D1728208

Larousse ENCYCLOPEDIA of ARCHAEOLOGY

Contents

Introduction

Opposite page: a warrior's head sculptured in terracotta from the site of Veii, one of the principal Etruscan cities. It dates from about the fifth century BC.

Frontispiece: the two mighty statues of the pharaoh Amenophis III at Memnon. The northernmost of these was famous in antiquity as the singing statue; the rising sun had a strange effect on the stone which made a sound like the twanging of a lyre. The emperor Hadrian is known to have heard it in AD 130, and Septimius Severus many years later. Some time after that there were repairs to the statue – and the singing stopped. No one knows now what could have caused it.

Today archaeology is in fashion. In the following pages we shall seek to establish the reasons for the somewhat belated popularity of a subject which was long regarded as tedious and even slightly absurd. It is, however, quite clear that today's public does not know the true nature of this science for which it shows an almost instinctive affection. Popular books and newspaper articles tend to present only archaeology's most attractive aspects and its most spectacular results. In contrast, those who might be termed semi-specialists – quite a number at present – often pride themselves on having knowledge of a branch of study which is particularly austere and even repellent, apparently deriving from it an increase of pleasure proportionate to the limited nature of the results.

Modern archaeology is subdivided into a considerable number of branches, differing widely in purpose and working conditions, while nevertheless sharing a certain number of basic methods in common. The present volume therefore comprises two parts. The first defines the discipline, tracing its development and indicating the principles which govern all research; it is the work of two classical archaeologists who have, however, worked in separate spheres with very different methods. To their satisfaction they found that, although they did not collaborate closely, neither contradicted the other on any basic issue. The second part of the book has been apportioned among specialists in various fields. Rather than give an exhaustive assessment of the archaeology in his or her field, each scholar has described the stages of the 'archaeological conquest' – a task made necessary by the fact that different schools of archaeological study are very far from being coeval and have advanced by widely separated paths – and then tried to select only those examples which appeared most appropriately to illustrate what investigation and study of monuments has contributed to our general understanding of ancient peoples and their history. These illustrations are extremely varied, according to the society under discussion; and reflect the authors' special skills and focus of interest.

Obviously, therefore, the reader should not expect to find in this book a complete picture of the ancient civilisations. Since a choice had to be made, it seemed preferable to pass over the medieval cultures, Christian and Islamic, Asiatic and European, although archaeology is used with increasing frequency to enrich our knowledge of them: some recent studies in medieval archaeology are, however, mentioned in the first part of the book. But these relatively recent cultures – very recent indeed, if one considers the total span of human development – are nevertheless being studied, especially from the literary records they have bequeathed us. Many of the buildings they erected are still intact and employed for their original purpose: between these cultures and our own there is no great gulf.

Other decisions may surprise the reader. The chapter devoted to pre-Columbian America is oriented less toward the study of the great 'classical' civilisations (Maya, Aztec, Inca) than to the investigation of the first traces of human activity in the New World. Here we are dealing with a particularly enthralling and exciting problem which archaeology alone can solve, and in which interpretations have had to be completely revised over the last few years.

We do not, then, provide a complete 'treasury' of world archaeology – which would be truly an impossible task to perform – nor a practical handbook for the beginner. Our intention has been to explain how a research technique has become the means of enriching and even revitalising culture and the study of man and how this has happened concurrently with the traditional disciplines, which it certainly does not seek to supplant, but beside which it can henceforward claim an honoured place.

Gilbert Charles-Picard

ARCHAEOLOGY at WORK

What is archaeology?

Roman art borrowed much from Greece, as is shown by this stone head of the Empress Livia Drusilla, wife of Augustus. The increasing realism of Roman portraits can be very valuable in identifying works of art and thus establishing dates.

An auxiliary of history The word archaeology (ἀρχαιολογία) means 'the study of everything ancient'. Unlike the names of other disciplines, evolved in modern times, this word was known in classical Greece, but to the Greeks it meant 'ancient history', and 'archaeologists' were a class of actors who specialised in themes based on ancient legends. The word was revived in a modified form in the seventeenth century by Jacques Spon, who was himself one of the first true archaeologists.

For a long time archaeology has been strengthening its claim to be regarded as separate from other related studies. At present it can be defined as 'an auxiliary of history'. That is to say, history in its true meaning is first and foremost founded on written records and sometimes calls upon more specialised studies to supplement the documentary evidence. Among these, archaeology has the widest scope. Whilst epigraphy is confined to the study of inscriptions, numismatics to coins and palaeography to manuscripts, archaeology covers all ancient artifacts, whatever their nature or size, from such colossal structures as Egyptian temples or Roman baths to the smallest and most insignificant potsherd.

Absolute truth The word 'auxiliary' implies a certain inferiority in the disciplines it describes. This, however, is becoming less acceptable nowadays, least of all to the archaeologists themselves, who often believe that they are in closer touch with reality than the historians or, in modern terms, that they are more scientific. In practice, the texts from which history is formulated shed light on only a minute fraction of the past. The human race has existed for hundreds of thousands of years, but written records are not found anywhere before about 3000 BC or even later; this skill was confined to a few people, and it was slow to spread. True history did not appear in Greece until about 500 BC and a little later in China. Archaeology, on the other hand, takes every trace of human activity into account. Moreover, documents give a particular interpretation of events. In even the most reliable (and the least frequent) cases, where accounts of actual eye-witnesses exist, they have obviously presented the facts in such a way as to reflect the most credit on themselves. More often there is only indirect evidence transmitted by numerous scribes who all more or less distorted the material handed down to them. Imagination often becomes inextricably interwoven with memory.

Scholars of the nineteenth century believed that textual criticism could enable them to recover the truth from beneath the embellishments of forgery, but today the finished critical work is often deceptive. It sometimes happens that archaeology is able to correct the errors of the annotators and to rehabilitate a discredited tradition. Archaeology's documents are authentic, and when they are 'read' correctly the past takes on new life.

Techniques are continually improving. For example, on a well conducted excavation it is usually possible to give a 'relative' date to the finds, but the methods employed by physicists to date radioactive carbon (C 14) have provided the archaeologist with a new line of attack which provides absolute dates. These facts make the conclusions of the scholar seem pitifully inadequate, when he is armed only with the resources of his intelligence and imagination to combat the distortions and omissions of the literary tradition.

Limited truth Although archaeology can, from time to time, recover precise knowledge and can thus take its place among the exact sciences, this knowledge is always severely limited. By taking into account the stratigraphy, the fabric and the decoration it can be stated, for example, that a certain vase or potsherd was made in Athens between 510 and 500 BC, and in some cases it is even possible to name the potter. But as soon as a shift is made from definite facts to general deductions the possibility of inaccuracy creeps in and increases in proportion to the scope of the speculations. The discovery of a vase on a site tells that the people who lived there had contacts with the city where the potter worked; but the exact nature of these contacts is much harder to determine. Starting with the vase itself it is possible to make a guess at contemporary events. If the vase was found in a particular structure, for example, it can help to determine the date of its erection or its destruction. Other circumstances which come to light during the excavation may make it possible to visualise how this came about. The presence of burnt matter indicates destruction by fire, which can usually be approximately dated by the pottery. If the burnt matter is found in several different areas of the site it can be assumed that it was destroyed during a war.

Opposite page: the Parthenon at Athens was constructed under Pericles between 447 and 432 BC and is the supreme example of the Doric temple. It has been used as a Christian church, a mosque and even an arsenal and survived many vicissitudes before its true worth was recognised. Much of the frieze was removed by Lord Elgin in 1806 and is now in the British Museum in London.

Carvings and reliefs often depict the way of life, behaviour and dress of the ancients. This bas-relief shows Babylonia and Assyria waging war. A city is being attacked with battering-rams and archers, and impaled prisoners can be seen top left.

The appearance of a very large number of artifacts belonging to one civilisation in another archaeological area indicates that there was an invasion. In a large part of western Europe, from Austria to Spain, cemeteries of underground tombs containing burnt bones, known as urnfields, dating from the beginning of the first millennium BC, are evidence of a major migration from East to West. About the middle of the same millennium the appearance of new tools and weapons, the most characteristic examples of which were found at La Tène in Switzerland, superseding those in the style known as Hallstatt, heralds widespread upheavals in the Celtic world.

When archaeology has only its own evidence to rely on, all the deductions belonging properly to the sphere of history are bound to be either very vague or purely hypothetical. If any book on prehistory is examined it will be found that along with a description of the finds from the excavation there is usually an attempt to interpret them in the light of other discoveries. There is an effort to find correspondences between ancient texts or folk legends and archaeological facts, or else a resort to linguistics to reconstruct the early forms of language. Unfortunately these attempts seldom produce a satisfactory result.

For instance, until the present day it has not been possible to reconcile the philologists' conclusions on the diffusion of the Indo-European languages with the archaeological facts. None of the theories attempting to explain the shift of the Maya population centres in the Central American forests, or their final decline, is really convincing. In another field, it is extremely difficult to interpret even the richest of decorated monuments when there are no legends to explain their meaning. The Minoan civilisation has bequeathed thousands of frescoes, figurines and engraved gems, but by themselves they cannot provide an explanation of the Cretans' religious beliefs and political institutions. If a precise picture of Mycenaean civilisation is beginning to be discerned it is largely due to the decipherment of the Linear B tablets, although they are nothing more than lists of property. Undoubtedly there will always be much controversy about the exact significance of prehistoric paintings. And what would the scholar not give for a definitive interpretation of the myths, Germanic or Celtic, illustrated by the reliefs of the Gundestrup cauldron?

History and archaeology, then, cannot be substituted for one another. True knowledge of the past can be claimed only when texts and artifacts, material objects and human testimony, illuminate each other. This means that only the briefest period, corresponding to a hundredth part of the existence of the human race, can be fully understood. However, it is comforting to recall that less than 100 years ago this period covered only a fraction of its actual extent, except for Egypt, and there were then only the vaguest, most erroneous notions about the ancient peoples of central and northern Europe, Asia and America. The very existence of mankind in prehistoric times was barely admitted by anyone except a few advanced thinkers.

From the collector to the scholar

Such progress as this could never have been achieved without a radical change in attitudes towards the relics of the past. For a long time they were studied only as curiosities or works of art, talismans with the power of evoking dreams. Then a few scholars began to use them as records. Still later the possibility of recovering large complexes enshrining the everyday life of the ancient world was discovered, and little by little they came to be preserved.

Relic hunters and collectors of antiquities

Amateur collectors of ancient artifacts, until recently known as 'antiquarians' (some famous learned societies still bear this name), doubtless existed from the Hellenistic era onwards. There were certainly many such among the Roman aristocracy. In Cicero's time the majority of noble Romans, starting with the orator himself, were already collecting works of Greek art of the sixth, fifth and fourth centuries BC, having

The Rosetta Stone was found by a French officer in 1799 near the Rosetta mouth of the river Nile. Its parallel inscriptions in three different scripts provided the French scholar J.-F. Champollion with the key to the decipherment of ancient Egyptian writing. The stone is now in the British Museum in London.

them copied when the originals were unobtainable, and establishing galleries in their houses which they opened to the public. Some of the temples and public buildings were transformed into veritable museums. The Greeks and the Egyptians were well aware of how to exploit the attractions their artistic wealth exercised over rich tourists (or such of it as they had been able to rescue from their rapacious conquerors). In the second century AD Pausanias wrote a description of Greece which is the forerunner of the Blue Guides or Baedeker. Romantic travellers wrote verses about the pyramids. But all these activities were still purely aesthetic. The Romans were very fond of debating obscure points of ancient history, but it is not known whether a single one of them ever thought of using an ancient monument to throw light on any of these disputed matters.

The triumph of Christianity destroyed dilettantism. Thenceforward, until the end of the Middle Ages, ancient artifacts were valued only where they had some religious connotation. The search for relics also led frequently to discoveries. One of the earliest and best-known was the alleged True Cross which was recovered by the Empress Helena, Constantine's mother.

An inscription in classical Greek capitals on a marble block at Epidaurus. Writings in which the ancients speak directly are of the utmost value to archaeology.

The dream of a Classical Golden Age In the fourteenth century a spirit not unlike that of the Roman collectors was abroad in Italy. Antiquities were sought for their beauty and as models for contemporary artists. 'In the fifteenth century,' wrote André Chastel (*Art et Humanisme à Florence*, p. 34), 'every humanist was a collector to some degree, and there was scarcely a studio which was not adorned with some statuettes and medals. ... One of Donatello's strong points was his first hand study of Roman ruins and artifacts, as well as his contacts with the Ancona Padua region where there was a flourishing trade in works of art, and where he gathered much useful information. Filarete mentions statues which he saw about 1435 in the possession of Donatello and Ghiberti, but the best-stocked studios were those of the Po valley. ... Interest in archaeology was not yet distinguishable (if, indeed, it ever was to be) from the cult of *mirabilia antiquitatis*. Poggio and Nicolini already regarded each object as a symbol: a vase, a statuette or a coin bearing an imperial portrait type was to them a species of talisman with power to stimulate the imagination. In Florence and the northern cities of Italy the highest value was placed on gem-stones, small sculpture and ceramics. The accent was still on the cash price. In the catalogues of the Medicis, Hellenistic and Sassanian vases, cameos, ancient intaglios and medals are enumerated and minutely evaluated, to the exclusion of fragments of sculpture.'

Modern Russian archaeology has shown that there was a settlement at Samarkand early in the first millennium, and it was a flourishing capital in Alexander the Great's day, long before the medieval caravans made it famous.

Archaeology has revealed at the fortress of Masada near the Dead Sea traces of the last stand of the Jews who rebelled against the Roman empire in the first century AD. When all hope had gone the defenders committed mass suicide rather than surrender. Here Professor Yigael Yadin (right), leader of the archaeological expedition, handles a bundle of forty papyri unearthed in the Cave of Letters.

This quotation admirably defines the Renaissance attitude to the ancient world. Unlike the Roman collector, the man of the *quattrocento* gathered works of ancient art not just for their beauty but because they evoked for him a way of life he would have liked to revive. His attitude was, in principle, similar to that of the modern archaeologist who uses ancient artifacts as a basis for a reconstruction of the past.

The Italian humanists' picture of the ancient world, however, was entirely intuitive and sentimental, rather than the rational and scientific view we try to form today. There was no attempt to recapture every aspect of ancient life. On the contrary, they systematically obliterated anything that did not fit in with their idealised notion of the Golden Age. By no means all the Greeks and Romans were thought worthy of recollection, but only the great thinkers and artists, whom they elevated to the status of demi-gods, divorced from the environment where they lived out their earthly days. Chastel convincingly demonstrates that all this activity was inspired by Platonic philosophy. The city of their researches was neither Athens nor Rome but the Utopian republic, the almost disembodied creation of the world of philosophy.

This passion for the ancient world was accompanied by total indifference to the fate of its monuments. More Roman remains were destroyed in the second half of the sixteenth century and the first half of the seventeenth than at any other time. No one had the least hesitation in doing this; the material of the demolished buildings was used for new structures. The fact which is most pleasing to discover today—that the ancients lived, had the same desires and needs as modern man and fulfilled them in much the same way—was not in the least interesting to the humanists, and even seemed to them to profane the celestial city on which their ideals were focused.

This idealistic worship of the past did not end with the Renaissance. In the eighteenth century a similar feeling inspired Johann Winckelmann (1717–68). For him ancient art, and particularly Greek art, for which he showed an almost religious veneration, was the symbolic expression of a splendid philosophical message. Winckelmann exerted an elevating influence, particularly on Goethe, which can be discerned in the second *Faust*. The Parnassian school of poets in nineteenth century France was to adopt a similar but less profound attitude. The views of the novelist Gustave Flaubert were original, stemming from the romanticism which turned to the Gothic north, the Middle Ages and the 'Barbarians' for the inspiration which over-exploited Greece could no longer supply. The Carthage in Flaubert's *Salammbô*, although it is based on extremely sound documentation, is also an imaginary city, the cruel splendour of which was displayed to shame the bourgeois vulgarity of Second Empire France. The controversy between Flaubert and Froehner, Keeper of the Louvre and a scientific archaeologist, is typical of these two contrasting points of view. Froehner's criticisms, it must be added, were not without a trace of pettiness.

At the end of the nineteenth century the beliefs of Louis Ménard, the 'pagan mystic', had a wide following among intellectuals and aesthetes. The resulting refined esotericism influenced many scholars, art historians, collectors and museum curators from that time until the First World War. Even today many great and

A revolutionary archaeological discovery was made at the turn of the century when Sir Arthur Evans unearthed, at Knossos in Crete, the remains of a magnificent civilisation which he called 'Minoan' after the legendary King Minos. This picture of Knossos shows the view across the storage magazines to the west court. Some of the walls still bear smoke marks from burning oil when the palace was destroyed by fire in the fifteenth century BC. Also to be seen are the massive jars (pithoi), in which the Bronze Age Cretans stored their goods, many of which were still standing in the palace when Sir Arthur Evans began excavating.

Valuable archaeological evidence is sometimes supplied by burial methods and by the gifts accompanying the dead. This corpse, which was found under a house floor in the Lebanon, had been placed in a large jar.

The existence of the Bronze Age cultures of the Indus valley was virtually unsuspected until the twentieth century. Their fine seals show animal motifs and a script which, so far, has not been deciphered. The upper seal shows an auroch bull seen in profile, standing before a censer. The bull in the lower seal is the hump-backed Brahmin type still to be seen in India today.

The development of modern professional archaeology The staff for these archaeological institutions was naturally recruited in the universities, where the science of archaeology gradually established – not without difficulty – a place for itself. Two opposing principles were in operation here: firstly, the tendency to early specialisation as seen, for example, in the United States, and secondly, the principle still in force in France, whereby only students with a sound groundwork of classical studies are accepted for this discipline. This means that the staff of the schools at Rome and Athens were chosen primarily from students of the humanities or history. The first of these methods produced excavators who were highly qualified in the most detailed technical processes, while the products of the second method were better qualified to interpret their discoveries. However, even with the aid of modern educational techniques a university can often provide only a theoretical training and the first contact the student has with the actual artifacts must take place in a museum.

The organisation of archaeological institutions has altered considerably since the beginning of the nineteenth century. In most countries this process was aided by the foundation of the great national museums. The most important of these is undoubtedly the British Museum, which acquired the Parthenon sculptures bought from Greece by Lord Elgin in 1816. Critics have sometimes regarded this as barbarously despoiling the most revered monument of the ancient world. Nevertheless, the removal of the work of Phidias to London, where it has been properly looked after, was one of those events which count as a fundamental breakthrough in archaeological studies. Until that time Greek art was almost exclusively known through Roman copies. For the first time one of its most important productions could be studied at first hand. A result of this was, admittedly, that the excavators subsequently concentrated almost exclusively on sculpture, and especially on classical sculpture. We are still trying to shake off the resulting prejudices.

The only remaining trace of it is the fact that today the study of Greek sculpture is the best developed branch of archaeology. If the study of pottery also seems to be approaching this perfected stage, this is due to the work carried out in the great museums of the western world. The reconstruction in Berlin of the Great Altar of Pergamum at the end of the nineteenth century revealed another aspect of Hellenic genius. The Louvre, founded on royal and imperial collections and supplemented by legacies, the most important of which was the Campana collection, has many masterpieces in its classical department, but the most impressive displays are those of the Egyptian and Western Asiatic departments.

France possesses also a most important prehistoric collection housed at St Germain en Laye. Chiefly because of the work of the Abbé Breuil, the development of the collection is still closely bound up with research. The present expansion of the great museums reflects the political vicissitudes of the countries which own them. It was only after the United States rose to be a world power that its institutions, such as the Metropolitan Museum of Art, New York, rapidly drew level with, or even overtook, their European counterparts.

Furthermore, the nations who have inherited the legacy of the ancient civilisations have had to take steps to prevent the export of works of art, which threatens to strip them of their cultural heritage. First in Italy, then in Greece, Egypt, Turkey, Syria, Lebanon, Iraq, Persia, and North Africa, museums have been established to complement the great archaeological sites. A decision has to be made as to whether it is better to keep the largest possible number of finds close to the site where they were found or, on the other hand, to reconstruct each culture in a great central museum. This is only one of the multitudinous problems facing modern museum curators. Without going into the question in detail, which would be beyond the scope of this book, let it suffice to say that the modern trend is to display only a few objects in such a way that the spectator can appreciate their aesthetic and historical value to the full, these public rooms naturally being supplemented by storerooms where the specialist can work on the whole body of the collection.

The revival of amateur archaeology Although the tendency of the last century and a half has been to confine archaeology to increasingly specialised professionals, a remarkable revival of amateur interest is now taking place. The reasons for this are many. It may be the reaction of the individual against the inhuman and tyrannical aspects of modern life – a sort of escape into the past comparable in principle, if not in practice, with the enthusiasm of the Renaissance, the dilettantism of the eighteenth century and the romantic yearnings of the Parnassian school for long-dead beauty. The great success of the works of C. W. Ceram, H. P. Eydoux and Leonard Cottrell have contributed to this trend, but they did not originate it.

Professional reaction to this unexpected influx of neophytes has been divided. Some regard it sympathetically and have established a generally cordial and productive co-operation. Others, however, evince either annoyance or distress. These 'new archaeologists', however, are mostly careful and painstaking, full of enthusiasm and ready to accept advice. As long as their work is supervised by an expert, which the law of most countries requires, they are a

When ancient buildings need repair modern archaeologists take great care to reproduce the original materials and methods as exactly as possible. Here bricks are being shaped to replace diseased masonry at Mohenjo Daro, in the Indus valley.

Persepolis. A procession of tribute-bearers climbs one of the stairways of the Palace of Darius.

useful supplement to the limited official teams.

Much more difficult than field work and needing a longer training is the interpretation of the results. The excavators often develop a possessiveness toward their discoveries and are very reluctant to hand them over for investigation by others. It often happens that good excavators make a serious mistake when they insist at all costs on publishing their finds themselves. It is equally necessary for the specialist to allow the finder the modest share of renown which is usually his only reward. It is essential to establish a system of mutual co-operation, which is easy enough to define in theory but much more difficult to put into practice. However, at the risk of seeming over-optimistic, it must be stressed that the acrimonious disputes between scholars and dilettanti which have so much delighted the onlookers are progressively dying out as methods become more scientific.

The scope of archaeology

It was mentioned at the ... ...ning that th... covered by archaeology is much larger than that of history. Every form of human activity is included in it, from the immensely remote day when *homo faber* first fashioned rude tools to a stated period the limits of which, however, are fixed only by convention and constantly under review.)

Such a vast field of activity inevitably needs to be subdivided, and since the early nineteenth century, i.e. since the beginning of scientific archaeology, no one has been able to claim an equal knowledge of all its branches, and no one has simultaneously studied Egypt, Greece, Etruria and Gaul, for example, as Caylus did. No one could, since the growth of the science has made it impossible for any one scholar to cover such an extensive area. Archaeology, then, includes numerous subdivisions, and the second part of this volume aims to give an overall picture of the whole discipline, but several omissions are inevitable. Everyone has his own preference, and does not care to depart from it. The Hellenist, for example, often dislikes working on Roman remains, some archaeologists confess to the opposite sentiment. Moreover, the excavator is very often seized by such an affection for one particular culture that he ends up by over-rating it and simultaneously undervaluing

Potsherds from the excavations at Agros in Greece being washed. Even the most unprepossessing piece of broken pottery may have an important tale to tell and must be cleaned, studied and recorded.

Detailed study of methods of flint working enabled scholars to establish chronological sequences long before modern techniques of laboratory dating were available.

all the neighbouring civilisations.

These prejudices can have serious consequences, since the remains of different cultures are often superimposed on one site, and it is therefore impossible to study the earliest remains without destroying the upper strata. Time after time valuable relics of medieval, Christian and Islamic buildings have been lost because they concealed classical remains. It is obvious that, since it cannot be done any other way, these unfortunate 'recent' buildings should at least be photographed in detail to preserve records of them, even if aesthetic effects have to suffer because of it. A French resident general in Tunisia, visiting the site of Dougga where the Roman forum is surrounded by a Byzantine fortification which obscures part of the Capitol, said to the director of antiquities, 'You ought to pull down the wall so that the temple can be seen better.' 'I'm going to pull down the temple so that the wall can be seen better,' retorted the archaeologist.

It would, however, be misleading to imply that archaeology is not a single discipline. Whatever the period he deals with or the country where he works, the 'detective of the past', as he has been called, employs methods based on the same fundamental laws.

The first, and most interesting, duty of the archaeologist is obviously the discovery of remains, usually buried in the earth but sometimes lost in the desert or overgrown with vegetation, or lying on the sea bed. But, whatever public opinion may believe, the archaeologist is not necessarily an excavator, an explorer, a diver or even a pilot.

The importance of interpretation In practice, the excavation is only the first step because once the remains are found they have to be made to speak. Furthermore, the processes of discovery and interpretation cannot be separated except in cases, unfortunately only too frequent, where intervention is necessitated by the threatened destruction of a monument. Otherwise the aims and method of the programme are dictated by what is known already and what the archaeologist is trying to find out about a particular culture. Excavation plays a part in the study of mankind's past comparable to the experimental method in physics and biology. The archaeologist is not free to dictate the conditions of his experiment as the physicist does; like the doctor, he has to be content with observations which chance puts in his way. On the other hand, he cannot remain a passive onlooker. Right from the start he must have a 'working hypothesis' about his finds; but the chief problem is to avoid becoming obsessed by this hypothesis and to be ready to modify it according to the results of his observations. The shrewdest and most experienced archaeologist rarely, if ever, finds exactly what he is looking for and often a new discovery wrecks a cherished and patiently constructed theory. Even eminent men have been known to lack the courage and humility to bow to the verdict of evidence, and have tried at all costs to support their suppositions with increasingly complex and far fetched explanations.

It will be seen later which are the best methods and techniques for avoiding such mischances and producing a satisfactory record of the excavation – a minutely detailed and easily interpreted record of the archaeological data. In fact, it should not be forgotten that, while the physicist can start his experiment all over again as often as he wants the archaeologist has no second chance if a site is destroyed in the process of investigation. It is lost for ever, and can only survive in the publications of its discoverer.

Principles of classification Fortunately, unidentified remains are not the only ones available to the archaeologist. Anyone who is seriously interested in the profession should not begin by studying such objects. He can *carry out his apprenticeship in museums where* there are numerous unpublished or inadequately studied artifacts. Many visitors who go to the Greek and Roman rooms at the Louvre by way of the Mollien gallery do not realise, for example, that the sarcophagi between which they are walking have never, for the most part, been seriously studied, or that many of them, including several works of considerable merit, have never been photographed. There is no better training for the beginner and nothing more

During the nineteenth century the 'Celts', as nearly all non-classical European cultures used to be known, were the object of a great romantic revival. This jewellery dating from the sixth century comes from the tumulus of La Butte.

useful to the advancement of knowledge than the publication of an entire series of such objects. It must, naturally, be a methodical publication including detailed analysis of each.

First of all, the would-be archaeologist must learn to observe and describe, which is not easy, then to pick out significant details which indicate the proper context for the object and permit the compiling of a typological sequence which can be developed into a chronological classification when a certain number of objects in the series have been dated. Thus the most valuable talent of the archaeologist, the visual memory, is trained. Faced with a new object, he has to compare it mentally in detail with the largest possible number of similar objects, noting the resemblances and the differences. Supported by these comparisons, which can be made by reference to a card-index (the automation now being applied to archaeology makes these findings more reliable), the investigator can answer a number of standard questions:

What is the object and for what was it used?
When was it made?
Where and by whom was it made?
How was it made?
What happened to it from the time when it was made to the time when it turned up in its archaeological provenance?
What information can it give us about the site as a whole? In particular, does it help with the dating?
If the object is not a straightforward utensil, has it any significance, and if so, what? Is it, for example, a religious object, something used for social purposes, a memorial of a particular event, etc.?
If it is a work of art, what aesthetic principles does it express?

In answering the last question, a judgment on values is permissible. It is, however, essential to be as careful as possible not to express a personal opinion, but one which could have been formed by a contemporary.

As can be seen, the last items on this questionnaire are more complex than the first. More extensive knowledge is necessary to answer them, and in every way the answer is more dependent on the personality of the investigator. The answers, too, are more open to dispute.

Archaeological evidence is all about us and can turn up in the most unexpected places. For example this picture shows a cobbler at Ismir in Turkey using the battered remains of a Corinthian column capital as a work table.

In practice this standard questionnaire can be applied to all traces of human existence, whatever their nature or provenance. It was drawn up with a view to a particular object – a relatively simple artifact of limited dimensions. However, the problems set by a monument, an architectural complex, a large sculpture or a painting are not fundamentally different. Naturally the fourth question dealing with the method of construction becomes both more important and more difficult. It is easier to find out how a craftsman works than to understand an architect's methods. When the monument is highly complex each part of it can be made the object of a separate enquiry, since they are not necessarily contemporary.

Various methods Research of this nature obviously requires very wide and diverse knowledge. Furthermore, archaeology recognises another subdivision, according to the methods employed, in addition to those based on the different civilisations. Scholars who specialise in the study of buildings cannot arrive at valid conclusions if they have not studied architecture. It is indispensable for an architect to take part in all the excavations where buildings are found. Conversely, experience has shown that an architect cannot undertake such projects alone. The first of the true archaeologists are those who have specialised in major arts such as painting and sculpture.

It is difficult to separate their domain from the history of art. Pottery experts are a more autonomous body, but even they lack the degree of independence enjoyed by numismatists in their study of coins. Epigraphists, who deal with inscriptions, come halfway between the archaeologists and the philologists.

The part played by these different disciplines varies according to which province of the archaeological kingdom is under consideration. At one end of the chronological scale are the prehistorians whilst at the other are the post-medievalists. The former have only the finds from excavations and the results of typological analysis to help them to reconstruct fossil man. But they, more than anyone else, have produced a set of absolute rules for the technique of investigating the soil. Classical archaeologists, who long regarded them as poor relations ('wanting to practise archaeology without having learned the ancient languages and very little acquainted with their own', to quote Salomon Reinach), have been obliged to learn from them.

Geology, palaeontology (the study of extinct life from its fossil remains) and ethnography are the sciences most useful to the prehistorian, and it is he who has benefited most from the new dating methods developed by physicists; dating by radioactive carbon (C-14) has revolutionised existing theories on the duration of the earliest stages of human development but is unsatisfactory for historical periods. When prehistorians have to fit into the framework of a university, they are drawn to the faculty of science as much as to the humanities.

However, even the most remote of our ancestors have left only traces of their material existence. When a study is made of cave paintings, Aurignacian and Magdalenian sculpture and megalithic architecture (i.e. consisting of huge stones) a certain amount of knowledge about art and religious history is naturally absorbed. An examination of the geographical distribution of certain typical artifacts and burial customs can enable even the archaeologist to reconstruct political and military events.

The prehistorian's methods remain valid for nearly all non-literate cultures, and for those whose language is not yet translated. However, such people were very often in more or less direct contact with literate cultures and objects coming from the latter permit the establishment of definite chronologies.

Rougiers, a model medieval excavation Medievalists, on the other hand, practised only one form of archaeology until recent times: this was the study of buildings and the plastic arts. Now they use excavations as well, and have

The harbour of Byblos, near modern Beirut in Lebanon, which has been in continuous use for at least 5000 years. Although tiny by modern standards, the harbour offered ample shelter to the powerful Phoenician fleets which set sail through the narrow entrance to explore the Mediterranean and beyond. They traded with the Scilly Isles for tin, and may have circumnavigated Africa nearly 3000 years ago. Much of their prosperity was founded on the export of magnificent Lebanese cedars, for which there was a great and lasting demand.

learned to make use of modern 'fossils' like domestic ceramics. These methods have greatly enriched the knowledge of periods like the Merovingian, from which there are few texts and even fewer large monuments. They also throw light on the life of the lower classes and the development of the economy. It is impossible to devote a whole chapter of this book to medieval archaeology, but the results obtained by Mlle Démians d'Archimbaud on the site of Rougiers in Provence, based on the account which she presented to the Association Guillaume Budé of Aix-en-Provence in April 1963, are of particular interest and provide an effective example.

This excavation comes under the heading of general investigations into the medieval life of rural Provence, inspired by the work carried out in England (under the auspices of the Deserted Medieval Village Research Group), and in Denmark and Poland. It will be observed that in both Denmark and Poland, which were untouched by Roman culture except for commercial and military contacts, medieval archaeology is a direct continuation of the prehistoric. In France as in England, on the other hand, there was a definite break, and where researches similar to those of Mlle d'Archimbaud are being conducted in the Paris basin by P. Courbin, archaeology indicates a return to a way of life very close to that of the Gallic period after the fall of the empire.

Thus in the region on the north boundary of the Ste Baume massif, which is where the abandoned village of Rougiers lies, the population which had settled in the lowlands during the *Pax Romana* returned after the invasions to occupy more easily defended sites. Several texts preserve further details of this development.

The excavations amply confirm this evidence. An Iron Age oppidum (fortified settlement) was established on the fortified spur of Piégu, protected on two sides by a sharp drop. The Ligurian Celts had managed to make it invulnerable by closing off the connecting neck of land with a double rampart. It was this ancient fortress which was reinhabited, probably in the fifth century. Indeed, immediately above the level containing pre-Roman pottery, a grey ware with stamped decoration was found, known by the name of Visigothic ware, which was actually used throughout the whole of southern Gaul in the fifth to seventh centuries. The new inhabitants reinforced the site by building a masonry fort over the Gallic rampart, with walls pierced by curious loop-holes, and they and their descendants lived there in peace for several centuries.

In 1150, however, the Sire de Signes, contending for the overlordship of Rougiers with the Abbey of St Victor at Marseille, seized the castle and destroyed it. The conflict was happily settled by the decision of the Vicomte de Marseille, who divided the property between the

A rock painting believed to be about 5000 years old at 'Diana's Vow' Farm near Rusape, Rhodesia. The large reclining figure in the upper half is a masked and bandaged body, probably of a chief or someone of rank, wearing an antelope mask and ready for burial. The figure below with raised knees is possibly a wife mourning him, but she could also be prepared for burial with him – in effect to follow him into the next world. The curved lines in the lower part may represent a river, a frequent symbol among primitive peoples for the barrier to be crossed before reaching the next world. Numerous other figures are shown, and offerings of food.

The Cauria (Corsica) dolmen has three supports on each side and a massive table-like capestone, the positioning of which must have been a great feat of engineering.

Jean-François Champollion. One of the greatest names in the history of archaeology. Champollion was a professor at the university of Grenoble at the age of nineteen. He was thirty-two when he published his key to the decipherment of Egyptian hieroglyphs and thus extended enormously our knowledge of the ancient world. He died at the tragically early age of forty-two.

Sire de Signes and the monks. The inhabitants were then able to rebuild another fortress on a ridge parallel to the site of the oppidum. A castle consisting of a donjon or keep and several towers guaranteed the security of the village and was itself protected by an enclosing wall. It has been possible to reconstruct the peasants' houses with their hearths laid directly on the earth, usually outside the dwellings. The assembled debris forms a picture of the household utensils: plenty of pottery and some glass, bone and metal objects. Animal bones provide some information about domestic and wild animals and throw some light on contemporary diet, whilst with pollen analysis it is even possible to discover what species formed the local vegetation.

After the middle of the thirteenth century the situation was safe enough to encourage a number of the farmers to venture into the lowlands again, where a better water supply enabled them to extend their cultivation. The fortress was not immediately abandoned, but it underwent modifications which indicate that its protection was becoming less and less necessary. The excavations show that this cautious and gradual descent to the lowlands was beginning at least half a century before it is first mentioned in the documents. It ended in the fifteenth century with the final abandonment of the fortress, and Rougiers became a lowland village for the second time.

From this example the kind of light shed by archaeological methods on economic and social history in particular can be seen. Hundreds of excavations like that of Rougiers are required before a serious synthesis of rural life in the Middle Ages can be written. The little that is known today is still enough to prove the unreliability of theories claiming to explain everything by means of a general philosophical doctrine of one sort or another. The same can be said of the classical world. For example, the great economic and social history written by the Russian historian M. I. Rostovtzeff, although it used archaeological data as much as possible, is completely out of date today. It is clear, too, that some archaeologists try to formulate rules to be applied to the modern world based on the laws which regulated the ancient communities. Archaeology, which makes possible the reassessment of these laws, and its progress is far from irrelevant to the conditions of modern

The critics of archaeology

Is archaeology in conflict with life? The foregoing conclusion leads directly to an enquiry into an essential point: are not the aims of archaeology fundamentally absurd? Its entire purpose, in fact, is to restore dead things to life. Nature, hostile to all waste, tirelessly reabsorbs the smallest remnants of anything which once lived and uses them again to produce more life. Archaeology, in attempting to preserve corpses and fossils, is working against nature. The peasant who smashes the marble sculpture or inscription which has jammed his ploughshare, and the contractor who has ancient walls cleared away so as to build on level ground, are acting in accordance with the laws of nature. Furthermore, these ancient cultures of the past which are being revived so laboriously had themselves scant respect for the heritage of their own forebears. It could even be said that they destroyed as much as they created. Pericles and his contemporaries obliterated many temples of preceding centuries and buried many statues, and Roman architects only erected their buildings after they had cleared the ground of all ancient structures. Medieval builders scarcely left any (earlier Merovingian and Carolingian) churches standing, and a number of their cathedrals in turn met the same fate at the hands of their Gothic successors.

We have already seen that the cult of classical antiquity did not stop the men of the Renaissance from destroying many buildings which had

survived in a good state of preservation until their day. This destruction, too, was more or less justified by necessity. How many others were inspired by idealism? Christianity, Byzantine iconoclasm, the Reformation, and the French Revolution all showed destructive zeal. The Russian Revolution has shown more moderation in this respect fortunately. Currently, many people deplore the consequences of liturgical reform in the Catholic church, and the effects of the Chinese cultural revolution can only be guessed at.

Wars bring in their train vast unintentional disasters which – although they are now kept to a minimum because of the opening they give for hostile propaganda – the compensating frightful increase in the destructive power of armaments has made worse. They also let loose the senseless fury to which in 1794 the Abbé Grégoire gave the name of Vandalism. Since then several attempts have been made to prove that the Germanic tribe from whose name this epithet derives deserved no more than any other the bad reputation which it has been given. In practice,

The archaeologist not only plays a vital role in discovery but also in preservation. A team of international experts assembled by UNESCO joined forces to cut the carvings and reliefs of Abu Simbel away from the rock and transport them to a new site where they could be rebuilt.

vandalism is just as natural as cruelty, and no nation is entirely free from it.

The task of preserving the relics of the past, then, entails not only a struggle against material forces but against perfectly natural and widespread trends of personality. It is not surprising that those who defend ancient relics are sometimes regarded, along with nature and wild life conservationists, as impractical idealists. Indeed the character of the scatter-brained archaeologist has often exercised the wit of satirists. If Molière's Vadius is more of a philologist, Edmond About brought back from his stay at the École Française at Athens, not a learned paper, but a caricature of an epigraphist who, mistaking the marks on a milestone for a classical inscription, provided an ingenious interpretation of what he took to be abbreviations incised in the stone in ancient Greek.

As it is hard to believe – at least in popular literature – that an intelligent man could waste his time on ancient stones, the antiquarian is often a smuggler or a spy in disguise in fiction, and in fact many archaeologists did work in the information services during the Second World War. Another manifestation of the dislike for archaeologists is to be seen in the *Voyage de Sparte* of Maurice Barrès; they are accused of lack of feeling, of taking all the poetry out of the past. Many comparable reproaches have been expressed in recent works which are, in brief, survivals of romantic idealism refusing to accept the disciplines of science. At the moment, however, these attitudes seem to be out of fashion. Archaeology and archaeologists, especially the amateurs, are enjoying popular favour. But there are greater dangers than simple ridicule.

First and foremost, the demographic development of mankind and the improvement of technical knowledge imperil the natural safe-deposits where the remnants of the past are kept. For example, , the mechanisation of agriculture results in ploughing going deeper, and this lays bare more ruins every day. Hundreds of Gallo-Roman villas have been discovered in this way in parts of northern France. In Tuscany the demand for land which has been virtually abandoned for 2000 years has revealed a horde of underground tombs which had hitherto escaped the notice of treasure hunters. We are all familiar with the drama of Aswan; the construction of the high dam which will probably rescue the Egyptian *fellahin* from thousands of years of misery nearly resulted in the submerging of Abu Simbel where Ramesses II had the greatest of all pharaonic temples carved out of the cliff.

Transformation in the urban life of the cities of Europe is threatening not only the medieval monuments and the seventeenth- and eighteenth-century settlements but even older remains, the existence of which is still unsuspected. In Paris a Roman bath was sacrificed in

Inca stone needs no restoration. The beautifully cut blocks at Cuzco in Peru are so closely fitted that even a knife blade cannot be inserted between them.

1935 to the extension of the Collège de France. Ten years ago in Rome on the Via Latina an unknown catacomb, the frescoes of which illustrated the beliefs of a strange gnostic sect dedicated to the joint worship of Hercules and Christ, was saved from destruction by the denunciation of a sacked workman. More recently in northern Tunisia the route of a pipe-line carrying oil from the Sahara crossed the site of an early Christian basilica. When a new road was being built at Carthage in 1960 workmen came across a pavement of the fourth century AD made of marbles and mosaics nearly 100 metres square.

There are many other examples, and fresh ones can be found almost every day by simply opening a newspaper. Usually the public authorities, the engineers in charge of the project and many private individuals do their best to see that these accidental discoveries are not destroyed. But there are few competent archaeologists; they are short of funds, and the precautions indispensable to methodical procedure prevent them from working rapidly. The halting of any great programme of work already under way represents a disastrous loss. Many remains cannot be moved, and a complete change of plan would be required to preserve them. Thus all sorts of troublesome problems can often arise. To avoid them each country would have to draw up in advance as complete as possible a list of its archaeological wealth, and this would be an enormous undertaking. Even when something is achieved, it can naturally include only known monuments.

It is true that modern techniques provide resources for more complete and rapid surveys. The Istituto Politecnico of Milan, for example, has used geophysical methods of investigation to discover several thousand buried ruins, recovered several thousand ancient objects and brought to light more than twenty painted Etruscan tombs, eight of which are of outstanding merit. These results, which extend our knowledge of Etruscan civilisation, were obtained in the space of four years. But even in Italy, which has the benefit of one of the most advanced archaeological services in the world, C. M. Lerici's defiance of convention has not been made without provoking some energetic resistance.

Archaeological problems and ethics

Traffic in antiquities Chance discoveries sometimes lead to illegal trading in antiquities, a trade older than archaeology itself. Already flourishing in the time of the Roman Empire, it probably reached its zenith in the Renaissance. Traffickers, however, must not be condemned

The Roman theatre at Orange has been renovated and modern seating, lighting and a stage added so that the ancient structure can still exercise its original function. It is one of the few Roman theatres in active use in the twentieth century.

out of hand. Many of them have been, and still are, remarkably knowledgeable, and some are accomplished scholars. Without them the museums of the countries which never possessed great archaeological wealth of their own could not have been established. And it seems only right that the cultural heritage of the classical world should not be exclusively confined to the nations inhabiting the areas where the culture originally flourished, especially as the new nations have made great contributions, both intellectual and material, to the progress of ancient studies. Moreover, traffic in antiquities can in practice only be supported by the break-up of old collections or by clandestine excavations. In the former case unknown pieces or objects thought to be lost are sometimes revealed. There are many private collections, particularly in Italy, to which the proprietors allow no access. The second case is much more reprehensible. Such excavations as these are obviously conducted without the slightest scientific care, and the provenance of the finds is falsified to avoid detection. In the Louvre, for example, there is a beautifully sculptured bas-relief representing the divinities Isis, Serapis, Harpocrates and Dionysus. Acquired in 1912, it was presented as coming directly from Egypt. The back of the stone bore a Christian epitaph in Latin. The stone, therefore, must have been reused towards the end of the classical period somewhere other than in the Nile valley, where the only languages were Greek and Coptic. An enquiry provided positive proof that it had been found in Tunisia, at a place called Henchir el Attermine. There is even a photograph showing it in the house of the colonist who first found it. Several scholars, however, regard this marble as a typical product of the Alexandrian school, and it is possible that it was taken to Tunisia from Egypt, if this took place during the classical period.

Some dealers are even unscrupulous enough to buy objects stolen from official excavations or museums. So as not to fall victim to this, the archaeologist must transform himself from a detective of the past to a detective of the present.

Forgers When a dealer cannot get hold of genuine articles, he is often tempted to make them, and the forgery business, too, has been in existence for a long time. In the first century

The vast necropolis of Cerveteri is one of the innumerable Etruscan sites which has benefited from modern archaeological methods. Tombs can now be located and examined by means of a periscope without excavation to determine whether they are worth further investigation.

AD the workshops of Athens were already producing pieces in the style of the classical period. At that time, and again in the Renaissance, copyists, imitators and forgers could scarcely be told apart: the same piece could be claimed as genuine or sold as a confessed forgery, according to the buyer. Some Roman men of letters amused themselves by adding inscriptions in the names of the great men of the Republic, a period from which remains were very scarce. A text found several years ago in a suburb of Carthage purported to be a copy of the victory dedication of Scipio Aemilianus after the destruction of the Punic capital. It was the signed work of a high official of the *imperial* administration. In the fifteenth century the young Michelangelo, when at Florence, taught himself to make forgeries like the celebrated Head of a Faun, the virtuosity of which was universally admired. Since no one yet regarded antiquities as historical records, no disgrace attached to such activities. The Italian taste for successful *combinazione* was to perpetuate something of this feeling until the present day. The great Belgian historian Franz Cumont tells how one day he had a conversation in Rome with a reputable carver of 'ancient' sculptures. Admiring his talent, he urged him to give up cheating and exhibit his own work. 'But I like making forgeries!' replied the artist.

During the Italian Renaissance deliberate copies in the classical style were considered an essential part of any reputable sculptor's skill. Head of a faun by Michelangelo.

There are many kinds of forgers. The more modest confine themselves to making modern pieces which could deceive no one but the most naive tourist. 'They ought to be encouraged,' remarked a famous archaeologist. 'These take the place of authentic pieces which *won't* be clandestinely sold.' At Carthage many 'Punic' lamps modelled on a genuine piece in the shape of a bearded head have been produced. There are so many copies that the 'potter' did not even trouble to fire them. If buyers had unwittingly put their purchases under the tap they would have seen them dissolve. These petty forgers are more picturesque than dangerous. It is quite another matter with the operators of large workshops, particularly in Greece and Italy, who now and then succeed in deceiving the watchful eyes of the curators of the largest museums. Here everything is conducted with the utmost care and every detail is based on an authentic piece. It is no longer possible, however, to make an exact reproduction of a known work, not only because the imitation can easily be spotted (this would not be proof of forgery, since there were copyists in ancient times) but because high prices are paid only for sufficiently original works. The forger, then, combines elements from several different models. And it is here that we find the possibility of unmasking him. An expert can demonstrate that these various elements belong to different periods and consequently could not occur together in the same work. However, proof of this kind is only possible in the case of pieces purporting to come from well-known workshops and following a style with a well documented evolution, like those of classical Greece. The ancient centres of production in the less civilised regions often combined contemporary styles with methods of older origin. The most successful forgeries have purported to come from these peripheral regions.

This was true, for example, of the Tiara of Saitaphernes, the acquisition of which was to cause the keeper of the Louvre nearly as many headaches as the theft some years later of the *Mona Lisa*. At the end of the nineteenth century several funerary mounds were excavated in southern Russia containing the remains of Scythian kings, as well as magnificent treasures. It was known that several of these tombs had been secretly looted, apart from the official excavation. Nobody, then, was surprised to see an extraordinary golden tiara make its appearance on the Paris market, with reliefs apparently from the hand of the best fourth-century Greek artists. An inscription in which none of the epigraphists could find a single suspicious detail stated that the piece had been presented by the inhabitants of a Greek colony on the Black Sea to the local prince, Saitaphernes, for protecting them from other barbarians. The tiara was bought for a considerable sum.

At this time the celebrated Bavarian archaeologist, A. Furtwängler, who was as well known for his learning as for his caustic tongue, announced that the tiara was a fake. This raised a great outcry in France. The German scholar was accused of professional jealousy, which was further exacerbated by the political rivalry between the two nations. The tiara was about to be published in a splendid volume under the auspices of the Académie des Inscriptions, *Les Monuments Piot*, with the co-operation of the greatest scholars in France, when disaster struck. A Russian goldsmith, entirely of his own volition, confessed that he was the creator of the masterpiece. The effect of this revelation on public opinion can be imagined. With deplorable injustice, however, the chief responsibility fell on Salomon Reinach, who had had doubts when the purchase was first discussed, but had made the grave mistake of joining too warmly in the controversy while the others were discreetly backing out.

An affair of this sort is obviously quite exceptional. It is several years since the Metropolitan Museum of Art, New York bought some terracotta statues of fighting Etruscan warriors which rapidly became famous. Some publishers, fascinated by their spectacular impact, chose them as illustrations for popular works, sometimes in defiance of the wishes of their justifiably suspicious authors. An enquiry revealed that the figures were undoubtedly modern enlargements of small bronzes. There is in existence a

A view of the Goreme Valley in Cappadocia in Asia Minor. The weird cave-like openings are actually churches, carved from the rock in the early days of Christianity. Such structures are an important part of man's past and as much the concern of the archaeologist as the most imposing ruin.

photograph of the warriors under construction in the workshop where they first saw the light of day.

One of the most famous forgeries in Great Britain was the complex of finds at Piltdown. In 1908 an antiquarian found, in a gravel pit in Sussex bones and tools which apparently belonged to a primitive kind of man. They were discovered with the remains of Upper and Lower Pleistocene mammals and 'Piltdown man' was supposed to have lived between 200,000 and 1,000,000 years ago. After decades of increasing scepticism Kenneth Oakley in 1949 showed by means of fluorine and other tests that no part of the skull was more than 50,000 years old. He went on to prove the whole complex a fake, composed of cleverly disguised fragments from an ape's jaw and a relatively modern human cranium. It was shown also that the stone and bone tools were faked and that the animal bones had been brought together from several sources. By exposing this forgery what has been called 'the greatest anomaly in the fossil record of human evolution' was removed. Detection of forgeries is one of the most thankless tasks which can fall to the archaeologist. In many cases absolute certainty is impossible. The Louvre has two bronze busts (found at Neuilly le Réal in the early nineteenth century) which for a long time were regarded as fine portraits of Augustus and his wife Livia. However, a German scholar pointed out discrepancies in the hairstyle of the empress and in the inscription on the pedestal which gives both the name of Livia (which she abandoned on the death of her husband), and the title Augusta (which she only received at this time). The two busts, then, must be forgeries made shortly before they were found during the Napoleonic era. There was nothing suspicious, though, about the circumstances of their discovery. Moreover, the forger never derived any profit from them, but this is not, admittedly, a decisive argument, since there are forgers working for 'the love of art'. Finally, the ancients themselves were often confused about the successive names and titles of this great lady whose life was so eventful.

It can happen that genuine pieces are the objects of unjust suspicion. Colonel Stoffel, the director of the excavation of Alesia, near Dijon, for Napoleon III, discovered in the trenches of the plain of Laumes, in addition to some Roman and Gallic weapons, a splendid silver cup with floral decoration which is still in the museum at St-Germain-en-Laye. A jealous courtier was mischievous enough to have a silversmith make a forgery which he placed where it would be seen by a number of people—on a billiard table in the château of Compiègne. Fortunately, the slander by implication was quickly discovered, and the recent publication of Stoffel's

Opposite page: the current fashion for ancient works of minor art has led to a growing traffic in forgeries. Small models such as this bronze statuette of a warrior from Sicily (opposite) are easily copied and even the greatest experts are sometimes deceived.

letters has amply vindicated the honour and conscientiousness of that officer.

Nowadays archaeology can enlist the aid of the natural sciences. It is true that for a long time forgers have known how to age their products artificially. In the Renaissance they used to be buried for a while before being offered for sale. There are some quaint recipes in existence, like that of the Egyptian scarab makers who, it appears, treated their work by making geese swallow it. Others took less care, like the German, who, having modelled a wax bust which he passed off as an Italian Renaissance work, left a piece of newspaper inside it. Museum laboratories nowadays make use of many methods to detect forgeries, X-ray examination, infra-red photography, chemical analysis of metals, fibres, etc. The traces left by modern tools can be discerned under the microscope.

Inexperienced or misguided excavators

Other enemies of archaeology act in good faith, but are dangerous all the same. These are the ones who are fired by undiscriminating zeal and set about a project which is either useless or beyond their powers. Since there is so much to do it is appalling how much time is wasted in tackling non-existent problems. Many sites have been irretrievably spoilt by badly conducted excavations. Most frowned on by the professionals are those who throw themselves into the study of the past from the standpoint of some preconceived notion, either political, religious or philosophical.

Many local enthusiasts have made fools of themselves by identifying their little town with some celebrated city of the ancient world. Even patriotism is a bad adviser when it leads to searching the ancient documents and texts for justification for contemporary doctrines or political claims. Italian scholars, for example, very rightly denounced the material and intellectual damage done by the archaeological policy of the Fascists, but Mussolini was far from being the only one guilty of this. In another sphere the legitimate curiosity roused by the mystical sects and religions – the worship of Isis, Druidism, the medieval heresies spread by the Templars and Cathars – such things have not only led to the publishing of many extravagances but have also inspired useless or even dangerous projects. The least tiresome result of these upheavals is the waste of money, already in short supply, which is available for the preservation and study of ancient monuments.

Archaeology and modern civilisation

In spite of its difficulties and dangers archaeology has become a comparatively popular study today, which indicates that man is determined not to lose sight of his traditions, whatever the changes in his way of life. It is altogether characteristic of this trend that countries which have recently changed their form of government are making a considerable effort, not just to preserve the relics of the past, but to have them more widely understood.

However, we can also discern among the most determined traditionalists an undeniable distaste for humanism as it was defined in the sixteenth and seventeenth centuries, and particularly for the academic study of the ancient languages. The great popularity of archaeology is evidently explained by its affinity with other aspects of modern culture. Like these, it makes

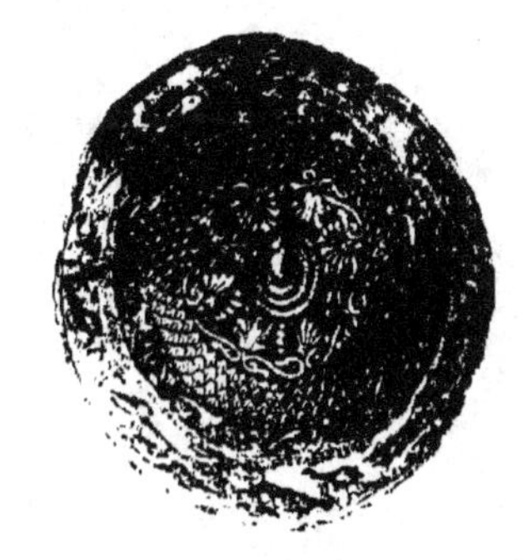

The Tiara of Saitaphernes was originally thought to be genuine and was displayed in the Louvre Museum in Paris. The revelation that it was a forgery caused a scandal in the archaeological world.

Opposite page: the work of the archaeologist may involve him in the life of past centuries—sometimes in the life of past millennia. In this corner of Asia Minor near Adana the ruins of a Frankish castle crown the distant hill, while the rock face on the left bears a Hittite relief of the Great King Muwatalli who lived from 1306 to 1282 BC.

more use of actual objects and representations than of written evidence. It uses methods of classification and deduction akin to the exact sciences, and practises a form of experiment. It shows a poor grasp of great events, but to make up for this is remarkably informative about social and economic conditions. It rarely deals with outstanding individuals, except insofar as they left their mark on society—a king, for example, is known by his portraits and his palaces—but it does bring us into intimate contact with a multitude of ordinary, humble people. Finally, its practice demands intellectual as well as physical hard work, and gives as many opportunities for the exercise of the power of observation as for those of deduction and memory.

Does this, then, indicate that archaeology is the branch of learning about the classical past which is destined to replace the study of the classical languages and texts in popular culture? It is true that a civilisation cannot be understood from a study of its remains alone: and this is equally applicable to both research and teaching. It is also clear that the great classical texts—Homer, for example—have a value and scope which go far beyond the context in which they were created, and that there is no substitute for the enriching experience of studying them.

However, although a classical education brings its pupils into intimate contact with great literature, on the other hand it leaves them almost totally ignorant of the artistic heritage of the ancient world, and gives a picture of life in antiquity which is both pompous and false, and distorts even the greatest writings. If we look at an eighteenth- or early nineteenth-century translation of Homer it seems ridiculous. The engraved illustrations have a certain charm of their own, but even the best of them take all the flavour out of the text they are supposed to illuminate. Nobody today knows Horace and Virgil by heart, but millions of people have visited the forum at Rome and the Acropolis at Athens, where many of the greatest Hellenists of the early part of this century never set foot. Even while admitting that a spectacle like *son et lumière* is not the best introduction to the sculpture of Phidias and the architecture of Ictinus, it must be agreed that the average man of today knows the ancient world from a different viewpoint from his forebears, and one which is none the worse for that. What is more, he has some knowledge of the non-Mediterranean cultures of which most earlier Europeans were ignorant.

At first archaeology was an elegant pastime designed to give pleasure to a privileged minority, and an auxiliary of the creative arts. Later it became pre-eminently a research technique practised by experts. Today, as we can see, it has become an essential culture medium and is strengthening its position as one of the fundamental modern humanities.

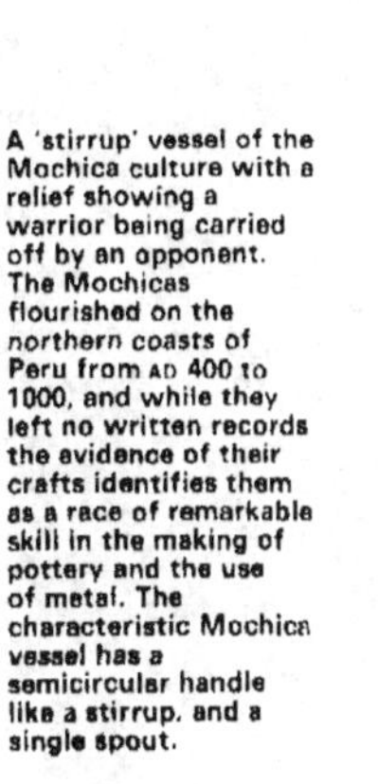

A 'stirrup' vessel of the Mochica culture with a relief showing a warrior being carried off by an opponent. The Mochicas flourished on the northern coasts of Peru from AD 400 to 1000, and while they left no written records the evidence of their crafts identifies them as a race of remarkable skill in the making of pottery and the use of metal. The characteristic Mochica vessel has a semicircular handle like a stirrup, and a single spout.

How monuments survive

The emperor Septimius Severus, who reigned from AD 193 to 211, a native of Leptis Magna. He conducted major replanning operations which completely changed the design of the city.

How does it happen that certain human artifacts escape destruction? It is easy enough to understand when they are tiny objects of hard materials like the dressed flints of prehistoric man, or potsherds. This kind of survival is, indeed, very valuable to the archaeologist because its presence on the surface of the soil leads to the disclosure of an ancient habitation site at first glance. In the course of an excavation such objects are frequently encountered and they furnish the basis for every stratigraphic classification and chronology. But if our ancestors had bequeathed us nothing more than these wretched remnants it is very likely that no attention would ever have been paid to them.

We have seen in the foregoing chapter that archaeology was born as a result of the emotional stimulus of the discovery of classical works of art, and secondly of the excavation of huge complexes where the life of the ancient world had been arrested as though by an enchanter's spell. Now we shall examine the conditions, clearly exceptional, in which this 'mummification' of vanished civilisations sometimes occurs. There are two cases in point: either some human agent has tried to save from destruction certain treasures which have temporarily or permanently ceased to be of use, or the preservation is the accidental result of a freak of nature.

Buried treasure

The human species does not lightly abandon great wealth. It sometimes happens that a man wishes to keep his memory alive among his descendants, and therefore sets up for their use the most durable monument he can devise. This feeling is particularly marked among rulers. The pharaohs of Egypt, the kings of Assyria and Persia, and the Roman and Chinese emperors succeeded in preserving remembrance of themselves by inscriptions and sculptures. It even happens that this sort of posthumous publicity infects ordinary people, who devote a large portion of their means to the building of a mausoleum for themselves.

However, this phenomenon is generally found only in those civilisations which combined great political power with a tendency to self-deification and command of a large surplus of money, at least as far as the ruling classes were concerned.

Nearly all peoples (at least since the Mousterian period, which goes back more than 100,000 years) have surrounded the remains of their dead with veneration expressed through offerings, often costly, and grave goods which they tried to render imperishable for as long as possible. This cult of the dead was not entirely disinterested. Certain individuals, usually chieftains, who were deliberately sacrificed in early times, were believed to exert a benevolent influence on the community, and particularly to ensure their economic prosperity. This explains why considerable expense was incurred to make sure of the eternal 'comfort' of these protectors.

This is a very different matter from the memorial tombs we shall shortly discuss, which are in themselves a form of publicity aimed at future generations. The mausoleum is made as eye-catching as possible to perpetuate its owner's glory, but the secret tomb is usually hidden, mostly underground. While the mausoleum contains very little apart from the remains of the deceased, the secret tomb is adorned with abundant grave furnishings to allow the dead man to continue as long as possible the way of life he led on earth. It is true that the most famous example, the pyramids of Egypt, could be quoted as a contradiction; but it seems that these artificial mountains were destined in the minds of their creators not to impress the living but to establish a link between the two divine kingdoms of the gods, the sky and the depths. Every care was taken, in addition, to ensure the inviolability of the eternal residences within them.

The innumerable secret tombs throughout the world are clearly a prime objective for the archaeologist. It was not without good reason that C. W. Ceram called his book *Gods, Graves and Scholars*. Some civilisations would barely be known were it not for their tombs. For a long time this was true of Punic Carthage, for example, until the discovery in 1921 of the 'topheth', the sanctuary where children were sacrificed. However, this was on the whole no more than a special kind of cemetery. This was also true of Etruria. We would also know very little today about some aspects of Celtic life without the great tumuli of Champagne, Burgundy, Switzerland and Germany. The persistence of a half-pagan funerary cult among the Franks has supplemented the work of Gregory of Tours with precise data on the armaments, clothes and ornaments of the Merovingians.

Opposite page: since the days of Schliemann the adventure of archaeology has evoked an immediate and enthusiastic response. The site of the Jews' last stand against the Roman power was Masada, probably the most famous name in archaeology in our time. The call for helpers in the excavation brought hundreds of volunteers from all over the world.

Four columns standing *near the Mediterranean Sea at Leptis Magna in Libya.* The city, which was the birthplace of the emperor Septimius Severus, lies about 120 kilometres east of Tripoli and the excellent condition of many of its buildings is largely due to their having been buried in sand.

In addition to the tombs, there are rarer hiding-places for objects intended for the gods. These are known as *favissae*, from the Latin word meaning a pit. They were used for the burial of cult objects which had become unusable for one reason or another, but which must not be profaned. The best-known example is the pediment of the old temple of Athena on the Acropolis and the kore statues mutilated by the Persians and subsequently buried in the filling of the terrace on the east of the Parthenon, where they were rediscovered in 1882 and 1885.

A change in fashion is sometimes enough to lead to the replacing of cult statues no longer considered sufficiently handsome. In Tunisia at Cap Bon several hoards of Punic terracotta statues were found, which had been relegated there in the imperial period and replaced by marble statues. Christianity borrowed this custom like so many others. Even as late as 1964 a nineteenth-century retable, rejected by a Gascon priest as unsuitable for the new liturgy, was ceremonially buried, not without outcry. It can also happen that someone discovers a deposit of objects piously protected from the iconoclastic zeal of profaners. In 1889 P. Gauckler discovered in a fourth-century house in Carthage a cellar completely sealed off by a mosaic, containing the images of a whole family of gods hidden there to escape destruction by the monks, a decree of the Emperor Theodosius having authorised the suppression of idols.

Foundation deposits are an offering made on the building of a monument to propitiate the gods and ensure the survival of the edifice. They are usually buried immediately below a wall or door. The Sumerians used to make copper nails especially for this purpose, to which they added the statuette of a tutelary god and an inscription giving the name of the founder and the function of the building. These deposits thus constitute a historical record of the utmost importance. The Phoenician custom was slightly different. The oldest monument hitherto discovered at Carthage is a sort of miniature shrine, a square chamber with a small dome surrounded by a kind of labyrinth. Under one of the walls was buried a Phoenician lamp in the shape of a beaked bowl and an amphora with twisted handles covered with geometric decoration. This vase, which was made in the Cyclades, dates the little shrine to the second half of the eighth century BC. It is the earliest evidence of the presence of the Phoenicians in Africa. The custom of placing foundation deposits, adopted by a great many peoples, has survived until the present day with the ceremony of laying the foundation stone of a public building.

Personal hoards Treasures, properly speaking, are always buried for a practical reason. In a time of trouble, war or revolution, the victims, forced to take to flight or fearing looters, hide their most valued possessions in the hope of coming back to find them once peace has returned. Very often something prevents them from recovering their property. Hoards of money or precious objects thus enable us to determine the date of great upheavals and to establish their extent. A vase unearthed at Tarentum in 1911 contained 600 silver coins struck in the sixth century BC by various Greek cities of Italy and Greece proper. It was probably deposited at the time of the internecine war which destroyed Sybaris in 510 BC. Variously sized hoards of darics (Persian coins named after Darius) were lost on Mount Athos and in Attica by Xerxes' troops. The Beni Hassan jar from Egypt is probably evidence of the invasion of Alexander and the insecurity of the Hellenistic period is shown by numerous deposits throughout Western Asia. Every one of the civil wars which preceded the establishment of the Roman empire or disturbed the peace led to the formation of hoards. It was, however, the Germanic invasions of the Roman empire more than any other event which forced the inhabitants of the devastated provinces to hide their wealth. In 1936 the great French numismatist, Adrien Blanchet, listed 528 Gallic hoards dated to the years AD 255–80. Since then this number has increased.

One of the most remarkable recent finds is the treasure of Kaiseraugst in the Swiss canton of Aargau near Basle, found in 1961. A bulldozer exposed a group of silver dishes, tableware, coins and ingots which had been buried in a box at the time of the Alaman invasion of AD 351. The date could be established with great precision because the raw ingots bore the mark of the usurper Magnentius who came to power in Rome in 350 and died the following year. The distribution of deposits enables us to trace the route of the Barbarian hordes and to chart the extent of their ravages. Several objects discovered in a bed of ashes in the ruins of the baths of Antoninus at Carthage proved that this superb building was systematically destroyed by the Vandals under Genseric. Hoards, thoroughly studied, faithfully reflect the economic fluctuations of the years preceding a disaster. By means of them we can measure the rise in the standard of living brought about by the *Pax*

Mesopotamian cities were often built of unbaked brick which easily disintegrates into dust again, adding greatly to the problems of the excavation, as is shown by this view of a corner of the Isin-Larsa city of the late third to early second millennia.

Romana. We can also understand the gravity of the devaluation which jeopardised this prosperity during the third century, when administration was revolutionised to deal with it.

These conclusions can even be extended to the realm of social life. For instance, J. Heurgon was able to reconstruct, by analysis of the jewellery found at Tenès in Algeria, several generations in the history of a high ranking Gallic family who entered the imperial service in Constantine's reign. They were transferred to Rome, promoted to higher dignities, and then driven out by the Gothic invasion. They fled to Africa and stagnated there until the barbarians invaded this last haven.

In spite of laws in many countries which have always reserved ownership either partially or completely for the state, there have probably been more clandestine than declared discoveries. Popular imagination is stimulated by these wonderful windfalls, and surrounds them with legends. In fact, historical personages often play a part in them.

A Roman *eques*, or knight, persuaded Nero to undertake excavations to recover Dido's treasure. This charlatan, who paid dearly for his failure, has many descendants. There was one who claimed that the treasure of the Queen of Carthage, hidden 'by magicians and high initiates', moved about under the earth with a

The pyramid of Khephren surrounded by the tombs of high officials and the desert seen from the summit of the Great Pyramid of Cheops (about 2650 BC) at Giza. The pyramids of Egypt are amongst the most spectacular monuments ever built and were intended to preserve the king and his treasures for eternity. However, the magnificence of these great sepulchres inevitably attracted the attention of tomb-robbers; virtually all of them were looted and the treasures of the pharaohs stolen and dispersed.

The Nabataean city of Petra, an important centre in Roman times, was built in an unapproachable situation among the mountains and deserts of Jordan. The top of the citadel is reached by flight of steps cut out of the rock.

certain amount of noise, and would have to be seized while in motion. Another found enough dupes to finance a project for pumping out an alleged tunnel under Lake Tunis which was said to contain heaps of gold and precious stones. Human credulity being boundless, scarcely a year passes without some visionary more or less seriously combing the ruins of a castle or church in search of the treasure of the Templars or the Cathars.

The question of survival

Some buildings are fortunate. Although the whole town is changing all around them, they pass through the centuries almost intact, and are adapted to the taste of each era, sometimes reduced to unworthy functions for a while, but then subsequently rejuvenated to their pristine splendour. It would be impossible to draw up a list of all these survivors, or to tell their stories, so we shall confine ourselves to one specimen – the Maison Carrée at Nîmes, the vicissitudes of which have recently been recounted by J. C. Balty.

This temple was constructed about 15 BC by Agrippa, son-in-law and trusted adviser of Augustus, to serve as an official cult centre for the Roman colony recently established in the capital of the Volcae Arecomici. At the beginning of the present era the function of the building was defined rather than modified. It was dedicated to the cult of the two sons of Agrippa, Caius and Lucius, whom their imperial grandfather had adopted as his heirs but who vanished from the scene in their early youth. In fact, the sanctuary dominating the forum played the same role as the Capitol in other cities, and it was actually known as the Capitol towards the end of the empire. The Visigoths who took Nîmes in 471 did not alter its status in the least. Attached once more to the royal domain, the former temple still housed the court of pleas of the bishop and the count in the ninth century. Contrary to the belief of seventeenth- and eighteenth-century scholars, it was converted into neither a cathedral nor a mosque.

In the first half of the tenth century the Maison Carrée, still known as the Capitol, was transferred to the *vicomté* and became an aristocratic fief. Soon after 1000 its owner, Canon Pons, built on to the original fabric a church dedicated to St Étienne; it was at this time that the intercolumniations of the pronaos portico – the front of the temple – were walled up. The cella, or main apartment, was not yet used for Christian worship. When municipal organisation was restored at Nîmes during the eleventh century the counts of Toulouse, to whom the town belonged, authorised the corporation to take possession of the ancient building which, having become a town hall, continued to fulfil, more or less, its original function.

Unfortunately in the sixteenth century the corporation saw fit to exchange the Maison Carrée, as it was beginning to be called, for a building near the clock tower. The new owner, Pierre Boys, built a two-storey house against the south façade, causing serious damage to the architecture.

In 1573 the Duchesse d'Uzès made a vain attempt to buy it as a mausoleum for her husband but this curious project came to nothing and the Sieur de St Chaptes, who had inherited the monument, transformed it into a stable at the beginning of the seventeenth century. Meanwhile, Nîmes had adopted the Reformed Church, and the Church of St Étienne of the Capitol, corrupted to Cap d'œil, vanished in the disturbances. The re-establishment of the Catholic faith brought comparatively happy results. Augustinian friars, looking for a new home, bought the property from the Sieur de St Chaptes in 1670, for the first time public authorities and private individuals attempted to restore the building to its original form.

Intendant Basville tried to buy back the Maison Carrée for the royal estate and, having failed to achieve this, he took advantage of the poverty of the friars to have the restorations subsidised by the province and the diocese. The work was entrusted to a skilful and conscientious architect, Gabriel Dardailhon. The church in the interior was cleared out and the original architecture restored as far as possible. The chapel thus formed an entirely separate construction with walls a hair's breadth away from the ancient walls. Unfortunately, it had not been possible to prevent the Augustinians from knocking unsightly dormer windows through the superstructure, nor from perching on the roof a bell-tower of 'very sorry aspect', as it was described by Vicar-General Robert, who had the unseemly addition dismantled in 1702. The Maison Carrée had at least recovered its exterior appearance.

But in 1765 new misfortunes overtook the building. Étienne Laurens de Carmagnole, a former clerk, 'claiming to have the secret of discovering treasures with a divining rod, and certain that there was a considerable one buried beneath the Maison Carrée, took advantage of the credulity of the friars and awakened their cupidity; desire for great wealth induced them to make terms with this visionary. It was public knowledge that they had agreed between themselves to excavate in the cella and that half the treasure should go to the convent, a quarter to the Brother cook who had fortunately procured the co-operation of Carmagnole, and a quarter to the latter'.

The Fathers wanted to conduct the excavation at night in secrecy but Carmagnole demanded an advance payment which even the Augustinians were not simple enough to give him. The trickster then denounced the enterprise to the Intendant by anonymous letter. This was an opportunity for the municipality to reassert its claims to the building, the matter having formerly gone astray in the labyrinth of judicial procedure. Plans were being mooted to move the Augustinians to another house when the Revolution broke out.

During the eighteenth century the Maison Carrée was scientifically studied by an eminent scholar, Jean François Séguier. It was he who first succeeded in resolving the question of the dedication. The bronze letters composing it had been torn off probably in the early Middle Ages, and the words could only be reconstructed by means of the holes left by the nails which once attached them. In August 1758 Séguier succeeded in restoring the inscription in the names of Caius and Lucius Caesar. It was only recently that E. Espérandieu was able to demonstrate that this inscription was preceded by another text relating Agrippa's contribution. The superimposing of the marks left by each of the dedications made this epigraphical problem unusually difficult. Séguier also conducted new restoration carried out by the son or grandson of Gabriel Dardailhon, with a subsidy from the province of Languedoc.

Khaznat Far'un, or 'Pharaoh's Treasury'. Petra. Protected from wind, rain and damaging sand-storms the façade, cut back from the rock face, has retained much of the sharp line of the original sculpture.

The Revolution put the Augustinians to flight and during the Terror the Maison Carrée was used as a granary. The Directory restored it to its rightful function by making it the centre of the departmental administration; during the Consulate the archives were deposited there. In 1806 the Augustinian chapel was demolished, allowing the interior of the cella to be seen again. There were soon plans for transforming the former temple into a museum, but this project was not to be implemented until 1816 and was completed in 1822. The vicissitudes of Agrippa's temple had now come to an end, and only the arrangement of the interior and the display of the collection have been modified to suit the taste of each successive epoch.

The baths of Antoninus at Carthage in Tunisia still bear the marks of deliberate destruction inflicted by the Vandals who seem, in this case, to deserve the reputation which was later bestowed on them.

The history of the Maison Carrée is very similar to that of all the ancient structures, temples, baths, triumphal arches or theatres which survive in the hearts of our cities. The temple of Nimes, in spite of its many tribulations, is nonetheless among the most fortunate, since the people of that ancient city have always regarded it more or less as the symbol of civic liberty which has rarely been altogether obliterated. It is because of this feeling of reverence that the sanctuary built by Agrippa is at present almost the only completely intact example of classical Roman architecture.

Buried towns

The fact that we have to dig to recover the remains of former life surprises a great many people. However, not all ruins are buried. On the contrary, ancient soil is often fretted and eaten away by erosion, but in this case the buildings resting on it deprived of foundation, inevitably collapse, and at best only debris is left. Even when the ground level has not noticeably altered, as in the case of the Acropolis at Athens, exceptional circumstances are required for unprotected buildings to survive for long the thousands of destructive elements attacking them. Indeed, in an inhabited area buried buildings are the only ones with any chance of survival. In the open spaces of the desert, however, important constructions can survive almost indefinitely, but even here, in the end the elements generally wear away projecting parts of the structure and silt up others.

The causes of burial can be many. When a site continues to be inhabited, the ground level rises by a natural accumulation of debris and rubbish, the depositing of which is especially plentiful in periods of unrest or cultural regression. Garbage collection is one of the first public services to suffer under an unstable government. In addition, contractors hardly take the trouble to sink their foundations deep enough, preferring, on the contrary, to lay them on walls built by more skilful forerunners. Many poor people have contented themselves with setting up their wretched homes in a great deserted building. Occurrences of this kind

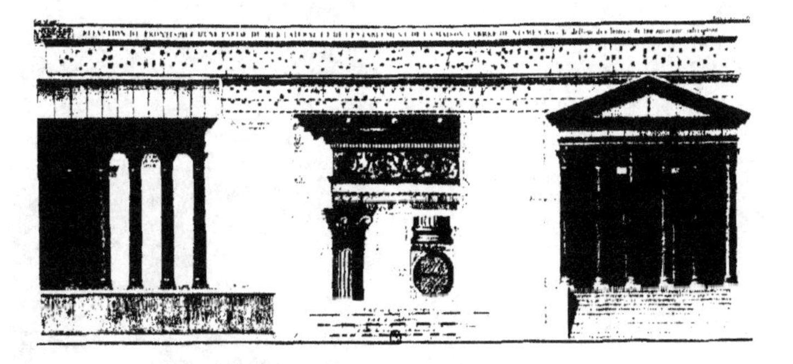

The Maison Carrée at Nimes is one of the few surviving temples of the age of Augustus. It has sometimes been misused, but the people of the city in which it stands have always valued it and protected it from complete destruction. Above: the inscription on the entablature which Jean François Séguier managed to decipher by means of the holes left by nails attaching bronze letters to it.

were almost universal at the height of the Middle Ages, and the practice survives to the present day. For instance, it was only necessary to move the village of Kastri to recover the sanctuary of Delphi.

On an uninhabited site fairly strong buildings situated in hollows (but not in the bottom of a valley, where erosion is too prevalent) get in the way of waterborne earth and windborne dust, and pile it up around themselves. The thickness and composition of the bed formed in this way are naturally very variable. It is rarely homogeneous, and is usually made of superimposed layers, some formed by natural deposits, others by human agency. When this site has been reoccupied after its destruction the people who settle there alter the ground level by building or farming, leaving traces of their passing each time. A building should never be cleared, therefore, without careful observation, photographing and drawing of the section through the soil covering it. The methods of carrying out this stratigraphy and its results will be discussed in the chapter on excavation techniques.

Some towns were violently buried by a disaster such as flooding or a volcanic eruption. The best known examples are Herculaneum and Pompeii. Herculaneum was engulfed in a river of mud consisting of ash and lapilli (small stones of lava) mixed with sand carried from the foot of Vesuvius by the torrential rains accompanying the eruption. This viscous torrent submerged first the outskirts and the suburban villas, and then crept into the streets, spreading as it went. Its movement was strong enough to push some of the walls down—for instance, those on top of the cavea of the theatre, where the audience sat—and to carry away some statues on its crest. Its depth varied in different places. Where it was confined between walls it rose to a height of twenty five metres above ground level. When it solidified, it formed a bed of soft rock, the *pappamonte*, which resembles the tufa of the Roman hills, but is not so strong.

Pompeii, on the other hand, was gently buried in a steady fall of windblown ash and lapilli which progressively covered the town with a blanket six or seven metres thick. This bed is stratified in superimposed layers and it is not so hard as the *pappamonte* of Herculaneum.

Dead cities of desert and forest An inhabited site often reverts to nature after man has briefly disrupted its normal course. Leptis Magna was originally an ancient Phoenician port founded about 650 BC on the Libyan coast about 125 kilometres east of Tripoli. At first it was built on a modest scale in the shelter of the estuary of a small neighbouring river, but after it had been taken over by the Romans the town flourished and grew as an outlet for the African trade. At the end of the second century AD one of its illustrious citizens, Septimius Severus, was raised to the imperial throne by the fortunes of politics. Wishing to put his native town on the same footing as the other great cities, he had an urban development plan carried out, the efficiency of which is acknowledged by the most advanced modern architects. He did not hesitate to divert the course of the river so as to be able to enlarge the estuary into an artificial harbour basin, or to establish in the dry river bed a majestic avenue leading to the new administrative centre consisting of a forum and a basilica of grandiose proportions.

Some centuries later, when the city was having difficulty in holding off the desert tribes who were determined to drive the Mediterranean peoples from this foothold, a violent flood broke the barriers which confined the river. The water returned to its old course, carrying with it huge quantities of sand which filled in the harbour and buried the walls and columns of the buildings up to their tops. Thus only the superstructures were lost, and in modern times Italian archaeologists have been able to recover them almost intact by clearing out the protective bed which saved them from looting.

The case of Leptis is similar to that of many other towns built on sites ill-adapted to the higher forms of civilised life. It needs only a relatively minor accident to cause the abandonment of the whole complex almost as abruptly as if it had been the victim of some major natural disaster. A military defeat normally causes only a minor break in the life of a well-situated city even when it is especially severe, but is often fatal to a cultural centre located in isolation in a hostile desert.

Palmyra never rose again after its conquest by the Roman Emperor Aurelian in AD 273. The fate of another Syrian 'caravan city' is even more typical. Seleucus, one of the generals of Alexander the Great, settled a group of Macedonian veterans at Dura, a strategic point on the upper Euphrates, and gave the colony the Hellenic

The South Viaduct, one of two built across the stream on which the Jordanian town of Jerash stands. It was the partial destruction or collapse of this viaduct at an unknown date, combined with the early disappearance of the North Viaduct, which prevented the Circassian settlers of 1878 from using the buildings on the west bank as a quarry.

Agrippa, who married Augustus' daughter and was one of his chief collaborators, had close connections with the province of Gaul and was responsible for many building operations, including the construction of the Maison Carrée.

name of Europos. It was laid out according to the geometric plan which the rationally minded Greeks favoured: in the centre was the agora, the civic assembly place so characteristic of the polis, or city-state, in contrast with the Near Eastern cities whose centre was religious or commercial.

Dura Europos outlived the kingdom which created it by three centuries. To its military function was added a cultural and commercial role which made it an essential relay station between the Mediterranean, Iranian and Semitic cultures. Unfortunately, in the third century the Sassanian dynasty of Persia aroused its subjects to fanatical nationalism. Dura Europos must have been one of the earliest victims of the bitter conflict between the empires of Rome and the shahs. About 260 the Sassanian King Shapur took the city where the Romans had quartered a garrison for more than a century. The siege was long and arduous; assailant and defender faced each other across scarp and counterscarp, where their desiccated corpses have subsequently been found. The city was abandoned and the sands covered it over.

In 1920 a French officer patrolling the frontier between Syria and Iraq saw rising from a dune the top of a wall covered with paintings. This sparked off a series of excavations financed by America and France under the leadership of Franz Cumont and M. I. Rostovtzeff, which led to discoveries important both for the history of art and for religion. Beside temples dedicated to the local gods were a sanctuary of Mithras, a synagogue and even a Christian church, the earliest known example. All these shrines were decorated with paintings showing an aesthetic concept entirely different from Greek classical art, and as indifferent to the exact representation of the human body as to that of objects, but imbued with mystic zeal which animated the followers of the different cults equally.

Other cities, it is true, were victims of neither accident nor attack, but gradually succumbed to a deterioration of the conditions which supported their existence. The land in Mesopotamia is scattered with high mounds where, at first sight, nothing seems to indicate a trace of human activity. These are, however, the remains of towns or palaces built of bricks laid on terraces of the same material. Here it is the material itself which has played the part of protective 'wrapping'. The walls crumbled little by little upon themselves, covering in a formless mass all the harder objects like carved stones or fired bricks, often marked with inscriptions. The mound, or 'tell', thus formed often became the site of a new settlement, which in its turn became a ruin. This cycle continued at Khorsabad until archaeologists dispossessed the most recent inhabitants.

Elsewhere it is vegetation which reconquers an area briefly held by mankind. The urban civilisations of the tropics often succumbed to assault by the forest and later on we shall look at the history of the Khmer towns and the Maya cities, suffocated one by one by the jungle which first cut them off from each other and then overwhelmed them. But if we are well informed about the relative decadence of the ancient Indochinese state and the shift of its metropolitan centres, the return of the peoples of Yucatán to a simpler way of life, particularly to living in huts, sets a problem which has not yet been satisfactorily resolved. Vegetation crept triumphantly over the buildings, eating them away and ending by fragmenting them, but the majority of the vertical walls offered no

A general view of Jerash from the Temple of Zeus, with the Forum, Street of Columns, South Viaduct and Temple of Artemis. Forums were usually built on a rectangular plan but for some unknown reason that at Jerash forms an oval.

Palmyra, the ancient city of central Syria, attained real importance only after Roman control was established about AD 30. Later in its decline Palmyra was taken by the Arabs and sacked by Tamerlane, and its ruins were forgotten until the seventeenth century. The swirling desert sand, kept in constant motion by the wind, accelerates the ruin, slowly wearing down the marble and stone carvings.

foothold and the main elements of the fallen structures, especially the roofs, stayed where they were, presenting the archaeologist with the pieces of something like an enormous constructional toy.

Great care is required, however, in clearing a ruin of clinging vegetation. The earliest explorers of the Maya sites limited their efforts to burning the forest, but the fire caused serious damage to the buildings. Moreover, the trees often held the masonry together, and when one fell, the other came down with it.

It is not generally known that ancient towns and buildings swallowed up by forests exist even in Europe. In the Paris basin in Normandy many Gallo-Roman shrines are hidden in the woods. At Venou in Cher J. Favière discovered under some brushwood the ruins of a village dating to the end of the Middle Ages, the earliest inhabitants of the site probably going back to the Gallic period, since there are burial tumuli *rising from the neighbouring fields*. The woods of France are less luxuriant and less formidable to the archaeologist than the forests of Yucatán and Cambodia. Nonetheless woods growing on ancient stones are a favourite haunt of poisonous snakes, and the legends of these venomous creatures guarding imaginary hidden treasure are not without foundation.

Having discussed different kinds of dead cities, it is time to examine a site and its excavation from the beginning. It will thus be possible to give a very precise account of how the archaeologist locates a dead city, and the way in which he can restore it to life.

Maktar Maktar is situated near the centre of Tunisia, on the same latitude as Le Kef to the west and Kairouan to the east, almost halfway between these two towns, i.e. about 160 kilometres south-west of Tunis. When the Carthaginians extended their dominion into the interior of Africa in the fifth century BC a fairly dense population inhabited this diversely contoured region (mountain chains, plateaux and inland plains) with its hot but healthy climate and luxuriant vegetation. These Numidians, a branch of the great Berber family, raised megalithic tombs very like the European dolmens. It is possible that they already practised agriculture.

In their land the Phoenicians founded cities *grouped together in districts. Maktar, established* on a plateau 1000 metres high, thus became the centre of a fairly extensive province numbering fifty villages. Disputed between the Phoenicians, the Numidians and the Romans, it did not suffer many wars because of its altitude which kept the main highways away from it. The language and culture remained almost exclusively Phoenician or Libyan until the end of the first century AD when Maktar was drawn into the great

This bronze jockey was found in the sea off Cape Artemisiun near Athens. It was made in the second century BC.

current of prosperity and modernisation flowing throughout Africa. The inhabitants, among whom many Romans had already settled, adopted the Latin tongue and customs, acquired citizenship for themselves and, about 175, for their city the status of a 'federated' colony of Rome. Punic traditions survived only in religious practice, and finally disappeared when the city adopted Christianity in the fourth century.

Although the end of the *Pax Romana* certainly brought troubles, no radical break can be seen in the life of the community until the height of the Middle Ages. In particular, most, if not all, of the citizens remained faithful to Christianity even after the Arab conquest. Nevertheless, conditions of urban life were becoming progressively more difficult. The fatal blow to the 1500-year-old city was struck by the Hillalian invasions of the eleventh century, which made a desert of the entire interior of the Maghrib. The site of Maktar was thenceforward unable to support human habitation until the second half of the nineteenth century. Only the name survived, attached to a market at which the neighbouring shepherds assembled every week.

The site of the ancient city, covering more than twenty hectares, is located on a line of ridges which divide the rainfall between two river valleys. Near their source these valleys diverge for a while before reuniting to make this plateau into a kind of island. The river Saboun on the north is very close, its steeply sloping banks defining the town limits. The river Ouzafa on the south is, on the other hand, three to four kilometres away. The slope, therefore, descends gently on this side, scarcely accentuated by several dry watercourses. Since the end of the ancient world ground level has risen an average of two metres in the eastern and western sections of this southern slope. These two protected areas are separated by a dry valley running more or less north–south where erosion has stripped off and even completely destroyed the ancient soil. The Saboun river valley has become deeper since the classical period, but without growing any wider on the town side. This valley is fertilised by the numerous springs rising on its slopes, although the surface of the plateau is arid limestone. The modern village, too, is situated on the slope to the north east of the ancient complex, which is largely deserted.

A number of buildings have successfully resisted the attacks of time. The most impressive is a large thermal establishment of the second half of the second century in the south-east sector of the town. These huge baths were constructed entirely of stone as a precaution against fire, and roofed with rubble vaulting bound with the celebrated Roman mortar. In the damp climate of western Europe this cement decomposes when it is buried; dry conditions, on the

The west portico and buildings of the Great Palaestra, Herculaneum. The Palaestra was larger than usual for the size of the town and included a large meeting hall and a cross-shaped swimming pool. According to legend Herculaneum, which lies at the foot of Mount Vesuvius on the Bay of Naples, was founded by Hercules on his return from Spain. In AD 79 it was buried along with Pompeii when Vesuvius erupted and volcanic mud flowed through and filled up every space—often leaving roofs and upper storeys in position. The ruins were first discovered in 1709 and by careful excavation it has been possible to preserve the houses as they stood.

The Hadrianic baths, Leptis Magna, were large and luxurious. They were dedicated in AD 126–27, and were probably exclusively used by men.

other hand, render it unbreakably solid. This is one of the reasons why Roman buildings are better preserved in hot dry countries like Spain, North Africa and Syria. The south-east baths at Maktar had two storeys, which was rather unusual. The underground structure, which has been investigated only partially, is absolutely intact. On the ground floor the great central hall with cold baths round it has lost the crown of the vault, but the walls round the base of the vault are still intact. The complex looks like a mound with two sturdy parallel walls emerging, crowned with masses of overhanging cement. The first thing that had to be done was to reinforce and restore these remnants of superstructure. Then it was possible to clear the three to four metres of fill (dust and sand) which had accumulated in the great hall. The baths and the mosaic floor reappeared, as well as the remains of makeshift structures built inside the Roman walls in Byzantine times, or during the medieval Islamic period. The excavation was then extended to the outer rooms, clearing notably a palaestra or gymnastic court surrounded by a fine colonnade with its arcades almost intact.

The town had another bathing establishment of almost the same date in the western sector. There, too, only a part of the vault had fallen in. The fill was even more important as erosion was practically nil in this part of the site. The ancient ground level was reached, on the average, four or five metres below the present level. Thus we were able to discover that in the fifth century the great hall of the frigidarium had been transformed into a church.

The visible ruins also included two triumphal arches. The one in the centre of the site, bearing a perfectly legible dedication to Trajan, turned out to be the main entry to the forum. The other, immediately overlooking the valley of the Saboun, marks the point where the highway from Carthage entered the town. Two tower-shaped monuments, one partially collapsed, the other proudly lifting its pyramidal crown, indicate the site of the main cemeteries on the outskirts. Finally, a kilometre to the east a row of arches indicates the course of an aqueduct. Besides these immediately recognisable structures, a quantity of scattered stones, often bearing inscriptions in Latin, Punic or Libyan, confirm the importance of the community which lived there for fifteen centuries.

These striking remains attracted the attention of an English explorer named Temple in the eighteenth century and were sometimes visited and reviewed, particularly in epigraphic studies during the nineteenth century. When the French protectorate was set up Maktar became the centre of civic authority, entrusted at first to an Algerian captain named Bordier, who first lived inside Trajan's arch, then in the great baths, and conducted the first excavations.

In 1893 he explored the temple of the Punic god Hoter Miskar, situated about fifty metres to the north-east of Trajan's arch. He chanced upon several Phoenician inscriptions, one of which, commemorating the construction of the sanctuary by a civic body, is one of the most important epigraphic records ever discovered in this language. A church built by a bishop named Rutilius was also discovered.

A study of the Latin inscriptions threw a remarkable light on social conditions in the city during the imperial period. An epitaph in verse dating from the middle of the third century recounted the exemplary life of a poor peasant. Beginning as a simple farm labourer, he became an overseer and at length rose to be owner of the estate by his hard work and thrift, and achieved a life of ease and civic honour.

Maktar, however, was very isolated. Even in 1920 it could only be reached from Tunis by a day's train journey followed by another day on muleback. The first archaeologist to make any protracted stay there, L. Châtelain, who was later Director of Antiquities of Morocco, was unable to clear more than a few scattered buildings in 1910 and 1912, the most important of which seems to have been a market.

These unprotected remains unfortunately sustained a great deal of damage during the subsequent years. It was, therefore, an almost untouched site which was examined in 1944. At that time the means were not available for the investigation of the largest buildings. A limited survey was therefore undertaken and a place with no signs of habitation was discovered to the south-west of the complex. Here the tops of several ancient walls could be traced on the surface, near to which two half buried columns were still standing. The first soundings revealed that the columns were firmly set on their bases, which in turn rested on an ancient pavement just under two metres below ground level. After a few weeks it was realised that the façade of a basilica, with a colonnaded court in front of it, had been discovered.

The earliest purpose of this building, used as a church from the fourth to the sixth centuries, was revealed only later by a very precisely dated inscription of AD 88 in the reign of Domitian. It was a kind of clubhouse for the youths of the district who received there a sort of physical, cultural and political education in a semi-military school. Extending the excavation of this 'college', the road which served it was found, running almost straight from north to south. This highway would probably connect with others, and work proceeded. Little by little the layout of the town centre was exposed, arranged round two large paved squares. The first, less than 100 metres from the college on the same road, already existed before Maktar became a Roman dependency, when it was inhabited only by Numidian and Punic peoples. The square was dominated by the temple of Bacchus, one of the tutelary gods of the city. There were temples to the two other patron deities, Apollo and Ceres, outside the walls. Later, when Maktar's population included a significant number of Roman citizens, Italian emigrants or enfranchised natives, they established a classical forum for their own use, the ceremonial entrance to which (the arch mentioned above) was dedicated in 116 to the Emperor Trajan.

For ten years Maktar was thus the subject of continual archaeological investigation and, to house the archaeologists and facilitate their work, a house and an excavation store were built. The work is still being continued from the Tunisian National Archaeological Institute, and during the last few years it has excavated the basilica of Rutilius and the amphitheatre from the clutter of modern buildings which obscured them.

Maktar is thus an impressive spectacle for tourists, particularly because of the exceptional state of preservation of the great baths, in the middle of a landscape where the harshness of the bare plateaux contrasts with the charm of the mountain gardens planted with rose bushes and raspberry canes. It is also, and no doubt will be for a long time to come, a veritable mine of inscriptions which enable the historian to follow, by concrete and precise examples, the integration of a typical African community into the imperial civilisation, and its adaptation to that culture.

Water as the guardian of the past

The seas, lakes and rivers have attracted the most daring men for thousands of years. Until the rise of modern industry in the eighteenth century, maritime trading was almost the only peaceful way to grow rich quickly. Trading, however, was very risky. The cargoes consisted of rare and precious merchandise: except in unusual circumstances it was not a commercial proposition to transport heavy goods of low value over long distances, even when they were necessities like corn. Furthermore, the vessels of the time were at the mercy of the elements, and shipwrecks were common.

A large part of the wealth of the ancient world, therefore, lies at the bottom of the sea. These treasures are comparatively safe there: objects submerged in deep water rest in peace in a gradually thickening coating of mud.

The Roman baths of the city of Maktar, which lay on a plateau near the heart of present-day Tunisia. The city flourished for 1500 years: it was founded by the Carthaginians and only died in the eleventh century AD when the Hillalian invasions brought ruin to a once prosperous area of North Africa. The Romans, in their period of ascendancy, made good use of the ancient city and evidence of their occupation is plentiful. Protected from vandalism by its remote situation, and from disintegration by the excellent climate, the masonry is exceptionally well preserved.

The magnificent early classical statue of the Greek god Zeus (or possibly Poseidon) was dredged from the sea in 1928 by fishermen off Cape Artemisium at the northern tip of Euboea.

A typical Mediterranean square-rigged ship carved on a Phoenician gravestone.

Among the water creatures are some species of worms and shellfish which make their homes in stone. Admittedly, water and the chemicals dissolved in it have a detrimental effect, especially on metals; but the resulting oxides and salts form a protective casing in themselves, which can easily be removed from the object when it is recovered. Finally, although many men have dreamed of treasure and of visiting sunken cities, and have gone to great lengths to find them, it is only in this century that a technique has been evolved which enables divers to work efficiently under water for long periods of time.

Lost cargoes: Mahdia There are many records of fishermen whose nets have drawn up objects of value. A curious bas-relief from Ostia shows the recovery of a statue of Hercules which was naturally regarded as miraculous and oracular. The first real underwater finds, however, took place on the east coast of Tunisia in the open sea at Mahdia from 1907 onwards. Some Greek divers who were fishing for sponges in these waters noticed some large cylinders sticking out of the sand at a depth of forty metres, which they took for the cannon of a sunken man-of-war. They were in fact marble columns. On closer inspection the divers discovered marble and bronze statues, the latter disfigured by thousands of wormholes, the former by calcareous accretions. They collected several and tried to sell them.

The Tunisian antiquities service got wind of this strange transaction, the source of which was revealed by an enquiry. The discoveries were officially taken over and work carried on until 1913. Interrupted by the First World War, they were resumed only in 1948 by the French team of Tailliez and Cousteau, who had meanwhile entirely revolutionised the technique of underwater exploration by replacing the heavy diving suit with light equipment allowing great freedom of movement. Father Poidebard, to whom we shall shortly return, took part in the expedition, which had its base on the vessel *Elie Monnier*. Unfortunately, the guide marks to the wreck, placed there forty years earlier, were no longer recognisable. To find the place again a whole new investigation was required, conducted by means of an undersea sledge, a kind of submarine aquaplane.

After five days of exhaustive searching the galley was discovered—220 metres from the place noted in the records of the earlier expeditions. The project was then taken up by volunteers from the underwater research club of Tunisia, led by G. de Frondeville. These investigations demonstrated that, contrary to all hopes, little was left in the galley. The cargo had been scattered on the sea bed without sinking into it or being silted over, and the larger part of it had easily been gathered up by the Greek divers. However, some very precise and interesting information was obtained about the ship and its cargo, which had previously been inaccurately described. It was not actually a galley, which is a long vessel powered essentially by oars, but a corbita (a word which, corrupted to 'corvette', later came to mean an entirely different type of ship), that is, a big sailing vessel, probably not decked, about thirty metres long by ten wide, with a capacity of up to 250 tons. The bulk of the cargo consisted of sixty columns, each one weighing just over three tons, which were wedged fore and aft in the hold in piles of five or six deep. In the spaces between them and on top of the columns the capitals were firmly stowed, together with blocks of marble, bowls, candelabra, bronze statues and furniture.

The home port of the ship could be accurately determined. Among the objects found were stelae engraved with Greek inscriptions, decrees and official records published in Attica, or more precisely, in the Piraeus. The corbita, then, came from this port. Four large marble bowls, candelabra of the same material destined for garden ornaments, and capitals decorated with griffin heads—all showed characteristics of Athenian work of the second and early first centuries BC. At that time Athens was going through a difficult period. For a hundred years it had enjoyed some degree of political prestige because of its alliance with Rome; but Rome was lacking in discretion and, moreover, favoured the oligarchic party.

At the beginning of the first century BC King Mithridates Eupator of Pontus called on the Hellenes to return to independence and democracy. He convinced a number of Athenian intellectuals, who were swept into power by the populace. In 86 BC, however, the Roman dictator Sulla defeated the armies of Pontus and laid siege to Athens. The city resisted courageously, but ended by surrendering and had to pay dearly for its impulse towards independence. Unscrupulous traders, who abounded in Sulla's retinue, descended on the city.

The Mahdia ship was one of those in which they carried off their loot to Italy. In its cargo were found not only cruelly ravaged treasures like the statue of Agon by Boethos of Chalcedon with its base carelessly broken off, but also some articles specially made for export such as the bowls and candelabra, which had only just come unfinished from the workshop. The hoard was probably amassed some years after the siege by a merchant rather than by a predatory army officer. Its destination was most likely the villa of one of those immensely rich Roman senators who absorbed all the wealth of the Mediterranean world after the conquest, and reproduced in their Italian residences the luxury of the Hellenistic palaces. The main structure of the villa was, no doubt, already complete, and its owner awaited only the arrival of the finest decorations.

However, the vessel carrying them was overtaken by a storm in the dangerous waters of Cape Malea, and was blown to within sight of the African coast where it foundered. The majority of the crew were probably able to put off in the lifeboat, and perhaps reached safety. Few human bones were found in the wreck.

This lost cargo contained at least two masterpieces: a bronze statue of a winged Eros-like youth, and a herm or pillar surmounted by a bearded head of Dionysus. Herms were usually set up in the palaestra to invoke the patron divinity of sport. The winged youth, who has been called Agon, is the spirit of competition. Winner of several races, he is shown placing on his head the filet which was the prize of victory. There is no other ancient work which symbolises the sporting spirit of Greece so well. Perhaps the bronzeworker who made it did not have creative genius of the highest order, but he was thoroughly versed in the school of the great masters of the fourth century BC. His Agon combines the sullen grace of the Praxitelean statues and the lively animation characteristic of Lysippus.

The artist's name is known also, for a signature was engraved on one of the bosses which takes the place of arms on the herm: 'Boethos of Chalcedon made this.' He is also known to be the maker of a famous statue of a child trying to strangle a goose, which was frequently copied. Flourishing in the third century BC, he founded a school in which his sons and grandsons followed him. We are exceptionally lucky to possess an original signed work by such a famous artist. If it had not been preserved for us by the shipwreck this group would, in all probability, have vanished into some medieval smith's crucible.

Submerged towns and sunken statues

Another bronze of even more outstanding power was preserved in similar conditions. This is the monumental Zeus, 2.60 metres high, striding forward brandishing a thunderbolt, which is the glory of the National Museum of Athens today (see chapter 11). It was recovered from the open sea in 1926 at the north cape of the island of Euboea, opposite the town of Histiaea. As at Mahdia, the statue was lost in a shipwreck. Other portions of the cargo were scattered on the sea bed which is forty metres deep around its resting place, and its sides still bore marks of the ropes with which it was stowed. The Greeks dedicated the northern cape of the big island to Artemis. It was here that they first engaged the Persians at sea in 480 BC, the prelude to their great victory over them at Salamis.

It would be pleasing if we could establish a link between this event and the statue, an outstandingly important piece from any point of view, which could only have been commissioned in exceptional circumstances. Zeus the Thunderer, creator and maintainer of the rule of reason against the forces of evil: who better to symbolise the struggle of the Greeks refusing to accept the yoke of the barbarian? The Artemisian Zeus is characteristic of the

Roman amphorae being salvaged by divers near Naples, Italy.

Bronze statue of a boy, probably by Praxiteles. It was recovered from the Bay of Marathon in 1925 and is now in the National Museum, Athens. The statue is slightly smaller than life size and the tousled hair has an encircling band with an upturned point at the front. This was a characteristic fashion of youths who competed in the athletic contests of the palaestra, in imitation of Hermes, their patron.

early classical style before the rise of Phidias. He has the serene strength of the Apollo from the Olympia pediment of exactly the same date, 465 BC, fifteen years after the battle—fifteen years of such lively cultural and political activity which changed Greece so fundamentally that the archaeologist should beware of insisting on the historical link, however attractive at first sight.

The catalogue of bronzes recovered from wrecks is already long and growing longer every year. After the youth of Marathon came the Anticythera jockey. Again, in the summer of 1964, a fine kouros emerged from the Hérault river at Agde, while other divers found the wreck of a Ligurian or Celtic ship laden with weapons and tools of the Hallstatt period on the same part of the Languedoc coast. Underwater investigations are not only useful in the field of art history, but also provide invaluable information about economic history. Two men in particular have contributed to the development of this branch of research in recent years: they are the Frenchman Fernand Benoit and the Italian Nino Lamboglia. Thanks to such archaeologists with their concentration on the study of pottery, and to such diving experts as the Taillez-Cousteau team, this has been one of the most growing branches of archaeology.

All the divers who spend their summers under water off the Côte d'Azur, Corsica and the Italian Riviera know that the sight of whole fields of amphorae on the sea bed is not uncommon. Some have discovered more valuable quarry than fish, which in any case quickly desert a region which has become the haunt of frogmen. A certain intellectual snobbery has helped to raise the prices of amphorae to phenomenal heights, and if the authorities had not controlled this traffic, which no legal measures could have foreseen, every single potsherd would have vanished from the north-west Mediterranean.

But why, it may be asked, should we deprive the holidaymaker of the triumph of fishing up and carrying off a badly made terracotta vessel as valueless in the ancient world as a plastic bottle to us?

Although each individual pot has only limited value—and there are always exceptions which only the expert is qualified to recognise—the total of the cargo provides extremely important information on the history of industry and trade. When a vessel is found loaded with new crockery for sale by the local agent of a firm, it is *a priori* virtually certain that all the merchandise was made at the same time at the same centre. Since there is always one datable piece, and since other chronological evidence can nearly always be found in the equipment of the ship itself, we have in this cargo an important sequence of dated pottery. After that a single pot, or even a fragment from this series, found on any site is enough to provide clear evidence of the date.

In recent years a dozen of these wrecks have been properly investigated in French and Italian waters. An outstanding find was at Grand Congloué, off Marseille. The research, conducted here under the direction of F. Benoit by the Cousteau team, recovered no less than 6000 pieces of black glazed ware made in Italy (in Campania, around Rome and in Etruria). The spread of this pottery during the last three centuries BC enables us to measure the economic consequences of the Roman conquests. The stamps on many of these pieces allow their date and provenance to be precisely established.

With them were a large number of amphorae which once held oil or wine, some of the wine jars bearing a seal inscribed in Greek with the names of magistrates of Rhodes. This confirms the great economic prosperity of the island which dominated the eastern Mediterranean particularly during the second century BC and contributed largely to the introduction of Roman influence there. Other amphorae bore the seal of Marcus Sestius, an Italian merchant with widespread contacts, whose dealings with Greece and Spain are recorded.

Such a project, while it sets problems for the specialist which need not be described here, amply justifies the report presented by F. Benoit to the Académie des Inscriptions in 1961: 'While excavations on land produce broken potsherds with no record of provenance

Greek sponge divers first discovered the sunken ship at Mahdia which contained vast quantities of Greek works of art—enough to fill five galleries at the Bardo Museum in Tunis. Far left: a bronze statue of a winged Eros-like youth. Above: a bronze herm by the Greek sculptor Boethos of Chalcedon which provides a rare signed work of the Hellenistic era. Left: head of Pan, or a satyr.

The head of a man whose body was discovered in a peat bog at Tollund in Denmark. He died by being strangled, probably as part of a ritual, and the rope was still in place round his neck. The condition of the head is evidence of the amazing preservative quality of the peat bogs.

The Gundestrup cauldron was discovered in 1891 in a peat bog at Gundestrup in Denmark. It is made of silver-plated copper and probably belongs to the first century BC. It was presumably used for ritual purposes and is important for the understanding of the religion in Celtic Europe and Scandinavia in the Roman period. It is richly decorated in high relief, on the inside with mythological scenes and on the outside with the images of unidentified deities.

and too often without a definite chronological link, the cargo of a ship is a slice of life. It preserves all the elements of an abruptly terminated moment of history—the construction and rigging of the ship, the living conditions on board, the merchandise and the pottery in use at a given date. It even provides evidence for contacts between two parts of the world, without misleading factors being able to creep in.'

In the same report it was also pointed out that forty wrecks containing amphorae have been found between Port Vendres and Monaco at a depth of less than fifty metres. An even more surprising fact is that no medieval ships, either warships or traders, have left any traces, probably because the cargoes have been completely silted over. This last statement says more than several lengthy volumes about the appalling deterioration in technology which overtook our ancestors after the fall of the empire, and is in itself an adequate comment on the navigational theories advanced about the alleged superiority of the Middle Ages over the ancient world, especially in the matter of land and sea travel.

Wrecks, however, represent no more than a fraction of the archaeological wealth hidden in the sea. From Brittany to Greece legends of sunken cities have been handed down. Although the legendary Ys in Brittany has never been discovered and will probably remain in the domain of myth, its Greek counterpart, Helike, which was overwhelmed by the sea and its ruins silted over with the alluvial deposits of the Peloponnesian torrents, has been located. In this case the catastrophe was attributed to the wrath of Poseidon, god of earthquakes and floods, which is entirely reasonable in view of the seismic nature of the earth's crust in the Balkan peninsula. The slight variations in sea level at various points on the Mediterranean coastline, however, have been enough to submerge areas which were all once habitable. This is particularly true of Fos, which is popularly regarded as the Provençal Ys. Houses which were still inhabited in the fourth century are now three metres under the surface. At Acholla, now known as Boutria, forty kilometres north of Sfax on the Tunisian coast, the L-shaped mole of the Roman port can clearly be seen beneath one's boat. At Baiae, an elegant seaside resort in the time of Cicero and of Augustus, an entire district of the town has been submerged.

Submerged towns are few, and none of them has been investigated, but it has been possible to relocate some important buildings under the sea. It was in this way that Father Poidebard, who is well known in another field as a pioneer of archaeology based on aerial photography, explored the ancient port of Tyre. More recently the Israelis have obtained significant results at Caesarea in Palestine where Herod ran a shipyard on very progressive lines. These investigations are still very difficult. Submerged blocks, eaten away by the waves and often overgrown with seaweed, can easily be mistaken for natural formations. Even when they are recognised it is very difficult to determine the date when they were deposited. At Baiae, nevertheless, Italian archaeologists have managed to apply the stratigraphical grid method under the water, enabling them to record the exact provenance of each object.

In recent years the same Campanian teams have also been able to practise one of the most attractive forms of underwater archaeology—the exploration of the grottoes on the shore which the great Roman aristocrats of the early part of the present era had sumptuously fitted up, not only with fishponds but with ornamental statues and mosaics. Tiberius was particularly fond of these *antra*: in the cavern at Sperlonga which belonged to him (it is situated not far from Gaeta) a group of monumental marbles illustrated the adventures of Ulysses in conflict with the Cyclops and Scylla. These statues, which have now been recovered and largely restored, were the work of the famous Rhodian sculptors Hagesandros, Athenodorus and Polydorus, who also made the celebrated Laocoön now in the Vatican. In 1964 P. de Franciscis, Superintendent of Antiquities of Campania, found a marble Neptune and Triton in the famous Blue Grotto of Capri.

Wherever the great maritime civilisations developed, the shores are rich in remains. This is true, for example, of the Scandinavian fjords. In the Roskilde fjord in Denmark five warships and trading vessels were scuttled in the eleventh century to stop pirates from getting into the inlet. They were refloated in 1962 and are currently being restored.

Treasures in streams, lakes and peat-bogs

Rivers and lakes also harbour treasures. Vast numbers of ancient artifacts lie at the bottom of European rivers and are sometimes picked up by dredgers. They are usually bronze weapons and tools of the Hallstatt period (about 1000–500 BC), indicating the intensive trade linking, by way of Gaul, the great Mediterranean nations with the British and Armorican tin mines. There are also remains of yet more wine jars, the staple Greek and Roman export to the Celtic world. In two months alone the dredging of the port of Chalon sur Saône recovered more than 20,000 amphora points. One of the finest of the Apollo statues of the severe style just prior to the time of Phidias is known from a copy found in the Tiber. It is scarcely necessary to mention the role played by lakes in the Neolithic alpine regions. Most of our information about the European civilisations of this period comes from the lake villages built on piles, more than 200 of which are recorded.

The part played by these inland seas was still important enough at the beginning of the first millennium to leave its mark on the Greek legend of the Argonauts. Lake Nemi in central Italy remained the centre of an astonishingly primitive cult until the end of the Roman Empire. It was long known that there were sunken vessels in these waters. To recover them the Italian government proceeded to drain the lake temporarily from 1926 to 1932. In this way two immense barges were discovered, seventy metres long by twenty wide, which had never actually sailed, but which still preserved traces of sumptuous decoration. The ships and the museum which housed them were unfortunately destroyed by the retreating Nazis in 1944.

The peat-bogs of northern Europe offer extremely favourable conditions for the preservation of organic matter and metal. It was the custom of the north Germanic peoples to throw their spoils of war into certain marshes to propitiate the gods, as well as the bodies of various executed captives. We owe to this custom the preservation of many works of art, the most outstanding of which is the Gundestrup cauldron. The reliefs which decorate this silver vessel provide a remarkable illustration of Celtic mythology. As for the amazingly well preserved bodies, like that of the Tollund man, who still wears round his neck the rope with which he was hanged but whose features have kept a serene, almost smiling expression, the sight of them rouses profound emotion. Wood and fabrics have not deteriorated in the least, and as a result the Copenhagen Museum has a collection of some of the best-preserved ancient garments in the world.

It can be seen by this survey, then, that whilst records of the many stages of the development of mankind have been preserved they are sometimes incomplete, and chance plays an important part in the preservation of the elusive fragments which contribute so much to our knowledge of life in the past. Many of the people who have been preserved for posterity died violently before their time by natural or human agency; their homes and cities were laid waste, and even if they could have foreseen the kind of immortality which was to compensate them for their decease it is doubtful if the prospect would have been sufficient consolation. Only a few privileged beings who generally lived better than their contemporaries lay preserved in the depths of tombs, until the day when the archaeologists disturbed their resting places and exhibited them in museums.

This inheritance which preserves the personalities of our ancestors demands the most scrupulous respect from us. To destroy it by carelessness or stupidity or for personal gain is both an act of impiety towards the dead and a waste of the patrimony for which we are accountable to our descendants. The responsibility resting on each human being who comes in contact with the relics of the past, be he professional or amateur, is naturally heavier than on those who have chosen other ways of life. The archaeologist is required to submit to a rigid discipline, the rules of which were slowly evolved and are always in the process of improvement. It demands uncompromisingly that he takes every care to preserve, because they are irreplaceable, the remnants of the past and he must extract from them all the information they contain.

The Oseberg funeral ship at the time of its excavation in 1904. The ship was well preserved in the soil and in the centre there was a burial chamber which contained the remains of a woman, faint traces of a second woman, and some superb treasures in carved wood.

How to locate a site

A figure found at Bouray near Paris, made of iron and cast bronze. The eyes, one of which is missing, were of inlaid enamel. Dating from the first century AD, the figure was identified as that of a god since it wears round the neck a torc, the Celtic symbol of divinity.

Among the most frequent questions asked by visitors to an excavation site are the following: 'Why did you decide to dig here—rather than elsewhere in the immediate neighbourhood? How did you know there was something to be found when there was nothing to see?'

These questions and many others like them credit the archaeologist with a sort of second sight in locating the undiscovered site he is investigating and, by attributing his choice largely to instinct, seem to exclude the operation of chance. There are, of course, many methods of discovering a site by rational means when an excavation is planned, and these methods have made striking advances in recent decades. It remains true, nevertheless, that in many cases chance alone has led to discoveries of major importance.

Chance finds

These partially accidental and involuntary discoveries known as chance finds, which are very often not confined to the object itself but lead to an excavation and thus the discovery of a site, are set in motion either by human agency or by natural forces. In fact, even natural erosion can expose the first traces of a lost site. Two elements share this function between them: wind and, more especially, water. Rain in the form of streams can wash away the bed of earth and expose a carved stone or a line of blocks, the characteristic sign of an outcropping buried wall. From year to year travellers and archaeologists very often note the appearance after winter of hitherto unseen remains in places they thought they knew well. Even on a site which was located long since and is already under investigation it is common knowledge that an ordinary shower can reveal such items as potsherds or small bronzes in the sides of a trench or a sunken road and, even better, it washes them and restores them to their true colours. A small bronze sphinx was discovered in this way after a rain squall by some visitors to the Argive Heraeum. Rain can also cause subsidence of terraces or even walls, revealing the secrets they hide. In the same way, but on a larger scale, a river which is finding new courses, or the sea eating away at a cliff-face, can bring prehistoric complexes to light. A fall in the level of a lake can expose previously submerged and unknown remains. It was in this way that the lake village of Zurich, for example, was discovered. This settlement was formerly established on the marshy shore of the lake and later submerged in the water, which preserved it. Like water, wind can free long-buried objects.

Mesolithic flint tools as well as more recent pottery have been found in the Sahara Desert right on the surface as if they had only just been deposited there.

Human activity, however, is more varied than that of the forces of nature. The search for the necessities of life, the planning of farming space or even simple games can lead to chance finds and subsequently to the discovery of an archaeological site. There are many examples of this process. It was a peasant returning from his field who discovered the corbel-vaulted tomb at Ras Shamra (Ugarit) on the Syrian coast, and in other countries many a peasant has felt his plough jam against the stones or the wall of a Gallo-Roman villa. In Malta the hypogeum at Hal Saflieni was discovered accidentally by workmen building new houses, whilst in Greece fishermen brought up in their nets the splendid bronze known as the Artemisium Zeus at the north cape of Euboea. Terracing and road-building operations have been responsible for numerous discoveries. This is not the place to discuss chance finds on already well-known sites such as, for example, the tombs discovered when foundations for a new house were being sunk in Athens and several other towns, or the discovery of a whole group of statues when a sewer was being dug in the Piraeus in 1959, one of the most remarkable of which was a sixth-century BC original bronze kouros. We shall confine ourselves here to finds which have led to the location of hitherto undiscovered sites. To the south of Argos a ditch dug by a bulldozer to conduct water from a nearby spring led to the excavation of a Geometric period cemetery (tenth to eighth centuries BC). Again it was a bulldozer which uncovered the first funerary urns ever found near the sacred lake of Sras Srang at Angkor in 1962 and made possible the location of the first known Khmer cemetery. In America the construction of motorways or turnpikes is followed by archaeologists conducting emergency investigations as the work proceeds.

A treasure of the La Tène period is known to have been discovered during the building of the airport at Llyn Cerrig, Anglesey, and a group of splendid silver coins was found under exactly

Opposite page: Machu Picchu, one of the few urban centres of pre-Columbian America to be found virtually intact. This detail from the ruins shows the extraordinary fine stone work where huge blocks were held together by the precision in fitting—no mortar was used.

Chance was responsible for the discovery of the Hypogeum at Hal Saflieni in Malta. This remarkable series of halls and chambers, cut out of soft limestone some forty feet below ground, was found by accident in 1902 by workmen digging a well. It was probably used originally as a temple, but its function later altered to that of a burial place, for many thousands of human skeletons were found together with personal ornaments and pottery. The hypogeum was in use about 2450 BC.

similar circumstances in Crete in 1954. A mechanical dredger working underwater at Spargi in Italy led to the discovery of a wreck.

Besides these peacetime exposures warfare, with its defensive trenches like those in front of the station at Syracuse, and its systematic destructions, like that of the Old Port of Marseille in 1943, can expose unknown sites or unknown sections of a recorded site. Finally, chance can lead to discoveries in completely unexpected places. For example, the cavern of Lascaux was found by two children chasing their dog. At Mari on the banks of the Euphrates it was the villagers who lighted on a Sumerian statuette. Many other examples could be quoted from every age and every kind of region.

Systematic investigation

Chance, therefore, still plays a part in the discovery of hidden sites, even if every archaeologist is not willing to admit it. This part, however, still fairly large in the discovery of artifacts, tends to be limited in the matter of sites as a whole. The word 'site' should be understood here in its broad sense, i.e. the location not only of vanished towns, but also of cemeteries, sanctuaries and even the land (cultivated or not) surrounding them, the lines of communication between them, such as roads or canals, and the fortifications ultimately defending them.

Indeed, it must be observed that even in the case of chance finds good luck on its own is not sufficient. A chance find is only a find in so far as it answers a more or less direct question, takes its place in a scheme of previously formulated hypotheses, and to some degree fills a want or expectation, sometimes very general such as the wish to understand the past, sometimes very precise. To take an example, the discovery of the funerary urns at Angkor by a bulldozer as just described was truly significant only by virtue of the fact that for a long time people had been speculating as to why no trace of a cemetery had been found belonging to this district of a million inhabitants, and regretting that for the most part very few complete vases had been found, but only potsherds. The same reasoning applies to all chance finds. You find what you are looking for.

The discovery of a site, then, requires some form of search for it. Schliemann wanted to find Homeric Troy. Generally, when an archaeologist is looking for a site his interest is centred on a specific region and, within this region, a particular period and, within this period, a particular problem. When he decided to dig at Lerna the American scholar J. Caskey was looking for an intact stratigraphical record of the early Helladic period (roughly the third millennium BC) to help him solve some problems which had arisen during earlier excavations on other sites. Of course it is not always possible to find that which is sought, but there should be a definite objective in view.

Once the problem of locating a site has been clearly defined a methodical enquiry into the solution becomes possible. The whole range of possible solutions, both old and new, are brought into play in surveying. The answer to the question, 'How do you find a site?' is, 'By different methods of prospecting.'

Prospecting and preliminary surveys

All the methods of prospecting, and they are many, are based on one fundamental principle: all remains which cannot actually be seen leave some kind of trace of their presence by virtue of the fact that they existed at a given moment. The whole point lies in knowing what sort of traces survive, and the best method of exposing them.

The methods of surveying can be listed under several headings: there is bibliographical research, the interpretation of aerial photographs, surveying at ground level, scientific methods and, finally, soundings. Since surveying is possible on several different scales (according to whether it deals with a cultural area as a whole, an entire country, a region, a particular site, a specific sector of a site or even a strictly limited part of this sector), the methods, too, reach different levels of precision. Surveying from books or aerial photographs, for example, can cover large areas as well as very restricted regions; surveying in the field can be conducted over a fairly wide extent of ground as well as within a given site. On the other hand, with some exceptions, it would scarcely be feasible to operate the scientific methods of detection over too large an area. The same is true of real archaeological soundings. To put it another way, although the most widely applicable methods can *a fortiori* be used on a small scale, the reverse is not true. Aerial photography allows for the prospecting of whole regions and is just as effective over an area of a hectare, but it would be difficult to envisage the use of electrical resistance or terrestrial magnetic surveys measurements over a whole valley, for example. On the other hand, the nature of the results of surveying on a large scale can be checked by methods proper to the smaller scale. Clues from written records can be confirmed or discredited by aerial photographs and these again by surveying. The conclusions of field archaeology in turn can be verified by geophysical methods of detection, and the results of these by trial trenches; every stage can be checked. The methods of prospecting form a coherent and graded whole in spite of their apparent diversity, and lead imperceptibly to the actual excavation.

This is why each method will be examined in turn. It will be seen that some of these methods have no bearing on the actual remains as tangible objects, but only on the picture which can be formulated from this information.

Literary research The most abstract but perhaps the most reliable, since it is the best defined, picture to be formed of a site should be sought in texts, maps and illustrations, since these documents can give unequivocal facts on the existence, and very often the name and the nature, of the site we are looking for.

The ancient authorities whose works have come down to us describe towns, sanctuaries, isolated monuments, roads and numerous other sites. Some of them, no doubt, were found long ago and are familiar, but others are known only from these references and have not been located. All kinds of documents can contribute to the search for a site, but pride of place naturally goes to topographical, geographical and historical works. The indications provided by writers like Herodotus, Thucydides, Polybius or Strabo, for example, are invaluable. Pausanias wrote a veritable guide book to Greece in the modern sense of the word. He not only recorded the traditions associated with this or that monument, but described towns and sanctuaries he saw with his own eyes, and even provided the distances from one place to another.

In 1950, for example, the École Française d'Athènes was working on the problem of the

Vertical aerial surveying provides a plan view of the ground. The camera points vertically downwards and the exposures are made at regular intervals whilst the aircraft flies along the lines of a plotted grid. Aerial surveying eliminates much tedious field work and often shows up features which would be indistinguishable from the ground.

town of Helike in Achaea on the south coast of the Gulf of Corinth; this town disappeared in the earthquakes of 373–372 BC and was submerged in the sea. The facts were well known, but the problem was to find its exact position. Pausanias, describing his journey from west to east, tells us that the distance from Aigion, the town immediately before Helike, is forty stades (seven kilometres) and, as he followed the coastline, this evidence is enough to provide at least an approximate location for the vanished town. It is noticeable that the references in the ancient authorities hold good not only for their own time but retrospectively to some degree for earlier periods, and even for preliterate ages, the traditions of which were later taken down in writing. It has been possible to compile a geography for the heroic period derived from the Homeric poems. Naturally, the ancient texts must be supplemented by inscriptions on stone or metal, which often permit the identification of hitherto unknown towns, or the locating of towns mentioned in other texts but not yet charted on the map. Coins should be classed as inscriptions on metal. They often bear the name of the town where they were struck and their provenance indicates its approximate locality.

If this principle works well for the ancient world it is even more applicable to more recent periods. The locating of medieval sites has benefited greatly from the variety of contemporary literary evidence. The deserted villages of this period have recently begun to arouse much interest. This type of research begins with the disappearance from the documents of the period (chronicles, hearth counts, ecclesiastical benefices and rolls, etc.) of the names of villages which had formerly been regularly mentioned. The accompanying details on the relative position of the vanished village with reference to others still surviving is the most

Some sites, despite clear evidence of their existence with buildings and monuments protruding above ground, were nevertheless still not excavated until modern times. This engraving by Giovanni Battista Piranesi shows the Forum in Rome in the eighteenth century. In the centre, partially buried, is the arch of Septimius Severus.

frequent means of charting its exact position.

Nearer to our own times the accounts of travellers recording places several centuries or several dozen years ago which have now completely disappeared are an equally valuable source of information. There is, for example, a mine of information on the Peloponnese contained in Abel Blouet's *Expédition Archéologique de Morée* of 1836, or from the reports of various nineteenth-century travellers in the East. Finally, just as the Homeric poems preserve even earlier traditions, so more or less recent transcripts of ancient legends, oral traditions finally set down in writing, can provide their share of information. The attribution of certain structures to giants, such as the Cyclops, can lead an archaeologist to a megalithic monument; furthermore, stories of demons and ghosts can indicate the presence of a cemetery. There are numerous examples. Few texts, when studied from this point of view, fail to lead to the desired site.

Maps The contribution of the texts is sometimes supplemented by cartography. No doubt the only really reliable maps are modern ones but, like texts referring to earlier conditions, maps often show something of vanished sites.

Ruins marked on ordnance survey maps and abandoned roads and footpaths can be significant. Surveys ancient and modern, or even itineraries without illustrations where these are lacking, can be very revealing. It can happen that a survey contains a detailed plan of a vanished village, even when the actual site has been levelled off. However, it is the detailed list of placenames preserved in the surveys which plays the chief part in locating lost sites. It is known that a number of placenames have précise meanings. An ancient habitation site south of Bolsena in Italy was discovered on a bare hill top which was still known as 'La Cività'. Mont Câtelet near Breteuil in the Somme district of France overlay a fortified camp. Names like Battle and Tombland speak for themselves. Blackheath could indicate a habitation site, in France *champ blanc* can mean a cemetery, and names beginning with Foss – a Roman road.

Finally, illustrations, which are most valuable for providing an idea of the appearance of a vanished site, can also help to locate it in the same way. From the plan of Nippur on a clay tablet to the bird's-eye views of medieval and more recent towns, from the blue print resting on the knees of Gudea's statue to the medieval miniatures, by way of views of besieged towns

The discovery of Troy, the ancient city made famous by Homer's poems, was one of the most romantic in archaeology. Despite the doubts of scholars Heinrich Schliemann, a German businessman, accepted the poems as literal truth and from early childhood was determined that he would find Troy and other Homeric sites. After several years study and travel he set out to do so, and conducted excavations there from 1871 to 1882. Nine successive cities or villages had occupied the site, the earliest dating from the Neolithic period. The seventh level was probably the Troy of Homer's period.

Raking light (note the shadows of the trees on the lower right) shows a circular feudal moat at the top of this photograph of Magny-sur-Tille. The ditch has been filled in, but it is still moist and appears as a darker ring. Shadow sites and damp marks are therefore both used here. The oblique line crossing the bottom of the photo from left to right is only the route of the modern aqueduct supplying Dijon.

such as the silver rhyton from Shaft Grave IV at Mycenae or the bronze doors of Shalmanezer III, and views of buildings on coins—all kinds of records can contribute to the identification of sites.

All these clues, whether they are texts, place-names, maps or illustrations must naturally be studied with a highly critical eye. Pausanias sometimes became confused and made mistakes, but experience has proved that this was not so frequent as used to be believed. Surveys are undoubtedly often inaccurate or vague or place names refer to too many Roman camps and roads. In spite of this, with judicious pruning these indications can be useful.

The limitations of bibliographical research lie not so much in what is said as in what is omitted. Many sites belonging to such early dates as the prehistoric and protohistoric, or even the historic period, have vanished without trace and are not mentioned in any of the surviving written records. In other cases the site is actually mentioned, but the details are insufficient and the lack of a plan is keenly felt. Bibliographical research, then, must only be used with reservations, carefully supplemented and checked.

Aerial photography After searching for the documents the archaeologist must look for the traces on and under the earth, revealed by aerial photography. Simple observation from the air, the forerunner and originator of aerial photography, which has existed since the invention of hot-air and hydrogen balloons, has rendered and still renders most useful services. This method, which far surpasses any results to be obtained by observation from a hill or some other high point, revealed the possibilities of aerial photography and makes it possible to grasp the lay-out and extent of sites which, seen from ground level, appear to be discontinuous and formless. Above all, it reveals pointers which cannot be seen at ground level. It is possible to see ancient roads and submerged structures from the air with the naked eye or with binoculars. But the possibilities of this method have been infinitely increased by photography, the only means of systematically exploiting it. Indeed, the observer cannot possibly see and register everything, and grasp all the implications of what he sees. Photography, on the other hand, can record everything, even details which are invisible to the human eye, so that they can be studied and compared. Surveying from the air is not the same as visual observation; it consists of the interpretation of aerial photographs.

Aerial reconnaissance from a captive balloon or a kite balloon gives the requisite perspective; but it is the horizontal movement of an aircraft (or even a helicopter or an ordinary balloon) which really permits systematic prospecting of whole regions and gives complete coverage of an area on a large scale. Closely overlapping photographs are taken automatically by a special camera so that nothing is missed and also to allow for subsequent stereoscopic examination. The area to be photographed is worked over in parallel lines, rather like a farmer turning his plough at the end of each furrow. The characteristics of the camera, especially the focal length of the lens, and the aeroplane's altitude are combined in order to work out the scale of the photograph. Oblique views, giving a more natural perspective, are suitable for previously located sites and supplement the indispensable vertical views. Finally, the use of filters (ultra-violet, for instance) and of special film such as infra-red makes it possible to discern traces which would otherwise be invisible, or extremely difficult to see.

Aerial photography shares similar principles with other prospecting methods: buried remains, even if invisible or nearly so at ground level, always betray themselves in one way or another. Aerial photographs react to this hidden entity in one of three different ways: they show variations in contour, in vegetation, or in colour.

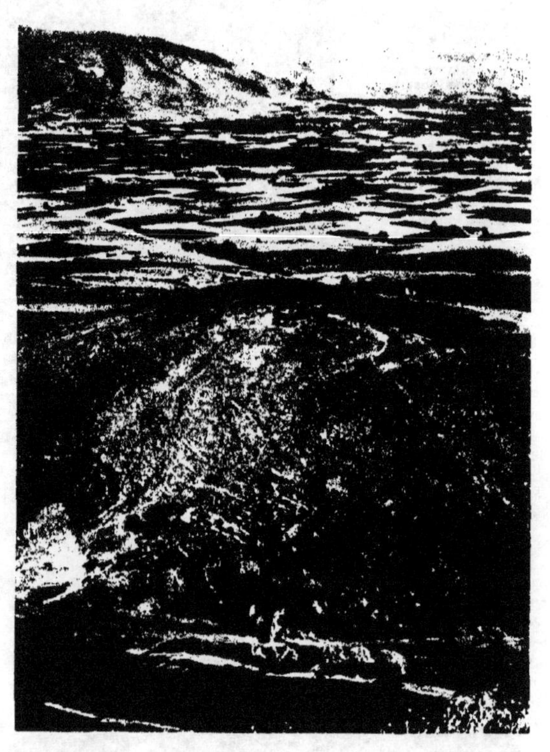

View taken from a height above Argos. The result is very similar to a low-flying oblique air photo. The slightest irregularities show up, and flattened structures are even more visible. Note the long shadows of the cypress trees, which show that the photograph was taken in the late evening.

Variations in contour Differences in relief, even very slight, caused by structures under the soil are thrown up by oblique lighting. An often quoted illustration is the way asphalt looks as though it is covered with lumps when it is lit up by the headlamps of a car. The same effect can be seen in fields by the light of tractors working at night. Aerial views, too, make use of the pattern of shadows to trace the outlines of buried buildings. Sites discovered in this way are known as shadow-sites.

Although these variations in contour are rare in countries where the ground has almost been completely levelled off for farming, some spectacular results have been obtained in England where the layout of some villages, especially those deserted because of the Black Death, stands out with remarkable clarity (e.g. Tursmore, Oxfordshire, or Burston, Buckinghamshire, to name but two). In such desert regions as parts of Syria the Roman *limes*, a third of which is actually buried, the caravan routes, and the frontier posts have been located in the same way. And the North African desert was the site of some brilliant discoveries by Colonel Baradez. Even beneath one metre of sand imperceptible undulations can be spotted in certain conditions, and elsewhere, when variations in contour have been flattened, buried structures betray their presence by the look of the vegetation or the lie of the earth over them.

Vegetation Crop-marks, or variations in plant growth, can occur in different ways according to the nature of the remains beneath. Above a wall, for instance, where water cannot readily penetrate, the growth of vegetation is thinner than over a ditch which is usually full of humus or refuse, and has good drainage. Consequently vegetation gives the reverse of the picture formed by variations in contour. Grain does not grow so tall where there is a wall underneath, and taller above a ditch, so that the relief appears in the 'negative'. In some cases, it is true, an opposite effect occurs, the lime in the mortar can fertilise the soil, and conversely the presence of flint in a ditch or the total dehydration of a mound make them unfavourable for plant growth. The height of the vegetation then faithfully reproduces the outline of the ruins.

Whatever the case, aerial photographs provide the necessary distance and an overall view similar to that of looking down from above at a carpet. The variations are named (from the kind of vegetation) crop-marks, weed-marks or grass-marks, but the principle is always the same. Variations in plant growth have led to the discovery of all kinds of remains, from protohistoric ditches to Roman villas and overgrown Roman roads in the desert.

Discolorations Where agriculture never existed or has been abandoned, hidden buildings and earthworks can betray their presence by the unusual coloration they cause. Here, too, the concentration of moisture in ditches, culverts, pits and wells or embankments has the effect, not of making the vegetation thicker, but of making the layer of earth above them a darker colour in contrast with the adjoining earth. These are called damp-marks, and they are a major contributor to crop-marks. Numerous

Right: crops do not grow tall above buried walls. Low sunlight shows up this irregularity and accurately outlines the plan of a small Roman villa. In this case, crop marks and inverse shadow are both used to aid the archaeologist.

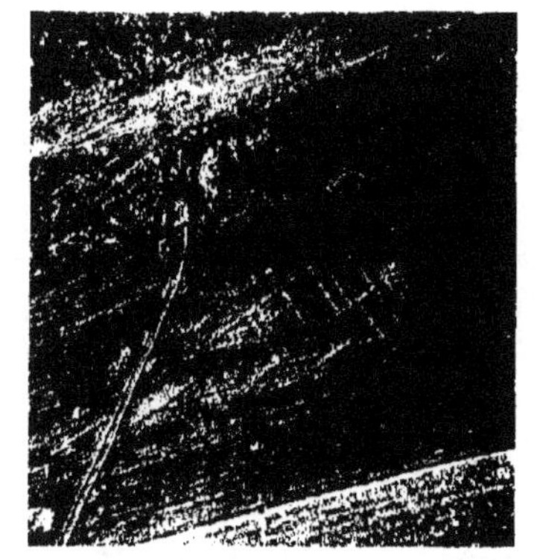

Below right: lucerne grows greener and thicker above ditches and post-holes, which are still damper than the surrounding soil. In this way two concentric ditches with the line of the palisade stakes inside can be seen perfectly clearly. On the left, two parallel rows of posts were also noticed. These were the false entrance. This photo was taken in June. The following year the enclosure was not visible because the weather was wetter than the preceding dry summer. A year later only the ditches were visible. Region of Genlis.

Below: vegetation is thin above buried walls and roads. This is the reason for the lighter stripes in the field, which outline the roads and blocks of houses (or *insulae*) of a small Roman town. The photo was taken in July, after a dry spell, above Nuits-Saint-Georges (Côte d'Or).

ditches, enclosures and roads (because of the ditches flanking them) have been traced in this way. Ancient land distributions, especially the famous Roman centuriations (a grid of lots of fifty hectares, to each of which a hundred colonists were theoretically allotted) have been discovered partly by means of this type of clue, such as the marks left by the irrigation channels in Tunisia, for example. In Indochina the highly developed system of canals linking the Khmer cities of Cambodia and harnessing the monsoon flood to irrigate the crops in the subsequent dry season was discovered in this way. Damp, however, is not the only discolouring agent. Agriculture, particularly modern deep ploughing, sometimes brings to the surface fragments of building material from constructions a little deeper, and spreads them around. They contrast with the surrounding soil and show up sharply on an air photograph. A typical example of this phenomenon is to be found in the Etruscan cemeteries where the tumuli which were once raised above the burial chamber with earth from the surrounding ditch were subsequently levelled, but still mark the location of the tomb with a lighter circle on the dark soil. In this case variations in colour are combined with damp-marks where the steps down to the burial chamber show as a darker radial line. The case of the 'fossilised' olive-groves in Tunisia has often been quoted too, where the Vandals compelled their prisoners to dig pits in alternate rows for planting trees, which show up perfectly on the tufa surface when seen from the air.

Combined with the true topographical clues supplied by aerial photography, such as long-distance alignment of sections of roads, thickets, rows of trees, etc. following the course of a Roman road, the discolorations, crop-marks and shadows which have been described comprise a more thorough and systematic survey than could be carried out on the spot. No new excavation nowadays can afford to neglect this preliminary investigation.

A special branch of the application of this method is underwater surveying (rather than undersea, since it is just as relevant for lakes and rivers). In this case the clues would naturally not be the same as on dry land. It is simply a matter of recovering submerged remains through a transparent medium. A well-known example is the discovery of the ancient harbour of Tyre by Father Poidebard. It has been possible to photograph submerged lakeside villages in the same way. However, underwater surveying is based on techniques which are fundamentally different.

How to choose the right time In all cases it is very important for the aerial photograph to be taken under the right circumstances. Great care is necessary in choosing the time of year, the weather and the time of day. It is fairly obvious, for example, that the shadows thrown by minor variations in contour are much more noticeable when the light is lower and more oblique. The source of light being the sun in the majority of cases, it follows that the photograph must be taken early in the morning or late in the evening. The use of artificial light, only possible over a limited area, calls for a night flight.

Dawn is the best time to look for damp-marks, the evaporation of the dew being faster on less absorbent soil. Aerial photography is particularly revealing after rain. The variations in drainage and drying out which occur, as described above, more slowly where there are ditches, can supply a veritable X-ray picture of the ground, but this effect is transient. A storm which flattens the tallest stalks can confirm the clues of the usual crop-marks. Thinning fog, white frost or snow melting more slowly above walls where the subsoil is very cold, more quickly above a spring or an ancient watercourse covered with a thicker blanket of earth, these provide a black and white negative, or positive, and can be used in practice to confirm the usual discolorations.

Variations in plant growth are more sharply defined at some seasons than at others.

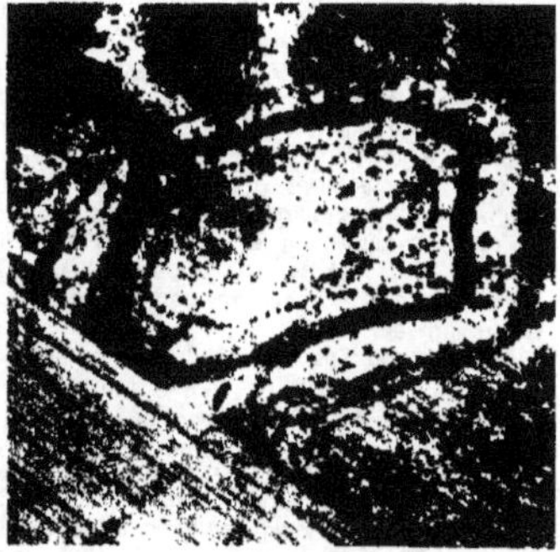

Sprouting corn, growing grain, some cereals ripening more quickly than others and hay drying more rapidly in some places indicate the presence of buried foundations or ditches. After the harvest weeds do not grow where the cornstalks were unusually tall and rich over an ancient ditch and so once more follow the course of the vanished structure. Wooded regions should be photographed in winter when the terrain is not hidden by foliage. A series of photographs of the same region taken at different seasons of the year, and in several successive years, provides useful confirmation and information by studying any temporary changes.

The interpretation of aerial photographs Aerial surveying, as the foregoing remarks will already have shown, is not a straightforward matter, it is an interpretation of photographic prints and this is a job for experts. In practice, one of the advantages of aerial photography over ordinary observation or photography is that it permits, by means of a sixty per cent overlapping of the prints, a stereoscopic investigation which restores or even exaggerates the contours. In this way small variations in contour or thicker growth of vegetation can easily be spotted. According to the nature of the problem in hand (habitation sites, cemeteries, lines of communication, etc.) the interpreter marks on a separate overlay of tracing paper all the points he considers significant. His working lamp substitutes for the light of the sun oriented from the same direction. Constant comparisons must be made between the map and the photograph, or between the photograph and earlier or later prints.

This method has its own problems and limitations. The possibilities of error are numerous. A cloud shadow can look like a significant damp mark or can blot out the slight shadows upon which the interpreter depends. The examples of 'negative' clues mentioned above can cause a road to be mistaken for a ditch or vice versa. It sometimes happens that at one point or another along a wall the foundation trenches provide favourable conditions for plant growth, and this too can suggest the presence of a ditch rather than a wall. Although it is easy enough to avoid confusing a modern trench promptly filled up with material differing very little from the surrounding soil, it is not so easy to distinguish the marks of a modern tractor, which look like the ditches at the side of the road, or perhaps the former site of a haystack which looks like an ancient house. An underground spring can make a damp-mark in a field which can play havoc with the interpretation. The threshing-floors of the Argolid, in the north east of the Peloponnese, round in shape but set up in the centre of a rectangle of stones, have been mistaken for ancient temples, since they are also usually oriented east–west.

Although aerial photographs sometimes make it possible to date superimposed structures, successive extensions of a town, or circular Bronze Age and square Iron Age enclosures (in this case, by applying a typology established by other methods) by relating one to another, the fact remains that mistakes and questionable chronology are common.

Finally, it must be noted that aerial surveying is sometimes unable to detect remains. In Sudanese Nubia, for example, J. Vercoutter excavated a necropolis which had never been spotted from the air. Furthermore, wooded or rocky ground is scarcely suitable for aerial photography; and, again, aerial reconnaissance is so difficult that it is necessary to begin by knowing *where* a site is to be found and studying only this area on the film. This was the case of the north Italian port of Spina, for example, where accurate photographs could only be obtained after detailed technical investigations. All this adds up to the fact that although aerial photography has produced some magnificent results and has numerous possibilities, it must be used in conjunction with other methods and checked by them.

Surveying at ground level The resources available for surveying at ground level are much more restricted than for air surveying; but ground surveying checks the findings of air photography as it does those of texts and maps. It adds to these findings and in some cases supplements an unsatisfactory result of aerial

Below: the corn was still green at Mirebeau (Côte d'Or) on 3 July when this photo was taken, but it is already short of water where there are buried walls. In the middle of a big Roman camp, a square with sides 100 metres long, perhaps the forum, can be seen. The two rows of light spots indicate column bases.

Bottom: Neufmoulin, 'le Mont d'Evangile' (Somme). Vanished ditches can be seen by means of the moisture in the fill.

Traces of a Roman villa 320 metres in length. Dressed stone walls are chipped by ploughing, and white fragments outline the plan on the muddy soil, which is damp in winter.

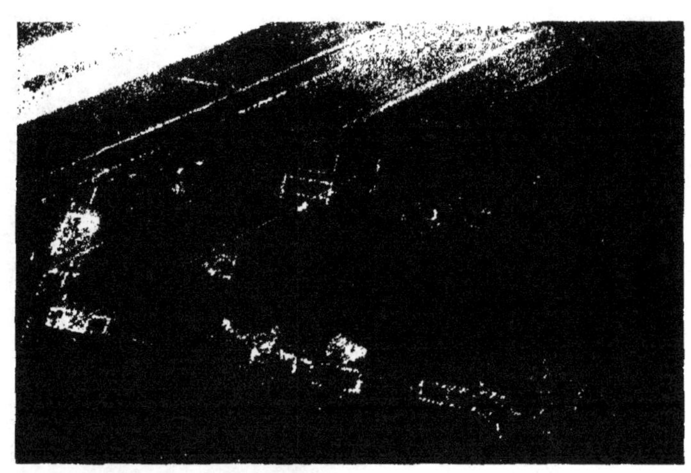

Aerial surveying. A Roman centuriation in Tunisia. The district was divided into squares with sides 710 metres long regardless of the contours, by means of the *cultellatio* process. Each square in turn was subdivided into smaller rectangular lots for the colonists. These lots can be seen clearly from a distance, although they have been ignored in current road-building.

photography. Field archaeology is the most traditional, and the most direct, form of prospecting, but although this form, too, was modernised some time ago, it is still irreplaceable because it is the first point of contact with the tangible reality rather than just the picture of the site under investigation.

In its most basic form surveying at ground level is simply carried out on foot. Before the building of the first Aswan dam the banks of the Nile were explored like this in strips five metres wide by twenty long. On the return journey the two teams of prospectors exchanged strips. Whenever possible the expedition is mounted on donkeys, mules or horses. Epigraphists scoured Asia Minor in this way. Nowadays Landrovers are used, as was the case of the Nile banks once more, when the construction of the high dam at Aswan was announced. The expedition is preceded by a study of the air photographs, and the vehicles are driven as slowly as possible. In some cases ordinary motor roads can be used, for instance when a Roman road cut at regular intervals by a modern highway is being investigated at certain specific points. In the case of underwater prospecting the glass-bottomed bucket which was formerly used in the early stages of a voyage and the old-fashioned diving suits are now superseded by the aqualung fed by cylinders of compressed air. For fairly extensive submarine surveys an underwater 'motor scooter' is sometimes used. Although they are employed for many other purposes than archaeological surveying, submarines can also be brought into service.

Whatever the means employed by the expedition, the method is partially the same as for air prospecting, but applied on ground level. The observer examines the relief 'tells', hills and feudal mottes or mounds—and small variations in contour, outcropping lines of stones, rising ground and depressions which may indicate the interior of a fallen house. He examines the vegetation. Crop-marks had been studied at ground level long before they were noticed from the air (the historian Louvet noted 'ancient tracks where the corn is not so high, which are places where ancient houses were built' to the south of Breteuil in 1631). He looks for soilmarks, for an ancient building, levelled

Vaux-sur-Somme (Somme). Aerial view of part of a Roman villa.

Below: Etruscan tumuli in the cemetery of Banditaccia at Cerveteri, and interpretive plan. Bottom: the black circles indicate foundations and the position of the entrance. The white circles are well-preserved tombs.

off by farmers, leaves a lighter patch which can be seen at a distance. He photographs what he sees. His survey must be as methodical as possible and must, if necessary, 'rake' the region in parallel strips.

If this type of surveying has neither the perspective nor the range of air photography, it makes up for it by being more thorough in some respects. The pedestrian can pick out various withered or dead trees in a regular plantation which probably indicate underground ruins; he can spot a hedge or a line of young trees growing on the site of an ancient wall or road which cannot be seen from the air. He can scrutinise modern walls and find reused inscriptions, architectural fragments and broken pieces of sculpture which have every chance of leading to remains not far away. Compared with the airman's perspective his close view has its uses. In the same way he can examine the natural or artificial stratigraphical sections exposed by the collapse of terraces, beds of streams, sunken roads, foundation trenches, water channels, etc.

The ground level observer's chief advantage is that he can collect facts which are not accessible from an altitude. He can pick up potsherds (after making detailed notes of their provenance) and he could do the same with bits of metal, glass, coins and bones. These objects provide evidence of the date of the buried remains which is impossible to obtain from the air. The pedestrian can visit local museums and private collections. He must not omit to make enquiries of landowners and farmers, who are in the best position to know what has turned up in their fields. The farm labourer may remember the trouble he had in uprooting stones which would be very troublesome in such a place, and may know about similar finds in the neighbourhood. All these clues must be noted down or put on a tape recorder. Without a thorough investigation it is impossible to understand the wealth of information which can be collected in this way - information which must be submitted to critical scrutiny, but is none the less precious. The field archaeologist checks the airman's findings, supplements them when they are insufficient and, by direct contact with the basic evidence of the site under investigation, sets the stage

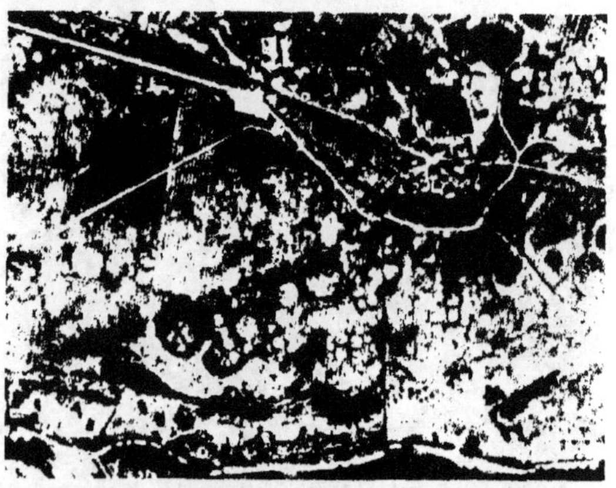

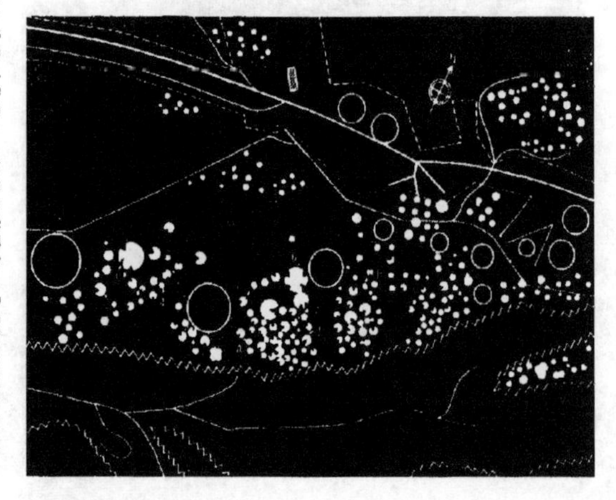

Prospecting by means of resistance measurements. Right: the instrument on its tripod and the five electrodes stuck into the ground at one-tenth of a metre intervals and connected to the transformer by separate wires. Below: diagram of the process and the curve obtained by it. Resistance drops above a ditch and rises above a buried wall.

for the excavation. Not more so than other prospecting methods, however, for this technique is not universally applicable. Grass, weeds and undergrowth sometimes completely conceal every potsherd and remnant of building. In this case other methods are necessary.

When texts are silent, aerial photography is blind and ground level surveying is paralysed, the burden falls on the truly scientific methods which have been brought into play and proliferated in recent years. These methods are basically physiochemical. They can be remarkably accurate, but they usually need expensive specialised equipment operated by experts. Like every other technique, these methods too have their limitations.

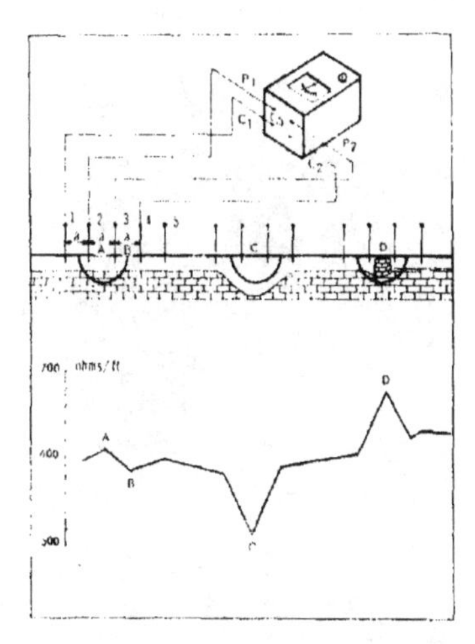

Electrical-resistance surveying Electrical surveying uses the same variations in humidity which already apply to the damp-marks (see above) detected by aerial photography or ground level prospecting. Moisture which makes plants grow also has the property of being a good conductor of electricity, which passes through the mineral salts dissolved in the water by a process of electrolysis. It follows that the wetter the soil, the more readily the current will pass through it, and the dryer the region, the poorer the conductivity of the soil. In pits or ancient ditches which collect the moisture in the earth, resistance to the electric current will be weak; conversely the resistance offered by a quickly drying stone wall or a footpath will be stronger. Simple quantitative resistance measurements and the equipotential line method are both based on the application of this principle.

With the equipotential line method a current flows directly from one electrode to another if the intervening substance is homogeneous, and in this case the equipotential (or 'equal voltage') lines are straight. If there are some remains hidden in the earth, on the other hand, the equipotential lines are disturbed. This method can claim credit for the discovery of Tepexpan man in 1947. It is, however, seldom used since the electrodes need to be several hundred metres long, as far apart as possible, and fed by a very powerful generator.

Resistance measurements are easier. Two electrodes (in this case metal rods) are needed to begin with, to measure the voltage of the electricity which is being used: two others, placed between the first two, measure the resistance. To eliminate the electrical charges which form between the soil strata and between the electrodes and the soil, the current should be alternating. If the earth between these two electrodes is abnormally resistant or abnormally unresistant, the measurements will show irregularities on the resistance graphs drawn to record them. These curves show clearly when there is a rise in resistance at the location of, for instance, a stone-built tomb, or a fall at the site of the ditch surrounding it.

In action, the four electrodes are spaced out evenly in a straight line and the resistances between them noted, a fifth electrode placed an equal distance away makes it possible to take a new measurement by means of a rotating switch without moving the three nearest electrodes. The first electrode is then moved alongside the fifth and so on. When the selected line has been completed the resistance along a parallel line is measured and so on. A comparison of the successive graphs obtained in this way may reveal a construction fairly accurately. The distance between the electrodes must be as nearly as possible equal to the estimated depth of the feature and thus eliminate superficial anomalies. Three hundred measurements an hour can be taken. This method was successfully applied to the detection of Etruscan tombs and again to the archaeological exploration of Purgi in central Italy, in 1962. The process runs into difficulties when it encounters natural pockets of clay or earth in the rock. Stony soil conceals remains and makes the placing of the electrodes difficult, and rain can falsify the results. Furthermore, pits of organic matter and burnt remains are hard to detect. This is not true of the magnetic process.

Magnetic surveying Magnetic surveying is based on an entirely different principle: the modification of the earth's magnetic field by the magnetism of various structures and types of buried remains. An elementary example of this process is furnished by the mine detectors used by the armed forces, but this instrument can only react to objects near the surface, which leads to confusion with recent and modern remains. An older process which has only lately been explained by science is the ancient 'sorcerer's ring'. Its vibrations arise from electric currents due to the penetration of water into porous materials. The magnetic field set up by these currents works on various points on the human body (now known to be located at the bends of the elbows) by a process which is still inexplicable. It follows that the 'witch' is sensitive not only to the magnetism of water but to every other form of magnetism. Experiments have been carried out with metals. The same process could operate in archaeological research with other weak magnetic fields.

One use of magnetism in archaeology, the thermo-residual technique, is based on the fact that clay which has been heated (for instance, in an oven or intense fire) acquires a magnetic force of its own, and retains it after it cools. In practice the clay contains varying quantities of iron oxides such as magnetite or haematite which cause its red colour, and their magnetic field aligns itself with that of the earth as the temperature rises, and remains aligned when the clay cools off again. The result is a localised magnetic disturbance in a given area, the intensity of which can be measured with a magnetometer—an instrument which expresses the relationship between the earth's magnetic field and the localised thermo-residual one. A systematic series of measurements taken on a regular grid covering an area will show irregularities due to the presence of a furnace, a burnt beaten earth floor, etc. These irregularities in the curves provide very accurate and detailed information; for example, the depth of the buried structure can be calculated. The exact centre of the structure is situated to the north of the point where the irregularity is most marked, at a distance equal to one-third of the depth. The size of the structure can also be estimated. It is obvious that this is a fairly accurate indication.

The process of magnetic measurement has recently been improved by the development of the 'proton magnetometer'. In practice, an ordinary magnetometer must be placed in a strictly horizontal position since it is the strength of the needle's deviation from the horizontal which is measured, and this always takes time. With the proton magnetometer, on the other hand, the measurements arise from a phenomenon which is only proportionate to the magnetic strength, and the instrument does not therefore need to be horizontal. Basically the proton in question forms the nucleus of the hydrogen atom. This proton can be compared to a minute magnetised needle revolving on itself. It tends to align itself with the magnetic field in which it is situated according to its own degree of magnetic power. It resists alignment according to the strength of its revolutions, and the result is that it goes into a spin like a gyroscope or a spinning-top which can stand up by itself and spin under the force of gravity. The speed or frequency of the revolution of the proton is proportionate to the magnetic strength to be measured. In practice, since the process is based on hydrogen atoms, a bottle of water is used. The bottle is wound with wire; when a current is passed through the winding it sets up a magnetic field along its axis, and most of the protons align themselves accordingly. When

Professor Lerici's motor drill.

Prospecting with a proton magnetometer. The assistant moves the bottle and the observer reads the measurements from the instrument.

Tarquinia, Italy. Professor Lerici at work on a site.

the current is sharply cut off the magnetic field disappears, the liberated protons realign themselves according to the local magnetic field and begin to gyrate again, and the combined magnetic impulse of all these revolving protons induces a weak voltage in the winding. It only remains to amplify this voltage to a convenient level and measure its frequency which is proportionate to the strength of the magnetic field.

There is, however, another side to the picture. A proton magnetometer is a very expensive piece of equipment. Volcanic rocks, modern pipelines and electric cables can cause errors and, as with resistance measurements, irregularities of the ground, modern pits and embankments can falsify the results. Nevertheless, the fact remains that magnetic measurements supplement resistance measurements and detect burnt material, baked brick walls, or even pits full of organic rubbish which have never been heated but still generate a magnetic force.

Sound wave surveying Acoustic surveys are traditionally carried out by banging the ground with an iron bar or mining probe which gives off a hollow or solid sound when it hits the ground and can guide the experienced user. It would no doubt be possible to develop more sensitive instruments. Seismic surveying, which consists of recording on a seismograph shock waves set up in the earth, can also be used, as with electrical currents, the diffusion of these sound waves is affected by buried remains.

Photographic probe with a miniature camera and flash equipment on the end. The regular rotation of the apparatus gives a panoramic view of the burial chamber.

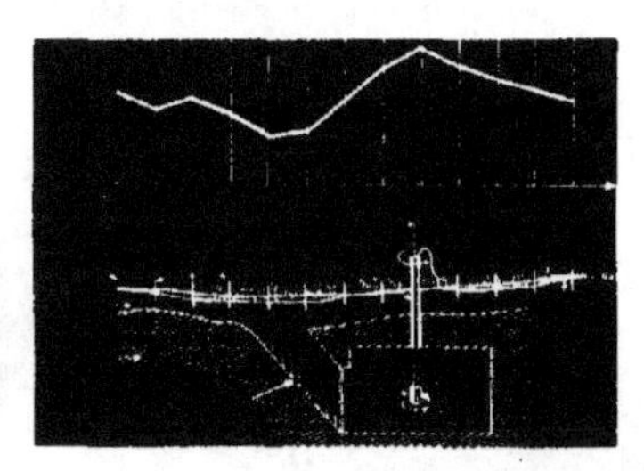

Chemical and botanical surveying Chemical surveying is chiefly founded on the analysis of phosphates. It is based on the fact that, unlike other mineral elements combined with nitrogen which are absorbed in the form of potassium nitrates, phosphoric acid, which is common in habitation sites, remains insoluble in phosphates of lime or iron. Quantitative analysis can reveal the presence of prehistoric habitation sites, as was the case in Sweden. It is necessary, however, to beware of modern fertilisers.

A similar method based on pollen analysis can also be used. The presence of pollens from cultivated crops can lead to the detection of human habitation.

The examples above illustrate the variety of detection methods available to the archaeologist when other signs are lacking. But here too, as in other processes, the risk of error is so great that nobody can afford to do without a practical check.

Trial trenches This ultimate check is supplied by true archaeological sounding and boring. Even in earlier days when they were the only known method of prospecting, soundings could verify or invalidate the excavator's theories. Nowadays, even though all the techniques described are available, they are still the moment of truth for, unlike other detection procedures, they try to make contact with the actual material at the site itself.

Some soundings are extremely delicate and restricted. These are soundings in the true sense of the word: that is to say, soundings carried out by means of some kind of probe. The simplest probes and augers are T-shaped rods driven in or screwed down by hand to the desired depth and then drawn up again. The presence of soil changes can point to a grave or habitation site.

Although this process is harmless enough above a house floor it can be imagined how destructive it could be in the case of a tomb, where such blind gropings could cause irreparable damage to the skeleton or the grave-goods. In recent years this method has improved strikingly. An Italian engineer, C. M. Lerici, has made use of probes powered by an electric motor which can bring up specimen cores from a depth of five metres. Some impressive results have been achieved with the help of an electric drill which reduces the risk of damage to a minimum by bringing up, not fragments of the object but photographs of it. This process has been applied with notable success to the examination of Etruscan chamber tombs. It must be preceded by a general survey, usually based on aerial photography, to pinpoint the exact centre of the chamber and avoid hitting the thickness of one of the walls. A tube holding a special camera and flashgun is introduced through the hole pierced in this way. It is thus possible to photograph an entire tomb and find out if it is intact or already looted and not worth excavating—always a costly process. The same result can be obtained with a prismatic periscope through which photography or inspection can be carried out from the surface.

The first stage of the excavation Before an extensive excavation begins a large number of soundings, each of which covers only a limited area, are made. Indeed, for a long time this method of surveying was virtually the only one in use and it is still practised today. Soundings were made more or less on the spur of the moment, perhaps according to plan on some occasions but also at random and completely

unsystematically. If the soundings produced no results, or none that interested the excavator, they were filled in and started again elsewhere. It was purely a matter of luck if they were even marked on a site plan, and this, unfortunately, still happens today.

It is hardly possible to do without soundings altogether – when there is no other way of finding out, for example, the depth of the archaeological strata or the general disposition of walls or above all, whether the remains under investigation are actually present, since they may be completely different from what was expected. But although it is useful, in some cases almost indispensable to make soundings, they must on no account be carried out in a haphazard way. It is important that soundings are made only the preliminary to an excavation; they must in no way jeopardise the ordered and methodical process all excavations require. The only way to achieve this is for the soundings to conform from the start to the plan of the forthcoming excavation, and to anticipate it to some degree. This means that before sinking bores or probes some idea must be formed of the plan of the eventual excavation.

This presupposes a clear picture of the problem to be tackled at this stage. If it is a question of locating and investigating a long construction such as a line of fortifications, a wall, a road or a canal, it should be cut, preferably perpendicularly; the trenches must be suitably oriented at the point indicated by the air photographs or other available means. If it is a question of an extensive complex like a habitation site, a cemetery or a sanctuary, the investigation designed to cover a given area can take various forms: parallel trenches at intervals (a grid), trenches in a staggered pattern (an interrupted grid), trenches sited to solve particular problems, etc.

The choice between these different processes must be dictated by the type of excavation being envisaged, since they must conform to the same plan. This system is of primary importance for the siting of the investigations by boring or sounding which precede the excavation in point of time. Moreover the parallel strips of aerial reconnaissance and ground level surveying, like the grid of electrical and magnetic measurements, all tend to favour a geometrical plan of excavation in the form of a grill laid out over the riddle to be solved.

These, then, are the methods available to the archaeologist for finding a site answering to the questions he has set for himself. Without ignoring chance finds he can make use of a large number of processes, from the study of texts and maps by way of the interpretation of aerial photographs, to the investigation of the soil by the latest scientific methods, which proceed from the general to the particular while checking and supplementing each other, and preparing for the actual excavation.

Saint-Jean-le-Froid (Aveyron), showing the first stage of the investigation of a French village which vanished in the Middle Ages. Two lines of trenches cross at right angles in the middle of the site. They give an idea of the whole site, which is covered with vegetation. In the foreground, the foundations of a defensive wall, built to protect the village in the fourteenth century, can be seen.

The excavation

The Mochica potters' remarkably explicit art encompassed most aspects of human experience. This one depicts a dead man wearing a nose ornament, which in the Mochica culture was usually made of gold.

A scientific technique

There are still, in the public at large and even among the archaeologists themselves, certain people who imagine that excavating consists of digging holes in the ground and rooting about at random in the soil in the hope of finding a masterpiece. The word 'excavation' is significant in itself. This quality is not confined to English and whether the words *fouille* or *Ausgrabung* or *scavo* or *askaphe* are used they all bear witness to the origin of archaeology - and to the stage at which it unfortunately all too often stops. This view of the matter leads, moreover, to confusion between the excavation and the preliminary survey and - as has been seen, the quality known as flair or luck in its most disastrous form.

However, for some time now enlightened opinion has accepted the fact that the part played by luck is diminishing, that the archaeologist is concentrating no longer on treasure but on remains which are apparently devoid of value or beauty and, above all, that methodical and painstaking procedure is more like an autopsy than a haphazard burrowing operation. Newspapers, holiday brochures, specialist publications and controversies have all spread the belief that excavations are absolutely straightforward and supported by a wealth of technical resources. Indeed, the prevailing impression is sometimes that arbitrary and gratuitous difficulties are multiplied for their own sake, for the sheer love of science. In short, the reasons for this method of proceeding are not always understood or realised, even by the professionals. Therefore, before describing how a modern excavation is, or ought to be, conducted it is as well to state the reasons why this activity has nowadays assumed such a distinctive form.

The object of the excavation In any attempt to explain this highly specialised organisation it is necessary to remember that the excavation combines several methods with one end in view, just as surveying uses a whole range of procedures in a highly specific plan. It was seen in the first chapter that archaeology was founded chiefly on the study of works of art, and thus had a limited purpose - the recovery of a statue or a beautiful vase, the location of a monument or an inscription. There is no written evidence from prehistory or protohistory and even with a historical period texts do not include all the information required. In the last analysis the sole aim of archaeology is, or ought to be, the compiling of history, if it can be stated at all. History must not be examined (as has too often and too long been the case) solely from the military, political, judicial, religious or artistic points of view, but must include its humbler and more fundamental technological, economic and social aspects. However interesting the work of art or the inscription, the history of the ancient world cannot be limited to that of buildings and their carved, painted or engraved decorations, or to the official records or registers preserved in inscriptions.

Nowadays the study of the forces motivating modern communities leads archaeologists to take an interest in the means of subsistence of vanished civilisations, their food, clothing, dwellings, the main raw materials available, the sources of power, technical knowledge and the products made, and transport and communications. Archaeologists study also relations of one community with another, the presence or absence of trade and coinage, knowledge of writing and politics and social organisation. Certainly monuments, carvings and coins, like inscribed and written texts, have long furnished a reflected picture of these conditions; henceforward, though, direct evidence will be sought.

Independently of the object of research as defined above, this invests archaeology with a distinctive quality in which chronology plays a fundamental part. In practice archaeological material does not usually furnish absolute dates, and, when it does supply them by luck or the application of recent scientific methods, they apply only to isolated finds which have to be assigned to the right context. It is therefore essential to know if any given finds are contemporary or successive and, when they are successive, which is the oldest and which the most recent.

All the historical conclusions to be drawn from archaeological finds presuppose that they have been classified in a correct chronological scheme. Unless this is done the wrong deductions can be made: a discovery can be ascribed to forgers, an influence discerned in what is really an innovation, and so on. The ultimate picture will be a false one. This is why it is so important to establish chronology. In archaeology chronology is founded on two separate and well-defined principles: contemporaneity and sequence. A certain group of metal imple-

Opposite page: after the topsoil has been removed more delicate tools have to be used in excavation. Here an archaeologist clears a double grave with a brush.

Uruk, the Erech of the Bible (and nowadays called Warka) is a large archaeological centre in Mesopotamia, about thirty-six miles north of Ur. It lies in an inaccessible region and has provided difficult working conditions for archaeologists. The picture shows the north-eastern side of the ziggurat in the temple precinct of E-anna. More than forty centuries are piled up at Uruk from the fourth millennium BC onwards, and in certain places there are as many as eighteen levels one on top of another.

ments, vases, potsherds, seeds and fibres, for instance, would have been buried at more or less approximately the same date, a certain building will, from its stratigraphy, be older or more recent than another. This is why the archaeologist attaches so much importance to establishing the 'stratigraphy', the sequences of successive complexes. This is the only way to *realise the full historical significance of the finds.*

The difficulty of such work perhaps explains why for a long time little interest was taken in the economic basis of the ancient communities, and why even recently it has been possible to complain about the lack of information on the economic history of the classical world. This may also be the reason why people have given up trying to establish the chronology of certain remains which are apparently inextricably interwoven. The feeling that these difficulties are insurmountable has perhaps had the side-effect of increasing the indifference shown to certain problems.

In fact, the domain of archaeology is richer and more expressive than could be imagined on the subject of dating as well as economic and social institutions. In the first place, a simple change of viewpoint can throw a new light on the finds, like a chemical submitted to an unusual reagent. It will be found that a tomb can supply information not only on funerary ritual but also on the clothing of the dead man, his rank in society, his age and the cause of his death, and that a certain religious scene can be informative on furnishings, foreign imports, armaments, etc. It is as if the mere act of posing the question produces the answer. There are, in addition, the more or less recent contributions of applied science; metallographic and spectroscopic analyses tell how a metal is composed, how it was forged and broken, and the temperature at which a vase was fired. Botany identifies the available species and the crops grown, and, in default of this, pollen analysis supplies this information and draws up a plan of the countryside with the help of geomorphology and climatology. Anthropology gives the composition of the population, organic chemistry identifies the most varied substances such as oil, dyes and paints. The picture which can be formed of the past is thus considerably extended. In the field of chronology there are results of carbon 14 and other methods which furnish dates in actual years, an achievement formerly thought to be impossible. The Palaeo-magnetic measurement process is already supplying even more precise information, and will continue to do so to an increasing degree. Nevertheless, in this field, too, the largest number of detailed facts come from the earth itself.

In practice, if the process of the formation of an archaeological zone is analysed it becomes clear that, contrary to what might be thought (and all too often was thought), it usually registers the progress of events in its own way. *How, then, is the soil of an archaeological zone made up?*

How archaeological strata are formed

Even today an idea of this process can be formed by observing what happens especially in underdeveloped countries, where living conditions, apart from a few minor improvements like electric light, the internal combustion engine and the radio, are not radically different from those which prevailed in the ancient world or the Middle Ages. The process can be equally well reconstructed by a kind of 'experimental archaeology'—when a house catches fire, for example, or a roof collapses or a ditch gradually falls in over a period of time. These phenomena, it seems, arise principally from the action of gravity which causes a floor or an object to remain where they are, and a wall to collapse, for example, combined with the rigidity, or at least the comparative rigidity, of matter. The result is that any structure or object stays where it is (or if it shifts it continues to occupy a volume of space) and, whatever changes it undergoes in the way of deterioration or compression, it survives in one form or another, sometimes barely visible or even recognisable, but none the less perceptible. Thus an inhabited level gradually becomes a stratum on to which small objects fall and are lost when they sink into its thickness, which increases because of the accumulation of refuse and dust. Should the site be abandoned, a layer of soil carried by rain and wind and enriched by the succession of rotting vegetation will form. If man reoccupies the site he will rebuild a second layer over the humus and the preceding floor, and so on. On sloping ground he has to build retaining walls and to fill terraces from the inside. Should the need arise he will have to repair a wall or reinforce or heighten it with stones or construction methods which can never be exactly like those used for the rest of it. He will be able to consolidate or pave the floor, or put down a layer of mosaic or cement which will cover the primitive floor without the site being abandoned in the intervening period. Finally, collapse, fire, warfare or earthquake can lay low what man builds up The fabric falls in disorder on the floor, covering any contemporary objects on it. At other times charcoal accumulates in a thick dark layer, more or less baking the beaten earth floor, for example, and causing a distinctive change in its colour. Thus, by successive accumulations, the future archaeological zone thickens and the ground level rises.

Intrusions This comparatively simple process can be complicated by the converse movement when objects become mixed in with earlier material, which is none the less perfectly comprehensible and very common. For example, the building of a wall will begin with the digging of a trench for the foundations. These various forms of trench penetrate deeper layers of earth or cut through them, and in these cases contemporary remains and evidence of the construction of the walls can fall into earlier strata. The same result can arise from ditch digging, refuse pits, wells and especially kilns (often half buried) or tombs. Finally, after the destruction of a building, new builders may wish to reuse the levelled foundation stones over again more or less on the spur of the moment. They will dig what are known as 'robber' trenches right up to the wall, and thus completely destroy it.

Whatever the case, this interference will be betrayed not by a rise in the height of the top layer but by a depression of some kind in this level in the region of the intrusion (as pits, wells and tombs are called). In addition, variation in the normal stratification is limited to certain points in this case. A far more extensive result will be obtained by levelling operations which will lower the whole of the most recent layer and thus destroy the older strata which are thus superseded. All these processes are familiar [illegible] and can be seen in action today wherever construction work is in progress. The result is that the sequence of strata of floors and the layers between them only seldom provide the sort of picture that could be likened to the successive pages of a book or even the limited irregularities of a *millefeuilles* cake. In practice the thickness of a layer is never even, being thinner here, thicker there and completely irregular further along. A layer of floor is never flat, but always slightly sloping so as to permit water to run off, or broken up by stairways and terraces which cause strictly contemporaneous surfaces to occur at uneven heights - for example, the inside floor of a house and the ground outside it. What is more, these beds and layers with their thousands of distortions are boxed in, some within others, since an archaeological zone, like nature, abhors a vacuum. Spaces full of air are rare, but they can occur in the case of certain tombs or at the top of a well when only the foundation has collapsed.

Traces of events It is possible for archaeological strata to form without leaving any traces. This is true of regions where the earth is *thoroughly homogeneous like sand, peat or silt*. All that man can do - the changes he brings about and the rubbish he throws away - is far from altering this consistency. On the other hand, as a general rule all the operations described leave their mark on the terrain where they take place. A fire will be indicated by a black layer, a destruction by a confused mass of material, etc. A stony floor will separate two layers of earth, a filling of soil brought from elsewhere will be different from the level it covers, and a foundation trench, even when quickly refilled with earth originally dug out of it, will not have the same texture as the untouched soil around it. A ditch will gradually fill up with topsoil which is darker than subsoil. A post-hole will preserve a carbonised stump.

It is this order of visible signs which will permit the observer, if he notices them, to reconstruct the progress of events. A certain kind of floor underneath another, for example, can only be older, whilst in order to build a wall or dig a ditch deeper layers may have had to be pierced and this intrusion can explain the presence of incongruous remains in an unusually early stratum. On the other hand, it is possible to discern a fill of earth from a much older zone behind a wall or terrace. If sequences can be made out it will be easier to see that objects are contemporary even when they are found

A modern excavation at Angkor in 1964. Square trenches (4×4 metres) oriented north–south and east–west are separated by bulwarks of earth one metre wide on which the stratification can be read. The trenches in the foreground are finished whilst excavation is continuing in the middle of the area. This was the first necropolis ever to be identified at Angkor (eleventh–thirteenth centuries AD).

Opposite page: volunteers, many of them students, at work on two famous sites. The upper picture shows work proceeding at Cadbury under a threatening sky, when there were hopes of finding positive traces of King Arthur's court. In the lower picture the desert sun burns down on the workers at Masada, the rock-fortress built by Herod the Great which was the last bastion of the Jewish zealots. It fell in AD 73.

Above, right: view of an eighteenth-century excavation in Greece, from *Ruines des plus beaux monuments de la Grèce* by Leroy (1758), Bibliothèque Nationale, Cabinet des Estampes.

Right: haphazard excavation. Slapdash techniques often lead to the loss of much valuable archaeological information.

Overleaf: the sheer size of ancient buildings was often the cause of disbelief when travellers returned to Europe and reported on what they had seen. An accurate description of the great temple of Amon at Karnak (ancient Thebes) will give the height of the 134 columns as 69 feet, and the diameter as nearly 12 feet. But the description is borne out by the photograph: the visitors are dwarfed by the immense structure, which was begun in the reign of the Pharaoh Thothmes I in 1530 BC.

Right: excavation as it was once conducted, and as it is still all too often done. The trench is shapeless, the edges are not sharp, debris is accumulating at the sides and the walls are not flat. The workman is knocking down the archaeological zone with his pick. Blows of of the pick have left deep vertical scars which obscure the stratification. No provision has been made for possible finds.

some distance apart but in the same layer, even when it slopes steeply. Likewise it is not surprising to find vases in a tomb, one or two metres down, identical to those found on the surface of the layer in which the tomb was dug. The whole secret lies in spotting the tenuous traces of these operations in soil which might have been thought to be homogeneous. This is the main object of the 'stratigraphic' method.

In this way the archaeologist can reconstruct the chronology of events and fill it out with all the information he has discovered in the course of his excavations. He must not neglect a single category, however humble, for it is in their interrelationships, their 'associations' as they are called, that their significance becomes evident. He will be able to reconstruct as absolute a history as possible.

The archaeological excavation, then, must be planned in such a way that all these associations and sequences with all their implications can be grasped.

Some systems of excavating

The immediate aim is to demonstrate the sequence of floors and layers (the strata) and the variations which form the intrusions. But the first step in an excavation is to strip off these layers and cause the stratification to disappear with them. Excavation defeats its own ends to some degree, as we have seen. A system must then be found to permit the (theoretical, at least) reconstruction of the vanished strata. Since the layers, constructions and intrusions represent volumes of space, it is not enough to reconstruct only two dimensional plans or sections. Both have to be available simultaneously, that is to say, the excavation must be regarded as three-dimensional. All modern excavations are conducted on this principle.

In former times, as can be seen even today though increasingly seldom, shafts used to be dug more or less at random according to the whim of each investigator. Alternatively, the whole area was stripped completely off down to bed rock. In contrast to this most modern excavations, including underwater investigations, are carried out within a horizontal grid. Standing baulks are left between trenches and these provide 'sections' - vertical cuts through all the strata. These contribute the third dimension, the grid (already presaged by the parallel strips of air reconnaissance, ground level prospecting and the electrical or magnetic grid) supplying the first two. That is to say, trenches are marked out with very little idea of what will be in them.

The topographical grid On some projects the grid is purely topographical; in this type of grid the intersection points can be marked on the surface by pegs (though this is not necessarily always done), for example, while the lines remain imaginary or are only temporarily represented by strings. The chief function of the squares of this grid is to locate structures or objects - the location may be designated, for example, square D18 or L7 - and the dimensions of the squares are consequently rather large (20×20 metres), which calls for subdivision into a second grid with a finer 'mesh'. In this method it is particularly noticeable that the placing of the stratigraphic sections is very flexible and usually does not coincide with the lines of the grid. They are always carried out at the place thought to be most promising, as

indicated by the nature of the remains already recovered; this may be, for example, along the axis of a house, perpendicularly to a wall, etc. The advantage of this system is its great flexibility, but it loses in thoroughness. The longitudinal and transverse sections are hard to keep in a straight line with one another. Baulks or partitions are preserved to demonstrate the vanished stratification, but the placing of these, too, is not necessarily regular. This was the system adopted by the American School in the Athenian agora and at Lerna, to name only one example.

The accumulative grid On other projects the dig progresses in a grid form in that first one square is excavated, then the next, then another, and so on. It follows that these squares are smaller (10×10 metres), and the stratigraphical sections coincide with the lines of the grid. It is thus possible to carry out continuous transverse and longitudinal sections very easily all over the site. In this method, however, it is necessary to lay out scale plans end to end, since the excavation of each square effectively destroys the section it brings to light. That is, although the grid is no longer imaginary it is still impermanent. The advantage of this system is that it supplies continuous sections which are themselves arranged in a grid, and exposes an increasing area of the site.

The Wheeler system The system which seems to be the best in comparison with the others, combining their advantages and avoiding their drawbacks, is that devised by Sir Mortimer Wheeler, which is now being adopted by an increasing number of expeditions. In this system the grid is no longer imaginary, and it can stay in position as long as it may be wanted without impeding the digging of an increasing area later. In practice, the grid, marked out by fixed wood or metal pegs, is overlaid with a second grid of

The ruins of a Roman temple of the second century AD were discovered during excavations for a new building near the Mansion House in London. The temple was dedicated to the god Mithras. An archaeologist works with a small hand trowel while excavators work in the background.

Opposite page: Machu Picchu, the Inca city which was not known to exist until 1912, when it was discovered by the American explorer Hiram Bingham. The city lies on the side of the steep gorge of the Uramba River and dates from about AD 1500. It was unknown to the Spaniards who destroyed the Andean culture.

Sir Leonard Woolley's excavations in the cemetery of Ur. Ur is in southern Iraq and was for many centuries the capital of Sumeria. It was mentioned in the Bible as the birthplace of Abraham and this reference was the only thing which prevented it from being forgotten.

Ministry of Works surveyors at work among the ruins of the Temple of Mithras in the city of London.

slightly smaller meshes marked inside the previous squares. A grid is thus obtained consisting of earth partitions separating the square trenches where digging is taking place. In this way the transverse and longitudinal sections are as easily obtained as in the previous system, but the progress of the excavations does not entail their immediate destruction. The baulks are to some degree continuous. There is therefore nothing to prevent these earth partitions being dismantled at any subsequent time. Apart from this, the whole site is clearly and indisputably marked out not only for the purpose of allotting responsibility, but to facilitate description. The square form of the trenches makes it possible to extend the excavation in any direction equally easily. As well as being clear enough for observation and photographing, the earth partitions (or causeways) allow easy and permanent access to all parts of the site, which simplifies both supervision and the removal of fill. Finally, it saves work, since not all the causeways are actually dug. The only drawback is that wherever the partitions intersect, the section is interrupted at that point, but this hiatus is generally very easily restored, and if a particular problem arises it is easy to pull down the relevant partition. On the other hand, the large number of stratigraphic sections meeting at right angles and continuing uninterrupted round the four vertical sides of the trench allows a precise three-dimensional reconstruction of the stratification of the trench. This is the finest system yet devised and the one that will be referred to in the pages that follow.

Laying out the grid

The excavation properly speaking starts with the laying out of the grid. At best this operation is preceded in some cases by surveying and making a chart of the disposition of the ground (a contour plan), or by the systematic removal of layers found to be superficial or unproductive or recently accumulated (ancient fill, etc.) by mechanical aids such as bulldozers. The laying out of the grid brings four independent variables into interaction: these are the derivation of the grid, its orientation, its scale and its naming. The derivation of the grid must conform to a physical boundary marked out on the maps so that the exact location of the dig can be charted. Apart from this, the base reference for the contours will thus also be accurately known. In practice, one of the intersections of the grid is matched up with a geodetic point with a fixed landmark, such as a milestone, etc.

The choice of the best orientation for the grid is a difficult and highly important matter. In fact, it is by far the most convenient if any walls found within the square trenches form an angle of about forty-five degrees with the partitions, because if they are parallel with the partitions there is a risk of their being too close to be examined or even excavated in the narrow corridor or 'well' formed in this way. It is also very useful to be able to name the four partitions 'north', 'west', 'south' and 'east' after the cardinal points of the compass. This makes immediate orientation possible at any point on the site, and facilitates description. It is irritating to

A Roman timber building at Richborough in Kent traced by the coloration of the soil.

have to deal with 'north east' or 'south-west' partitions: these mixtures and combinations lead to frequent mistakes.

The scale of the grid is the measurement of the side of each square trench. There is no advantage in making these sides too big, because in a huge trench small constructions may be a long way, if not cut off, from the stratification to be seen on the partitions. On the other hand, the trenches must not be too deep for a given size; they must not be 'wells'. In normally firm ground the side of the trench (or 'square') must be equal to the estimated depth to virgin soil. Here, preliminary soundings will have supplied the archaeologist with information on both the nature of the ground and the depth of the archaeological zone. To give an example, a square with sides four metres long is very often suitable. It follows that, the squares being separated by earth partitions one metre wide, the grid marked out by the pegs will have meshes of 4 metres+1 metre=5 metres, which is a convenient measurement. Four of these squares form a 'hundred-metre square' (i.e. a square whose sides are ten metres). Finally, it only remains to name the boxes of the grid with letters or numbers. The choice between letters or numbers for the horizontal and vertical axes of the grid is dictated by the natural features of the site. In practice there are only twenty five letters available (capital I is not used as it can be confused with Roman numeral I), but there is no limit to numbers. If a very long site is being investigated, then letters will be used for the width, numbers for the length. If the site is extensive in all directions, various other devices can be adopted, such as zA, zB, zC, etc. or aA, aB, aC, etc.

When these decisions have been taken, the next proceeding is to place the pegs in position, wooden ones in loose earth, metal in hard ground. Above all, if the site is hilly a surveying device like a theodolite, for example, must be used for accurate positioning of the pegs which are to serve for marking out the plan of the trenches.

The plan of campaign

At this point the plan of the excavation must be decided, that is to say, the plan of the trenches which will actually be carried out within the grid, and the order in which they will be executed. The plan of the excavation varies according to circumstances. If the site has only been broken into by the preliminary probings the problem is to investigate the terrain, no longer in the limited manner of prospecting, but systematically, and this can be done in several different ways.

The site can be bisected by two lines of trenches which cross at the estimated centre of the complex. If it is a fortified mound or a Middle Eastern tell, the centre, which is not necessarily the highest point, can be investigated by means of a number of trenches (3×3=9), the ramparts can be cut at the site of any gate, and the two sectors of the dig subsequently joined up. Plenty of plans are possible according to the

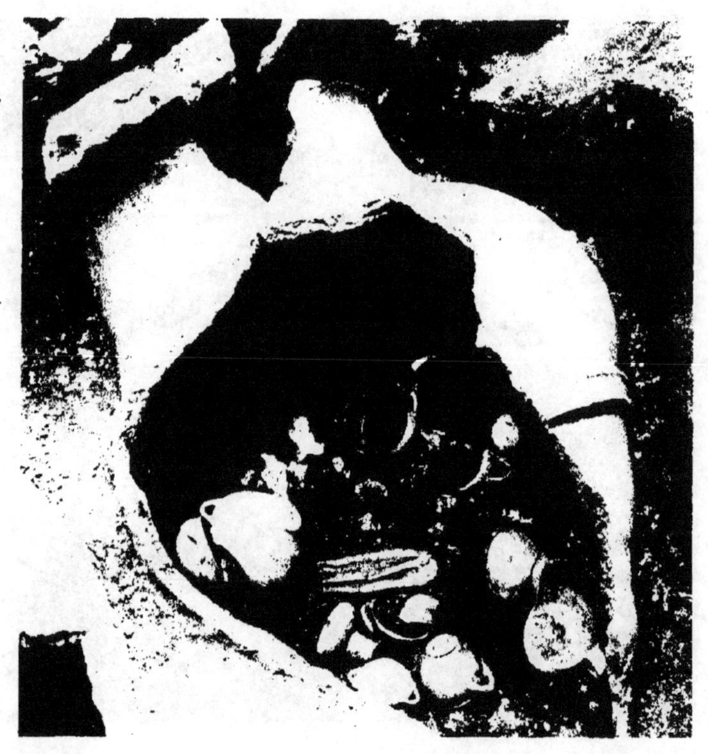

Inhumations in a large jar (Greece, eighth century BC). The breakage of the jar has enabled the worker to see its contents as they have been for the last eighteen centuries. In the middle are the bones of two bent legs. Analysis of skeletal material shows that at least three successive corpses were buried in this tomb. The vases were broken by the introduction of the new bodies. It is probably a family tomb.

circumstances. The essential thing is to have a plan and stick to it. In the special case of a huge palace or Roman villa, the grid can be adapted to the arrangement of the rooms and courtyards; in the case of a tumulus, it is divided into four quarters or quadrants. If, on the other hand, the site has already been investigated the problem is made easier by the earlier discoveries, which suggest the most productive directions and points. This is the state of affairs after the second campaign of the excavation. When there are several distinct sectors being excavated some distance apart, it can be useful to link them together by a long narrow trench, discontinued as necessary, which establishes the chronological relationship of a given layer in one sector with another layer in a different sector. It goes without saying that this plan is only a framework for the excavation and when something interesting is discovered – the first tombs of a cemetery, for example – the investigation can be extended as far as necessary in that direction; this is one of the advantages of a grid.

An important element in the planning of the excavation is the choice of the place for dumping the fill. In practice, the first result of any excavation is naturally the removal of a more or less considerable quantity of earth and stones. Debris must not be allowed to accumulate at any place which may be the scene of a future excavation, for it would then have to be shifted again. It is advisable, then, to move it to a safe distance, and in any case, not to pile it up too close to the trenches, as all too often happens. In addition, if the excavation has to be filled in again at the end of the campaign the debris must not be too far away. Above all, it is essential to avoid carrying fill up the slope from which it originally fell. The system, which consists of refilling every trench immediately with the earth from the next trench, has some drawbacks. The removal of debris sometimes demands the construction of a narrow-gauge railway or an

In the foreground is an oval rubbish pit dug in the stone floor, which has just been emptied. In the background, on the earth partition, can be seen a section of the stone floor which was cut by the foundation trench of the wall, the end of which still survives. The level belonging to this wall is higher, at nearly the same level as the top of the present wall. This type of observation is of primary importance on a dig. Many historical conclusions depend on it.

approach road for trucks and dumpers; barrows have to circulate, as do the basketmen, on the partitions or causeways. The essential thing is to choose a place everyone knows, not too near and not too far.

Whatever the dispositions adopted in the excavation plan, it is of primary importance not to excavate all the planned trenches simultaneously. The excavation of the largest number of trenches must, in practice, be guided by the first trench which acts as a sort of pilot trench. This trench will give a preliminary idea of the stratification and confirm the survey soundings; it will verify the general direction of the walls and will do much to facilitate the excavation of subsequent trenches. The placing of the first trench, then, is also important. This is one of the times when there is still scope for the famous 'flair', although it is not absolutely essential. Method will do the rest.

In practice, the marking out of the first trench, and all the subsequent ones, will be conducted in the following way: within the square of the grid as shown by the four pegs a smaller square is marked with nails or iron rods and strings, so that these strings are half a metre from the imaginary lines of the grid. As a result, the pegs are located at the intersections of the forthcoming causeways, in the middle of each causeway.

All the labour of setting up the project, which can be arduous and protracted, must be completed before the first stroke of the pickaxe. But if the dig has been properly organised, the excavation will move forward steadily and function smoothly.

The excavation

The first blow of the pickaxe is not as exciting as might be thought. In the framework thus prepared, the sequence of events is largely foreseeable, the scientific outcome is more or less ensured and the only surprises are the advents of really exceptional finds. The pilot trench can be dug fairly rapidly and perhaps less carefully than those which follow it. Its primary object is to supply as quickly as possible a picture of the overall stratification at a precise point on the project. Indeed, it would be extremely unsystematic not to make sure at once of the number of strata, their nature and their complexity. Special attention should also be paid, while making the necessary observations on the floor structures and the first traces of material to be seen, to digging the trench perfectly straight so that the partitions are flat and vertical. The workmen are guided by a plumb-line at the corners and even in the middle of the sides. The partitions are dressed with a trowel so as to smooth off the marks of the pickaxe, which obliterate the strata. If a stone is stuck in the partition it is left jutting out, but its intersection with the face of the partition is very carefully cleaned with a trowel. Over hanging walls can be cut back. It is necessary to identify virgin ground (not virgin soil) at the bottom of the trench, i.e. before all human occupation, and thus devoid of all industry or signs of hand-made remains, and, in order to be sure of this, it is essential to dig deeply enough into the unproductive zone to be certain that there is no older occupation level underneath.

Even if a very thin or almost invisible layer has not been noticed, or a shallow pit has been overlooked or the location of a wall has been misunderstood in the first trench, the mistakes can be rectified if the sides of the trench are properly preserved, for these 'sections' offer to the observer a record of the superimposed layers, or stratification. (This is the phenomenon which will allow the overlooked floor, the unidentified pit, the foundation trench and the true ground level to be spotted.) The observer must therefore always know how to recognise the clues supplied by the earth and stone partitions, he must know how to read and interpret the stratification.

During excavations at the Tower of London for the foundations of a new Jewel House to contain the Crown Jewels, traces were found of the first Tower. This was a massive earthwork constructed by William the Conqueror. The old mound is of darker earth in the strata.

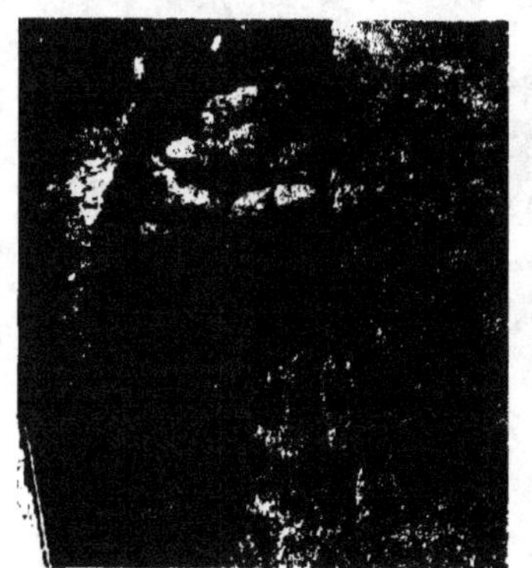

Example of a 'salvage trench'. The trench to be seen in the foreground was once filled by the foundations of a wall. All the stones of this wall have later been salvaged except for those in the background. The empty trench was filled up with black earth like that which occurs on the present ground level, which contrasts with the surrounding yellowish clay of the substratum (it can be seen on the vertical wall at the back). The original position of the wall can be accurately determined.

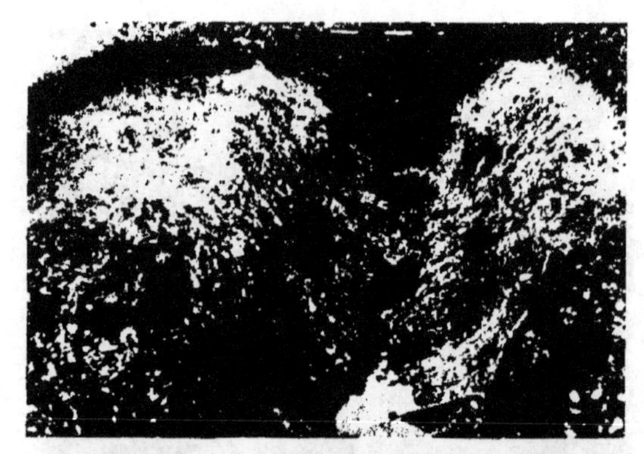

Experimental archaeology. Above: an ancient ditch recently dug out. Below: the same ditch a year later. The corners are rounded and the bottom has partly filled up. Traces of this process are often encountered in excavating.

The language of the earth

This is, in practice, a real language, or better, a dead language with a vocabulary and grammar, a morphology and syntax of its own. Moreover, it has to be translated. A dictionary of stratification could be compiled to indicate what a given manifestation in the section should mean. Thus the beaten earth where people have trodden will look like an almost invisible line along the section, but it is always distinguished by a difference in colour between the layer underneath and the layer above it. A paved floor will leave a much thinner and more irregular line of pebbles on the vertical partition than might be expected. A stone flagged or plastered floor or a mosaic will naturally be much easier to identify. A carpet will show as a thin carbonised black line with a few woollen threads sticking out. The levels separating the floors and layers, being often thicker, will be cleaner. A layer of vegetation will include roots; a level formed during a period when the site was deserted will consist of heavy rich dark soil. An occupation level will be equally dark, but spotted with refuse, potsherds and the remains of bones. A fire layer will be very black, or sometimes grey but scattered with bits of charcoal. A destruction level will show as a confused mass of material such as stones and plaster, sometimes mixed with nails from the woodwork. Virgin soil will be clear. Walls seen in section will look like a heap of stones, not so well bonded as they really are, but dressed on the outer face. The foundation trench, which links the ground level contemporary with the building to the base of the wall, will be distinguished from the older layers it passes through by a fairly clear line, and by its more or less chaotic fill. In the same way, ditches, drains and pits of various sorts, particularly those of graves and wells, will contrast with the surrounding earth. Every structure will have a corresponding mark on the side of the section.

The partition, however, has other uses than the identification of two dimensional tangible facts: it also demonstrates their chronological relationships. A level immediately above another will obviously be more recent; on the other hand, everything belonging to the same layer in one way or another will be contemporary. The essential principle is that an object belongs to a particular layer, rather than how high it is. An object is not necessarily more recent because it was found higher than another.

If the principle were always true according to which the higher an object is found, the more recent it is ('upper is later'), all the layers and floors would have to be horizontal and without variations, and, as has been shown, this is scarcely ever true. In sloping layers it may happen that a sherd typical of an early level will be found above one typical of a later layer. Similarly, much more advanced pottery may be found in the bottom of a well than in graves at a higher level which were cut through in making it. For the same reason refuse which has fallen or been thrown down a pit or a disused silo will be found in a deeper level than contemporary objects found on the surface from which the silo or the pit were built. Exactly the same is true of material from a grave. Only a few examples can be given but examples of these inverted stratigraphies are numerous on any excavation; they are enough to show that the archaeologist needs to be skilled at this form of intellectual gymnastics. He has to be able to produce a 'simultaneous translation' of the 'copy' presented to him by the stratigraphic partitions.

The application of the Wheeler system in Greece. The topographical grid is represented by metal pegs five metres apart. Trenches measuring 4×4 metres have been dug between these pegs. The stratification can be read on the walls and can be easily followed from one trench to the next. Notice the neatly swept causeways. Since the excavation must be filled in again, the fill is stacked nearby.

Developing the excavation

The clues garnered from the pilot trench will now allow the excavation to be extended fairly confidently into adjoining trenches. In fact, the four immediately adjacent squares are only separated from the first by an earth partition a metre thick. It follows that there is, as a general rule, very little likelihood of a complete change in stratification only a metre away. It is far more probable that it will be identical. It is therefore possible to foresee within a few centimetres at what level a certain layer or floor already noted in the first trench will reappear in the new one.

However, the new trench is big enough at a scale of 4×4 metres for the stratification at its furthest end (1+4=5 metres away) to alter, not fundamentally but significantly; thus a very thick burnt layer in the pilot trench may grow thinner further away from the heart of the fire until it finally disappears almost completely. Or again, passing from the inside to the outside of a house, the ground level, however strictly contemporary, will not necessarily be on the expected level. This phenomenon is even more marked in the case of a terrace or a flight of stairs.

To be able to excavate a new square with complete confidence it is necessary to ascertain whether or not there has been any change in the sequence or nature of the strata. Therefore a pit 1×1 metre, or a small control trench 1×2 metres is dug in one of the corners furthest away from the pilot trench. This cutting is sunk one metre or half a metre deeper than the rest of the trench, and usually precedes its excavation. It is thus possible to note in advance any alterations in the recorded stratification at this point.

However many precautions are taken, they can never be proof against closely localised variations such as a random pit, a hole in the ground or a mound. Such things can completely falsify the excavation of the entire trench if they are not noticed. Remains which formerly fell into the bottom of a pit may be assigned to the deeper level below, and since they really date to the following, later period they will lead to a mistaken dating of the deep layer. It is therefore essential to avoid such confusions, and this can only be done by means of a thorough investigation of the whole square.

The aim is no longer, as it was with the pilot trench, to establish the vertical sequence of the layers, but to clear them one by one horizontally over a sufficiently wide area. The next step, therefore, is the progressive removal of the surface of the whole square by scraping and planing away the earth: the layer is peeled off. The expert workers who carry out this task have learned to recognise variations in the hardness, texture and colour of the earth where their implements are at work. The archaeologist, for his part, notes patches of different colour which may indicate a ditch, a grave or the opening of a well in the earth. He picks out changes or unusual survivals in the material, particularly the potsherds. It is better to dig too deeply than not deeply enough, for the mistaken inclusion of older elements does not falsify the chronology. This technique is indispensable for the correct investigation of a thin layer. If the layer is thicker the best procedure is a very small working face where several centimetres but not more can be removed at a time (nothing can be worse than for the working face to include several layers at once). If the layer is very thick several successive cuts through it are made. It is always necessary to break up and rake through the earth so as not to overlook a single find, however small, or a single coin, and if necessary to sift or riddle the earth which has been dug out.

The excavation of intrusions of all kinds must be carried out at the same time as that of the layer to which they belong, i.e. starting from which they were originally made. In this way all the material assignable to any one period is collected at the same time. Furthermore, intrusions are usually limited and form closed deposits which are particularly useful for establishing chronology. Care is necessary in keeping material from a foundation trench or a half-

A section pegged for three-dimensional recording at Maiden Castle, Dorset.

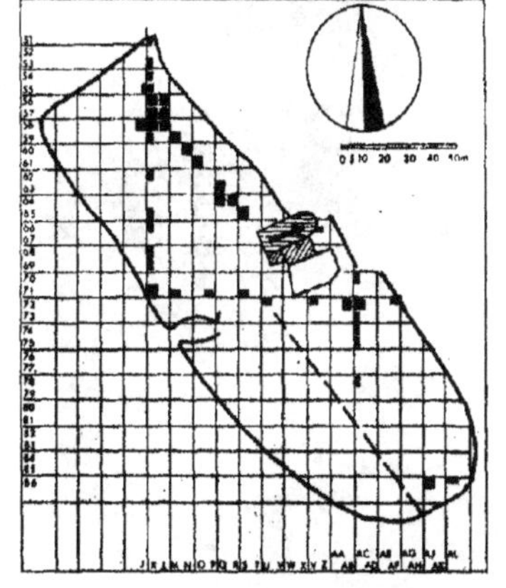

Exploring a site. The trenches are arranged in a pointed shape, first from north to south (above), then from east to west, and finally from north to south again. The grid is accurately oriented north–south. The trenches are rectangular at first (2×4 metres), then extended into squares (4×4 metres) wherever the excavation yields interesting finds.

Cross-shaped investigation of a site. A north–south line of trenches crosses an east–west line in the middle. The excavation is extended in the west in the area of some houses of the village under investigation. The grid is oriented roughly north-west by south-east so as to cut across the walls at an angle. One of the pegs of this grid is lined up with a local landmark. The trenches of the east–west line are square (4×4 metres) while those of the north–south line have been selected so as to pass through the north gate of the church.

Centre and lower: these two plans show the development in the next two seasons of the excavation shown above. The investigation of the village has been completed and that of the cemetery surrounding the Romanesque and the present churches has been started. The south gate and the area between it and the church have also been explored.

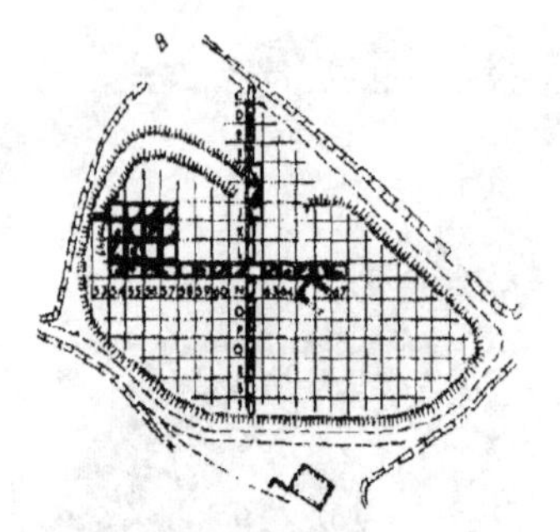

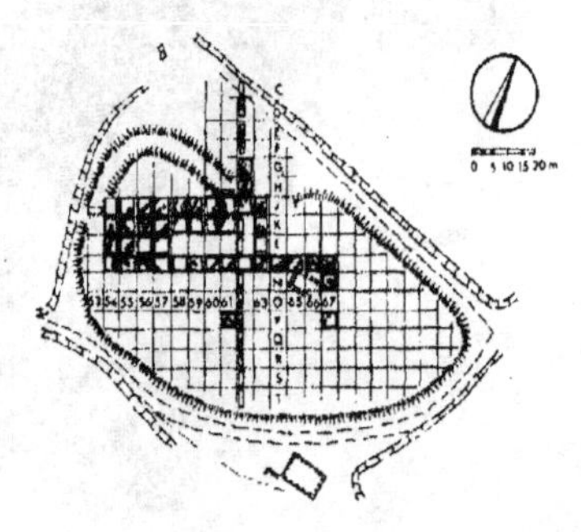

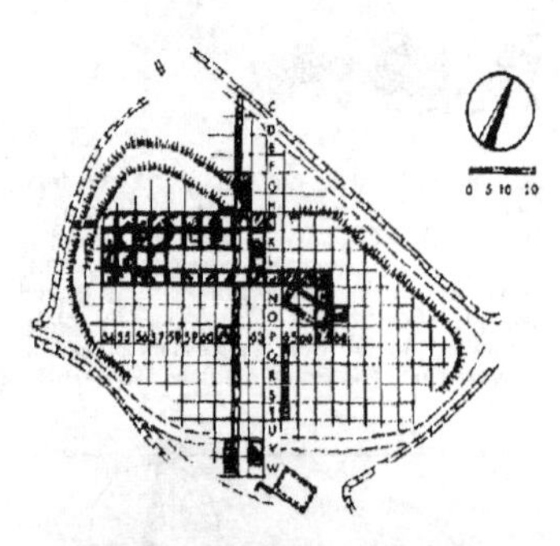

buried kiln separate from the finds from the older levels around it; for this purpose a small earth wall is kept to act as a divider. A pit can be the object of a stratigraphical investigation in itself, for it may not have fallen in all at once. It is divided into two or four sections, each of which is separately excavated. The same procedure is followed for the contents of a jar found standing upright in the earth where it was buried, if it is possible to clear one side of it, or if the mouth is wide enough. A well must be similarly excavated by sections (for example two metres wide) in cases where the contents are stratified, on the chance that a jug or a seal stone might have accidentally been dropped into it before it collapsed. The most interesting, and often the richest, intrusions occur in graves.

Graves The excavation of graves demands a special technique of its own. This process reproduces the stratigraphical investigation of trenches in miniature and in more detail. A grave is usually first apparent by the opening of its shaft, whether or not it is furnished with a stele or some other kind of marker. The filling of the shaft must be separately removed. The actual tomb appears either in the form of a simple cavity hollowed out in the earth, with or without a coffin, or in the form of a dressed stone cist (in which case the first thing to appear is the stack of horizontal slabs which formed the cover), or in the form of a jar laid on its side or set at an angle. (Chamber tombs and beehive tombs are another matter.) Tumuli contain individual graves of the same type as the earlier ones. The burials can be by inhumation or cremation. Usually the grave is at least partially filled up with earth and is cleared by very careful removal of this. A skull, the body or the mouth of a vase may appear as the knife (which should be either blunt or made of wood) scrapes away the fill, which is hardly distinguishable from the earth surrounding it.

The experts who carry out this work try to remove only the earth, leaving everything in the grave in its place: not only vases, weapons and the skeleton, but the most fragile gifts (hens' eggs, seeds, wooden objects, etc.) and the most delicate goods (beads, fabric, vegetable and organic matter, even saltpetre). Whenever necessary a technician strengthens the offerings or bones (using, for example, polyvinyl acetate), taking great care not to flood them, which might destroy periosteal remains, which surround the bones, or pollens in the earth. In cases where

Bottom left: pilot trench or control pit. In a corner of the trench a square (1×1 metre) is dug below the surrounding level before the main trench is made. It serves as a pilot for the excavation and indicates what will be found in the trench. A wall of earth is made to isolate the material from the pit temporarily but carefully from the finds in the rest of the trench, which do not belong to the same horizon.

Far right: section of a defensive ditch made to protect a medieval *castrum*. It has a clearly identifiable V-shaped outline. When the *castrum*, which was probably wooden, was replaced by a fortified castle, the ancient ditch became useless and was filled in with various layers of fill, one after another, which can be seen in the section.

Munhatta, Israel. Extending the excavation. The cleared area in the centre of the photograph with its buildings and ditches has been excavated in previous years. In 1966 a series of new squares was sunk round this excavation. A square is just being cut, and others are already being excavated. Note the strings between the pegs of the topographical grid and those which mark out the parts which have actually been dug. The numbers 17, 18 and 19 on the north–south line of trenches are written on labels attached to the causeways. The earth is removed in a trailer.

the grave has been used for one or more later burials, or has been disturbed or looted, it must be even more meticulously investigated, since it is sometimes possible to reconstruct the course of events in the most minute detail, with all the characteristics of personality and ritual they reveal. Graves are clearly one of the most spectacular aspects of excavating.

If walls are encountered as well as layers and floors, they are left *in situ* as far as possible. When the excavation goes deeper, shelves and pedestals of earth are left to support them. If they are so situated or entangled that they make deeper digging impossible, and they are neither handsome nor significant enough to warrant preservation, they have to be dismantled and removed. When this is done the course where they meet the partition of the trench is left in place as a record. This demolition is sometimes productive and reveals sherds, coins or reused inscriptions. In other words, stones have to be removed just as often as excavated earth.

When the four squares adjacent to the pilot trench have been dug, the stratification is re-checked on all the partitions, and then the four squares at the corners of the first ones are started. These are no longer guided by one partition, but by the two neighbouring sides of the trenches adjoining it. It is therefore possible to dig these new trenches with still more security, although this does not do away with the need to make pits or control trenches at the outer corner, as before. When these are dug, an area of three squares by three is available, or nine adjacent squares, which cover a total area of nearly 200 square metres (when the square measures four metres per side). It is then possible to halt the excavation of all the squares at the same layer to get an overall view of the layer. If, on the other hand, the material is not entirely unfamiliar, the stratification can begin to be translated into stratigraphy—dated interpretation.

The excavation can then be extended in the direction which seems the most promising, judging by the finds from the first nine trenches.

At the end of the campaign it is very easy to investigate the causeways in their turn, if this seems advisable, or to excavate parts of them, leaving the baulks (1 × 1 metre) where each of the pegs is located. Excavating the causeways gives an extremely accurate control, since their stratification is very minutely recorded by means of the partitions on each side, only one metre apart. Thus each layer in turn can be removed in one piece with all the finds in it. The continuity of the excavation, particularly that of the buildings, is thus finally established, but only after it has yielded up every possible scrap of information.

Recording the data

These data are, in fact, the sole justification for all the foregoing practical labour; and on any project a part of the time and trouble at least as important as the actual digging is devoted to recording the finds of all sorts which may have been collected.

The most practical aspect of caring for the finds is the safe storage of all the remains which are found during the course of the excavation: potsherds, objects or fragments of metal, glass, bone, and wood, coins and various utensils. Every one, without exception, must be cleaned since incrustations of earth sometimes hide the painting on a potsherd, and, since it is the most recent specimens of the material which give the best idea of the date under investigation, it is

Gradual dismantling of the causeways of the grid. On the right are two square trenches which still have walls. In the middle of the dig, most of the walls have been removed after being drawn and photographed. At the back on the left, some evidence survives. On the left, lower parts of the walls have been preserved. In the middle, the buildings have been cleared.

necessary to ascertain which these are. Then the body of the material can be studied from a firm standpoint.

A large number of finds can sometimes lead to storage problems. On the excavation site, boxes or linen bags are required for groups of pottery found together, strong paper bags for isolated or infrequent sherds, small bags, preferably transparent, for small finds or coins; made-to-measure boxes are used for some types of finds, such as swords. Specimens of soil (for geological or microscopic analysis), organic matter and charcoal (for carbon-14 dating) are put in sealed tubes. The specimens for magnetic measurement are coated with plaster on which the north point is scratched. Bones are steeped in polyvinyl.

There would be no use in gathering all these objects without enough information to give them meaning. The essential information is their stratigraphical position—the identification of the layer, the level or the floor where they were found. If necessary, this basic information is enough. But very often it is not definitely known if a certain sherd, coin, or other object should be attributed to a given level or especially to a given floor.

In this case, and in a more general way as a check, it is necessary to determine the position of the object, and this requires the measuring of its three co-ordinates. Its location on the horizontal plan is given by the distance between the object and the two or three nearest pegs. The depth can be measured by extending a string horizontally along the partition of the trench so that the height is, or can be, more accurately taken. If the case requires it, all these measurements can be furnished by a theodolite or level permanently established on the site. A measuring stave set vertically at the object in question makes it possible for this implement to measure both its distance and its depth as well as the angle of its direction with the magnetic north, and this is sufficient for pinpointing the object in three dimensions. These measurements, it should be remembered, are completely meaningless by themselves; they are only significant when they make it possible to refer the object in question to its exact place in the stratigraphy, in the layer or pit—even if this was

Excavating a tomb. The workman is using a delicate tool to clear bones surrounded by earth. In order to remove the earth without disturbing the bones and other objects, it is sometimes necessary to leave some earth underneath as a support. The brush (by his feet) is as useful as his knife. There are two skeletons on top of each other in this tomb. The lower and older skeleton was laid out in the bottom of the tomb. Its bent legs can be seen to the right of the figure. The uppermost and later skeleton was placed on top of the other one across the tomb. A cup full of meat was set on its right shoulder.

not identified during the excavation – or on the floor where it was actually deposited. These measurements ought to have been recorded from the start of the dig onwards, and the archaeologist's first care is to choose a number of points of reference on the site for this measurement. This information is transferred to a ticket marked with the name of the site, the date, the identification of the square, the three co-ordinates and, lastly, the number of the layer, the description of the object, etc. This ticket must always remain on the object, always accompanying it on its journey to the central depot, to its cleaning, restoration and storage, until it finally gets its accession number, which is usually impossible on the spot before cleaning and identification. The archaeologist spends a good proportion of his time writing out these tickets, because he is the only one who can do it correctly.

Locating structures rather than objects requires the drawing of plans – contour plans for earth or stone floors, and ground plans for architecture. Length, height and thickness of walls, columns, doors or doorways, etc. are measured and marked on a sketch which will ultimately be properly drawn or directly reproduced on the desired scale, and this will obviously take more time.

Orientation and angles are obtained by triangulation. It is sometimes advisable to check these measurements with a theodolite, and then hand over the work to a cartographer or architect who will draw the final plans, elevations and sections. While the dig progresses several successive plans of the different structures superimposed on each other in the same trench can be made. This is why it is important to identify the different walls and other elements on the plan, not just by a shower of confused letters which may lead to misunderstanding, but by directions, i.e. orientations (the terms 'right' and 'left', always relative and inadvertently reversible, should be as strictly banned, as also the words 'high' and 'low' or 'upper' and 'lower'), or by technical descriptions (e.g. 'north wall', 'brick wall', etc.). The juxtaposition on a scale drawing of plans obtained in different squares provides the overall plans of each stage of development of the site.

Typical example of a finished trench. Parallel walls of an atrium portico (late Roman). The base of the corner column is still in position, but the doorstep, curiously placed in front of the wall, the stone next to it and the drum before it have been kept in place by an earth support. Crossing the bottom of the picture is a much older wall but the circular pit which just cuts the line of this wall is of later date.

The excavation of a trench. A workman on the right is cleaning the shaft of a child's tomb, a simple hole in the ground. The slabs covering it are leaning against the side of the trench behind the workman. Beside him, a digger is piling the soil on to a small ledge (lower right) from whence it will ultimately be removed. On the left, a small tomb indicated by its shaft has been cleared. It was dug from the level on which the man with the shovel is standing, and is older than the child's tomb which is being cleared.

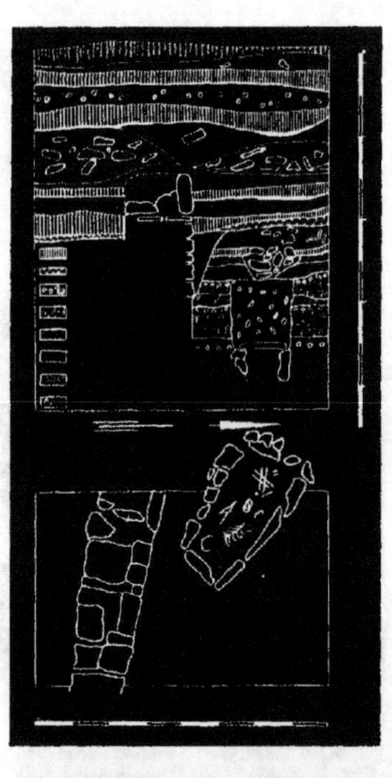

Stratigraphical section of a trench 4 metres wide and 4·1 metres deep. (The key on the lower left of the upper diagram reads, top to bottom: burnt layer, cement, rubble, clay, sand, earth, brick, gravel.) Beneath the surface level, a first burnt layer was found on a stony layer. Then a second burnt level over a thick layer of rubbish. A third fire, fairly thin, over black earth belongs to the wall shown on the plan below. Beneath that, a fourth fire over a cement floor belongs to the same wall as before but in its original ancient state (the wall was rebuilt after this and before the third fire). This is shown by the foundation trench on the right of the wall. The excavation was then confined to the right of the wall. A fifth very thin burnt layer belongs to the destruction of a small wall shown in the photograph at the bottom of the page. Below it, three floors of beaten earth in turn were pierced by the digging of the tomb constructed right at the bottom and therefore contemporary with the first of the three floors. The diagram relates to a Mediterranean tomb site (2000–1600 BC), and the last two floors are older than the tomb. Virgin soil is shown at the bottom of the diagram.

Stratigraphy

The stratification to be seen on the well smoothed earth partitions can be 'fixed', at least provisionally, by knife blades, but this is not necessary and adds one more inevitable burden to the interpreter. The drawing of the sections is executed, like all the other drawings, by squaring up. The verticals are provided by *plumblines'* or by a *single plumbline moved* along a metre at a time (or, in very complex zones, a half metre at a time), and the horizontal is supplied by the string, adjusted to the vertical with a spirit level or again with a theodolite, which has already been used during the dig for measuring depth in locating objects. The *resulting section should be as clear and informative* as possible. If the drawing is not 'representationally' executed, e.g. in colour, and if symbols are used to show a layer of rich earth, fill, a burnt layer, a plaster floor, etc., such symbols should be as informative as possible. The layers should first of all be numbered from the top, *not from the bottom, because it is possible* that still deeper layers may be found later on, or in another sector. It is better to distinguish too many layers or sublayers to begin with and reassess them later according to a study of the finds.

The juxtaposition of the stratigraphic sections from a series of trenches supplies transverse or longitudinal sections for the whole site. Juxtaposition of successive sections from the four sides of the trench will give a panoramic view of the trench, and thus the sequence of the layers which may have been removed during the dig can be restored in theory and in practice.

Part of the preceding stratigraphic section shown in the upper diagram (from the lower right). Under the cement layer at the top of the photograph, from whence the cutting of the foundation trench of the wall continues diagonally, is a small truncated wall. Beneath it are the pebbles filling the shaft of the tomb and the slabs on top of it.

Photographing the site

Photographs always call for a kind of interpretation different from drawings and it is necessary for photographs then to obey certain rules in recording archaeological sites. For views of the whole area, aerial photographs can be used, or a camera with telephoto lens can be suspended from a captive balloon or a kite balloon (making sure that the cable or wire does not get between the camera and the subject) can be used, or it is enough to take views from a neighbouring hill, steeple or photographic tower.

For closer, more detailed views, the natural objectivity of the camera calls for careful preparation of the subject. The object to be photographed between the stones of a wall must be thoroughly cleared of earth and of visible debris. All modern implements, clothes and shoes must be removed. If some photographs include a measuring rod to give the scale, it ought to be unobtrusively placed parallel with the side of the photograph, not across it. In any case, at least for a close-up shot (this is not true of a panoramic view of a palace, a citadel or a monumental gateway), the scale may perhaps be given by including a workman or a tourist. The grass at the edges of the causeways, at the ends of the partitions—and all over the site to begin with—must be cut. The choice of the angle from which the photograph is to be taken must be carefully made so that a person studying the photograph understands what it is meant to convey, and there must be no contradiction between the picture and the caption. Anything which is actually horizontal must look horizontal; anything vertical must look vertical.

Lighting must be worked out in the same way. The facade of a wall or inscription can be taken by photographing obliquely, against the sunlight or by using a moving light. As a general rule, apart from attempts at special effects, very dark shadows should be avoided since, of course, nothing can be seen in them: a soft diffused light is preferable, and this can be obtained by using a diffusing screen, or by waiting until the sun is off the subject or until a cloud softens the sunlight. If necessary, a wide-angle lens should be used to include the whole of a subject if it is not possible to move back far enough from it. The immediate foreground should be as sharply focused as the background. As a general rule, both black-and-white and colour photographs should be taken. The surveying in plan and in section can be done by photogrammetry (reliable photography from which accurate measurements can be taken). Ciné film should record almost all the dig: that a certain well or grave may turn out to be empty is just an occupational hazard. While the photograph records the details of the site, it is just as necessary to record the technical details of each print such as the date and time it was taken, not for its own sake but so as to be able to tell if one photograph was taken, for example, before rather than after another.

There can never be too many photographs. Indeed it might be said that there are never enough. Very often they are the only record of something inevitably destroyed in excavating, and they are all that many people will ever see of the dig.

Finally, all this information must be entered in the excavation journal or notebook. The archaeologist must write down everything he observes during the excavation, all the remarks, ideas and theories (even the unfounded ones), which occur to him in 'brainstorming' sessions. Long after the dig is ended these observations often play a fundamental part in deciding between possible answers to problems he never even thought of on the spot. Nobody ever takes enough notes or makes enough sketches. The archaeologist must write down everything he gathers, wheat and tares alike.

Young people, with suitable supervision, can make a valuable contribution to archaeology. Here students from secondary schools and universities are seen at work at Kunszallas in Hungary. When a sand-pit was opened the remains of an Avar settlement from the fifth and sixth centuries came to light. Numerous graves, personal belongings and jewels were found.

Equipment and personnel

The project as a whole requires the use of a large amount of different equipment, from bulldozers and dumpers to pinpoints, by way of picks, trowels and tweezers. The pickaxe is no longer the hallmark of the archaeologist. It should be added that on sites a long way from anywhere a dig-house for living, cooking and working very often has to be rented or built, and a workshop and store organised. All this equipment has to be operated by an increasingly numerous and skilled staff: not only labourers and foremen, but technicians (surveyors, architects, photographers, caterers, etc.) and the archaeologists themselves. Every part of the dig should be permanently supervised by a competent person. The director of the dig very often needs the help of a public relations expert. The days are gone when a lone archaeologist reigned over a host of unskilled workers. Devoted to its ultimate objective (the compiling of as closely dated and complete a history as possible) and armed with its method (stratigraphy), the modern excavation is totally professional, and the techniques of the amateur are no longer employed.

Establishing dates

"N·R·LOOSE E.B

Lekythos (oil jar) of the sixth century BC showing warriors in combat. The movement towards a simpler style with all the emphasis on the figures is already apparent.

How an object or a monument is dated

The establishment of chronology is the primary objective of the archaeological scholar. It is because he can date the evidence he discovers and studies, sometimes very precisely, that he can attain limited but none the less absolute truth. On the reliability of his dating depends some of the accuracy of the deductions he will form from this evidence and then pass on to the historian. A specific example drawn from an investigation will show then the step is taken from strictly factual chronological conclusions to hypotheses based on them.

Underwater work south of Cádiz resulted in the discovery in 1925 of a large bronze statue of a man wearing a cuirass. The place where it was recovered appeared to be the site of the famous sanctuary of Melkart founded by the Phoenicians from Tyre when they settled on this part of the Iberian coast, and later submerged in the sea. The statue certainly represented a Roman general, but which one? The head was completely gone, but the decoration on the cuirass was intact. A bearded mask symbolised the god Oceanus; on the breast were two griffins on either side of a candelabrum; on the abdomen an acanthus calyx spread its scrolls, reaching to each side. These details could be compared with those on some 350 known ancient cuirassed statues (a systematic catalogue was compiled in 1959 by the American scholar C. Vermeule). Analysis of the general form of the armour and accessories made closer comparisons possible. In the end it was decided that the Cádiz cuirass was definitely Roman work, probably cast in Spain itself, but it showed some typically Hellenistic characteristics. This hybrid type of cuirass was fashionable during two periods: the second half of the first century BC and the first half of the second century AD. During both these periods there were Roman statesmen who were closely connected with the Roman town of Gades, the modern Cádiz.

Julius Caesar, still only a humble quaestor, received a promise of a dazzling career from the ancient Phoenician oracle. Later on, his right-hand man, or prime minister as he would now be called, was an extremely rich citizen of the town, named Cornelius Balbus. In the year AD 97 a Roman citizen from Spain, Trajan, became emperor; he attributed his success to the protection of Melkart, identified by the Romans with Hercules. His nephew Hadrian, who was to succeed him, also emphasised this family devotion, which underlined the legitimacy of his claim. So three rulers could be the subject of the Gades bronze—Caesar, Trajan or Hadrian. Archaeological evidence, however, suggests an early date and consequently points to Caesar.

Such reasoning is obviously invalid unless the point of departure is absolutely certain. This leads to the problem of ascribing a period to an object. When this has been explained then the more difficult case of a building, the dating of which derives from the study of several different elements, will be examined. Objects themselves can be divided into four categories: those inscribed with the exact date of manufacture; those without a dated inscription but with a shape or decoration indicating their age almost as clearly; those which can only be assigned to a period by analysis of the material, and finally, those remains which cannot be directly dated, but can be assigned to their context through their association with pieces from the first three groups in the same archaeological strata.

Objects dated by inscription

Objects dated by inscription enter the realm of archaeology's sister study, epigraphy. It places at the disposal of archaeologists texts by the thousand, worked on all kinds of materials such as stone, terracotta, metal and even gems. The authors of these texts are no less varied, nor are their purposes. There are many humble epitaphs; but there are also religious, historical and political texts, and even genuine works of literature in the case, admittedly exceptional, of some clay tablets found during excavation in the Near East. Not all these writings are dated, but at least the most important ones, the official documents, are.

Chronological systems of the ancient world Unfortunately the archaeologist is up against a serious problem already, the problem of translating ancient references into terms of current chronology. The classical world often dated by natural events, particularly astronomical sightings. 'For the peasant,' wrote E. J. Bickerman (*La Cronologia nel Mondo Antico*, Florence, 1963), 'the sailor, the soldier, etc. it was naturally very important to recognise the changes in the seasons as they occurred. Over

Opposite page: mosaic floor of a Roman villa at Volubilis, Morocco. Some buildings can be dated by the style of their mosaic floors, which follow a recognisable sequence of development.

Classic Maya characters from Palenque. It is known that most Maya inscriptions of this type relate to calendrical calculations, but the script has not yet been satisfactorily deciphered.

the generations the sky had been observed and the custom had arisen of linking the seasons as well as the principal events of the farming year to the *rising and setting of a few of the most* brilliant stars. For instance, the period between the rising and the setting of the Pleiades was the time when navigation was possible. Furthermore, astral phases were taken for the causes or signs of changes in the weather.

'Before the civic calendar and along with it, there existed throughout the ancient world what is known as the agricultural calendar, i.e. marking out the course of the year (and of the night) by the position of the stars. The easily recognisable stars which can be used for this end, such as the Pleiades, Orion, Arcturus and the Lyre, were few. Thus Hesiod relates that the time for the vintage begins with the morning rising of Arcturus (at that time, 17 September), the time for navigation ends with the evening setting of Orion (20 November), and the spring begins with the evening rising of Arcturus (24 February), etc.'

This essentially practical system of dividing up the time is well adapted to the permanent and uniform elements in human life. It is a step towards the creation of more or less advanced calendars, and these are continually being perfected. But it is no help with the computation of properly historical dates, i.e. the pinpointing of exceptional events which only occurred once. It was a long time before anyone thought to relate these events to a fixed point, and at first they were only vaguely linked to one another. 'Before this war,' Thucydides says (I, 24), 'the Epidamnians banished the oligarchs.'

Scarab recording the marriage of the pharaoh Amenophis III to Tiye. The inscription is found on the base.

The establishment of regular monarchies and the invention of writing, which are related according to C. Lévi Strauss, very naturally encouraged progress in chronological calculation. Every sovereign kept his records, partly for religious or economic reasons, but also no doubt, to increase his own prestige. The human race, or at least a few parts of it, then moved on from the prehistoric period, when life was always cast in the same mould and no single individual left his mark, to the historical period when every generation, even every year, stood out in contrast with its predecessor.

Thus computation by regnal years and dynasties, the custom throughout most of Asia and in Egypt, was evolved. When various Mediterranean peoples abolished monarchy rule for life in favour of temporary magistracies, the name of the magistrate superseded that of the king. This was the system used on the majority of classical inscriptions, the only official records of the Greeks and Romans. It is naturally impossible to make use of a text dated to the rule of a king or eponymous magistrate unless a complete list of the dignitaries is available. However, these rolls were kept in unbroken succession and have often survived. When they are lost (e.g. the *suffetes* of Carthage) or when their authenticity is suspect (e.g. the Roman consuls of the fifth century BC), there is no truly historical record of the community.

For a long time Egyptian history was obscured by uncertainty about the order of the dynasties which were listed in the reign of the Ptolemies by Manetho, a half Hellenised priest, and confirmed by various inscriptions, documents and papyri. His lists, acceptable enough for periods of peace and unity, became highly

dubious for times of unrest when several pharaohs were recognised at the same time. For instance, Manetho often recorded the reigns of rivals as if they followed one after another, when, in fact, they were contemporary. In this way the two conflicting chronologies, the short and the long, arose, and they differ by several thousand years. The interval between the Twelfth and the Eighteenth Dynasties was computed by Manetho as 1600 years, although it was no more than two centuries. It was possible to correct this by the astronomical observations—those essential aids to the accurate functioning of the calendar—which are interspersed throughout the chronicles. For instance, the heliacal rising of Sirius (Sothis to the Egyptians) was noted in the fifth regnal year of Sesostris, the fifth king of the Twelfth Dynasty, and the reference in the official calendar enables the archaeologist to establish that this took place in 1874 BC.

Parallel to the civic year there was a religious chronology based on the periodic recurrence of certain festivals. About the beginning of the fourth century BC the Greeks decided to avoid the confusion arising from the proliferation of cities and eponymous magistrates by using a system of dating which was marked by the names of the victors in the chief Olympic events. The system of dating by Olympiads evolved from this; the games occurred every four years, with each one numbered in sequence. The system was revised at the beginning of the Hellenistic period, which brought great advances in the rationalisation of chronology. Increased international contacts and the idea, then dawning on the intellectuals, of the interdependence of the history of different peoples stimulated the creation of simplified systems which could be applied to huge territories. In this way the concept of the era was developed.

The first was the Seleucid era begun on I Dios (a Macedonian month at the end of summer) 312 BC by Seleucus I, who was then overlord of Mesopotamia. Babylonian astronomers were the first to achieve a calendar with a year of fixed length, obviously an indispensable necessity for the successful functioning of the era system. The Seleucid era lasted much longer than the dynasty, although the latter finally disappeared early in the first century BC the system was still in use in the East as late as the Middle Ages. Other eras were instituted by rival dynasties and independent cities.

Contrary to popular belief, however, the Romans never dated *ab urbe condita* ('from the founding of the city'), from 21 April 753 BC. This foundation date, calculated in the first century BC by the scholar Varro and slightly correcting the calculations of Fabius Pictor, was not by any means universally accepted. Until the sixth century the consular lists remained the only official historical determinant. In a great many inscriptions referring to the emperors, however, it is the imperial titles which give the most reliable information. The emperors' *tribunicia potestas* (which gave them the same rights as the former tribunes of the plebs) was regarded as an annual magistracy, regularly renewed, usually on 10 December. The number given in all the texts containing the complete title is therefore equivalent to the regnal year.

It is remarkable that among ancient peoples the Maya of Central America achieved quite

The science of time-measurement was carried to an extraordinary degree of sophistication by the great Central American cultures, who devised a calendar which was accurate to within one day every six thousand years. It was carved by the Aztecs on a stone disc, thirteen feet in diameter, which was discovered in Mexico City on the site of Tenochtitlán. The outer border contains two serpents representing time, within which are sun's rays. At the centre is the sign indicating the present era, and preceding ages are indicated by the four arms, which also bear the 'motion' sign. The twenty day-names surround the central symbol. The stone was carved just after AD 1502.

The columns and superstructure of the hypostyle hall in the temple of Amon at Karnak are covered with beautifully carved hieroglyphic inscriptions, many of which supply invaluable evidence for dating the building. The building is nearly a quarter of a mile long and the work of extension, rebuilding and decoration went on for nearly two thousand years.

Roman imperial coins with their details of the emperor's tenure of office supply accurate information on dating, and objects found with them can be dated by association. Above: the Emperor Caracalla, son of Septimius Severus, whose reign was infamous for its cruelty. Opposite: the Emperor Diocletian who strengthened the state and empire. However, his economic policies were less successful and he also persecuted Christians.

independently a really advanced chronological system, remarkable for its attention to small details as well as the implications of larger areas of time.

Fossilised chronology: scarabs and coins

A dated inscription is generally found carved on a large stone block which was once part of a public monument, which is as much as to say that this unequivocal method of solving a fundamental problem is not always available to the archaeologist. He is therefore particularly interested in small finds which convey precise chronological information and are, moreover, very common. In this category are Egyptian scarabs and coins.

Scarabs are a kind of amulet and seal characteristic of pharaonic civilisation, but imitated by the Phoenicians. They consist of a small hemispherical piece of glazed faience or hard stone carved to resemble a dung-beetle, the insect which the Egyptians regarded as the incarnation of the sun god (they believed that he pushed the solar disc across the sky as the scarabaeus rolls its ball of dung). The first scarabs were funerary amulets to promote the resurrection of the dead man, then they became seals with the owner's name engraved on the flat lower face. In the relatively late Saite period of the Twenty-sixth Dynasty (663–525 BC), the inscription often named the reigning king or a great pharaoh of the past. The scarab thus becomes a historical record of great value. A scarab bearing the name of Amenophis III but certainly made long after the reign of this pharaoh, who reigned about 1400 BC, is perhaps the earliest evidence of the Phoenicians as far west as Lixus on the Atlantic coast of Morocco.

As to coins, which are known to have appeared in the seventh century BC, they are the province of numismatics, a highly specialised and very precise study. The first known pieces issued by the Ionian seaboard cities, and then by the island of Aegina, already bear a sort of crest of the issuing city. 'Issued by private enterprise, coinage soon became a state institution; it assumed a different character according to whether power was despotic and centralised, as in Persia, or shared throughout democratic cities, who used their coinage to display their individualism and their mercantile spirit. Overlying the whole is an echo of superstition' (J. Babelon, *La Numismatique Antique*).

Corinthian hydria (water jar) of the sixth century BC showing mourning for the dead Achilles. Minute study of the development of vase painting has enabled scholars to establish the sequence of styles.

Darius of Persia was the first ruler to use his own portrait on coins, if indeed the figure of the archer on the darics can be interpreted this way. Until the Roman conquest there was a real abundance of coinage throughout the Mediterranean world. Nearly 1400 towns and 500 sovereigns with their own mints are known. Apart from the 'type', as the main central design is called, these coins bore only a short legend—generally an abbreviated form of the name of the issuing authority—when they had one at all, and no date. Nevertheless they convey numerous historical facts. The vicissitudes of Athens in the seventh and sixth centuries BC are reflected in the *Wappenmünzen*, or heraldic coins on which C. T. Seltman recognised the emblems of the Eupatrid families, and then in the rival issues of the Alcmaeonidae and the Peisistradidae, one distinguished by a wheel, the other by a horse. In the fifth century BC the distribution throughout the Mediterranean of 'owls', Athena's symbol which often appeared on Athenian coins, attests the political and mercantile supremacy of Athens.

After the reign of Alexander the information to be derived from coins becomes considerably more accurate. In a monarchy the obverse of the coin now usually bears the ruler's portrait, a striking resemblance engraved by a great artist. The inscription consists of the name of the monarch, his title of *Basileus* and frequently his surnames. Sometimes, especially in Egypt, a number denotes the regnal year. In a republic the name of the magistrate in charge of the issue is fairly often given. On the reverse famous monuments, statues and temples are frequently shown. These representations are extremely valuable as archaeological illustrations. On suitable occasions a special event was commemorated: for instance, Victory crowning a trophy on the reverse of the tetradrachmas of Seleucus I Nicator symbolises the Battle of the Ipsos which raised him to the throne, while an elephant stands for the defeat of his rival Lysimachus.

The spread of the coin-using economy throughout the Mediterranean world was linked with that of Hellenism. Rome only began to issue the bronze *as*, a surprisingly archaic piece, at the beginning of the third century BC, but the rapid progress of the Roman issues was accompanied by astonishing numismatic advances. After much discussion, it is now known that the stabilising of the Roman system coin

Krater (wine-mixing bowl) signed by the painter Euphronios. The custom of inscribing names of potters, painters and handsome popular favourites on vases has enabled scholars to establish sequences in the various workshops.

Etruscan polychrome amphora showing scenes of hunting and chariot racing, with fabulous animals. For a long time Etruria was believed to be the source of all painted pottery in the Greek style.

Amphora (wine jar) of the sixth century BC by the Nessus Painter. This artist, who worked in Athens, is named from the frequency with which he shows the centaur Nessus (on the neck of the vase). Another favourite subject, a gorgon, occupies the body of a vase.

cided, within a few years, with the victory over Hannibal. In the second century and most of the first century BC the reverse types chosen by the magistrates, who belonged to the senatorial nobility, were inspired by the pretensions and ambitions of the leading families. Otherwise they commemorated events of the past and present, sometimes with astonishing precision in view of the small size of the field.

Among the most remarkable coins of the 'peripheral cultures', without a doubt, are those of Gaul, which are as valuable for the information they convey on the various clans and their chiefs as for the aesthetic concept, so different from classical Greek canons, that they demonstrate.

One consequence of the rise of Augustus was a complete transformation of the monetary system. While a number of independent mints existed, particularly in the east, the coinage of the empire as a whole was in practice centralised under the control of the ruler. Its historical character was further accentuated, partly by the legend which perpetuated the ruler's title, but chiefly by giving the number of his terms of tribunician power and hence the year of each issue, and also by the choice of type. That of the obverse was always the sovereign's portrait after the Hellenistic fashion. The reverse, constantly renewed, illustrated in the most graphic detail the political programme and achievements of the reign. It should finally be added that variations in the metal alloy make it possible to follow the economic fluctuations of the times.

It is clear that coinage became increasingly historical from its first appearance until the end of the ancient world. There was, moreover, a very clean break at the end of the fourth century BC, showing that at this time there was a fundamental change in ways of thought. Titles, portraits and reverse types suddenly lost their individuality and, instead of demonstrating the monarch's position with regard to the world he governed, they show nothing but his relationship with divine providence founded on immutable metaphysical beliefs.

It can be seen from this brief summary that a comparatively small proportion of ancient coins, Hellenistic and Roman, supply precise information, particularly dates at a glance, which can instantly be used by the archaeologist. The value of this source, however, has been considerably increased by the researches of numismatists. It should be noted that these are primarily collectors. They are chiefly concerned with compiling practical catalogues of the hoards of coins at their disposal. For the pre Roman series this classification is based essentially on geography, and secondarily on chronology. For the Roman series, the classification was based for a long time on the likeness of the obverse. The coins of the Republic were classed by the families of the issuing magistrates and those of the emperors, within each reign, by the legends on the reverse in alphabetical order.

These were the principles adopted in the two great French works of the end of the last century, by Ernest Babelon for the Republic and Henri Cohen for the Empire. This method allows the rarity of the piece, on which its market value depends, to be swiftly determined, and simplifies the arrangement of coin cabinets. Furthermore, numismatists usually worked on collections formed from the Renaissance onwards by royalty and the great aristocrats, who were not concerned with the proven ance of a coin, but only with its beauty and its price. An entirely revolutionary viewpoint has been adopted in the present century, the leaders being the keepers of the British Museum. The fundamental principles are now the place of issue of a coin, its date and its composition. It has been possible to reconstruct the history of the different mints by studying coins and their variations. Coins with few distinguishing features have been assigned to their precise context in hoards along with well-known pieces. At the same time numismatists are increasingly interested in the actual historical information to be derived from a study of these pieces.

Coins, then, no longer seem to be such obedient witnesses of former events, ready to answer to the call of the archaeologist: he has to know how to make them speak, and the task is no less arduous in this case than in that of apparently dumb evidence like pottery.

Mythological scene from the handle of the François Vase. This famous krater names not only the makers, Cleitimas and Ergotimos, but most of the figures in the crowded action covering both sides.

Pottery

Every time a new branch of archaeology is examined similar problems present themselves. At first objects attracted the attention of dilettanti because of their beauty; collectors and dealers set to work to classify them, often for purely practical reasons. Finally it was observed that a history of the ceramics industry could be derived from this classification, and that this provided valuable information on history in general. The study of pottery, among others, developed in this way. It could be said that this discipline was born in the eighteenth century because of the discovery in Etruscan tombs of a number of vases which were as outstandingly elegant in shape as in decoration.

The classification of Greek pottery The French scholars of that period, Montfaucon, Caylus and their Italian counterparts Buonarota, Gori, Guarnacci and Passeri first believed, naturally enough, that these vases had been made in the region where they were found. A Paris antique dealer still preserves the memory of this theory in the name of his shop, 'The Etruscan Vase'. Under the influence of Winckelmann, their origin was subsequently sought in the Greek colonies of southern Italy. However, the progress of research in Greece itself after the beginning of the nineteenth century demonstrated that an abundance of painted vases were to be found there. Albert Dumont, the first director of the École Française d'Athènes, proved by means of a study of the signatures inscribed on certain pieces that a large number had definitely been imported into Italy from Greece. By a natural process of reaction a tendency arose during the second half of the nineteenth century excessively to minimise the part played by the workshops of Magna Graecia and Etruria.

The great German expert, Otto Jahn, in the preface to his *Beschreibung der Vasensammlung in Munchen* ('Description of the Vase Collection in Munich'), published in 1854, would allow them to be the origin of only a few decadent series. He subsequently had to modify this opinion, which was deservedly contested by other scholars of the period. A fundamental truth nevertheless emerged: in the sixth, fifth and most of the fourth centuries BC the main centre of production was firstly Corinth, but this was rapidly overtaken by Athens, which dominated the market for a long time. It was during this period that the craft reached an

Wine pitcher from Rhodes decorated with bands of animals. The tapestry-like effect of the closely massed figures and decorative motifs owes much to Western Asia.

astonishing peak of perfection typical of the spirit of a people who could never bear to use an ugly vessel, even for everyday purposes, and whose genius allowed them to create works of art using only everyday materials and methods.

In the last quarter of the nineteenth century, pottery experts concentrated chiefly on distinguishing the various production centres at work in Greece, and on reconstructing the history of each workshop. The essential devising of a system of terminology was taking place at the same time. It was fortunately possible to discover the terms used by the Greeks themselves, and thus the catalogue in this chapter was compiled. All these terms are named after the function of the pot.

'I have often noted,' wrote F. Pottier (*Catalogue des Vases de Terre Cuite du Musée du Louvre*, III, p. 608), 'the essentially functional character of Greek vases, and this cannot be emphasised too strongly, since there is a widespread prejudice in favour of attributing a purely aesthetic value to them. There were neither knick knacks nor large ornamental vases in the ancient homes of Greece and Italy.' Actually Pottier was mistaken here with regard to the Romans, who used to display vases purely for ornament.

'Even the subjects chosen by the painters show that painted pottery kylikes, oenochoae, kantharoi, skyphoi and rhytons were actually used as drinking cups at banquets, as well as kraters for mixing the drinks, amphorae for oil and wine, hydriae and loutrophoroi for water, and lekythoi, aryballoi and alabastra for perfumes and unguents.'

Because of the very fact of their functional nature, these forms persisted for a long time, although not without development, and this provides the main basis for chronological classification. Every one of these forms, moreover, includes a number of variants which often have specific names: this is the case of the dinos, an ovoid vase set on a separate base from the body, and the pelike, a pear-shaped variant of the amphora which is only found during one limited period. The general form of the body and the type and position of the handles altered from one period to the next. Nevertheless, it is not upon these variations that chronological classification is founded. The decoration provides a more useful basis

The work of restoring pottery found at the agora in Athens requires much patience and care.

In this respect, conditions were exceptionally favourable. The art of the Greek potter was always lively and creative in practice. It was forever producing something new in accordance with logical and easily followed laws. From the time of the earliest discoveries it was noticed that certain pots had a painted decoration consisting of human figures, some of which were in red on a black background, others in black on a red ground. Other pots were decorated chiefly with exotic beasts and fantastic monsters arranged in a regular frieze like those in the art of Western Asia of many periods. Finally, another class was distinguished by geometric ornament, the painter stylising even silhouettes into simple shapes when he wanted to show people or animals. De Witte in 1865 was the first to establish a chronological classification applicable to all these groups which is still, broadly speaking, valid. He distinguished nine series:

Pottery in a primitive style dating back to the tenth or twelfth centuries BC.
Oriental style pottery from Corinth of the seventh century BC.
Black-figure vases of the sixth and fifth centuries BC.
Red figure vases of the fifth and fourth centuries BC.
White-ground vases of the same date.
Gilded vases of the fourth century BC.
Relief decorated black glazed vases of the same date.
Italiot (that is, from the peninsula of Italy) vases, believed to be [illegible]dent forms of the third century BC.
White-on black pai[illegible]ases also of the third century BC.

These categories based on the evolution of style have been confirmed and corrected from observations made during excavations. Unfortunately, until the end of the nineteenth century very few Greek cemeteries were properly excavated, and tombs, more than anything else, are the chief sources of vases. Most of those which found their way into collections were unearthed by peasants, who hastened to sell them to antiquarians. One of the first cemeteries to be investigated in a satisfactory manner was the Dipylon at Athens, where the finest examples of Geometric pottery were found in 1871 and then in 1891.

If any modern work on Greek pottery is examined it will be seen how far de Witte's classification has advanced in the last hundred years. The sequence of the main series is as follows.

	BC
Mycenaean (late forms known as sub-Mycenaean) up to	1075
Protogeometric	1075–950
Geometric	950–710
Protocorinthian	750–620
Early Corinthian	620–595
Middle Corinthian	600–575
Late Corinthian I	575–550
Late Corinthian II	550–500
Early Attic (6 classes of black-figure, 1 class of red-figure)	600–480
Attic red-figure, severe style	480–460
Attic red-figure, free style	460–430
Attic red figure, florid style	430–400
First fourth century Attic style (few known examples)	400–375
'Kerch' style	375–330
Black-glazed Attic (pre-Campanian)	450–300

A part of the Research Laboratory at the British Museum, showing the equipment required for the synthesis of benzene which is used in the process of radiocarbon dating. All the great museums have elaborate workshops and laboratories in which a vast amount of work is done. One of the most important tasks is the dating of objects; in this the advance of science has greatly assisted the archaeologist.

After the end of the fourth century the workshops of Greece were in increasingly close contact with those of Western Asia, Egypt (especially Alexandria), and above all, Italy. The traditional methods of manufacture changed fundamentally. Painted decoration vanished (the last red-figure vases were made in Apulia at the end of the fourth century BC). Vases with stamped or incised decoration were preferred, or vases with relief ornament imitating metal.

It is very necessary to emphasise that the dates given for the beginning and end of a style were chosen primarily to crystallise the idea, and are not as completely inflexible as really historical dates. While it can be definitely stated that Pericles was *strategos* (military commander) of Athens from 443 to 428, it is barely permissible to say that the potter Erginos and the painter Aristophanes, who worked in the same studio, were contemporary, even though this example comes from a particularly well-documented period. In spite of this vagueness, the fact remains that the chronology of Greek pottery is a source of information of the utmost importance to archaeologist and historian alike. Certain periods such as the Geometric and the Oriental style, are even named historically after the type of pottery in use at the time.

As to the uses of this admirable precision instrument, it is not possible, unfortunately, to summarise here the work for which such brilliant scholars as Edmond Pottier and Sir John Beazley are renowned. Let it suffice to say that there is continual cross-reference between analysis of painting and the results of excavation. An excellent point of departure for the peak period of Attic pottery (second half of the sixth and first half of the fifth centuries BC) is provided by inscriptions. These preserve the names of workshop owners, of certain painters and of some particularly popular young men, some of whom later became prominent political figures. By means of these writings, which Jahn was the first to list systematically in 1854, will be seen the emergence of that unhappily rare element of ancient art, the personality of an individual artist, like Douris or Brygos.

The Pantheon, the best preserved of all the ancient buildings in Rome, was erected in 27 BC by Marcus Agrippa, son-in-law of Augustus. It was burned down and later rebuilt by Hadrian who ordered that the name of Agrippa be restored to the pediment, thus causing confusion to later scholars who found it difficult to believe that nothing of the original structure survived.

Opposite page: Baalbek, city of the sun. There was a city on the site already ancient when the Romans arrived but it is their legacy which makes the ruins a matter of wonder. The upper picture shows a romantic lithograph by David Roberts whose book *The Holy Land* appeared in 1842. The huge columns in the distance can be seen in brilliant sunlight in the lower picture; they are the remains of the Temple of Jupiter Heliopolitanus and soar over sixty feet, (the columns to the right are part of the Temple of Bacchus) The six columns are all that remain of fifty-four which surrounded the statue of the god.

In 1906 Pottier perfectly explained the route followed by scholars when making use of this information.

'Historical or not, named figures on vases can still help to establish an important chronology of potters' names. M. Hartwig has very convincingly shown that two different vases bearing the name of the same youth must be contemporary within ten years. It follows that the potters or painters who inscribed these youths' names on their vases must equally be contemporary. These correspondences are valuable, and M. Klein has arranged displays which show at a glance the groups of artists belonging to the same period. Thus the name of Hipparchos links Epictetos and Paidikos, that of Leagros links Kachrylion, Oltos, Euxitheos and Euphronios.

'It is also possible to distinguish different periods in the life of the same artist. The vases naming Leagros are among the earliest works of Euphronios, and those mentioning his son Glaukos are naturally twenty or thirty years later. ... Furthermore, we know that Eucheiros was the son of Ergotimos, one of the makers of the François Vase, and that Tleson and Ergoteles were sons of Nearchos. This is therefore a means of differentiating between two generations of artists and, consequently, a great many works. In fact, Nearchos is related to Amasis in his style of painting. Tleson, in his turn, trained a whole series of painters of small cups' (*Catalogue des Vases Antiques du Musée du Louvre*, III, pp. 711–713)

The study of the development of such ornaments as palmettes, rosettes, etc. has also furnished the basis of a relatively chronological series, i.e. the determination of the order in which the various types of vases succeeded each other.

These dates have been obtained by means of excavation. The investigation of the cemeteries of Greece proper (for example, the Kerameikos at Athens) has provided the answers to many questions – among others, the exact date of the beginning of red figure pottery in the last years of the sixth century BC. Excavations conducted on the sites of Greek colonies, especially those in Italy, have been particularly informative on the classification of archaic work (early Attic). The historians have preserved the foundation dates of these settlements. The earliest pottery found there should, in theory be contemporary with the arrival of the colonists. It has been possible to discover, however, tha[t] some towns which prided themselves on thei[r] very ancient origins could not produce an[y] evidence in support of their assertions.

An interesting example to the contrary is th[e] Sicilian town of Megara Hyblaea, a small po[rt] situated twenty two kilometres north of Syra[-]cuse, which was investigated from 1949 on[-]wards by F. Villard and G. Vallet. According t[o] tradition the Megarians arrived to settle ther[e] in 728 BC. Herodotus tells us that the town wa[s] destroyed in 483 BC by Hieron, tyrant o[f] Syracuse. It was rebuilt at the end of th[e] fourth century BC by another Syracusan tyran[t] Agathocles. Finally, Livy tells us that its ul[ti-]mate destruction was brought about in 212 B[C] by the Roman troops of Marcellus at the tim[e] of the famous siege of Syracuse.

This time the excavation confirmed t[he] historians' accounts. Large quantities

The peristyle of the House of the Golden Cupids, Pompeii. The building was named after the decoration of one of the cubicles – cupids engraved on gold foil and placed under glass discs. The west end of the peristyle is raised (left) like a stage and many of the reliefs and hermae in the garden carry theatrical masks. The house was owned by Poppaeus Abitus and is a fine example of the refined and theatrical taste of the Neronian period.

Opposite page: another picture by David Roberts is the coloured lithograph which gives a very strange view of the desert city of Petra. Trajan incorporated the city into his empire and the inevitable theatre was built there. The ruins of Petra became known in Europe in 1830, from the reports of the French traveller Léon de Laborde.

Protocorinthian pottery of the last quarter of the eighth century BC proves that the Megarians really did arrive about 720 BC. Hieron's destruction is attested by the almost complete absence of sherds of the fifth century BC, although the abundance of pottery of the sixth century BC indicates the prosperity of the little community. The Hellenistic reoccupation is clearly shown by stamped black glazed ware of the Campanian type and by the coins which allow the buildings of the reigns of a later Agathocles (319–289) and Hieron II (270–215) to be accurately dated.

It should be noted in passing that these correspondences are very important in the study of pottery. The finds lying under the floors of houses can, in principle, be assigned to the end of the fourth century BC at the latest. Finally, the excavations have added a chapter forgotten by the historians to the story of Megara. The inhabitants came back to rebuild their homes after the departure of the legions, and the town survived in obscurity for another two hundred years, until another force came and wiped it out altogether. The discovery of a hoard of forty-seven silver coins in a jar connects this event with the civil wars of the second Triumvirate.

Mexican pottery Greek pottery has certainly been the most valuable to historians, and the best exploited from this point of view; but whenever a nation was endowed with enough creative instinct to impart a distinct character to its humblest domestic utensils, the archaeologist can use them as the basis of classification. This is true, for example, of China and pre-Columbian Mexico and Central America. The latter case is particularly interesting since here the archaeologist is dealing with a civilisation which left practically no decipherable written records.

Mexican pottery provides for the archaeologist material which is both rich in variety and can be precisely classified. Writing in the catalogue to an exhibition of Mexican masterpieces at the Petit Palais, Paris, in 1962 F. Gamboa observed that 'Maya pottery evolved from simple shapes and monochrome colouring. During the Mamom period beginning about 850 BC, dishes and flatbased pots, jars and jugs with striations reminiscent of gourds were made, as well as clay statuettes with perforated eyes. During the Chikanel period, which lasted until the beginning of the Christian era, bichrome vases with negative painting and al fresco decoration appeared. The shapes became more and more varied. Next came the little-known age called the Matzanel period (until 380). The Tzakol period (300–600) produced bichrome and polychrome pottery, and dishes with three or four feet as well as pots with different bases similar to those of Teotihuacán. The Tepeu period (600–950) was distinguished by polychrome vases decorated with human figures, animals and ritual scenes, vases with relief ornament, dishes worked in intaglio with notched bases and rims, and moulded clay figurines.'

At this point the methods of dating the pieces which are rightly considered to be works of art should be examined, but in doing so the ill defined frontier between true archaeology and art history is being crossed. Never

Doric columns of the Propylae, the main gate of the Acropolis, Athens. In classical architecture there are five so-called 'classic orders' of columns – Doric, Ionic, Corinthian, Tuscan and Composite. The Doric was the earliest to develop in Greece and was used for most temples and the Parthenon.

theless, every archaeologist ought to be an art historian as well, and the reverse is also true, at least as far as the ancient world is concerned. A valid judgment about the date of even a masterpiece ought to be based essentially on practical criteria such as details of garments and jewellery, for example, before allowing aesthetic judgments, which are always subjective, to intervene. It must be admitted, however, that the true province of archaeology lies above all in articles in common enough use to warrant the formation of general conclusions about the society as a whole, or at least the greater part of it. All handicrafts, and especially pottery, depend on this. The appearance of the home, with or without furniture, ought equally to be included in it.

Pottery for everyday use

The poor articles, which are naturally in the majority, have nothing to recommend them and have been scornfully rejected by generations of antiquarians, must also be considered. Even today the feelings of the uninitiated, watching the archaeologist picking out shapeless sherds of coarse, badly fired pottery, taking endless care with knife and brush, are divided between astonishment and pity. However, there is something to learn even from the remains of these wretched pots.

First of all, there are objects of a distinctive shape. Indeed, the commonest cooking pot in the earliest home is rarely the same as that used two generations before. To tell the truth, typological development is very slow when manufacture is exclusively confined to village craftsmen who, except among the most inventive peoples, stick to repeating ad infinitum the models learned in their apprenticeship. Progress becomes more rapid, however, when there is large-scale production aimed at a slightly wider market. Bases and handles, less breakable than the bodies of pots, are precisely the elements which are most easily modified. 'A simple fragment of the foot or base of a pot, in combination with the shape and the technical characteristics, can sometimes prove to be a chronological clue, possibly vague, it is true, but nonetheless useful,' wrote J. P. Morel in a treatise on the black glazed pottery of the Roman forum and the Palatine published in 1965.

Everyday metal objects Among the metal objects, tools change very little because their shape is dictated by the requirements of efficiency.

Some of the saws, hammers, and other tools dating from the Bronze Age are very like those of today. Weapons and ornaments, on the contrary, change from one generation to the next. The chronology of the Celtic civilisations has been established by means of these variations. The safety-pin, for instance, more gracefully known to archaeology as the 'fibula', is in itself the subject of a classification scheme. It spread from Italy throughout central Europe and Gaul; rare north of the Alps in the Bronze Age, as in the early Iron Age (Hallstatt I), it showed a great variety of types in Hallstatt II.

J. Déchelette (in *Manuel d'Archéologie Préhistorique Celtique et Gallo-Romaine*, II, pp. 854 ff) distinguishes three classes in this period: fibulae with a plain bow and a unilateral

A capital used on a Greek Corinthian column decorated with acanthus leaves and scrolls. These were later developed by the Romans and Renaissance architects.

spring; fibulae with a plain bow and a bilateral spring, and fibulae with a coiling bow and fibulae without a spring. J. J. Hatt has shown that after Caesar's conquest of Gaul the shape of fibulae represented in sculpture can be dated. In the time of Augustus a fibula in the form of a dolphin and another with a disc in the middle are found. The mother goddess of Naix, one of the most remarkable Gallo-Roman religious sculptures, wears a fibula similar to those found in Germanic tombs of the time of Claudius and Nero. Under the Byzantine empire the fibula became a badge of military rank and, by extension, of the political office attached to it. It was the fibulae from the treasure from Ténès in Algeria which supplied J. Heurgon with the basis on which he was able to reconstruct the history of an aristocratic family in the fourth and fifth centuries.

Dating from radioactive and magnetic material

There will always be a mass of material to which typological methods cannot be applied but which is datable by other methods; in the case of carbonised organic material those of physical science are helpful. When alive all organic substances contain a fixed amount of radioactive carbon (isotope 14). After death this is dispersed by radiation at a uniform rate and it is possible to calculate when death took place (if it was during the last 50,000 years) by measuring the residual quantity of radioactive carbon. It is true that this method, disputed by some, includes a margin of error which varies according to the age of the specimen; for example, for substances some 2000 years old this margin is of the order of 200 or 300 years.

This alone makes the method unsatisfactory for classical times onwards because documentary records usually furnish a more precise date. Moreover, the rate of disintegration can be accelerated or slowed down if the specimen has been subjected to a variety of chemical reactions – if, for instance, it has been exposed to rain – and it is necessary to take specimens from uncontaminated places and keep them in non-metal containers. Despite all precautions, unreliable results are often obtained, and it is almost impossible to arrive at an acceptable result without carrying out a number of control experiments at the same site.

But with all these drawbacks, the method is of great importance to prehistoric studies, for, since the earliest days of the discipline, workers have searched eagerly for a way to translate the relative chronology based on the classification of Stone and Bronze Age material into absolute dates. The several centuries' margin of error of the carbon 14 method is insignificant in dealing with cultures with an extremely slow rhythm of development, for instance, the last Magdalenian period lasted for at least 1000 years. It is possible to experiment with specimens from many different sites belonging to the same stage of cultural development, and a comparison of the results can eliminate error. It is true that carbon dating cannot be applied to material more than 50,000 years old. But at that point a similar technique, based on the disintegration and transformation of radioactive potassium into argon (A40), can be used for material more than a million years old, and this method has allowed sites of *Homo habilis* to be dated.

Up to the present, it would be true to say that historical archaeology has benefited very little from the methods devised by atomic physics, but the techniques are being improved every day. Not only are physicists working in their own field, but archaeologists have sometimes thought of looking for carbon samples in unexpected places. One of the most serious difficulties they have encountered is finding the relationship between an ancient hearth and the remains of building and furnishings around it. It can always happen that visitors to a ruin lit a fire there.

There are yet more processes which can be tried. Professor Thellier of the Faculty of Science of the Sorbonne has been researching into the magnetic properties of terracotta for some years. Every pot contains minute particles of iron which form an infinitesimal magnet. Firing causes the field of this magnet to become fixed, and this field has exactly the value and strength of the earth's magnetic field at the time and place of firing. As the variations of the earth's magnetic field follow a regular curve, there is a possibility of determining when firing took place. Up to the present Professor Thellier has concentrated on reconstructing this curve by measuring the magnetic field of firmly dated terracottas, but once this variation is measured it will certainly be possible to turn this process to account in archaeological research.

There is therefore, hardly any trace of human activity which cannot state its age if it is properly interrogated. This justifies the extreme care taken by modern excavators not to overlook anything and the precautions they take in collecting the evidence. It will be seen, then, how much more complicated the task of the archaeologist has become, and how much heavier his responsibilities are growing as techniques become more precise and scholarly. Indeed, it is absolutely impossible for one man to carry out personally all the investigations that have been described. It must be added once again that the most difficult problem has not been examined, since the case of comparatively simple objects only has been considered. Subjects such as buildings demand a combination of studies.

Dating public buildings

If the houses in which people live and the public, civic and religious monuments of an ordinary town are examined, even a cursory enquiry cannot fail to show rapidly how complex their history is. The date when they were first built is comparatively easy to determine; it is often indicated by an inscription on monuments, and even private houses sometimes have the year of construction and the architect's name carved on them. Subsequent alterations, however, are another matter. Maintenance of the structure alone obliges each generation to add its contribution to the work of the builders. Someone often tries to enrich their appearance with new decorations, or to adapt them for different functions from those originally planned, or, on the other hand, to give them back their original form. These vicissitudes are not always easy to discern.

Accounts, inventories and dedications

The career of classical monuments was no less chequered; and the archaeologist rarely has an unbroken series of records at his disposal to reconstruct it. This good fortune, however, is sometimes the lot of the architectural historians especially those who deal with Greece. At Delos, for instance, parts of the accounts of the *hieropoioi*, magistrates in charge of the administration of the sanctuary during the period of the island's independence (314–166 BC)

Constructions can be dated by their style. An early form of Greek building is shown in this fragment of a polygonal wall on the south slope of the Acropolis. Later, with the development of the more refined classical style, rectangular blocks were employed as shown by this picture of the southwest wall of the sanctuary of the Telesterion at Eleusis. Mortar was not used in either case.

have been recovered

'The accounts, with several fragments of estimates, refer to a number of buildings, the majority of which existed from the beginning of the third century, or about 280 BC, and deal with a vast number of projects of varying importance from the placing of a keystone, the periodic revarnishing of the woodwork and the replacement of a few tiles to the construction of a new temple or a great stoa (colonnaded building). Unluckily, only a few haphazard parts have survived. It should also be remembered that the accuracy of these documents is only moderate, even when they are not mutilated. The wording is more or less abbreviated, they rarely indicate the location of a building, and still more rarely do they describe distinguishing features.

'The inventories drawn up when the island became independent have the same shortcomings, and they deal with only a small number of buildings. The Athenian inventories (in 166 BC Delos fell into the hands of the Athenians, who established a colony there) are much more useful. Apart from the usual offerings wreaths, jewels, vases, paintings, etc. statues, hermes stelae, doors and windows are mentioned. They list the principal monuments of a sanctuary, referring to the parts of each temple. Finally, they include, along with three public buildings (the *ekklesiasterion* or assembly hall, the *prytaneion* or senate house and the *gymnasion*, the forerunner of the Roman thermae), nearly all the temples and *oikoi* (shrines) of any importance, and these buildings are not set down at random, but grouped by districts. It is nearly as good as a guide book' (R. Vallois, *L'Architecture Hellénique et Hellénistique à Délos*, p. [illegible])

In the absence of accounts and inventories, the date of a building's construction may be indicated by its dedication. Even when this inscription is only partially preserved, epigraphists can often restore it. There are, admittedly, many misleading inscriptions. Names of founders who became the victims of political or religious revolutions either during their lifetimes or after their deaths (e.g. the heretic pharaoh Akhenaten and quite a number of Roman emperors) were chiselled out. Ambitious potentates unscrupulously annexed the works of their predecessors. Ramesses II usurped many buildings this way, and the same practice earned Trajan the nickname 'herb pellitory' (*herba parietaria*, a plant which grows on walls). By an absolutely contrary whim, Hadrian caused the name of Agrippa to be restored to the pediment of the Pantheon in Rome, which he completely rebuilt, and it was a long time before archaeologists could believe that nothing survives in the temple as it stands today of the work carried out by Augustus's son-in-law.

Dating objects in Pompeii presents no difficulty to the archaeologist since it is known that Mount Vesuvius erupted in AD 79. This picture shows a commercial bakery in the town. Many fascinating details of humble life which are never mentioned by contemporary authors were preserved because of the disaster.

The history of the most famous classical buildings has been preserved by writers, but it is not always easy to reconcile their references with the archaeological data. The name of the original founder is often preserved, even when every trace of his work has disappeared, which is precisely the case of the Pantheon. However, the chronicles of certain particularly renowned buildings are often accurate and detailed. It is known, for instance, that the Temple of Apollo at Delphi was reconstructed twice after its foundation in 548 BC: once between 514 and 505 under the patronage of the Alcmaeonidae, who had been banished from Athens, and subsequently from 370 to 340 after it had been destroyed by an earthquake in 373. The new temple was built by subscription, its sculptures being executed by two Athenians. It survived until Roman times.

The evolution of building styles and decoration The detailed study of monuments whose vicissitudes are recorded provides a basis for researches into buildings overlooked by the historians and lacking inscriptions. Their age can be worked out by analysis of the architectural structure and decoration, and also by means of finds recovered from their foundations. In the case of Delos, for example, members of the École Française d'Athènes began by dividing the buildings into categories according to the construction of their foundations. As a check on the accuracy of this classification they studied the shape of the bronze or lead staples joining the blocks together (the Greeks only used clay mortar, and in important structures they fastened adjoining blocks with metal), the system of the walls, the methods of marble cutting, the pro-

The House of the Faun in Pompeii is one of the most beautiful examples of an ancient dwelling to have survived from ancient times and shows in both plan and decoration considerable Hellenistic influence: Far left: the coloured marble impluvium, in the centre of which stands the statuette after which the house is named. Left: the bathrooms and kitchen from the second tetrastyle atrium. The passage at the far side of the peristyle leads through to an even larger peristyle with a garden.

portion and shape of the mouldings, etc. These researches have made it possible to subdivide the earlier groups and to determine the order of their sequence and the transitional phases between them. By combining the results thus obtained with the epigraphical and historical data, particularly the inventories and accounts mentioned above, very accurate results have been compiled.

Sites affected by some historical or natural disaster, if they have a well attested date of building and reconstruction, naturally provide the safest basis for architectural chronology. This is the case with the Greek and Roman colonies whose settlement was recorded by the historians, and with cities which were abandoned after being destroyed by an invader. All the buildings of Pompeii and Herculaneum naturally antedate AD 79.

By combining - with endless precautions - the results obtained by many different branches of research, it is possible to establish general laws, applicable to a whole culture, governing the evolution of its architecture. In Greece, the Mycenaean citadels were built in the 'cyclopean' style, i.e. using large rough blocks with the interstices filled with smaller stones. During the same period there was another, more polished style of building known as hammer-dressed work, in which the blocks are roughly finished, but not yet geometric in shape. The polygonal style, which was in use as late as the fourth century BC, was used especially for the retaining wall of the terrace at Delphi, and employs stones dressed to fit closely together, either with obtuse and acute angles (sharp polygonal) or with curves (blunt polygonal). In the trapezoidal style the blocks, usually parallelepiped in shape, often have interlocking projections. Finally, in the classical Greek style which was used for important buildings, the blocks are perfectly rectangular and often exactly the same size, the vertical joints falling precisely in the middle of the block in the course below. This is known as isodomic masonry.

Roman masonry is differentiated from Greek by the use of cement (except in a few very important buildings in the grand style), which was probably learned from the Carthaginians at the time of the Punic Wars. At the beginning of the Christian era a very distinctive style, known as 'reticulated' work, consisting of small lozenge-shaped stones, was often used. After the middle of the first century, however, baked brick, which was still unknown to Vitruvius, came to be the main building material of the Roman empire, either by itself or in combination with courses of small stones intersecting it in regularly spaced bands. It should be noted that Roman bricks are often stamped with the date, which is particularly useful for establishing an accurate chronology.

Among the most useful decorative elements to historians are the capitals and mouldings on the entablature. The earliest Doric capital was comparatively high, but gradually tended to grow shorter. The relationship between the height of the echinus (the moulding which forms the curved portion of the capital) and the total height of the capital is of the order of thirty five per cent to thirty-four per cent in buildings of the first half of the fifth century BC,

thirty-three per cent in the Parthenon, and drops to twenty-seven per cent in the fourth century BC (G. Roux, *L'Architecture de l'Argolide au IVe et au IIIe Siècle*, p. 410). The Doric column of the earlier temples tends to be squat, that of the fourth century BC much taller. After the Second Persian War, the architects of Athens enriched the entablature with a number of mouldings on the architrave, the frieze and the cornice. These innovations were not yet adopted in the Peloponnese. The architrave was the same height as the frieze in the fifth century BC. It was generally narrower in the fourth.

In Roman architecture, where the Corinthian order is predominant, artists enjoyed great freedom during the Republican period. A certain stabilisation of forms took place in the reign of Augustus, particularly from 20 to 15 BC. The acanthus of the capital, which until that time had looked like a spiky thistle, almost metallic in its hardness except where it was replaced by soft slightly indented leaves, took on a distinctive charm during this period. The foils are usually divided into rounded leaflets shaped like spoons or olive leaves. The cornice is enriched with a hitherto unknown element, the modillion, a sort of small bracket supporting a sharply projecting overhang. During the Empire architects tended to enrich the decorative mouldings with floral ornaments and even human figures, especially masks, with baroque exuberance. Leaves and figured decoration even invaded the column shafts. Composite capitals, a combination of Ionic and Corinthian, appeared towards the end of the first century. Very often the capital was ornamented with figures, the vitality of which to some extent foreshadows Romanesque sculpture.

Byzantine architectural decoration derives from Asia Minor and Roman Syria, where the Hellenistic tradition had been preserved and enriched with elements borrowed from other parts of the Empire. We find the spiky acanthus again, for instance, on the magnificent capitals adorning the earliest Byzantine churches, particularly those of Theodosius II (early fifth century). Through this intermediary, antique forms were introduced into Islamic art, which is classified by precisely this element, very obvious in the early stages but progressively diminishing.

The analysis of architectural ornament is just as fundamental for establishing the chronology of medieval Western Asian buildings, although in this period there are far more written records available than there are for the classical world. It should further be noted that the memory of Greek forms stayed alive throughout numerous variations springing from the minds of the artists; thus the crockets of Gothic capitals are the distant descendants of Corinthian scrolls.

Dating individual houses

For a long time classical archaeology was only interested in public buildings. Indeed, in the majority of ancient civilisations, the private houses, small and built of perishable materials, left very little discernible trace. Prehistorians

The House of the Faun contains some of the most famous mosaics in the world. Here are seen details of Alexander (left) and Darius (right) from a mosaic depicting the Battle of Issus which decorated the floor of the peristyle and dates from about 300 BC. Both Alexander and Darius are treated realistically rather than flattered.

and protohistorians have been the first to recover, by very arduous techniques, the hut foundations consisting of marks left by the wooden posts which supported huts made of branches. This type of research is very advanced nowadays, particularly in the German, Scandinavian and Slav countries, and is beginning to be applied to the more highly developed cultures. M. Cserny, Professor of Egyptology in the University of Oxford, has recently published the details of a workers' village discovered in the Fayyum. Our knowledge of Greek houses in the archaic and classical periods, however, is still very inadequate. There is not nearly enough material available until after the fourth century BC, i.e. from the period when a rising standard of living and the development of individualism endowed private houses—at least those of the richer classes—with a monumental character comparable with that of public buildings.

The evolution of domestic architecture
The archaeologist who investigates a Hellenistic or Roman house does not have available the easiest dating media which he meets with in a public building. There is no dedicatory inscription, contrary to modern custom, the architect did not carve his name or the year he built it on the structure. The identity of the owner is often in doubt. Even at Herculaneum, where everything is just as it was at the time of the eruption, there has been much dispute about the names of the inhabitants of various houses.

To determine the construction date, essentially archaeological media are often used. The chief of these are analysis of the finds and the structure of the masonry. Although private houses were naturally less carefully built than public monuments in general, the broad rules affecting the one can be more or less applied to the other. Thus the following styles were distinguished at Pompeii:

1. A pre-Samnite period dating to the sixth and fifth centuries BC. The building was carried out in large rectangular limestone blocks.
2. The first Samnite period (fourth and third centuries BC): masonry of small, less regular blocks appears beside the big limestone blocks, bonded by large vertical toothing in a style of Punic origin. Volcanic stone was used as well as limestone.
3. The second Samnite period from 200 to 80 BC, characterised by the use of large tufa blocks.
4. Finally, the Imperial period, distinguished by wider and wider use of baked brick, sometimes associated with reticulated work, sometimes alone.

Cave canem, 'Beware of the dog'. A mosaic in the entrance of the House of the Tragic Poet, Pompeii.

Mosaics From the end of the fifth century BC the Greeks, reviving a technique formerly known to the Minoans, began to pave the floors of their houses with pebbles set in plaster. By using different coloured pebbles they soon carried out geometric or representational designs which, on the whole, have the main characteristics of their other artifacts. Many of these 'pebble mosaics' have been found in the ruins of the town of Olynthus, sacked by Philip II of Macedon in 348 BC. More advanced pavements, made in the same way, decorated the palace of the kings of Macedon at Pella—one of them possibly represents Alexander himself struggling with a lion—and of Antigonus Gonatas at Palatitsa in northern Thessaly, dating to about 280 BC.

During the third century some craftsman thought of replacing the rough pebbles with squares of stone cut so as to fit exactly (tesserae). Thus the genuine mosaic was born, the earliest examples of which are found in Delos and at Pompeii, in houses built soon after the foundation of the Roman colony in 80 BC. In one of the latter, the House of the Faun, the world-renowned mosaic depicting one of Alexander's battles with Darius of Persia was found, along with eight other decorated pavements. Such floors are veritable pictures, exactly reproducing great masterpieces of painting. The majority of the Pompeian houses, however, had much simpler floors with a plain white ground decorated round the edge and in the middle with geometric motifs. This austere style, which contrasts strikingly with the richly decorated walls, lasted until the destruction of the cities near Vesuvius. However, several mosaics datable to the last years of these cities show a more elaborate and fantastic decoration, including several human figures.

Italian floors of the second century, the best collection of which is found in Ostia, were still carried out in black on a white ground, but along with the geometric motifs, stylised plant forms became increasingly common, as well as human figures and animals which came to form real pictures with the human figures shown in silhouette. This style was imitated in several provinces. In North Africa, however, a school of mosaic artists appeared at the beginning of the second century who returned to polychrome mosaic in the Hellenistic style.

Their methods of composition, however, were similar to those of the Italian school; like them, they put great emphasis on the non-representational elements (geometric patterns or stylised floral ornaments) of the design, which always left a large area of the white ground showing beside the graceful lines traced on it. These compositions are therefore as light as they are elegant. In the Greek-speaking countries, however (the most important group was found at Antioch on the Orontes), mosaic artists concentrated on representational designs, often drawn from literature, rather than simple abstract motifs with little variation. Finally, an original school appeared in the last half of the second century in Gaul and Germany.

Related to the Italian school, they lost no time in separating themselves from it, not only by their adoption of the polychrome style, but especially by their preference for heavier motifs surrounded by wide bands and plaiting, which more or less smothered the background. This style spread out during the third century, giving rise throughout the Empire to floors entirely covered with rich but very crowded polychrome decoration reminiscent of tapestry. Polychrome representational pictures are associated with these abstract designs (the black and white style finally died out about the middle of the century), the artists moving gradually further away from the realism typical of classical Greek painting. They made no attempt at perspective, and to give a feeling of the third dimension the figures were scattered about the field according to basically decorative laws. Stylistically these mosaics are already close to the Christian pavements which appeared on church floors during the fourth century.

The decoration of a house is not of course always contemporary with its construction. When the dwelling is inhabited for several generations, its occupants often redecorate it according to current taste. As a result it often happens that ancient floors or painted plaster, or at least recognisable traces of them, were preserved underneath a new coating, and it is partly by means of these traces that it has been possible to reconstruct the chronological sequence of the various styles. As to the furniture, ornaments, etc., they generally belong to the last occupants, but certain valuable pieces may have been kept for a long time: in this way a house at Volubilis, the mosaics of which were laid in about 230 BC, contained two bronze busts, one of which was a portrait of Cato of Utica, probably executed during his lifetime, and the other, which shows the features of none other than Hannibal as a young man, either an original work dating to the third century BC, or an Augustan copy.

Mosaics at Gaser in Libya. These were discovered by a farmer digging his land and represent some of the best preserved early Christian mosaics that have survived.

Restoration, exhibition and publication

Apollo, as the Romans of the late empire depicted him. The representation of divinities in the ancient world changed century by century and is a common guide for the archaeologist. This languid young god – wherever he was found – could never be confused with the formidable Greek original.

Contrary to the general belief of the uninitiated, the archaeologist's work is by no means done once the excavation is terminated. It could almost be said that the most important, and sometimes the most difficult, part still remains to be done. When he searches for traces of the ancient civilisations deep in the earth where they are buried, the archaeologist necessarily destroys the site itself, the tangible expression of a moment of the past. This petrified slice of history only survives in his memory, and its only reflection is in the notes and photographs he takes. It does not belong to him, it is his duty to pass it on, first to his colleagues, some of whom may use it to demonstrate facts which he himself is not altogether capable of deducing, and then to the general public at large, since they need it for their instruction and aesthetic pleasure.

This duty should be done first by publishing, which should be prompt and accurate, and then by lecturing and by writing books, pamphlets and articles for the informed general reader rather than the expert, taking care not to distort or abridge the subject too drastically. The archaeologist should also, however, take care of the site and the buildings he has discovered. He has been obliged to damage the former and deprive the latter of the protection built up around them by nature.

The conservation and restoration of objects

Conservation in the field The first stages of restoration are, in fact, inseparable from the actual excavation. Many remains are in such a fragile condition when they are found that they would be instantly destroyed by exposure to the atmosphere if their delicate fabric were not immediately reinforced. Others are actually destroyed already, but can be revived to some degree by modern methods: this is true of nearly all organic substances such as flesh, fabric, wood, etc. In the section on the excavations of Pompeii it will be explained how, since the middle of the nineteenth century, Italian archaeologists have known the method by which to take casts of the corpses of people and animals, plants, wooden architectural elements and furnishing which imprinted their shapes on the bed of ashes which engulfed the town. Nowadays some remarkable salvage operations are being conducted, like the reconstruction of the robes of the Merovingian Queen Arnegonde, whose tomb was discovered in the crypts of Saint-Denis in 1959.

By means of X-ray photography, M. Fleury and A. France have succeeded in separating from the mixture compounded of the queen's body and her clothes enough fragments of textiles to determine that she was buried in a chemise of fine linen, an eastern gown of purple silk, and a red silk tunic encircled by a wide belt. Her foot-wear was made of an exquisitely soft leather. This sumptuous outfit was adorned with embroidery and jewels, one of which (a signet ring) revealed the dead woman's identity. The majority of these treasures would naturally have been lost if the excavators had not proceeded with the utmost caution, and if they had not had the use of laboratories equipped with all the resources of physics and chemistry. It is unfortunately certain that even at Saint Denis, other equally interesting grave groups have been lost because the investigators were not careful enough, not only in the nineteenth century but even more recently. It should be stated in this context that too many amateurs imagine that the excavation of graves can be conducted by beginners, since it does not demand very much in the way of physical work. This is untrue: it is, on the contrary, a very delicate job requiring both great care and much equipment.

The restoration of paintings Another particularly delicate class of remains includes paintings. The action of light and damp on walls painted with frescoes quickly destroys them. This perishability was recognised in the Renaissance, but until the present day the means used to remedy it were not only inadequate but even harmful. At best, the ancient painting was copied by an artist. Records like those preserved in the archives, of which there are very many, are still extremely useful to the art historian – for example, the golden dome of the *Domus Aurea* of Nero is known to us only from a water colour by François de Hollande. It is, unfortunately, almost impossible for the most conscientious draughtsman to avoid 'interpreting' the model he is copying. The majority of the figures in ancient paintings copied in the seventeenth and eighteenth centuries are in the contemporary style of the artist's customary models.

Opposite page: the theatre of Herodes Atticus before restoration. This second-century building is no longer a striking ruin at the foot of the Acropolis but a fine theatre in regular use. See also page 139.

The walls of Nineveh and part of the remains of the great city, which was first investigated during the great age of Near Eastern exploration in the mid-nineteenth century. This picture captures very well the apparent confusion of a famous site. The archaeologist's task is to make order of the evidence and extract valuable information from the result.

The plaster with the parts of the composition regarded as the most interesting (i.e. almost exclusively the representational pictures) was next detached from the wall without the slightest scruples about destroying the unity of the ancient artist's composition. The 'fine pieces' selected in this way were transported to a particular gallery or museum, usually not without first being handed over to the mercies of a restorer, who attempted partly to safeguard the best preserved parts by varnishing them over, and partly to restore any lost parts. It need hardly be said that by the time he had finished the work was quite different from its original form.

These, then, were the methods by which the collection of wall paintings in the Naples museum, on which the majority of classic works on Greek and Roman paintings are based, was formed. As to the remaining decorations left *in situ*, they were abandoned to a slow but inevitable death by the combined attacks of the elements and the vandalism of tourists. Numerous Etruscan, Roman and medieval frescoes have perished in this way.

Happily, methods have radically changed. For one thing, special devices nowadays can reproduce with absolute fidelity not only the lines but even the colours of ancient paintings, and the practice of removing the decoration has been abandoned, except in cases where buildings are in places where preservation is totally impossible, such as desert sites like Dura Europos. As an illustration of former misdoings and the progress made since then, no better example could be quoted than the paintings of the Odyssey from the Esquiline the outlines of which were recovered by Signora Anna Gallina, a student of the School of Ancient Art of the University of Rome, under the direction of Professor R. Bianchi Bandinelli.

In 1847 and 1848 the city council of Rome ordered the excavation of the ruins of an ancient house on the Esquiline built in the 'reticulated' style which was characteristic of buildings of the later republic and early empire. One of the walls was found to be covered with particularly interesting frescoes, which were soon recognised as illustrating the story of Odysseus. The Fine Arts Commission of the council first had the paintings covered with wooden panels to avoid deterioration: then they called in a painter named Pellegrino Succi and commissioned him to 'remove and clean' the paintings.

Meanwhile, revolution broke out in Rome and a republic was proclaimed. Succi, who had already removed two of the paintings, kept them in his own home until the return of Pope Pius IX. He then took them to the Capitol museum. At the same time (September 1849) the excavation was resumed under the direction of the famous Canina. Five more intact paintings and half a sixth were found and removed by Succi. On 22 July 1850 the whole series was bought for the city of Rome for a price equivalent to about £1125. A little later

These three pictures show the stages of restoration of a Bronze Age cauldron found at Feltwell in the county of Norfolk, England. The first shows the fragments, all that were found. The next shows the painstaking reassembly of them: note how the missing portions have to be allowed for. Finally, with fibreglass supports, the fragments can be displayed as something remarkably like the original cauldron. The work was carried out in the British Museum.

they were offered by the corporation to the pope, who had them deposited in the Vatican museum, where they were entrusted to a new restorer named Ciuli.

Fortunately fairly accurate facsimiles of the frescoes were made as soon as they were discovered and one of the pictures from the series found its way into a private collection, which was bought several years later by the state-owned museum of the Baths of Diocletian. This picture, very fragmentary and broken in several places, was not restored in the nineteenth century. Succi and Ciuli employed a very different style and technique from that of the Roman painters. It will never be known if they painted truly *al fresco* on wet plaster, according to the method illustrated by a curious Gallo-Roman bas relief from Sens, or on dry plaster. In any case, the Roman painter's style, like many others of the second and fourth centuries, was related to modern impressionism. Figures and objects were suggested by blobs of colour and touches of light, with blurred outlines. This style was utterly incomprehensible to artists of the nineteenth century brought up in the traditional schools, as can be seen by examining the ancient paintings in the Louvre, many of which were restored during this period. The figures restored by Succi and Ciuli, on the contrary, are surrounded by a clear unbroken line, incised in places, and volume is indicated by shadows.

From 1956 to 1959 the technicians of the Vatican museum attempted to remove the nineteenth century restorations from two of the paintings. The stucco used to fill up the gaps was carefully removed, and they tried to dissolve the modern tempera additions, first with a solution of dilute hydrochloric acid, then with a basal wash. The results were unfortunately very disappointing. The work only survived in a few coloured parts and more or less polychrome patches. In fact, except where it had been covered by the work of the restorers, the background of the ancient painting proved to be disastrously fragile because of its extreme thinness and susceptibility to chemical reaction. The directors of the Vatican museum wisely decided not to pursue the treatment any further.

Nowadays the restorer is absolutely forbidden to add anything whatsoever to the ancient painting. His work must be aimed solely at strengthening its background. Two methods are actually employed by the Istituto di Restauro in Rome: one, the *strappo* method, consists of detaching the plaster from the wall, and is used especially for medieval paintings; for ancient paintings the *stacco* method, in which the layer of paint is removed from the plaster, is preferable. The procedure for this is similar to changing the canvas of an easel-painting; i.e. after cleaning the painting a very fine gauze is stuck over it, then a stronger one. The background is then removed and very gently scraped off, leaving nothing but the skin of colours. The latter is then placed on a new background, the nature of which varies according to whether the work is going to be replaced in the building it once adorned or kept in a museum. In the former case the new ground is separated from the wall by a layer of air to protect it from damp. The colour is also sealed off from the atmosphere by a clear varnish after the removal of the gauze. When the composition is disfigured by gaps they can be filled in by a drawing, so long as this is distinguishable at a glance from the original.

The remains of a mammoth discovered near Madrid by workmen preparing a housing site. A deposit of sand yielded the head bones of young female, the size of them indicating a creature of enormous proportions which died by drowning about 80,000 years ago. The work of conserving the bones with cellulose and plastic solutions was carried out on the site. When the structure was sufficiently toughened the bones and the tusks, nine feet long, were lifted by a crane and are now housed in a museum.

The restoration of mosaics and pottery

Restoration of mosaics is carried out by similar methods. It is true that the cubes of stone forming the coloured layer in this case are very much less perishable than paint, but glass paste cubes were also used, and these are dangerously fragile. The weak point of a mosaic is the plaster in which it is set. This is highly vulnerable to changes in temperature, particularly if it happens to be exposed to frost. Water seeping into it can split it, and then it is vulnerable to plants growing in the cracks, and accidental or deliberate damage done by people or animals. It is appalling how many visitors succumb to the stupid temptation to make off with a tessera as a 'souvenir'.

As soon as a mosaic is found it must be removed. This operation is carried out by first glueing a strong fabric on the outer face, then

A similar technique to that described on the facing page is used on a Roman bronze bowl found in England. The first picture shows the condition of the bowl after the metal was cleaned. Then a synthetic resin was used to brace the almost-complete bowl to enable its presentation as an excellent example of its type. The work was carried out in the British Museum.

A fine mosaic floor on the island of Delos. Further damage to mosaics – after their centuries of burial under earth and rubble – is often inflicted by tourists on the hunt for souvenirs.

detaching the pavement and its support and stripping off the layer of plaster. At this stage the fabric with the tesserae glued to it back-to-front can be transported and stored for quite a long time. From now onwards all that need be done to repair the pavement either *in situ* or in a museum, is to prepare a fresh bed of plaster (which should have a layer of air between it and the earth if the mosaic is going to stay out of doors) and to lay the fabric with the tesserae on this and detach it as soon as they are firmly set. This technique has been improved in Germany and it is now possible to remove the mosaic on a light plastic support which makes it easy to handle. While it is being raised, broken or lost tesserae can be replaced one at a time by new tesserae of the same material. This operation, which is comparatively harmless in non-representational designs, must be used with the utmost caution in the case of illustrative pieces; in such cases a barely perceptible alteration is enough to change the expression of a face, for instance, completely. When the pavement has major gaps they can be filled in by line drawings executed directly on the plaster ground.

The repairing of broken pottery is comparatively easy. In the case of vases, the pieces which fit together are glued and supported by sand until dry. The missing parts can be reconstructed in dental plaster.

The treatment of metals In the case of metal the problem of preservation is extremely difficult. Pure metal is not stable under natural conditions and as soon as it leaves the founder's hands it tends to combine with the oxygen in the atmosphere, or with acids. With the exception of gold, the only metal unaffected by deterioration, every metal object found in an excavation has been subjected to more or less advanced oxidisation or transformation into salts. Even after the first cleaning, this reaction starts all over again and the glass cases of the best-equipped museum are no impediment unless, indeed, they are filled with an inert gas like nitrogen. The restorer must wage continual war to preserve the object. The first stage, which should follow the excavation as closely as possible, consists of scraping to find the 'skin' of the object.

'Everything external to the original surface must be removed, everything internal preserved.' The rule is simple, and also defines that confused term "patina". This word, banished from the experts' vocabulary, has always been and

A collection of Greek pottery raised from the sea bed at Bonifacio in Corsica. Careful cleaning and restoration will be carried out in museum workrooms.

Opposite page, above: Castlerigg stone circle in Cumberland, with the hump of Blencathra behind. There are forty-eight stones in the ring, which served some religious or ceremonial purpose for the people of the north-west in prehistoric times just as did the more famous circle at Stonehenge for the people of the south-west.
Below: Stonehenge, the great prehistoric stone circle which stands on Salisbury Plain in Wiltshire, England. The outer stones are the later, the inner original ones having been hauled all the way from South Wales. Stonehenge dates from 2100 BC.

still is misused. It ought only to be used of artificial changes applied to the surface of a metal to accentuate its contours and, by extension, to the mineralised coat of carbonates and sulphates which forms on bronzes exposed to the air, but it should never be applied to oxidised objects found in the earth' (*Archaeologia*, 1965, 6, p. 11).

After cleaning, any damage spoiling the original appearance of the piece should be repaired. Only damage which seriously impairs the object's aesthetic value should be repaired – for instance, the Hercules of Bordeaux, who had lost an eye when he was first discovered, now has a whole face; damage must be left alone, however, if it was deliberately inflicted by the ancients themselves (in the performance of a ceremony, for instance), or even if it resulted from a specific historical accident. For example, the hand of the same Hercules, mutilated in an ancient fire, has never been repaired. Restorations must be obvious, for honesty's sake, but only to close examination by experts. Finally, it is essential to stop, or at least to retard the corrosive action of the atmosphere on the piece. This action is specifically the result of the development of chlorides on the metal in the presence of damp air. Chlorides can be removed either by a chemical process (in this way, however, part of the metal is also stripped off) or by neutralising the action of the damp air by impregnating the metal with a suitable product. All these operations should be carried out with caution and discrimination. What is more, they are not always necessary, since many objects are practically stable in normal conditions. Metal + oxides + environment form an equilibrium. The object must naturally be properly protected to preserve this balance. It is the conservationist's job to find a place where it will be secure in every sense.

The treatment of stone The case of stone sculpture, especially marble, is quite different from that of metal. The material is theoretically less susceptible to chemical reactions, on condition, of course, that it is not exposed to bad weather, or the droppings of animals, which are a source of corrosive acids. In this respect there is hardly any problem at all unless the work is out of doors; and even then it does not run much risk in a hot dry climate without pronounced variations in temperature, which can split stone.

On the other hand, carvings which have been exposed to heavy rains, and (especially in large towns) an atmosphere contaminated with chemical effluvia and microbes, are triply threatened: by physical erosion, by chemical erosion and by veritable infectious diseases carried by microbes or minute plant life. Cleaning is not enough to preserve them, and what is more, it destroys the surface of the stone. It is therefore usually necessary to remove these carvings to the shelter of a museum. This sets serious problems when they form part of the decoration of a building. The best copy, a cast in whatever material looks most like the original stone, is still sufficiently different to endanger the effect of the building. There is sometimes no alternative; and it is comforting to remember that statues from cathedral porches, for instance, can be seen better in museums, where they are safe, than in their original position. It is still desirable for the museum to be as close as possible to the shrine, so that the visitor can go on there with the impression of the building

Opposite page: the tomb of the Persian king Darius, carved from the rock face at Naqsh-i-Rustem. The Achaemenid dynasty fell before Alexander, and Persian glory might have died for ever. But 700 years later the Sassanids revived Persian greatness and one of their reliefs, on the lower left below the tomb, shows the humiliation of Rome herself: the Emperor Valerian kneeling before the victorious King Shapur.

The extent to which restoration is desirable, or even permissible, has always been a source of controversy among archaeologists. In the case of pottery, here being examined in the museum of Salonika, the problem is relatively simple.

Restoration work. In the Baia Mare museum in Rumania the fragments of an elaborate vessel of the Hallstatt period, found at the Lapus necropolis, are reassembled. Missing pieces are painstakingly simulated by the artist. The vessel is believed to have been used in a cremation ritual.

Repair and restoration of pottery discovered during the excavation of the ancient Agora in Athens.

One of the laboratories *in the Ancient Monuments* section of the *Ministry of Works* in London. Better *methods of preservation and conservation continue* to be sought.

At a museum in Brussels a carved wooden statue is given a *longer lease of life* by immersion in a bath of wax.

still fresh in his mind. The solution adopted at Strasbourg, where the museum has been housed in the Maison de l'Œuvre Notre Dame, home of the brotherhood who have looked after the maintenance and structure of the cathedral, could be quoted as a model for the museum is arranged so as to be as pleasing to the archaeologist wishing to study the carvings in detail as it is to the amateur who is chiefly in search of aesthetic enjoyment.

Statues unearthed during an excavation are sometimes seriously damaged. The problem is to decide how far it is permissible to restore their deficiencies. This type of surgery does not raise any questions of principle when it is possible to reassemble all the scattered fragments of the original. But an archaeologist requires a highly developed sense of judgment and an excellent visual memory to succeed in this search. An excavation often produces part of a sculpture dug up much earlier. The torso of the Winged Victory of Samothrace was found in 118 pieces by the consul Champoiseau in 1863. Twenty-three years later the same diplomat discovered the base, shaped like a ship's prow, on which the goddess stood. But it was not until 1950 that a French archaeological expedition added Nike's right hand, minus three fingers, to the puzzle. This hand cannot be replaced because the arm is missing, but it has provided the answer to a long-disputed archaeological problem.

A coin of Demetrius Poliorchetes, one of the *Diadochoi* of Alexander the Great, struck in 306 BC, shows Victory poised on the prow of a galley. For a long time it was believed that it represented the Samothrace Victory but the angle of the statue's neck (the head is missing) suggests that she could not have been blowing a trumpet like the Victory on the coin; what is more, it seems that the freedom of movement of the statue, the treatment of the drapery, and above all, the integration of the work into the great architectural composition which it crowned, all point to a later date, about the end of the third or the beginning of the second century BC. Stylistic judgments, however, have little validity in a period like this, where the artistic heritage is so sparse. The hand could not have held a trumpet; open and extended, its sole gesture is an announcement of triumph. Any connection with the coin of Demetrius is therefore ruled out, and it is generally believed that the group commemorates a victory won by the Rhodian fleet in the north Aegean *c.* 200 BC.

Restoration without alteration From the sixteenth to the nineteenth centuries, antique statues were greatly sought after by kings and noblemen for the adornment of their castles and parks. These collectors would not have dreamed of displaying a damaged work – many people even today grumble at the noseless heads to be seen in museums and accuse the archaeologists of negligence when they cannot find the missing pieces. They always repaired marbles in an attempt to restore them as nearly as possible to their original appearance, but their main care was to conform to the current idea of beauty. This operation, moreover, was often conducted under an erroneous notion of the subject's identity. Cunning craftsmen excelled in fabricating a *collage* of ancient fragments joined together with bits of the raw material. Even in 1811 the king of Bavaria ordered Thorwaldsen to restore the decorations of the Temple of Aphaea on Aegina, which he had just acquired. The work of the famous Danish sculptor was very skilful, but it has radically transformed the statues into completely new works. The results are hybrids which appear in most works on Greek archaeology as authentic representatives of early fifth century sculpture. Thorwaldsen's additions are now being removed.

While deprecating such excesses, it is often impossible to display debris as it emerges from the excavation. When a statue is broken in several places which do not fit together, it is not at all a good idea to set them on a plinth side by side. It is a matter for the personal judgment of the curators and restorers, who have to find a happy medium between the requirements of scholarship and those of aesthetics. Some time ago in Italy it was customary to fill in the missing parts of statues with a translucent plastic substance. The restorer could not be accused of trying to deceive by passing his own work off as authentic, but the discrepancy of these curious conglomerations excited more laughter than admiration. This 'progressive' practice has wisely been abandoned.

There should be no illusions about the fact that it is completely impossible to restore an ancient work of art to its original condition.

Even if exactly the same technique as the original artist's is used, the simple fact that the physical environment has disappeared, and that the intellectual motivations behind the making of the monument were completely different from our ideas, would prevent us from viewing it with the same feelings as the people of its own time. This is another aspect of the fundamental problem and difficulty of archaeology: it tries to ignore the passing of time and to go backwards through the course of history; but time cannot be abolished, nor history effaced. The monument seen today is not the work of the original artist. It has been transformed, deliberately or accidentally, by a thousand go-betweens. The restorer and the exhibitor do not reverse this process, they only add to it.

Earlier restorations, now considered objectionable arose from the spirit of the time, which regarded ancient remains not as historical records but as talismans for evoking dreams. Their authenticity was not in the least important. The attitude nowadays is more complex and, on the whole, less consistent. 'To restore an object,' wrote A. France-Lanord, 'is above all to restore its meaning to it.' Like the Italian professor Cesare Brandi, he regards the archaeological specimen as a personality in its own right, or in any case, as a 'social phenomenon with its own origins, which are full of historical and temporal meaning', with 'powers of irradiation'. All this is doubtless true, but the problem arises because of the fact that the 'personality' of an object does not exist in itself, but only in the mind of the student, and is different for the archaeological historian, the restorer, the exhibitor or the collector, who may be an artist, or a 'middle man' of varying degrees of discrimination. The archaeological historian is trying to discover what the object was like when it was new before studying it. The restorer, who is half-way between artist and historian, also concentrates on its value as a source of information, but from a more practical viewpoint. He would like to know the methods of the artist who created it, and to restore the object's definitive meaning, but he is also interested in the vicissitudes which have meanwhile transformed it, and he hopes that the end result of his work will please his personal taste. The exhibitor, be he museum curator or private collector, will concentrate chiefly on the latter aspect, and will take care to have the object placed in the setting he considers suitable for it. The amateur will not notice the setting at first, but will look at the object for his own personal gratification. These various demands can contradict each other.

A. France-Lanord gives an example of such dilemmas in the case of the excavator or museum curator who comes to him '. . . with a mass of shapeless debris and a drawing of the object as he would like to have it. There have been occasions when I had had to explain that the object was, to all intents and purposes, "dead".'

There is another, yet more serious problem: in some cases, certain absolutely genuine characteristics of the object which have been toned down or obliterated by time are repugnant once they are restored. Thus there is no doubt that the majority of ancient marbles and friezes were painted. This colour is tolerated in archaic work, which somehow preserves it better than others, such as the pediment of the Hecatompedon, for instance. Similarly it is acceptable for baroque pieces, but many people would be revolted at the repainting of classical

A statue is raised from the sea bed at Baia, in the Bay of Naples, by a team of skin divers. The recovery of a work of art, whether from the sea or from centuries of interment in earth, is the first stage of a long process which must take place before it can be shown to the public.

Modern development does not always destroy the relics of the past. A fine bronze statue of a *kouros* (youth) was found in the Piraeus, the port of Athens, during road-building operations.

The magnificent Hellenistic statue known as the Winged Victory of Samothrace dates from the second century BC. It stood originally in the sanctuary of the

statues which would give them a 'chocolate-box' effect. Indeed, no curator has dared to attempt this experiment, except on casts which are not exhibited.

The restoration of buildings

Most of the problems just examined recur in an accentuated form in the case of architecture, and particularly in large urban complexes. They vary, however, in the case of buildings which have always stood above ground and those which, after being completely lost, have been brought to light again by archaeology.

In the first case the basic elements of the ancient building have only been preserved because it was re-used, though obviously for entirely different ends from its original function. The story of the Maison Carrée at Nîmes, told in the second chapter, is a good example of these *vicissitudes*.

The Parthenon was used as a church in the Middle Ages, and as a mosque during the Turkish occupation. In the nineteenth and twentieth centuries most of the buildings which were previously transformed in this way have been relieved of the superfluous structures spoiling their appearance. These restorations do not usually rouse much protest, although they often necessitate the deconsecration of a religious shrine. A well known example is Hagia Sophia, which the Turkish government converted into a museum, authorising the cleaning of the fine Byzantine mosaics which were hidden under a coat of plaster while the church was being used as a mosque. However, Maurice Barrès castigated the destroyers of the 'Frankish tower', built in the fourteenth century on the south wing of the Propylaea in Athens, in his book *Voyage de Sparte*. Schliemann took the initiative in this demolition, which obviously explains the reaction of this nationalistic writer in part. The regrets expressed by many lovers of the old city of Rome when Mussolini had the imperial Fora cleared seem to be better justified. In this case, indeed, the operation necessitated the destruction of several churches and palaces of the seventeenth and eighteenth centuries of indisputable value and interest, and obliterated a whole district full of picturesque life.

The dangers of modern life It is usually the requirements of everyday life rather than those of archaeology which threaten the remains of medieval cities, Renaissance mansions, and neo-classical and baroque palaces. Every day the newspapers publish the mutual recriminations of the conservationists and developers who are keeping up a running battle in the matter of town planning which has not been settled with regard to the other arts. There are,

'The clearing of the Propylaia', from a painting by Sir Charles Fellowes. When the Greeks secured their independence from the Turks the Acropolis became the scene of intense activity – the beginning of the attempt to salvage what remained of the former glory.

A print of the *Parthenon made in* 1797 shows the accumulation of dwelling houses and the Turkish mosque built after the explosion of the powder magazine stored there in the seventeenth century. All non-classical structures have now been removed.

The tiny Ionic temple of Nike (above) on the Acropolis of Athens was demolished during the Turkish occupation. Luckily the ruins were left where they had fallen and the temple has been largely restored after painstaking study (above right).

unfortunately, still too many town councils who cannot see that the preservation of the ancient monuments of their city is desirable for more than aesthetic reasons, and that it often offers better returns to keep a narrow street flanked with beautiful house-fronts than to open up a wide trunk road for the convenience of speeding motorists. Architects nowadays try to modernise a district without changing its external appearance, to make old houses more comfortable and restore them at the same time. These projects are unfortunately, very expensive, and it is not easy to think of a practical use for former aristocratic residences designed for a completely different society from ours. The same difficulty applies to castles in the country. It is true that they can be used for vacation centres and social groups, but such bodies are not always in sympathy with their surroundings and sometimes cause thoughtless damage. Even churches still used for worship often seem to be ill-adapted to the modern liturgy. The creation of museum villages and districts is an acceptable alternative; but it can only be done in exceptional cases, and the complexes produced in this way often have an unpleasantly artificial and sophisticated atmosphere.

These problems, like so many others discussed, bring out the fundamental paradox of archaeology: its efforts to do battle with time. There is no definitive solution to these problems, either. It is plain that all the architectura remains of the past cannot be saved. In man cases, some have to be sacrificed to others. I churches have been destroyed to recove temples, how many cathedrals bury ver interesting ancient buildings forever beneat their foundations? When a derelict building i a danger to public health, no one but a fanati could demand its preservation at any cost. It i obviously necessary to choose and settle th question, taking all the aspects of the probler into account; and only the government is in position to do this.

The Nubian temples A dramatic examp which roused world-wide public opinion oc curred in Egypt in the case of the high dam Aswan, which is typical of the fundament problem of archaeology. The rescue of th Abu Simbel temples, the necessity for whic was incontestable, could have been effected various ways, but it was beyond the Egypti government's means. The matter was th taken up by UNESCO; this was the first tir this organisation had taken charge of archaeological project. The funds were secur by international collection, but unfortunat this did not realise the necessary sum estimat by the experts. This shortage of funds had be taken into consideration by the committee

charge of making the choice between the different salvage plans submitted. Some of these plans suggested keeping the water out of the sanctuary by building a coffer dam. Others proposed the removal of the temples, or their reconstruction on a new site above water level; the latter plan was actually chosen and put into practice. It had the obvious major disadvantage of uprooting the temples from the site for which they were designed. Furthermore, a reconstruction, even if it is conducted with every possible precaution, does not restore the original building, and can produce only a copy made of the ancient materials. The 'personality' of the building is in great danger of failing to survive this major surgical operation. It is true that it is always better to operate on a sick man, even at the risk of crippling him, rather than to abandon him to certain death. What is more, the Abu Simbel temples were not the only ones threatened by the building of the Aswan dam. The lake would have to submerge an enormous area of Egypt and the Sudan, now deserted but once the home of some astonishingly highly developed cultures. The investigations undertaken by Polish archaeologists there revealed the almost totally unknown art of Christian Nubia, which held off the Islamic conquest until the twelfth century. The frescoes from Faras now in the Warsaw Museum are as good as the best Byzantine painting.

Some methods of architectural restoration If permission has been granted for the preservation of a building the means have to be found and the right architect engaged. The technical problems of restoration arising at this point are fairly close to those that have just been examined for movable remains. Until recently the architect used to see restoration in very simple terms: it was a matter of replacing the lost or badly damaged parts of a building. If there was enough money, nothing stopped him from restoring it to its original condition.

Not so long ago the American School of Archaeology rebuilt the portico constructed in the second century BC by King Attalus II of Pergamon on the eastern edge of the Athenian agora. The greatest pains were taken in this case to reproduce faithfully the Hellenistic construction and decoration. The quarry which provided the original material was reopened and every block was carved after an antique model. However, the result is not acceptable to everyone. Most visitors think this large new building as cold and dreary as the palaces built for King Otto by his Bavarian architects in the last century. The psychological reasons for this attitude are complex: the ruin in its fallen state keeps a romantic charm which can be appreciated without going so far as to emulate Volney's enthusiasm. The chief drawback of the Stoa of Attalus as it is now is that it is the only restored

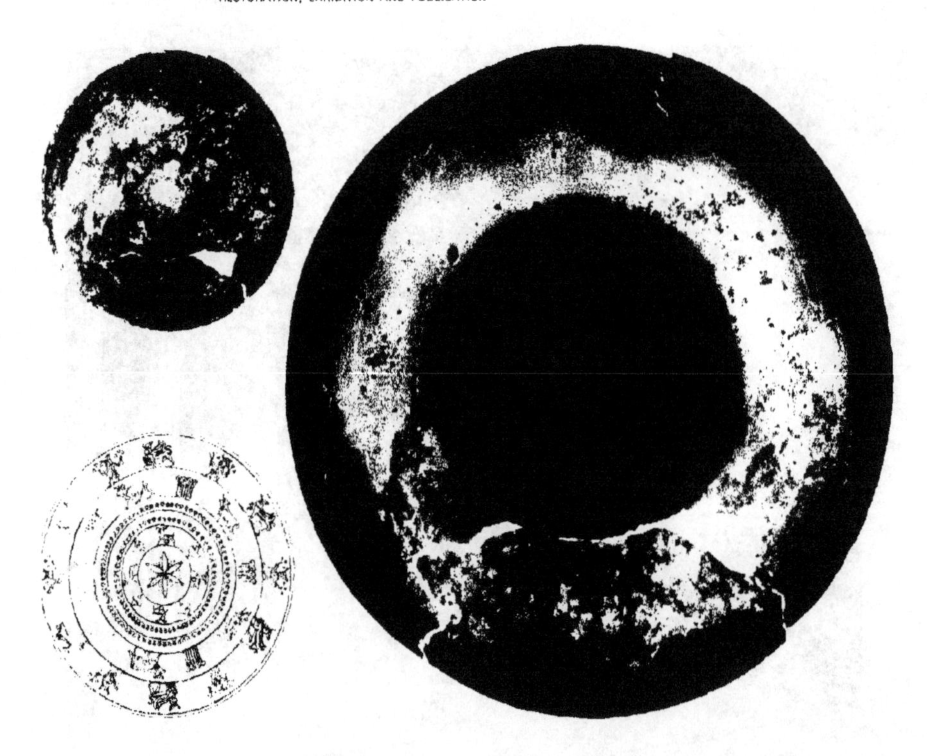

A shallow bronze bowl, found in the ruins of Nimrud, in the condition in which it was delivered to the British Museum, is seen in the top left picture. The large one shows what radiography revealed: the incised decoration, dimly seen, leads to the drawing in the third picture. The bowl, fully restored, could be seen to be Phoenician though found in an Assyrian city.

building among the ruins of the ancient site. When it has aged sufficiently, perhaps it will be a passable counterpart to the Hephaesteum. It is, moreover, particularly difficult to build a successful Doric colonnade.

A less ambitious form of architectural restoration simply consists of setting the fallen portions of the ancient structure back in their places. This is known as *anastylosis* (re erection of columns). One of the first operations of this type was carried out in 1835 by the German archaeologists and architects Ross, Schaubert and Hansen. The temple of Athena Nike (the Victorious), which had stood intact on the south west bastion of the Acropolis in Athens until 1676, was demolished by the Turks, but they left the remains where they were, and there were enough left to reconstruct the shrine. After the earthquake of 1894, which shook up the remains of the Parthenon, that most celebrated of Greek temples was in its turn the object of a discreet anastylosis conducted by N. Balanos until his death in 1930. The Erectheum, already restored in 1837 by Pittakis and a little later by the Frenchman Paccard, also owes its present appearance to N. Balanos. There are many other operations of this nature which could be described, but one of particular merit which deserves mention concerns the theatre at Sabratha, in Tripolitania, which was restored by an Italian team led by the archaeologist Guidi and the architect Vincifori. Three superimposed storeys of colonnades had to be re-erected in front of the stage wall, which curved into three deep semicircular bays. The baroque effect of this restoration is highly pleasing. It is regrettable only that it is supported behind the stage by concrete buttresses, but these are fortunately invisible from inside the 'pit' (cavea).

The name given to this kind of operation is explained by the fact that setting up the columns and entablature again is usually sufficient to restore the aspects of a Greek temple. Roman architecture is another matter again, since the main elements of the building upon which the architectural balance depends are the walls and the vaults, which are generally made of bricks and mortar. This is remarkably durable in a dry climate, but because of this very fact, it can sometimes happen that certain surviving parts of it in high positions have no support. The foundations must be strengthened, as they have often been weakened by the removal for reuse of bricks and the large stone blocks which form the ties. This was done by A. Lézine at the great baths at Maktar in Tunisia, for instance and in some of the rooms in the Antonine baths at Carthage. Some parts of the vaults and walls were only held in place in these buildings by the rubble that had filled up the rooms, the masonry having been removed for exploitation in the galleries after the building was buried. The restoration naturally had to be completed before the fill was removed.

Another difficult problem was set, again in Tunisia, by the amphitheatre at El Djem. This 'colosseum', nearly as majestic as the one in Rome, stayed intact until the seventeenth century. But since it served as a fortress for the inhabitants when they rebelled against the Bey, one of these rulers had a great breach opene in the wall with his artillery. Now, just as the gallery round the Colosseum at Rome was roofed by a continuous ring of vaulting, the comparable part of the African amphitheatre consisted of a series of radially-disposed rooms, each one entered by one of the arcades in the façade, and each roofed by an individual vault, the thrust of which was taken by the adjoining vaults. The opening of the breach resulted in a concentration of pressure at the edges of the gap. This pressure, which tended to force the structure apart from then onwards, was fortunately comparatively weak because it was largely taken by the pillars between the arcades. After 200 or 300 years, nevertheless, the effect began to be felt, and disaster would have followed if restorations had not been carried out. L. Poinssot, the first director, had the broken pillars on each side of the gap replaced by concrete models, disguising this material with a facing taken from the large blocks lying around the building. Unfortunately the metal tenons securing this facing proved to be inadequate. A. Lézine, who took up the task after the Second World War, then decided to face the reconstructed parts with whole blocks which he brought from the quarry at Rejiche, which supplied the Roman builders.

The humidity of the oceanic climate causes English, French and German archaeologists problems unknown to their Mediterranean colleagues. When Roman masonry has been buried for a long period of time in really damp soil and is then dug up, the plaster impregnated with water disintegrates if exposed to frost and thaw. 'At the end of the terrible winter of 1956,' wrote A. Audin in *Lyon, miroir de Rome dans les Gaules*, 'I remember standing at the foot of an ancient wall when it was touched by the first March sunbeams. A silky rustling struck my ear: suddenly thawed, the mortar was running gently down the walls.'

After restoration, properly speaking, comes exhibition—a process which can hardly be separated from it in the case of buildings.

Presentation of buildings and sites

Visiting an archaeological site is not always a pleasure. Often the visitor, attracted by a famous name and a few promising posters, finds himself faced with a waste ground where, among thistles and long grass full of restless animal life, he sees a few derelict stone walls at the bottom of holes where rubbish is piling up, some column-drums and capitals, some scattered *tesserae* and fragments of painted plaster (all that remains of the frescoes and mosaics), and a carved stone on which the lettering of an inscription or the contours of a relief are gradually wearing away.

It would be better not to excavate, or to re-inter the ruins, sooner than end up with a result like this.

Custodianship and public information

Every cleared site should be protected, administered and made to pay its way. The first step is to have the ruins fenced in and guarded. It is not easy to find the right custodians. Admittedly there are usually plenty of applicants for the job, but the chosen candidate will also have to act as a tourists' guide. As his education is usually rather limited he may fill up the gaps from his imagination, which is often only too flourishing.

This folklore has clearly defined categories arising from the popular misconceptions of the ancient world which the visitor brings with him. The Romans, for instance, were fierce conquerors who relaxed between wars at decadent orgies, indulging themselves on high days and holidays by throwing a few Christians to the lions. Every house is thus regarded as a house of ill-repute, and the visitor's attention is sedulously drawn to the obscene motifs frequently drawn or carved on the walls, without mention of the fact that these are usually talismans for averting the evil eye. At Leptis Magna in Tripolitania, where every street corner has its sign displaying a phallus, the visitor might well feel overawed at what looks like an insistence on the former inhabitants' virility. In fact, the signs boast of nothing—they were purely prophylactic. Everyone who visits the amphitheatre will be shown the martyrs' prison facing the cages of the lions destined to devour them. In fact, the victims of these beasts were mostly criminal offenders. Other custodians demonstrate even more imagination. One in the Ile de France, while he took very good care of the area entrusted to him (a temple associated with a theatre and some baths), explained to tourists that this was a rest centre for the troops of four legions quartered in the vicinity. Few visitors know that there were practically no troops in Gaul, and that from the beginning of the second century the whole of the Rhine frontier was patrolled by exactly four legions. It would have been better to have told them that such monumental complexes were common in the French countryside and they allowed the Gallo-Roman peasants in sparsely urbanised districts to have the benefit of the hygiene and the entertainments typical of life in imperial Rome, as if they too were city-dwellers.

The site of Bulla Regia, one of the finest in north-west Tunisia, was recently guarded by an African who became so passionately fond of it

In a developing urban environment, ancient buildings with outstanding architectural merit but no modern function sometimes present a problem. This has been successfully solved in the case of the Hôtel Carnavalet in the Marais district of Paris by converting it into a museum. It was once the home of Madame de Sévigné.

A striking aerial view of the Roman amphitheatre at El Djem in Tunisia. It was in danger of collapse after being used as a fortress in a rebellion in the seventeenth century, when the great breach in the walls was made. The reinforcements carried out after the Second World War were made with stone from the same quarry which supplied the original Roman builders.

that he himself came to be known as Bulla Regia, and almost forgot his real name. The inhabitants of this town, the climate of which is stifling, used to retire during the dog-days to underground rooms, the decorations of which are well-preserved. Bulla Regia would point out a 'French mosaic', then a 'German', and then an 'Italian'. Eventually he and his visitors would get to the 'Greek mosaic', which was actually bordered with the kind of meander known as a Greek key.

However picturesque the guides' rigmaroles may be—they already existed in the ancient world and some of their nonsense has survived—it is obviously much better if the visitors are correctly informed. The best way of doing this is to place at their disposal a pamphlet compiled by a competent expert. In Italy the Ministry of Public Education has published a series of *Itinerari e Guidi* in more than a hundred parts describing very clearly and accurately all the big sites and main museums. The drawback in an enterprise like this is that it is apparently not a good enough commercial proposition to interest a private publisher. Moreover, the rules governing the administration of public funds in some countries make the publication of a work of this kind very difficult. This could be a matter for national archaeological societies, the formation of which would be desirable for several other reasons. In the meantime, it is admittedly possible to find adequate information on plenty of sites. But the centralisation to be found in Italy is extremely convenient for the visitor, who is sure of not being misinformed by the booklet he buys.

Another procedure currently being used in some countries consists of recording an explanatory talk in several languages. A speaker operated by inserting a coin is made available to tourists in the building or at the entrance to a site.

Should archaeological sites be landscaped? It is absolutely essential to clear a site of undergrowth. Trees often displace stones, and even weeds can cause damage to floors. Moreover, brambles and thistles may scratch visitors or tear their clothes, and grass growing on stonework often harbours snakes. When a stone building in a good state of preservation is built directly on rock, like the Greek temples on the Acropolis, it would obviously be absurd to try to decorate them with greenery, which would, in any case, have little chance of survival. Athena's olive, piously replanted near the Erechtheum, is the only plant allowed on the sacred hill, but many ruins owe the essence of their attraction to the vegetation enveloping them. Even the most austere archaeologist could not wish the Farnese gardens to be uprooted to get at the foundations of Tiberius' palace. Nobody regrets that the Trémeaux family laid out their park over the ruins of Tipasa. On the contrary, there is an increasing trend among present-day archaeologists to turn their excavation sites into parks once the investigation is completed. Some cross-grained persons insist that no more should be done than the reconstruction of the ancient gardens, as has been done—and uncommonly well done—at Pompeii, for instance.

But planting trees and flowers among ruins is not just a fancy, nor even a means of making a visit there more attractive by disguising the depressing effect of the waste-ground, it is essential, especially for sites in an urban setting. By making a 'green space' of the site the demands of health and town planning can be satisfied, and a defence is provided against anyone who accuses the antiquarians of monopolising building land. Judiciously arranged and controlled greenery and flower beds can enhance the smallest remains of a wall and the most insignificant mutilated carving. Plants can follow the ground plan of a lost building which it would be dishonest to reconstruct. This device has long been used, for instance, to recall the church of Port-Royal which was razed by the order of Louis XIV. In the park of the Antonine baths at Carthage the walks follow the ancient streets, only the paving and the drains of which survive. Here too the cisterns of vanished houses have been made into little museums displaying the carvings and mosaics from the site. Plants always give life to ruins; the mimosa and bougainvillia at Leptis Magna add considerably to the attractions of the site.

Son et lumière This need to reanimate ancient remains takes many different forms nowadays. It is a good idea, in principle at least, to recall the great events in their history by means of *Son et lumière*. Some of the theatres have been back in use for a long time. The *choregia* at Orange produced by Mounet-Sully have been widely imitated. Attic tragedies, performed in the theatre of Herodes Atticus in

Athens and the theatre at Epidaurus in a modern Greek version fairly close to the original, exercise an emotional impact on the audience which film producers have been able to give a world-wide application through such films as *Electra*, shot at Mycenae in the ruins of the palace of the Atreidae. Not all the improvised summer productions, certainly, have the same majesty, but all these attempts, even the least successful, are manifestations of the current interest in the classical world mentioned in the first chapter of this book. This interest may be less serious and profound than that of our ancestors, but it has at least the advantage of reaching many who were formerly debarred from education.

The ruins of the temple of Zeus at Olympia are typical of the school of archaeological thought which rejects restoration. Mistakes are avoided and strict accuracy is sometimes preserved at the expense of visual clarity and popular appeal

The museums

Museums, originally founded to gratify the classical interests of a few great noblemen, are now open to all and are much used in education. Their function, admittedly, far surpasses that of archaeology for they deal with subjects as diverse as modern art, folk arts and crafts, and science. There have been attempts to compile general laws applicable to all these institutions, and this is the object of museum studies, which cannot be regarded as belonging exclusively to either art, history or archaeology. In a work on archaeology, however, it is impossible to pass over the problem of conservation and display of movable objects after examining these questions, particularly with regard to buildings

Museums devoted to individuals and single sites It could certainly be said, from the archaeological point of view, that to deposit an object in a museum is a lesser evil rather than an aim to be pursued for its own sake. The ideal solution would be to retain all the finds on the actual site so as to keep the petrified section of past time as complete as possible. It is perfectly obvious that this solution is only possible in theory as it would expose the most valuable monuments to inevitable destruction, either by the forces of nature or by vandals and thieves, whose depredations cannot be prevented by even the most careful supervision. Moreover, many ancient objects have not reached us via excavations, many of them were

Only the highly trained expert eye can obtain much information from the tumbled masonry of the temple of Zeus at Olympia. The minimum of clearing has been carried out, but otherwise the site has been left as nearly as possible as it was found.

Even the most careful restorations, if over-zealously pursued, can have unfortunate results. In repairing the château of Pierrefonds, Viollet-le-Duc believed he had exactly reproduced the original medieval structure, but the total effect is unconvincing.

The use of infra-red light is a frequent aid in museum laboratories. The first picture shows a medieval manuscript as it originally appeared. The second shows that a great deal of the original writing had in fact been erased. British Museum.

separated from their proper environment several hundred years ago in very obscure circumstances. Others have been discovered recently, but out of their context, which has not been scientifically studied for one reason or another, or even recorded. This is true of innumerable chance finds at which no archaeologist was present.

In these circumstances, several systems can be suggested for the display of collections. One consists of an artificial reconstruction of the archaeological environment where the object was found or (not by any means the same thing), the historical setting for which the object was designed and in which it 'lived'. This solution is easily adopted if the museum is a sort of shrine devoted to the memory of a single individual.

One of the earliest of these attempts known concerned the emperor Augustus. The historian Suetonius, who lived a hundred years after that ruler, relates that his house, still quite a modest home, had been preserved intact on the Palatine alongside the far more luxurious palaces of his successors. The furniture and personal possessions of the sovereign were exhibited to the curiosity or reverence of visitors. Although modern man has theoretically abandoned the practice of deifying great men, there are still plenty of shrine-museums of the same sort. The palace of Versailles is one of them, and there are many more devoted to great statesmen like Washington and Napoleon and to saints, writers and artists. Even genuine archaeological museums sometimes fall into this category. That of Châtillon-sur-Seine, for example, is completely permeated with the spirit of the mysterious Celtic princess whose remains were found surrounded by her treasure beneath the tumulus at Vix.

The Châtillon museum can also be quoted as an example of an excavation museum. In these, all the finds from a site are reunited close to the place where they were discovered. The idea is pleasing in that it allows the visitor to supplement his impression of the ruins almost immediately with that of the decorations displayed close to the site. At Olympia, for instance, or Delphi, he will very naturally see the statues and metopes of the great temple of Zeus or the friezes of the Sicyonian and Siphnian treasuries in terms of their setting in the Altis or the Sacred Way. Cathedral museums like that of Strasbourg mentioned above produce the same effect. The spirit of each site thus emerges very clearly, and it can be seen how it dictated the work of the artists of each successive period in their wish to follow the line marked out by their predecessors, while conforming to the taste of their own age.

Where there are no major works of art a museum can perhaps display a collection of the modest implements of daily life, which creates an attractive atmosphere. At Châteaumeillant in the Cher district the simple collection in a noble house of all the pottery of the period of the Roman conquest admirably evokes the way of life of Vercingetorix's contemporaries, especially their economic dependence on Italy, which obviously preceded and promoted their political and military independence. Excavation museums, however, raise a number of practical questions which are not easily solved. The chief difficulty is that of security when small but valuable objects are displayed. The style of exhibition, too, is a problem. In many collections formed at an earlier date, objects which have never been properly restored are gradually disintegrating, piled up in utter confusion in glass cases. An expert can nearly always find some object there which its finder did not understand, and these are often very interesting. For many reasons the creation and administration of these private museums ought to be controlled by some nearby public authority, and local feeling should not be allowed to oppose the transfer of outstanding pieces to a regional important centre, either with the object of protecting them from internal and external danger, or of allowing them to be more greatly appreciated.

National museums: exhibition halls and students' rooms The big national museums are, in fact, the only ones capable of presenting the whole artistic evolution of the chief periods by grouping a number of major works representing them. Modern museum practice has abandoned the custom of heaping all the wealth in their possession in the public exhibition halls. Only the objects of outstanding quality or originality are now displayed. Each one must be set at a sufficient distance from its neighbours for it to be examined by itself from several angles. The hall forms a veritable artistic composition in itself, and this is composed by taking into account not only the effect of each piece, singly and in groups, but the architecture of the building, which is often undistinguished. Naturally this presentation ought also to fulfil the requirements of art history, i.e. to group together works belonging to the same period and the same school. In older galleries ancient statues were deliberately associated with modern paintings, as is still done today when it is a question of showing a personal collection in a private home. These medleys have not been completely abandoned in museums where the 'palatial' character is uppermost, like the Prado in Madrid or the Apollo Rotunda, the Percier and Fontaine Rooms, and the great gallery on the first floor of the Louvre. Finally, in the interest of public instruction, information should not be confined to individual labels explaining the objects, even when they are precise and detailed. A notice in each room should give the basic outlines of the history of art of the period shown in it.

The requirements of archaeology, aesthetics and instruction are not easily reconciled. Some of these have permanent needs, others vary according to current tastes. Some curators unhesitatingly accept that exhibitions should be up to date and changed at regular intervals. The fact that the object has been totally divorced from its original setting lends it a different meaning from that originally intended by its creator. This displacement is perhaps no bad thing in the case of Greek statues which, even in the ancient world, were regarded as self-sufficient and which, from the sixth century BC onwards, manifest their creators' ideal of virile strength and balance, but also, to a far greater degree, their transcendental aspirations. Even architectonic sculptures which have been removed from their context, like the two metopes from Olympia taken to the Louvre by Blouet, still convey their essential message, because they express clearly and vigorously, in extremely restrained language, confidence in human energy and faith in the rule of reason. But styles of exhibiting currently in fashion often enhance perspectives and details which the ancient onlooker would not have noticed, or which he would have considered totally unimportant, by the use of more or less artificial lighting. The use of artificial light in dark rooms or for visits after dark is still accepted with reservations.

When small objects like coins are displayed, it is a good idea to focus the spectator's attention by showing him, along with the object, enlarged photographs which can be works of art in their own right. But care must be taken not to reduce the original to a pretext. An exhibition of Gallic coins arranged according to these principles several years ago gave rise to a lot of dubious theories—for the most part imaginary—about the spirit of Gaul. If a spurious value which they never originally possessed is thus imparted to certain objects, others are sometimes stripped of the qualities on which their distinction depends. Many adverse judgments on Roman art have been based on sculptures and paintings which are admittedly mediocre in themselves, but might have been effective elements of a composition. If the statues from the flowerbeds at Versailles and the trophies crowning the dado of the palace were set up in a museum, they would give a poor idea of seventeenth century French sculpture.

Where museums are concerned there exists once more the old conflict between aesthetics and history which has dominated archaeology since it began, and which it would be a mistake to think finally settled. If the researcher is not completely at his ease in the exhibition halls, he can enjoy more freedom in the laboratories, libraries and storerooms. It is still necessary, of course, for these sections to be properly organised, and this is not always done in French museums. England, where the British Museum is a fine research centre, and Germany are better provided in this respect.

The arrangement of the storerooms is an important part of modern museum studies. Formerly, nearly all the collection used to be on exhibition, but nowadays three-quarters of it is often kept out of sight. It is an essential duty for the curator to make sure that these pieces

A statuette of a bull overlooks a shallow pool in the house of a rich citizen of Pompeii. The modern trend is to leave as much of the archaeological material as possible in its original position.

A much older Greek theatre than that of Herodes Atticus is the fourth-century BC one at Epidauros. The remarkable condition of the original fabric and the beautiful surroundings make it the perfect setting for performances of the classical Greek plays.

are safe and readily accessible to students. It is possible to divert some of them to associated museums or even to place certain of them in public buildings as ornaments, though this is a dangerous solution since they may be damaged. It is absolutely essential to keep complete series in the storerooms, which must be large, well-lighted and comfortable enough to work in. If there is not enough space available, it is better to maintain old fashioned exhibition halls and minimise the risk of breakage and theft.

Publication

There is no need to emphasise the urgency of the need for publication. Leaving an excavation or a chance find unpublished is nearly as bad as pillaging a site or demolishing a building. Unfortunately many archaeologists are guilty of this. There are two reasons which explain, without in the least excusing, this negligence. Firstly, overwork: nearly everywhere the meanness of the authorities imposes archaeological retrenchments on directors, and to a lesser degree on the leaders of expeditions- a task beyond their means.

Secondly there is the fear of making mistakes which makes scholars reluctant to publish their discoveries. They want their work to be impeccable, with nothing further to be desired in the finished product as it leaves their hands. It is very unusual for anyone, however learned, to possess all the knowledge needed for a full interpretation of an ancient monument. The progress of the discipline alone renders the acquisition of such knowledge daily more difficult. Every year several thousand articles, notes and books appear on the subject of classical archaeology alone, and these works are published in an increasing variety of languages.

In the nineteenth century many learned papers were still written in Latin, but this practice has gradually fallen into disuse, and the efforts of the partisans of 'living Latin' have not yet produced any really appreciable result. But let it be said, once and for all, that it is hopeless to embark on an archaeological career without at least a reading knowledge of German, French and Italian, and further problems are presented when faced with works in Slavonic languages, which are becoming increasingly numerous and important, and Israeli publications in Hebrew. Fortunately, a synopsis in one of the more familiar languages is usually attached to these works. Reviews, the most important of which is *Fasti Archaeologici*, published by the International Union of Classical Institutes at Rome, compile very useful bibliographies. A considerable effort to keep up to date is needed in spite of all this. No one person nowadays can hope any longer to master all the special studies like epigraphy, pottery and numismatics, which are each developing increasingly complex methods.

The excavator who wants to publish his findings promptly runs a great risk of omitting from his bibliography an important article published in a foreign country or of unwittingly transgressing the rules laid down by the pundits of some specialised discipline. Some censorious spirits exercise a positive reign of terror in their own field and reduce it to a desert.

Fear of a critical onslaught and the idea that a real scholar has no right to make a mistake often condemn many scholars to silence and infinitely delay the publication of essential monuments, which will only be incompletely described after their discoverer's death, when the circumstances of the discovery are forgotten.

The varieties of publications It is obviously essential to combat this state of affairs, and all that is needed is to distinguish clearly between provisional and definitive publications. The former must, above all, be prompt. New discoveries must be announced within a year. The report need only contain a precise description of the site and the buildings, and this task can be shortened by plans and photographs. What is more, if the author is qualified to interpret the discovery he can suggest his ideas briefly, but it is not essential at this point. Administrative rules nowadays generally require such reports to be submitted by everyone who has a state subsidy or even official authorisation to excavate.

In France these reports are published by the directors of archaeological districts in the review *Gallia*. Some archaeologists rebel against this obligation, and some of the most eminent, taking their stand on their own importance, have succeeded in freeing themselves from it. This reticence is usually dispelled by the fear of seeing some old colleague claim the credit for their discoveries. The principle of scholarly ownership is, in fact, very powerful in most scholars, and rouses as many strong feelings as material ownership. This is a perfectly legitimate and natural attitude in itself; it is very easy to understand why a man who has devoted his whole life to demanding studies in return for scant recompense, or an amateur who has often paid part or all the costs of his research from his own pocket, should wish to keep at least the modest fame attaching to such works. This kind of moral ownership is unfortunately very difficult to define. Firstly, except in the case of a very limited discovery, the real author is not easy to pinpoint: when the head of an expedition or service entrusts a site to a colleague, usually not so highly qualified as himself, when he gives him advance instructions and guides him throughout the course of the work with advice, which of the two should take the credit for the discovery?

It often happens, too, that other scholars have worked on the site before, and although they are no longer taking an active part in the excavation, they are still interested in it. Have not they, too, some share in a success which often results from the approaches they instigated, or which seems to be the logical product of the work they put in hand? This problem, which arises in most disciplines, can hardly be solved, however inadequately, except by mutual agreement.

Once the chain of command has been established, how far do the director's rights extend? Can he claim to keep the excavation records to himself until he is ready to make use of them, or ought he to put them at the disposal of other scholars, some of whom may be better qualified than he is? There are many awkward problems to settle, and they are for ever raising bitter controversies. In practice it is becoming increasingly apparent that the definitive publication of an important site can only be carried out by a whole team; the best rule, which has been adopted for most of the Greek sites (Delphi and Delos under the direction of the École Française d'Athènes, Olympia under the Deutsches Archaologisches Institut, Corinth under the American School, etc.), appears to be the publication of separate issues, each dealing with one category of discovery reviewed by an expert.

Archaeological publications are expensive, and those of a completely academic nature can only appeal to a limited readership. Comparatively brief reports are generally included in the reviews organised by the official bodies such as academies, institutes or missions, or subsidised by the academic research centres

Phallic symbols, often used in a respectable context as amulets against the evil eye, seen in a more graphic role as signs of a house of ill-fame in Leptis Magna.

View of the restored seating and original *skene* wall of the theatre of Herodes Atticus during the course of restorations.

Coins and medallions are difficult objects to display. In even the finest arrangement some details will be obscured by tricks of light or refraction. Many museums show accompanying photographs to offset this. The pictures here show, left, a sestertius of Nero (AD 37–68) and, right, a bronze medallion of Septimius Severus (AD 193–211).

Opposite page: Minerva, the Roman goddess. A Roman column capital from Khamissa and now in the Musée Gsell, Algiers.

The first picture shows a fragment of a thirteenth-century chasuble, the decoration and embroidery almost obliterated by time. The second, taken under infra-red light, gives a good idea of what the original robe must have looked like. British Museum.

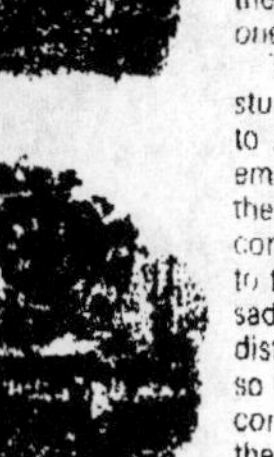

Local archaeological societies nearly all publish a bulletin. It is not desirable, however, for these periodicals to be too numerous. They would have trouble in surviving at all, and they might include too much third-rate material on too great a variety of subjects. Their circulation is necessarily very restricted. As to monographs, many of which are doctoral theses, they are usually published in the main official periodicals.

In addition to the strictly academic publications, there is nowadays an ever increasing number of popular books and periodicals. This development, which arises from the recent popularity of archaeology, is not always approved by the experts. They accuse the authors of such books and articles sometimes of incompetence, sometimes of encouraging by fulsome flattery amateurs who may be deterred by such publicity from submitting to the necessary academic discipline, and sometimes of making research sound like a romantic adventure when it is really extremely hard work. These criticisms may often be justified. But the public certainly has the right to know what is going on, and it would not be possible to raise the funds for excavating and the upkeep of the remains, or for negotiating changes in town planning to save some ancient monuments, without first rousing popular interest in archaeology. If the professionals want to prevent their work being presented to the public incompetently or inaccurately, they have only to do the job themselves. Or alternatively, if they do not think they can express their ideas clearly and readably enough, they should choose a 'ghost writer'. In any case, it is unforgivable for them to shut themselves up in an ivory tower and allow no one but the chosen few disciples into it.

Turned towards the past by the object he is studying, the archaeologist is thus brought up to date again, not only by the necessity for employing the most recent techniques, but by the influence he can and should exert on his contemporaries. He should not confine himself to the fine and noble duty of acting as ambassador and advocate among them of their distant ancestors, who are not altogether dead so long as their memory survives. He must also contribute to the formation and development of the spirit of our times. One of the most influential branches of thought of this age originated with a paleontologist and prehistorian. Certainly, Father Teilhard de Chardin was as much a priest and philosopher as a scholar; he had genius and literary talents which would have enabled him to transmit his message by means of any medium. But even without exceptional gifts it is possible to contribute effectively to the building of the new humanism which is now regarded as indispensable for happiness, so long as the rules of scholarship are scrupulously obeyed, and the selfish temptation to hoard the fruits of these studies is resisted.

Whenever a fraction of the human race enjoys a period of comparative peace and harmony—as was the case in the first centuries of the Roman empire, for instance—it remembers its past and formulates the rules of its order from this knowledge. One of the reasons for hope is that nowadays, apart from ways of understanding nature and controlling our environment infinitely surpassing the knowledge and techniques of our predecessors, we have a far more extensive, complete and accurate picture of history than they had.

The RECOVERY of the PAST

Prehistoric archaeology

Female figures with grossly exaggerated contours are a common feature of Paleolithic art. The Gravettian 'Venus' of Laussel is distinguished by the crescent-shaped object she holds in one hand.

Food gatherers

Definition and methods By the simplest of its numerous definitions, prehistoric archaeology is the reconstruction of mankind's most distant past before the invention of writing. Instead of beginning history with 'our ancestors the Gauls', or dating our origins back to the Homeric age or the Biblical fourth millennium, the prehistorian plunges into the bewildering depths of the geological past. This is still part of history, however, and in the absence of written evidence the prehistorian has to make use of many more sources of information. He also has to collect together all the clues which are independent of writing in order to compile a valid restoration of the past and extend the boundaries of history, for his basic objective is the history of the human race.

Implements of dressed stone, flint, quartzite or hard limestone, bone tools such as points or harpoons, durable ornaments, the monumental constructions known as megaliths, paintings, engravings and carvings on the rock walls of caves, clay models in the depths of an underground passage and sherds of pottery are frequently the only materials from this far distant history. More indestructible than manuscripts, these finds allow us to go back still farther to the very dawn of our species. If some human being had roamed the face of the moon 2,000,000 years ago, the prehistorian would be able to find traces of him.

He, and he alone, can go back to the distant origins of mankind. But however indestructible the records, they are still difficult to interpret. A few dressed flints, the polychrome bison painted on the great ceiling of Altamira, the piled up spheroids of El Guettar, the bifaces of Sidi Zin and the spear of Torralba have neither the eloquence of the Code of Hammurabi nor the precision of the Essene manuscripts from the Dead Sea; that is to say, the archaeologist has to combine a large number of extremely tenuous facts and clues and to look beyond this evidence and analyse its environment. In these studies the context is even more important than it is for classical historians. With bone burins (tools with bevelled points) the prehistorian can associate kitchen refuse, the bones of animals and the slender fish bones of the salmon species to reconstruct the diet of the Magdalenian hunters of the tenth millennium. Analysis of sediments, study of the impressions of plants on volcanic soil, and minute, miraculously preserved grains of pollen will give the archaeologist an idea of climate and vegetation. He can reconstruct the geographic environment, and in his ceaseless quest for this multifarious evidence he paints a more complete and detailed picture of the human condition than written documents sometimes provide. Even the distribution of this evidence in the archaeological strata can suggest human occupation, living or working quarters more than 2,000,000 years old. Finally, archaeological research into human remains—a brain pan or a jaw-bone sometimes happens to survive—can deduce the evolution of beings different from ourselves, the forerunners of our species. The study of these beings immeasurably extends the past of the human race.

The difficulty of interpreting the evidence is lessened by its great abundance and extreme disparity. A section of Old Stone Age (Paleolithic) life covering 50,000 years is sometimes more rich, detailed and expressive than that provided by a medieval document. Prehistoric archaeology often yields more copious and varied historical material than history itself. Deprived by definition of the end product, it still remains the broadest picture of the most distant cultures.

Until the last few decades the prehistorian's chief problem was the chronological classification of the innumerable remains. The relative positions of the objects in an archaeological zone—these Aurignacian burins being more recent than those Mousterian scrapers buried two metres lower—was for a long time the only dating available, and completely relative. Interest in stratigraphy was such that it inspired historical excavations, sometimes to the exclusion of all others. When the sedimental elements and the provenance of animal remains are known it is often possible to extrapolate a comparative date, to attribute the Sidi Zin bifaces of the Acheulian culture to the same period as the *Elephas atlanticus*, to recognise the Mousterian scrapers of Neanderthal man as being coeval with the great cave bears, and to attribute the end scrapers to the Magdalenian culture, so rightly known as the Reindeer Age.

Research into the structure of the atom and the stages of its disintegration has now supplied an absolute chronology. The precise measurement of the rate of disintegration of carbon in organic material can give dates up to 50,000

Opposite page: a small ivory head of a female figure with delicately carved features and stylised long hair was found at Brassempouy in France. It belongs to the Aurignacian period.

Detail of a prehistoric rock painting from southern Rhodesia. The graceful kneeling figure shows pronounced negroid characteristics.

years ago, with a margin of error of a few centuries, and often a few decades (deriving from carbon-14). The transformation of radio active potassium into argon makes it possible to go several million years beyond this point, while strontium gives an absolute date for the oldest rocks on earth. The chronology of the world, like that of the human race, is in the process of achieving the precision of the historical periods. The original gulf between history and prehistory is becoming narrower. There is the same synthesis of methods, history adopting for its own ends, and often much to its advantage, the modern scientific methods of prehistory. In fact, they are both concerned with the history of mankind, although they apply different criteria to the evidence under observation, and if the prehistoric evidence is more tenuous, at least it cannot be falsified by the subjective quality of a text. The flaked pebbles of Australopithecus were never worked over for posterity, as have been the memoirs of many famous men.

The pebble cultures

The origins of the earliest human industries are lost in the bewildering depths of immensely distant time, henceforward calculated by millions of years, and this chronology is without doubt the most astonishing and revolutionary contribution of prehistoric archaeology in recent years. The Olduvai Gorge south-east of Lake Victoria in east Africa is scattered with an impressive deposit of Quaternary material—in a very broad sense of the word—more than 100 metres thick. The earliest volcanic tuff at the bottom, resting on a rocky basalt foundation, reaches as far as the Aeolian sand of the last millennium, and forms the richest stratigraphic succession in the world. It includes 100 metres of deposits containing stone objects, bones of extinct animal species, human and, at the bottom, prehuman remains.

In a series of low strata thirty-six metres thick Dr L. S. B. Leakey and his colleagues found rich lacustrine deposits of fauna habituated to a hot damp climate of permanent forests. A study of sediments and a detailed examination of the skeletal remains showed certain discrepancies which could not be explained by any natural causes, and these suggested the intervention of mankind. Prehistoric archaeology often consists of the investigation of such discrepancies, which frequently reflect past human activity.

In a limited area excavations brought to light exceptionally abundant skeletal debris, remains of small vertebrates (tortoises and fish), and early mammals of a large size, the bones of which were broken to expose the medullary canal and the epiphysis now lost. There were remains of food, accompanied by rough stone tools of lava and quartzite, from rudimentary implements made from natural stones by chipping off one or two flakes by knocking them against other pebbles, to very carefully worked pebbles shaped into polyhedra and scrapers with one face flaked off (choppers) or both the original faces removed (chopping tools).

The great Magdalenian bison from the cave of Les Eyzies (left) shows several individual characteristics such as the tuft under its chin and the exaggerated hump which suggest that art had progressed, by this period, to representational portrayals of specific subjects.

The Abbé Breuil's reproduction of a wild boar painted on the wall of the cave of Altamira (right). From its earliest beginnings until its great flowering in the Magdalenian period, cave art was predominantly concerned with the animals upon which all human life depended.

The whole body of these finds represents a 'living floor' of prehistoric man. On this little island, newly emerged, lasting perhaps only a brief dry season (storms and floods would soon destroy skeletal remains), mankind occupied this floor, worked his pebbles in the earliest industry, and captured his first easy quarry (his prey would be young, slow and almost defenceless), the broken bones of which he left behind. At the time of writing this is the oldest prehistoric level in the world.

As to its date, the recent process of measuring the disintegration of radioactive potassium into argon gives a chronological value of 2,300,000 years. Some calculations suggest only 1,850,000 years for the bottom of level I and 1,230,000 years for the top. The University of Berkeley suggests 1,750,000 years for the earliest living floors, with a margin of error of five to seven per cent (or a chronological bracket of 1,870,000 to 1,630,000 years). Von Koenigswald dated a piece of basalt from the bottom in a Heidelberg laboratory and suggested 1,300,000 years. Whatever the number may be, one fact is indisputable: this, the earliest evidence of human activity found so far, goes back much more than 1,000,000 years and approaches the 2,000,000 mark, if not actually passing it. Mankind must thus look for his mysterious origins in the distant Tertiary geological period.

After the Second World War the Abbé Breuil had the temerity to suggest 800,000 years for the earliest tools found at Abbeville in the alluvial deposits of the Somme. Current science is even more daring than the reasoned and reasonable ideas of the Abbé.

There is much to be learned from Olduvai. A skull, a tibia, a fibula and a few teeth suffice to reconstruct the prehuman being responsible for these crude tools, who has been named *Kenapithecus Wickeri*, or *Pre-Zinjanthropus*, since higher horizons have also produced tools associated with the remains of Zinjanthropus, the Australopithecus of east Africa. The average length of the European tibia is from thirty-four to thirty eight centimetres. That of the Iturian pigmy is twenty-nine to thirty-one centimetres and a chimpanzee's twenty to twenty-two centimetres. The tibia of Pre Zinjanthropus measures twenty-eight centimetres. He was beginning to stand upright, a decisive step towards humanisation. Discoveries in 1963—most of a skull and an almost complete jawbone—led Leakey and his colleagues Tobias and Napier to regard this fossil as a true representative of the human race, whom they named *homo habilis*.

The first stage of the human race is thus represented by this wave of *homo habilis* and Australopithecus, the creators of the pebble culture. Remains of Australopithecus are found not only in various levels of the Olduvai Gorge but in the Sahara north of Lake Chad, and in Bechuanaland in southern Africa. Does this mean that Africa is the cradle of mankind? The answer to this question is too much at the mercy of new discoveries. The geographical extent of the pebble culture is wider than the area delimited by the skeletal remains of Australopithecus. The pebble culture has been discovered in a cave on the French Riviera. Not all of Asia has yielded up its secrets, and it must be pointed out that the unusual position of east and south Africa makes them geographical cul de sacs. The cradle of the world has perhaps to be sought in the immense Afro-Asian crescent which extends from the Himalayan ranges to the great African rift valley.

The biface users

The crudest tools in the world, choppers and polyhedra, did not change for several hundred thousand years, but more developed forms gradually came to be associated with them. It is not surprising to find this mixture of primitive and more advanced forms in the same archaeological context. The equipment of any twentieth-century workman often includes the pick and sledgehammer as well as the pneumatic drill.

In the eponymous site in the Somme valley at La Port-du-Bois in Abbeville, the Abbé Breuil identified a primitive industry later than the pebble culture (he suggested 800,000 years) consisting of tools worked on both faces with a stone hammer. They were made from igneous rock and have slender edges. Earlier authors used to call them hand axes but nowadays they are known as bifaces. They were the all-purpose tool, useful for scraping, boring, pounding, cut-and-thrust work, and every other purpose, and are associated with abundant flakes. Much later, these bifaces were roughly shaped and fined down with a stone hammer but they were smoothed and finished by supplementary working with a wooden hammer. They are characteristic of a new culture known as the Acheulian, after Saint Acheul, a suburb of Amiens. In the Middle Acheulian period and especially in the Late Acheulian, the biface acquired an efficient shape and functional aerodynamism which became a technical art form. An obvious aesthetic pleasure accom-

Another example of Magdalenian art, from the caves of Niaux in the Pyrenees, is the bison shown with a spear wound in its side. The painting was found at some distance inside the cave and some authorities see this kind of prehistoric painting as a form of hunting magic.

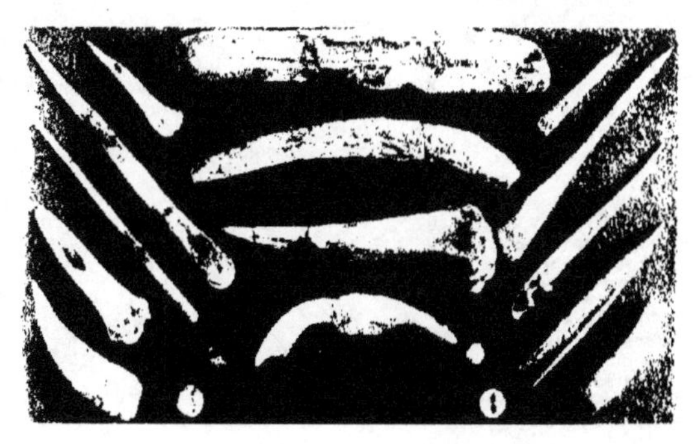

Bone implements found in the dolmen of La Lauzère. Animal bones, roughly worked with flint tools and given a fine finish by polishing on a smooth stone surface, were in common use from the Upper Paleolithic period onwards.

Dr L. S. B. Leakey, whose researches in the Olduvai Gorge in Africa have made enormous contributions to studies on the earliest formative stages of the human race.

panied the creation of these tools, which became comparatively lighter at this time. Some Acheulian bifaces are thus worked on large flakes, and these flakes are carefully retouched to produce an increasing variety of shapes for more and more specialised functions which have not yet been identified.

The Acheulian culture also covered most of the old continent and it reappeared in a deep level of the cave of Castillo in Cantabria in northern Spain, eighteen metres below modern ground level, associated with the cold-weather fauna of the penultimate glaciation. But in the heart of Castilla la Vieja, in the province of Soria, on the sites of Ambrona and Torralba, Acheulian bifaces were accompanied by the remains of hot-weather fauna such as *Elephas antiquus*. The site of Torralba represents an encampment of elephant-meat purveyors on the shores of an ancient lake. Bifaces made of limestone, quartzite and flint were found during Professor Powel's investigation, with femur ends and the large molars of the early elephant. Since the fauna has been preserved, future discoveries of some skull or jawbone may complete the picture of the craftsmen of Torralba.

Professor Arambourg found what at present are believed to be the originators of the Acheulian culture, in the heart of Maghreb, on the site of Ternifine-Palikao. This magnificent site in the Mascara district also represents a lake-shore culture living in favourable conditions which attracted not only a rich and abundant fauna but also mankind. The implements there include choppers, polyhedra and bifaces, and their interest is considerably heightened by the presence of plentiful fauna allowing the relative dating of the site: *Elephas atlanticus* (which superseded *antiquus* in Africa), *Tichorhinus simus* and *Equus mauritanicus*.

In 1954 and 1955 Arambourg found three human jawbones and a fragment of a parietal at Ternifine. Absence of chin, cheekbones rising low and broad from the jawbone and larger teeth than those of the present show that these jawbones are distinctive of a wave of new people, *Archanthropoids*, of whom these *Atlanthropoids* of Ternifine are the prototypes. Such were the makers of the Abbevillian and Acheulian bifaces. Under various names, this wave of population covered the ancient world – *Mauer Mann* in Germany, *Sinanthropus* at Chou-kou-tien not far from Peking, and *Pithecanthropus* in Java – although corresponding tools are not always known.

As to the sites yielding only Acheulian tools, they are innumerable. They are frequent on the shores of ancient lakes like Ain Hanech on Lake Kara. Surveys on the site of Sidi Zin in the Kef district of Tunisia in April 1965 enabled archaeologists to discover and define the Acheulian horizons of the site investigated by Dr Gobert in 1950. The bottom of the site proved to be an ancient floor composed of a mixture of closely trodden pebbles, over which lay a rich and fine industry of choppers, polyhedra, bifaces and hatchets worked in the local limestone and pebbles from the shore.

The clearing of wide areas of the ancient floor allows us to hope for the discovery of dwellings and, as bones of atlantic elephant and gazelle are preserved, perhaps for anthropological confirmation of Ternifine. From now onwards, horizontal stripping must succeed

The Abbé Breuil, doyen of prehistoric studies. The validity of his suggested dating of ancient material, made without any aid from modern scientific techniques, has repeatedly been confirmed by current laboratory tests.

the vertical attack on the layers which only defines the relative chronology of the site – which is now known – and does not allow the actual forms of human occupation to be discerned.

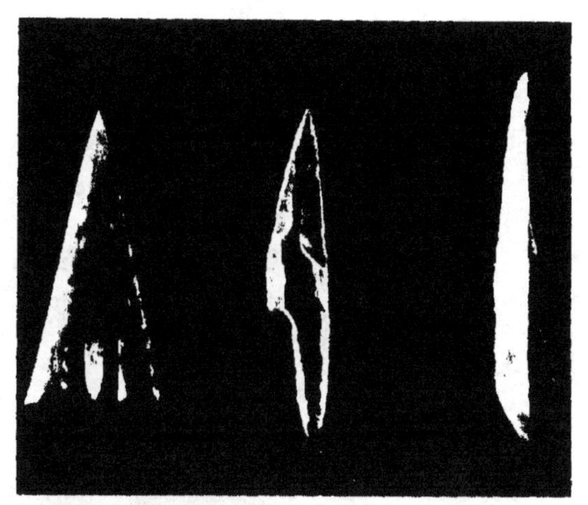

Three examples of prehistoric forms. Left to right: a flint arrowhead from Morbihan, a shouldered point from Charente, and a blade of Aurignacian type.

Flake culture

Prehistoric evolution is not susceptible of rigid classification, and the hermetically sealed cells of recent years, in which the well-defined cultures used to be imprisoned according to certain distinct races, are breaking up almost everywhere. At present, one of the most radically reassessed eras of this evolution is without doubt the period between 100,000 and 50,000 years ago.

Contemporary with the last stages of the Acheulian biface industry, a new method of working appeared which was distinguished by the systematic cleaving of flint nodules previously roughed out and prepared by preliminary working. By means of these prepared striking platforms, the craftsmen produced quasi-industrially some magnificent ovoid flakes, regular in shape, known as Levalloisian, which are ten to fifteen centimetres long. These flakes, still fairly light, are the Levalloisian points which are found with a variety of tools. The Levalloisian culture saw one of the most important technical revolutions in flint-working, a revolution which has been identified by the work of Professor F. Bordes.

Incomplete or poor Levalloisian, sometimes ignorant of the way to prepare nodules or obliged to be content with more modest raw materials, the Mousterian industries (from the site of Moustier in Dordogne) are basically distinguished by the point and the scraper. Variations of origin, territory and material make it possible to distinguish many branches such as typical Mousterian, La Quina Mousterian, Acheulian Mousterian, denticulated Mousterian and Charentine Mousterian.

The human remains associated with these industries are extremely varied. They form the third wave of population, the *Paleanthropoids*, which includes the Steinheim skull, the Swanscombe skull (still associated with Acheulian products), the Fontéchavade skull with the crude industry known as Tayacian, and especially Neanderthal man, who originated the typical Mousterian and Charentine types. The Mousterian culture and Neanderthal man between 75,000 and the thirtieth millennium are the best documented. From the end of the last interglacial period onwards they occupied the whole of the last glaciation (known to geologists as the Würm Glaciation), the period of severe cold during which the great cave bears lived.

Mankind then abandoned the great plateaux he had formerly roamed, to seek refuge in the west in the many caves in the chalk and limestone of Europe. Le Moustier, La Quina, La Ferrassie, La Chapelle-aux-Saints, and Arcy-sur-Cure are Mousterian sites in the depths of caves or hidden under rocky overhangs. Winters were hard and the new fauna included cold-weather species such as the mammoth, the blue fox, the marmot and the snow-mouse (in levels in the Suard cave at La Chaise-de-Voultron in Charente). A hunting economy finally became predominant, the earlier resources of food gathering having given out as the climate grew harsher. The hunters of Fontmore (Vienne) ranged over a territory with

Professor E. Jacobshagen of Marburg University holds the skull of a Neanderthal woman brought to light near Kassel by the floods of August 1956. It is about 120,000 years old.

a radius of two to five kilometres. Neanderthal man of La Quina in Charente in the Upper Mousterian period easily dismembered the victims of their hunts – reindeer, horse and ox. Flint points could be fitted with handles. Two spears, like the Mousterian bone spear from Castillo, made it possible to kill a variety of game which their tools dismembered, sliced and cut up as cleanly as a modern butcher's steel knives. The Mousterians of Grimaldi and the Observatoire cave in Monaco systematically worked innumerable stags' horns.

Neanderthal man, the most distinctive of the Paleanthropoids, was already a man in spirit. A sense of the hereafter appears in several burial rites. The dead were buried in pits dug by the living, who contributed weapons and offerings of food to the departed (graves at La Ferrassie and Mount Carmel in Palestine). At Regourdou not far from Montignac-sur-Vézère, a tumulus

The earth banks and stone circles of Avebury Ring stand in and around the houses and fields of the village. The complex, which dates to about 1800 BC, covers over twenty-eight acres.

of large stones protects a grave, and nearby is a pit containing the skull and bones of a bear under a huge slab of rock weighing 800 kilogrammes, one of the earliest known megalithic ceremonies.

At El Guettar, south of Gafsa, 8.3 metres below modern ground level, a Mousterian floor yielded a regular cone of sixty spheroids (75 centimetres high by 1.3 metres in diameter) full of long flint flakes, near a spring possibly connected with a water cult. At Toirano, in a Ligurian cave, as well as Neanderthal footprints in the clay floor of the passage, small balls of clay were found which seem to have been projectiles for bombarding an animal-shaped rock.

However tenuous they may be, all these facts suggest the origins of belief, still troubled and mystical, but possibly the first sign of the human spirit. The fact that Neanderthal man died out without any direct descendants certainly does not settle the question.

The Swanscombe and Fontéchavade skulls in particular, sometimes known as pre *sapiens*, play the most direct part in the formation of the new wave of peoples who formed the *homo sapiens* of the Reindeer Age, but many anthropological and geographical pointers are still lacking before this complex and flourishing human evolution can be defined with any certainty.

Skeleton of a Neolithic man from the chalk pits of Pantin in France. It was found in 1848.

Eleven graves believed to be 4000 years old were found on a building site near Magdeburg in 1960. A skeleton is examined *in situ* before removal for further study.

The great hunters of the painted caves

New chronological perspectives Although the absolute chronology of the food-gathering cultures (formerly Lower and Middle Paleolithic) goes back more than 2,000,000 years, carbon 14 dating brings us surprisingly close to the men of the Reindeer Age. They are represented by cultures using light tools consisting of flakes and splinters which Breuil named *Leptolithic* (light stone). These men, 'men like ourselves', according to the Abbé's famous statement, 'neither more stupid nor more intelligent, neither handsomer nor uglier than we are', achieved the most wonderful artistic revolution the world has ever seen. Faced with this new development, the prehistorian steps down in favour of the prehistoric art historian.

The first rudiments of representational art occurred in the Chattelperonian (from Châtelperron in Allier) and Aurignacian (from the cave of Aurignac in Haute-Garonne) periods, associated with very varied flint and bone tools. A lower level at Arcy-sur-Cure suggests 34,000 years ago for a Chattelperonian horizon which yielded a curious pendant, one of the earliest pieces of jewellery in the world. The typical Aurignacian material of La Quina in Charente is dated to 29,000 years ago, and the Aurignacian at Willendorf in Austria, celebrated for the discovery of a heavily formed 'Venus' statuette, goes back 29,800 years. The first known engravings (those traced on clay with the fingers representing bovidae in the Cantabrian cave of La Clothilda), and the earliest paintings (the negative hands surrounded with colour on the walls of the cave of Gargas in the French Pyrenees, the deer carried out in red dots at Covalanas in Asturias) belong stylistically to the Aurignacian phase which appeared about 30,000 years ago.

Somewhere around the tenth millennium this great artistic revolution took place in the West, exactly where its cradle had been, with the last expressions of Magdalenian art (from La Magdalène, a district of Tursac in Dordogne). The Late Magdalenian of the cave of La Vache in Ariège occurred 9850 years ago, according to a carbon-14 dating by Columbia University, and 10,500 according to a more recent dating by the University of Gröningen (the discrepancy between the dates obtained by different laboratories using different methods, and without doubt a better analysis of residual carbon in recent years, is reassuringly small). This dating of the La Vache material establishes the chronology of the portable works of art, and of a polishing tool bearing confronted ibexes identical to the wall paintings of the neighbouring cave of Niaux.

It is only in the last few years that archaeologists have dared to hope for such precise dates, and before the Second World War the prehistorian was reduced to subjective estimates. We should pay homage in passing to the Abbé Breuil, whose estimates often fell within the limits designated by carbon-14.

If he were working now, however, the Abbé would have to change the title of his magnificent and irreplaceable work on prehistoric art entitled *Four Hundred Centuries of Wall Painting*. Western art did not really begin until the thirtieth millennium, to end in the tenth. This is no more than 200 centuries.

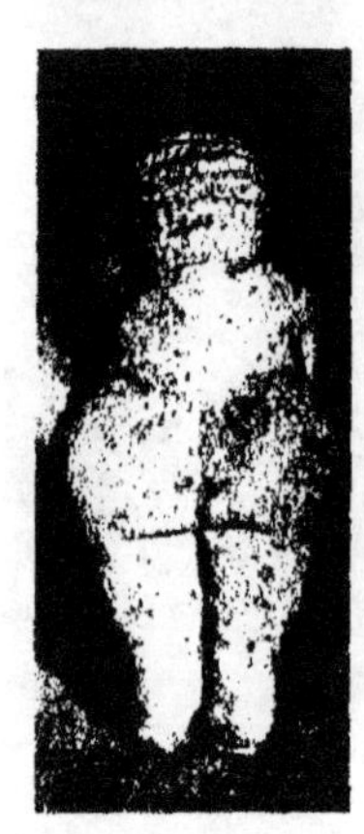

The famous statuette known as the Willendorf Venus, from the site where it was found in Austria. Of the Aurignacian culture, it is nearly 30,000 years old.

The cultures of the Reindeer Age

The cultures of the Reindeer Age in its broadest sense correspond conveniently to the Abbé Breuil's more technical definition, the Leptolithic. Tools, in fact, assumed new forms and a lighter appearance, a new step towards efficiency. The basis of the tool is still the flake obtained by progressively working down a flint nodule all round the striking platform. With long narrow flakes and splinters, the craftsmen produced an extensive range of tools such as fine end scrapers, burins with a dihedral corner suitable for cutting, drawing or engraving, and backed blades reinforced with a few deft retouches. There are some real spear heads and magnificent carving knives. Along with these stone implements, abundant and varied bone tools, which are even richer and very well documented, were developed. In the deeper levels of the caves, bone is preserved better than in older open air sites like Torralba and Sidi Zin.

Here, too, are bone points, lozenge-shaped points with the base split into a single or biconical bevel-edge, and in the Magdalenian horizons, extremely rich hunting gear is found, including spears with an elongated terminal neck, harpoons with a single or double row of barbs, fish spears, bird arrows and spear throwers for projecting and aiming. They are all very light in weight. Such an abundance of tools makes it possible to define the stages of the formation, development and evolution of the Leptolithic cultures more accurately. The terms used to define it are also numerous, but their number, like their relative diversity, ought not to obscure the profound unity of the period between 30,000 and 10,000 years ago in the West.

The first millennia of the Leptolithic are occupied by the archaeological horizons known as Chattelperonian and Aurignacian, the last by the Gravettian (from La Gravette in Dordogne) and Magdalenian horizons. They all,

The dolmen of Locmariaquer, Morbihan, with its huge stone slabs, is a typical example of megalithic building. These structures show a new sense of the hereafter embodied in the permanence of massive stone work, and indicate a concerted communal effort.

however, used the flint flake and its derivations, with an overall preference for flakes with a cut-away back in the whole Perigordian group, a term which sometimes includes the Chattelperonian and Gravettian cultures.

An original civilisation, meanwhile, remained completely distinct from the others; this civilisation was to some extent intrusive between the other two. It is known as the *Solutrian* period from the rocks of Solutré in Saône-et-Loire, and it produced an abundance of foliate points, 'laurel-leaf' blades, 'willow-leaf' blades, notched points, and arrow heads with tangs and barbs made by the bifacial rather than the flaking technique. Solutrian arrow heads can bear comparison with those of the end of the New Stone Age at the turn of the third and second millennia. The accurate and thorough investigations of Professor L. Péricot on the site of Parpallo (in the province of Valencia) were needed before these arrow heads could be attributed to the Solutrian period. This culture of foliate points and bifacial technique is found in middle and central Europe, where it could have originated, although many African cultures are technically related to it.

The Solutrian period marks the high point of flint working in the West. The laurel leaf blades reached a length of thirty-five centimetres to a few millimetre thickness, with magnificent retouching carried out with a bone retoucher. In a later phase the earliest bone needles with a manufactured eye were made of bone splinters smoothed and finally polished on small sandstone polishers with a very fine surface.

Painted clay female figurine from Strělice in Moravia. Such figures are Neolithic in date, but they show the same exaggerated forms as the earlier stone examples.

Living conditions The twenty millennia of the Reindeer Age saw appreciable changes of climate, which was predominantly cold during the retreat of the ice-cap around the middle of the eleventh millennium. The cold-weather fauna of the tundra, with the reindeer, the musk ox, the wolverine, the blue fox and the lemming, alternated with the fauna of the cold steppe country where the saiga (small rodents like the hare) roamed. Mountain species like the ibex and the chamois came down from the Alps and the Pyrenees. The great game trio of the Leptolithic period remained the reindeer, the horse and the bison. The more spectacular powerful mammals, now extinct, were characteristic of this cold-weather fauna: *Rhinoceros tichorhinus* and the mammoth (*Elephas primigenius*) adapted to the cold by its long thick pelt. This mammoth, with its long and terrible curved tusks, lived on sedges, wild thyme and the yellow Alpine poppy, bitter ranunculus and gentian, and all the other plants which have now deserted the Aquitanian lowlands to persist in the Alpine and Pyrenean climate which suits them.

These harsh climatic conditions (the study of pollens confirms the evidence of the fauna) induced mankind to take refuge under overhanging rocks, and in caves whenever he could

Another view of the huge Bronze Age stones at Avebury in Wiltshire. The stones are sarcen, a particularly hard form of local sandstone.

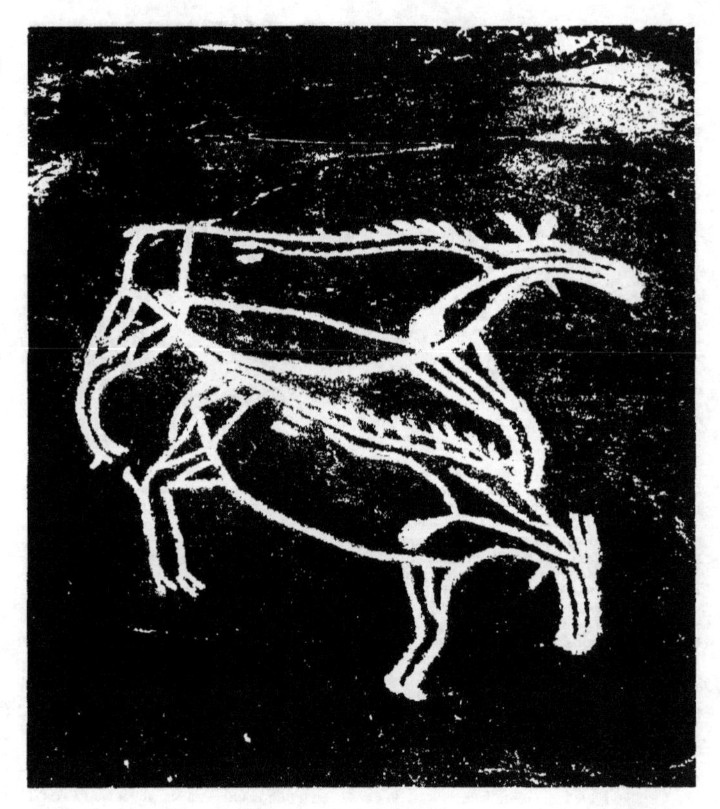

Rock drawing from Norway of two animals (perhaps elks) mating. The magical character of the picture is attested by the 'life-lines' representing internal organs running from the animals' heads.

find them. He fitted them up with low walls and protected himself from the cold *outside* by screens of branches or frames of wood or bones supporting hides. He constructed floors of carefully fitted paving stones.

In the cave of La Vache at Alliat, several large pebbles gathered from the banks of the nearby torrent, the Vicdessos, form individual hearths holding logs of willow and pine wood, some charred fragments of which still survive. In 1955 (L. R. Nougier and R. Robert's campaign), a large pit with sides more than two metres long by forty centimetres deep full of pebbles was found. This pit represents a communal hearth, a true 'tribal hearth' where ptarmigan were stewed by means of frequently renewed heated stones.

These creatures represent ninety-five per cent of the bird remains from this magnificent late Magdalenian site. The diet was supplemented by game mammals, principally ibex. The reindeer was rare because its withdrawal had already begun, and it continued to be rare in mountain sites, being a lowland animal. To these basic foods of ptarmigan and ibex they added the chamois, the stag, the hare, rarely the bison and the horse, and the nearby river supplied plentiful trout.

Stone and bone tools and kitchen refuse from the hearths of Alliat, Bruniquel and the cave of La Garenne at Argenton-sur-Creuse (excavated by Dr J. Allain) all combine to paint a clear and detailed picture of a hunting economy.

Animal art It is not surprising that the art is essentially and magnificently zoomorphic, since early man was faced with a way of life closely linked to animals and drew the essentials if not the whole of subsistence from them. Friezes of mammoths are found at Rouffignac, stags leaping across the river at Lascaux, polychrome bison on the arresting relief of Altamira, sculptured horses at Cap Blanc, and bison modelled in clay at Tuc d'Audubert. And there are also thousands of animals engraved on portable objects—reindeer horns, shoulder-blades and shin-bones of bison, and slender bird-bones bearing many lively engravings of reindeer, horses, ibexes, rhinoceros and the whole bestiary of the period.

Western prehistoric art is much more easily explained as an expression of everyday hunting life rather than sexual hostilities recently postulated and briefly fashionable.

If one thing is surprising it is the actual origin of art. It started between the thirtieth and the tenth millennia in the favourable triangle of Aquitaine. There were several reasons for this. The paradise of the great Leptolithic hunters began at the Charente and covered Aquitaine, which is sheltered from the cold strong continental winds by the Massif Central and open to the soft Atlantic climate via the south-west facing valleys of the Pyrenees. It extended as far as the barriers of the Pyrenees and Cantabria, which are also rich in shelters and overhangs. In the great cold and windy plains of central Europe, devoid of natural shelter, life could only be bitter and hard. In the vast funnel of Aquitaine where the whole of Eurasia converged and ended up, the hunting communities found plenty of refuges and crowded in there, following step by step the gathering together of the game. Aquitaine was a huge natural catch-net and this dual concentration of men and game was the great opportunity of Western art.

It will be seen that human progress was so slow as to be hard to measure. It is not known how many million years elapsed before the first anthropomorph passed from using his bare hands to hands extended by a splintered pebble. A million years was needed for the advance from dressed pebbles to bifaces, and rather less for the achievement of points and scrapers. To a superficial observer the two million years between the beginnings of *homo habilis* and the dawn of the great hunting civilisations might seem empty. They are, however, marked by endless experiments, and this human progress although interminably slow, remained the work of a tiny population. Only a few thousand people inhabited the whole of the West during the culture of the biface users. The innumerable bifaces found on the African continent, often even on the floors of the Reg and the Hamadas in the dunes of Tihoda and Ravi, ought not to give rise to any illusions, since they have survived for more than 1,000,000 years. They suggest discouragingly slow progress, achievements barely won, lost again, rediscovered and long forgotten, since contacts were difficult and problematical from one person or one generation to another.

In the Magdalenian period the first increase in the human population took place. Professor Péricot suggested the figure of 500,000 for the Iberian peninsula. The figure is probably too high, and the figure of 500,000 for Gaul is more reasonable, working from different criteria. But on a world scale the numbers remained of the same order. This population was concentrated in the melting-pot of Aquitaine, and the contacts between the tribes and generations transmitted techniques and beliefs. The conditions necessary for the development of true civilisation were present, and the genius of a few hunters was enough. This progress can only be explained as the genius of the few being transmitted to the many. For the prehistoric hunter, for whom the chase was life itself, indeed the only means of supporting a difficult existence, the first rudiments of self-expression still remained an action. Imitating the claw marks of the cave bears whose coeval he was, he scraped the marly limestone or the clay floors with the marks of his own fingers. One day there miraculously arose from these accumulated lines the likeness of an animal he knew, according to Professor Pittard, as a child sees his daydreams materialised in the clouds. From the active and still undefined character to the conscious character which distinguishes and definitively accentuates the resemblance, the leap to art was accomplished. The animal was *created*.

In its first stages art was a true act of creation. The dynamic concept of this art explains most of its surprising features, the walls of Les Combarelles, the engraved pebbles of La Colombière, and the deepest part of Les Trois Frères with its fine engravings all bear, heaped up and superimposed, images of the horse, the rhinoceros, the bison, the reindeer of the chase and the daily encounter. These animal forms are so many creations, picture realities, since art – and this already is art – is action.

In the eighteenth century the lines of standing stones at Carnac still included more than 4000 menhirs. The Age of Enlightenment dispelled the superstitious fears by which they had been protected for so long, and there are now less than 2000 left.

A rocky corner, a hollow, a piece of stalagmite naturally suggested some animal. The resemblance was deliberately accentuated and underlined. Perhaps Neanderthal man of Toirano felt some premonition of this.

Animals transfixed with arrows, which comprise ten per cent of the figures in the black chamber of Niaux, add the action of their destruction to that of their creation. In the cave of Montespan the great horse moulded and incised in the clay is covered with fingermarks and spear wounds. Some of the fingermarks are later than the spear-blows, others earlier, since a spear-blow obliterates a fingermark. Bare hands and spears were levelled at the representation of the animal in a frenzy of destruction. After the earliest engraving (the still-awkward oxen of La Clothilda), after the earliest paintings (the still stiff deer of Covalanas and La Haza with their red dots), the creative urge grew stronger and more original. Then community rites appeared, heralding the sorcerer artist and the priest-artist, creator or destroyer according to need.

The vast caves and deep passages, wide rocky overhangs and natural walks under the entrances became cult areas. Prehistoric art was not so much the underground art of the Western caves, although caves have preserved these paintings and engravings better than anywhere else, but rather, animal art. There the animal was king and god. If the narrow passage of Les Combarelles seems to be the guardian of a secret, if the Trois Frères cavern yields only its vision of the 'horned god' with his head crowned by stags' antlers after long dark windings, this is unavoidable because of the subterranean nature of the place. On the other hand, the carved friezes of Roc de Sers, La Chaire à Calvin and Angles-sur Anglin allow plenty of space for a crowd of people.

The vast rotunda of Niaux and the great Breuil Passage of Rouffignac, enlivened by a wonderful monumental frieze of rhinoceros and mammoth, also have room for mass participation. The great religious ceremonies and even the closely related theatre probably originated at Cap Blanc à Angles, Laussel or the dome of Niaux ten or fifteen millennia before the Dionysiac rite. Some more modest themes reveal a no less profound meaning, already singularly close to modern thought. On the cornice of Las Monedas a black horse seems to be leaping from a dark abyss gaping at its feet. At La Pasiega another black horse rises from the depths of the earth, like the mare in foal of La Pileta, just as the great brown horse rises from the life giving cleft at Lascaux.

The animal world emerges from the generating shadows, the dark mouths of mother earth. This was probably the chief contribution made by the discovery at Rouffignac in 1956. Indeed, the monumental friezes, for example that of the sacred way with its five engraved mammoths, supply much new information: the feeling for pyramidal composition, harmony and creative rhythms, and the force of the confronted mammoth *Leitmotiv*, an opposition very common in mobile art which was to remain a basic theme of Western iconography and was not slow to be adopted in the East.

The great ceiling of Rouffignac crowds its fifty figures one upon another in a magnificent example of picture-reality. But why this ceiling, and only this one? This was the result of a decision. It was the only ceiling at Rouffignac which could be approached by a tunnel-shaped underground passage in the second level of the cave, which then climbs under a pillar decorated with paintings, one of which shows a human figure, and sweeps towards the third level of the depths, where roll the mysterious watercourses of historic legend.

From 1575 onwards some writings allude to 'paintings' and mention this shadowy stream. As Delphi arose as a result of a fissure in the rocks of the Phaedriades, the animal temple of Rouffignac resulted from the underground opening into deep passages coiled in the bowels of the earth, the Earth Mother, creator of men and of the animals from whom all life was created.

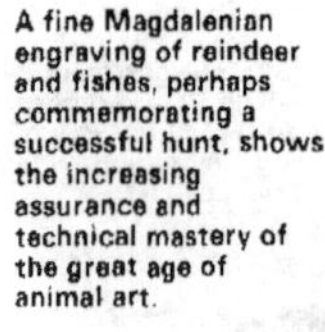

A fine Magdalenian engraving of reindeer and fishes, perhaps commemorating a successful hunt, shows the increasing assurance and technical mastery of the great age of animal art.

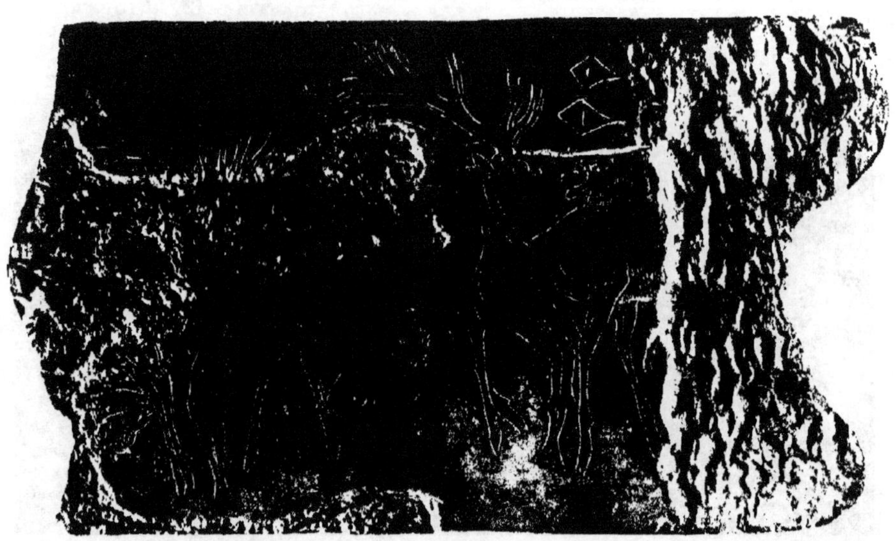

The decisive turning point of the human race, the Leptolithic period, with its apogee in the Magdalenian era between 14,000 to 13,000 and 10 000 to 9000 years BC, contributed new techniques, an unrivalled animal art, and thought and belief singularly close to our own 'classical' thought – so similar indeed that it is justifiable to see its distant origin here. For the first time in the history of the world, the West made its mark on civilisation. Perhaps no other period was so progressive, so civilising and so disinterested. From its cradle in the West animal art spread throughout the world, increasing its extent with the passing of time, with the antelopes engraved on the rocks in Norway, the paintings of eastern Spain, the engraved buffalo of the Atlas mountains, the long horned cattle of Tassili in Algeria and the rich Saharan paintings, to reach the farthest bushmen with its last flowering.

Mobile art such as statuettes and hunting weapons attest a veritable Leptolithic diaspora in the Far East. By way of the Bering Strait, which the glaciation had converted into an isthmus, hunters reached the Yukon valley and spread over the great American plains. In the marshes of Texcoco, not far from Mexico, the earliest American, Tepexpan man, lived 8000 years ago. Not far away at Santa Izabel in Iztapan lay a mammoth, its hind limbs still in place, sunken in the bog. Obsidian foliate points were used for cutting up game. Long before Columbus and Erik the Red, Leptolithic hunters discovered America.

The Middle Stone Age

The impact of the artistic revolution is such that an immense vacuum seems to follow it, and authors speak freely of a period of regression. The truth is more subtle, and prehistoric archaeology, deprived of high points but reaping the advantage of this extensive era, makes use of precisely this unbroken series of human experiments, to lessen the minor setbacks and retreats which affected the upward climb of progress.

After the Magdalenian, climatic conditions grew easier, the grassy steppe produced grain, and here and there trees grew thickly. From the developing forest glade, the West passed to thick forest of mixed growth in the fifth millennium. The withdrawal of the reindeer to the north remains typical of this intermediate period, the Middle Stone Age, or Mesolithic. But the reindeer's departure did not leave the West empty; the deer took its place alongside the wild boar. The ninth to fifth millennia form a real Deer Age to follow the Reindeer Age. A similar change took place in the caves on Mount Carmel in Palestine; the gazelle replaced the fallow deer.

Mesolithic economy increased in variety. The traditional game was supplemented by gathering fruit and berries, now possible again, and by collecting shell-fish. During the Mesolithic millennia the whole world followed this practice: snails in the Pyrenees, from Cantabria to Mazd'Azil and at Santimamine, and shellfish from the sea forming the Scandinavian kjökkenmöddings (kitchen middens) and the Brazilian Samaquis (shell middens). In the eastern Maghreb the Capsian sites near Gafsa can be recognised from a distance by large patches of ashes and considerable accumulations of snails. The eponymous site of El Mektar is dated to 6450 BC.

Mesolithic man no longer submitted to the rigours of an entirely hunting economy like his Magdalenian ancestor, and this variety of resources ensured his more regular survival. Similarly, a major economic turning point was the gathering of grain. The whole world took up this practice almost simultaneously. On the Iranian plateaux and in the caves of Natoufian in Palestine the first flints set in a deer horn handle were found. These are the earliest sickles, the flints of which show a typical polish caused by repeated friction against the corn-stalks. They date to about 7000 BC. In the West in the area of Rouffignac, in a region of the industry known as Sauveterrean (from Sauveterre la Lémance in Tarn-et-Garonne) C. Barrière found a whole series of flint knives also showing polish caused by corn-stalks. The layer was dated 7005 BC by carbon-14, and several polished knives were also found in the next layer below it. In the same millennium maize was growing on the high Peruvian plateaux in the Paracas district.

It is not surprising that animal art disappeared once its underlying magical and religious motivation was gone. The picture reality, the living substitute for the coveted prey, was no use to the snail and corn gatherers. Art did not totally disappear, as used to be thought. At La Borie del Rey, L. Coulange found an engraving of a deer like animal on a piece of limestone, and C. Barrière found a polishing stone, again at Rouffignac, with hatched geometric decoration, both of Sauveterrean type. In the shelters of eastern Spain a different art form appeared, to some extent free from religious overtones, with dynamic scenes of hunters pursuing herds of cervidae. Narrative art flourished here, telling posterity the story of their exploits by means of pictures. A historical tendency was already growing, and this grew stronger during the millennia of the New Stone and Bronze Ages, to end in stylisation. The thick statuettes of the western Aurignacian and Magdalenian periods appeared again several millennia later in Anatolia, as well as at Hacilar and on the shores

Prehistoric hands outlined in paint are superimposed on sketchy drawings of bison in the cave of El Castillo in Santander. The paintings, which clearly have a magical function, were executed in the late Aurignacian period.

of the Aegean. They lost none of their symbolic significance, representing mother-goddesses or fertility idols. Finally, the Mesolithic tools preserved the Perigordian and Magdalenian cultural traditions: these tools were flakes and splinters rich in a wide variety of points, spear heads and harpoons. The composite tool is a considerable advance on the one-piece harpoon. Under various names such as Sauveterrean, Tardenoisian or Capsian, tools often assumed reduced geometric forms, and demonstrated new efficiency, often being industrially produced. With its greater variety of resources and wider economy, the Mesolithic period set the stage directly and effectively for the great demographic and economic revolution of the New Stone Age.

The Neolithic revolution

During the New Stone Age, or Neolithic period, man became a stockbreeder and farmer, producing food. Thenceforward he built from the 'earth'. This was the great turning point of prehistoric evolution, which took place around the sixth and fifth millennia, with similar dates for most parts of the globe, since this transformation overtook the whole world. It is in Western Asia that the cradle of agricultural life has been sought and found, with the first grain-growers (cereals grew wild there already).

A corn cycle evolved, with preparation of the earth by scratching with a ploughshare, sowing first in holes, then in lines, reaping the grain below the ear with a sickle set with flint teeth, and grinding in mills with flat millstones. This type of economy barely evolved in the Mediterranean area, and threshing with a *tribulum*, a large board armed with flint teeth, as practised on the threshing floors of Castilla la Vieja, the Maghreb and Syria still follows the Neolithic tradition.

From this time forward a study of traditional techniques can shed light on the last part of prehistory, since they are so close to those of our own times.

The first domesticated species, the dog and the goat, appeared at the same time as the earliest harvests. The herdsmen of the Iranian and Saharan plateaux played an active part in transforming the nomadic life of the former hunters into a pre-agricultural pastoral life. The drying-up of the climate between the thirtieth latitude and the tropics about the fifth and fourth millennia led to a general descent by the upland pastoral people into the plains, where there was now more water because of the great rivers. From this concentration of population in the 'fertile crescent', the well-watered lowlands of the Tigris, Euphrates, Indus and Nile, with its regularly flooded valleys, arose the need to till the soil and domesticate new species of animals. From the clay of the river banks came pottery which superseded and imitated the hides used by the hunters and shepherds. Sun-dried bricks made their appearance. Traditional archaeology now takes over these problems, although writing was not yet invented and we are really still dealing with the prehistoric period.

The Neolithic movement in the West remains the province of the prehistorian. The agricultural movement there appeared to take place on a more modest scale, although the chronology was very similar. The stage was set for the change to peasant economy throughout the world, even in America where the Chalca culture at the site of Chicoloapan represents a Neolithic village with dwellings, hearths, mills and grain harvests in complete ignorance of pottery between 6000 and 3000 BC. This site is the New World equivalent of Jarmo.

In Europe the dog appeared simultaneously with the northern Maglemose marsh cultures as the first domestic species about the sixth

Many megalithic works have suffered from vandalism or been dismantled to make way for farming and development, but the capstone of the fine dolmen at Midlaren in Holland is still in position.

millennium, and it was probably about the fifth and fourth that the second wave of domestication can be justifiably attributed, with the little ox of the peat-bogs and the boar of the marshes found in the grey soil of the Martières swamps, the lower terraces of Somme, and the Le Havre district.

After cultivating the first fields, the Neolithic farmer attacked the encroaching forest which he had to stop and even drive back in order to reclaim his territory. These clearing operations were the major preoccupation of the Neolithic period. They called for a new tool, massive and powerful, and the traditional reinforced tools only survived for hunting. The Neolithic wood-cutter was armed with a saw, an axe and a pick. These were the 'peaceful weapons' of the Campignian cultures. In Europe the Campignians (from Campigny, a district of Blangy-sur Bresle in Seine Maritime) spread to Belgium, came down the Oise valley and settled in the wide loess plains of what is now the Ile de France with its fertile, easily tilled soil.

From Mesopotamia, Asia Minor and the eastern Mediterranean a second Neolithic wave spread up the Danube valley. These were the 'loess farmers' (Danubians or Omalians), craftsmen who produced a rich pottery with globular shapes decorated with strips and chevrons applied to the vase known as Ribbon ware. These Danubian farmers also reached the Ile de France and are dated to 3400 BC at the site of Cys-la-Commune in Aisne.

A third Neolithic wave reached Europe from Egypt by way of the Mediterranean seaboard and the islands: it was rich in a variety of pottery, vases with round bases ornamented with multiple impressions made on the clay before firing, especially marks made by a cardium, a small, very common shell (this is the so-called cardial ware). This pottery is found at Tassili des-Ajjer on the north African seaboard, and from Mediterranean Morocco to Spain, Provence and Liguria. A plain pottery without decoration superseded it, known as Cordaillod ware; it is typical of the site of this name, a lake-village of Neuchâtel. A later technique, also Mediterranean in inspiration, added fine hound's tooth or chequered markings in alternate squares to these vases. This was to be the pottery of the settlement at Chassey (Sâone-et-Loire). These last Mediterranean currents completed the Western Neolithic movement, contributing more extensive domestication including the goat and the sheep and the technique of making axes from hard rock by rough hewing and polishing.

In the Ile de France these new currents met the strongly entrenched Campignian farmers. The result was a technical and cultural fusion forming the Campignian Neolithic, which was based on massive stone axes, picks and flint saws to which hard stone axes came to be added, and an extremely varied pottery became common from this time onwards. Carbon-14 measurements give dates of 2800, 2600 and 2400 BC. This fully developed Neolithic occupied the middle of the third millennium. At this time Europe was undergoing a powerful demographic expansion accompanied by an agricultural transformation. It can reasonably be estimated that the population of Gaul increased to ten times its former size between 3000 and 2000 BC, rising from about 500,000 inhabitants to perhaps about 5,000,000. The exploitation of agricultural lands permitted this expansion, but demographic demands made it inevitable.

The resources of the soil had to be everywhere more intensively exploited. Villages multiplied, first widely scattered on the edges of the high ground, later more and more crowded in the interior as the soil became depleted. Deforestation accelerated and the thickly populated countryside took on a familiar aspect, with great wheat producing plains, sometimes separated by residual patches of woodland which

Some of the largest known menhirs are found at Locmariaquer in Brittany. Many such stones have been broken up to obtain building material, but others are too big to be profitably attacked.

A fine aerial view, taken in high summer in 1948, of the site of the Neolithic camp at Windmill Hill in the county of Wiltshire, England.

were all that remained of the former forest of the period of Atlantic climate during the fifth millennium.

Searching for raw materials on the surface, particularly flint, no longer produced enough to supply the demand. The Campignian peasant became a miner, extracting flint from pits dug in the chalk and from deep tunnels which he reached by means of shafts. On a site like Spiennes in Hainault the flint mines curiously foreshadow the coal mining of the eighteenth century which was developed a few kilometres away at Mons. From Hainault to Aveyron towards Mur-de-Barrez and from the chalky uplands of the Oise region to Mont Ventoux, throughout the whole Campignian area, shafts and tunnels allowed the extraction of flints which were flaked and polished in neighbouring workshops like Tussencourt, Girolles Châteaurenard, Murs and Malancenes, and later in the Grand Pressigny region. With this first industrial and mining civilisation came the accidents and diseases of mining (tuberculosis caused by the fine chalk dust) which were to be grave social problems in modern times.

The Ile de France, with its diverse and complex economic activities, had received many people from the east and south, and in its turn it sent out colonists to carry the Neolithic culture still further. Settlements, techniques and ways of living swarmed out one after the other. The Thames valley, geologically the counterpart of the Ile de France, was then colonised by Campignian migrations. Windmill Hill, the eponymous site of the British Neolithic culture, was a small agricultural community encircled by ditches which played the part of enclosures for the troops on guard. Causeways crossed the ditches to provide access to the cultivated areas. These ditches dug in the clay formed pits for the extraction of the indispensable flint. The flint mines of Grimes Graves in Norfolk attest this Campignian influence as late as the beginning of the second millennium.

By the end of the third millennium the main features of Neolithic agriculture and crafts in Europe were established. Agrarian structures were built up and almost all the land exploited. Forests respected by the Neolithic peoples were often equally respected by the Roman and the monk.

The social and megalithic revolution of the second millennium The Neolithic inhabitants of the Mediterranean area stayed close to their caves for a long time, taking shelter there from the heat rather than from the cold. The exploitation of agriculture freed mankind from living in caves and encouraged him to settle on the land. At Los Millares in Andalucia the huts of the agricultural village were grouped together at the very end of the spur. The graves, for a long time hidden in a cave or mountain fissure, thenceforward formed a necropolis. The tomb was communal, consisting of schist slabs supporting low walls topped by a corbel vault. The burial chamber was reached by a narrow corridor intersected by doors cut in the schist slabs. The monument is

From a Stone Age village in the Orkney Islands. The interior of one of the huts at Skara Brae.

really an artificial cave crowned with a tumulus of stones. In the Mesolithic period at Teviec, a small Morbihanese island, stone-built tombs had already appeared. From that time onwards the same idea was passed on in a multitude of different forms.

The hypogeum of Antequera is an expression of Megalithic thought, with its burial chamber more than 25 metres long by 6·5 metres wide, rising to a height of 3·3 metres. The foundation slab weighs nearly 320 tons, and the 31 slabs of the monument add up to 1600 tons. Elsewhere small rubble masonry was used for the tomb, as in the huge cairn of Barnenez, which conceals an impressive series of tholoi. The tholos of the island of Carn in Ploudalmezeau north of Les Côtes du Nord is dated by carbon-14 to between 3270 and 1500 years earlier than the tholoi of Mycenae.

Many of these burial monuments built of large or small stones raised up or dug in, dolmens, covered walks and hypogea, in Palmela, Provence, Pantalica and San Muxaro date to the third millennium, and the latest of them reach the second. The hypogeum of Mournouards III, hollowed out in the chalk of Mesnil-sur-Oger (Marne), dates to the late eighteenth or early seventeenth century BC. These hypogea were re-used for generations, and later on, new tombs continued the tradition. This funerary ritual, with its monumental hierarchy, reveals a social revolution which the evidence of such metals as copper and bronze also attests. Chieftains directed the construction of vast monumental complexes and the common people, submissive and disciplined, carried out their orders. Only a strongly hierarchic society could have raised skywards the great menhir of Locmariaquer with its weight of 348 tons, or aligned the thousands of menhirs of Carnac and the great stone circles of Avebury and Stonehenge.

Dated at least to 1800 BC, the first site of Stonehenge already included a great circular causeway in the Windmill Hill tradition, the orientation showing that the site was dedicated to sun worship. The trilithons of the horseshoe enclosure were set up about 1600 or 1550 BC. On an enormous pillar twelve metres high three shapes were deeply incised: two were axes of the bronze type and the third was a dagger of the same metal. These are early Bronze Age shapes and the same forms are known in Greece about 1600 BC.

Similar engravings on the schists of Mount Bego, on the steel blue rocks of the Camonica valley, show bronze hatchets and daggers, but also representations of ploughshares and chariots, mountain cabins and agricultural buildings, and thousands of bovidae and horned animals (more than 40,000 at Bego), and multiple solar symbols in a completely picturesque illustration of an agricultural and pastoral economy of the Bronze Age.

But on the rocks of Naquera mounted warriors are to be seen armed with long lances and javelins. The Iron Age was coming and already, in the Oglio valley, the first Roman legions were on the march.

Western Asia before Alexander

A vessel in the shape of a two-headed duck from the Hittite capital of Boghazköy in Anatolia. The Hittite empire was a major power in Western Asia from about 1400 to 1200 BC.

Western Asia: the meeting point Between the Caspian and Mediterranean Seas and the Persian Gulf lies the region of Western Asia, which is also mistakenly known as the Near or Middle East; an area which in fact includes several nations Iran, Iraq, Turkey, Syria, Lebanon, Israel and Jordan.

At first sight there are striking contrasts between these various countries. Their geography, climate and way of life differ in many ways, and range from the snowy heights of Iran and Turkey to the heat, sometimes excessive, of the desert lowlands of Iraq and Syria, the uneven rocky coast of the Mediterranean to the damp, swampy shores of the Persian Gulf, and from the forests of Pontus, the orchards of the Lebanon valleys, to the bare Syrian desert.

Despite so many strongly marked differences, Western Asia forms a single entity because of its geographical location. It is situated in the position of an essential junction between the three continents of Asia, Africa and Europe. This exceptional position has influenced the history of the people who settled there from the beginning of the ancient world.

Western Asia is a crossroads, a way station and a meeting point. Enormous mountain ranges like the Elburz, Taurus and Zagros mountains or formidable barriers like the Amanus, Ansariya and Lebanese highlands never seem to have prevented men from finding ways through, and these are often still in use today. The extreme diversity of this immense complex has made trading essential. At a very early date the inhabitants of regions lacking certain commodities made haste to find places where they could obtain these essential supplies. The obsidian trade, and transport inland of pretty sea shells, are attested by archaeology from the Neolithic period onwards. Later on came trade in metal wood and stone, bringing even more truly momentous consequences. From the economic necessities which led to various exchanges arose the unceasing need for maintenance and inspection of the highways, and the interpenetration of influences, in defiance of natural barriers and artificial frontiers set up by mankind. This is probably the reason why the various countries of Western Asia were traversed by cultural and technical movements even before the establishment of the earliest of the higher civilisations on the face of the earth which originated in Mesopotamia – the valley of the Tigris and Euphrates – during the fourth millennium BC.

This exceptional position at the meeting point of three continents, and these multifarious contacts from the earliest times give Western Asia a unity which is apparent on an overall scale to the geographer, the economist and the historian; the archaeologist, too, recognises it.

Practical and synthetic archaeology Archaeology, however, has two aspects: it is first and foremost a practical field study of the smallest remnant of material concerned with mankind's past, and then a thorough investigation in the silence of offices and laboratories of the elements collected in this way. This is the province not only of the archaeologist, but of the historian, the epigraphist, the pottery expert, the paleobotanist and an increasing array of other experts, since the discipline is advancing swiftly by enlisting the aid of the modern 'exact' scientific methods. This chapter is intended to show a few of the guises which the researches of the first category can assume for the archaeologist when he investigates, for example, an ancient site. On this level the connections so clearly existing in Western Asia are superseded for the time being by more limited problems. From the start the field archaeologist is tied down to a specific patch of ground under a political and administrative jurisdiction which is no less clearly defined, and he has to work with the approval, protection and assistance of the local authorities. This is why, having explained the reason for uniting these seven countries under the same name in the same chapter, it will be necessary to consider their archaeology under two separate headings: firstly, by tracing the successive stages in the archaeological discovery of preclassical Western Asia, showing how it was influenced by the current political conditions, it will then be possible to return to Western Asia as a whole to examine a few branches of study and a few problems of contemporary archaeology.

The history of archaeological research

Early travellers The tangible traces of the preclassical civilisations of Western Asia, by which is meant the cultures created by the people who flourished there before the conquest of the Achaemenid empire by Alexander the Great, were, for all practical purposes, as though

Opposite page: the remnants of Achaemenid might. These columns, which marked the entrance to the Apadana in the great festival city of Persepolis, stand forty feet high and bear witness to the power of Darius the Great who built the city in the fifth century BC. It was destroyed by the victorious Alexander in 331 BC.

The Zagros mountains of Iran. Great natural barriers such as these and the Taurus range were *no* deterrent to the races who crossed and re-crossed Western Asia looking for desirable lands in which to settle.

obliterated from the face of the earth for centuries. All that remained to be seen in a few widely dispersed places were a small number of reliefs carved on cliff faces, stone columns or staircases. Ancient Western Asia, by choice or necessity, used sun-dried or baked brick for building, which rapidly crumbles back into dust, and the collapse of the abandoned cities has formed artificial mounds which are now known as *tell, tepe* or *huyük* according to the country where they are located.

The history of these Asiatic lands was almost entirely lost except for a few stories and descriptions in the works of classical travellers and historians like Herodotus and Diodorus Siculus and a few passages in the Old Testament. Pilgrims visiting the Holy Land were very well informed, but this did not prevent Rabbi Benjamin of Tudela, travelling as far as Mesopotamia in the twelfth century, from recognising the accuracy of local tradition regarding the location of Nineveh.

The seventeenth and eighteenth centuries

Travellers explored these lands in the seventeenth and eighteenth centuries. There was, for example, the Bavarian scholar Leonhard Rauwolf, and the Frenchman J. B Tavernier. Nevertheless we have to wait until the eighteenth century before any desire to observe accurately and understand the ancient world can be discerned among men then known as 'antiquarians', who included diplomats, travellers and businessmen.

In 1617 Don Garcia Silva Figueroa, the Spanish ambassador to Persia, identified a mass of ruins in the province of Fars in Iraq as Persepolis by means of a description in Diodorus Siculus. The Italian traveller Pietro della Valle explored Western Asia from 1616 to 1624; in 1621, also at Persepolis, he copied (admittedly rather inaccurately) an inscription carved in stone in hitherto unknown characters, which he introduced to Europe. He was also the first to use a pick for excavating, notably at the site of Babylon, and he returned to Rome with 'curiosities' collected on his travels, among which were inscribed bricks. Pietro della Valle did not establish the connection between the characters traced on pieces of clay, the first Western Asiatic antiquities to reach Europe, and those of Persepolis; this was not done until 1700 when Thomas Hyde wrote as follows: '*dactuli pyramidales seu cuneiformes*' ('pyramidal or wedge-shaped signs'). Since that date their name has not changed. They are still called cuneiform.

Jean Chardin (born in France but naturalised British) took advantage of his commerce in precious stones and rich gold work to visit some parts of Western Asia. The story of his travels in Turkey, Persia and other places was published in 1711 in three volumes. Rock cut tombs in the neighbourhood of Persepolis and remains of architecture and reliefs from this demolished royal residence were described here with the utmost care and illustrated by many fine drawings, superior in both quantity and quality to those which had so far accompanied travellers' tales. A copy of an inscription was appended. Chardin added also, 'I daily perceived that I was finding more truth and beauty in divers passages of the Sacred Books than ever before, since I had before my eyes the natural and moral objects to which these passages referred.' This reflection was prophetic: it foreshadows the resurgence of Biblical studies to which the archaeological rediscovery of Western Asia was to lead.

Other 'curiosities' made their appearance in the eighteenth century, and after della Valle's bricks came the collections of the European amateurs. In France, for instance, the Comte de Caylus owned two cylinders and an alabaster vase, details about which he published in 1752. The figures shown on them he claimed must be 'Persians', since they resemble those on 'the monuments described by Chardin'. As to the inscription engraved on the vase, it suggested the following ideas to the Abbé Barthélemy, which he imparted to Caylus. 'Its characters are the same as those of the writing on the ruins of Persepolis, which are characters made in the form of a wedge or nail.'

In the same period collectors of antiquities were travelling in other countries. Some of them had the intention of studying Graeco-Roman remains in Asia Minor in 1764. From this time onwards, others understood that astonishing remains of lost civilisations lay hidden beneath the ground in Western Asia. Joseph de Beauchamp, the churchman and astronomer, passed some time in Baghdad and subsequently published his account of his investigation of the site of Babylon in the *Journal des Savants* of December 1790. Like Pietro della Valle before him, he had hired Arabs to excavate there. Between 1781 and 1786 Beauchamp collected inscribed bricks and small finds and brought them to France. With the aid of the Vicar-General of Baghdad it was possible to begin to understand how the great cities had disappeared, laid low by human agency as well as the elements. 'There are huge mounds which were used, and are still used, for building Hella, an Arab town of some ten to twelve thousand inhabitants.'

The beginnings of archaeological research

With the nineteenth century archaeological research began and gathered momentum on a vast field which was divided into two by the existing political frontiers. On one side was the Ottoman empire, in which Asian Turkey was divided into provinces roughly corresponding to Syria, Lebanon, Palestine and Iraq, on the other, Persia. There were two reasons for this acceleration of archaeological work, widely different but complementary and mutually beneficial from the archaeological viewpoint. These were the interest aroused among European 'antiquarians' by Pietro della Valle's publications and those of Chardin, Beauchamp and many others; and the intrigues manoeuvred by Napoleon in the hope of overthrowing the British ascendancy in India. These considerations strengthened British desire to secure positions in Persia in order to reinforce imperial communications.

For the same reason the French were induced to show a variety of interests. In 1851, after a long expedition, the painter Eugène Flandin and the architect P. Coste wrote, 'The French government wished to do for Persia what it had

The ruins of Persepolis as they appear today.

Early travellers in the east, astonished by the remains of mighty unknown cultures they found there, brought back reports and drawings. Chardin, who explored Persia in the eighteenth century, published a surprisingly accurate view of Persepolis.

WESTERN ASIA BEFORE ALEXANDER

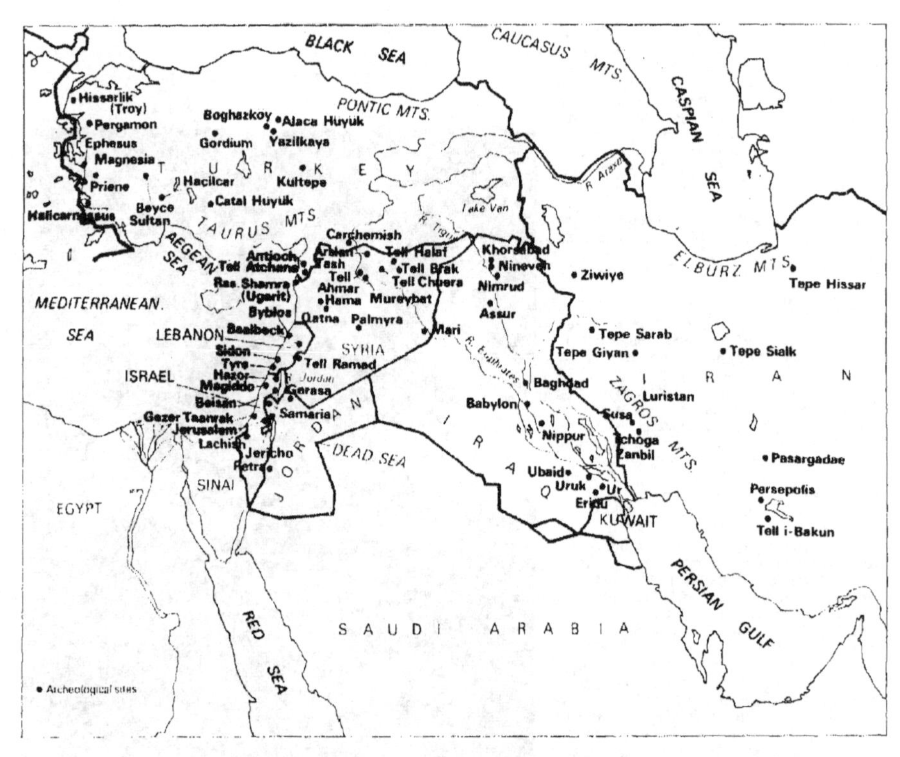

done for Egypt and Greece. It realised that the monumental history of this country was complementary to those of the other two, and felt that there was a void to be filled ... ' Thus expeditions of English and French scholars, counterparts of the British agents of the East India Company played a decisive scholastic role in Western Asia.

These accelerating archaeological studies assumed a rhythm and character dictated by local conditions, but on the whole they reflected the great political events which divided their course into three major phases:

Before 1914 (division of Western Asia between the Ottoman and Persian empires)

Between the First and Second World Wars (British and French mandates provisionally replacing the dismembered Ottoman empire *in some regions*).

After the Second World War (besides Persia, now known as Iran, which had preserved the *imperial form of government, the national* bodies mentioned above came into being each in its turn)

An inscription in Persian cuneiform writing, from the east stairway of the Apadana in Persepolis.

Back to the past For less than 200 years knowledge has been advancing without interruption. Buildings have reappeared, civilisations have been rediscovered, writings deciphered and forgotten languages reconstructed.

Man's return to his own history began with *comparatively recent periods, and from there it* went backwards in a sort of count-down towards the distant prehistoric ages. In the process archaeological methods have gradually improved. The discipline took shape while it advanced. Its two fundamentals became stratigraphy and typology in cases where written evidence was sparse or absent, and in addition, new techniques developed by the modern sciences have been adopted.

It may be asked if such alterations have wrought a change in the basic principles of archaeology. To begin with it was little more *than a search for rarities conducted with little* regard for the recording or preservation of the discoveries. Little by little the archaeologists became conscious of both the importance of the smallest remains in any attempt at reconstructing past history, and the fact that detail which is not recorded at the time of excavation is lost for ever. It is, in practice, not so much a matter of a change in viewpoint as variations in human personality, allowing, of course, for the means available during each period. The history of each age shows, indeed, that some men entered upon their work in the capacity of men of letters, others from more selfish motives.

Claudius James Rich, who in 1841 was the first to write a memoir on the site of Babylon, is typical of the first of these categories: 'I announce no discovery, I advance no interesting hypothesis . . . I shall therefore confine myself to a plain, minute and accurate statement of what I actually saw.' Later on, in 1864 after his expedition to Phoenicia, Ernest Renan remarked, 'If the aim of an expedition is to bring back the largest possible number of finds to the galleries, the coast of Phoenicia would be the worst choice in the world. But France is well aware that the system of expeditions financed by museums or (which comes to the same thing) the custom of certain museums of having expeditions conducted on their behalf has serious drawbacks, we have kept to the sound principle that the aim of a learned expedition is not the service of idle popular curiosity, but the advancement of knowledge. While attending to the interests of our collections, I have never believed any investigation was wasted if, without producing any portable finds, it helped to shed light on some problem or to clear some monument.' The best archaeologists of the present day have exactly the same aims as Rich, and none of them would deny Renan's statement.

W. K. Loftus provides an example of the other tendency. Exploring the sites of Lower Mesopotamia, he actually admitted that he was seized by 'a nervous desire to find large important museum pieces with the smallest possible outlay of time and money'.

Iran

Persia, officially known as Iran since 1935, which celebrated the 2500th anniversary of the monarchy in 1961, has often remained on the sidelines while political upheavals have overtaken the other countries of Western Asia. It is true that geographically speaking this land occupies a place apart. Opening to the north on an inland sea, the Caspian, and to the south on the Indian Ocean, ancient Persia consisted chiefly of high plateaux, often desert, of remarkable aridity, bounded by mountains: the compact mass of the Azerbaijan range in the north continues in the Elburz, and the Zagros to the south extending along the Persian Gulf. Life here was governed by the presence of water and the excessive variations in the climate. The harshness of these conditions gave Iran a distinctive character in the past, and has preserved it up to the present. In addition, although it is bounded by the Soviet Union, Turkey and Iraq, it is also in contact with the Far East by virtue of its common borders with Afghanistan and Pakistan. This position midway between Europe and the Orient is perhaps symbolised by the romantic name of the 'silk route', and partly explains the destiny of Iran before the Suez canal and the invention of the aircraft.

Only the standing columns of Persepolis and some rock walls ornamented with reliefs commemorating the great Achaemenid and Sassanian kings preserved the astonishing history of the departed past above the ground. Since the end of the eighteenth century, as we have seen, travellers and then scholars began to take an interest in the inscriptions on monuments. This was the first step towards the decipherment of the Persepolitan scripts and the reconstruction of the dead languages. The discoveries of the Dane Carsten Niebuhr, about 1778, followed by his publications and those of other pioneers, showed that there were three different kinds of writing. These three languages were decisively established by Grotefend at the beginning of the nineteenth century, while the Persepolis inscriptions, to which others were continually being added, were the object of studies conducted in an atmosphere aflame with intellectual achievement throughout the whole of the first half of the nineteenth century.

Many of the secrets of Mesopotamian history were revealed when the cuneiform script was deciphered in the nineteenth century. A closely written clay tablet in the Louvre shows the script at the height of its development.

This epigraphic victory was entirely scholastic, since Iranian archaeology was non-existent during this period. In 1884 investigations were *conducted at Susa under the leadership of* Marcel Dieulafoy. A French mission was established there in 1897, and has continued work there without interruption ever since. This work, however, was carried out in the manner of the times, using remarkably primitive methods.

Modern archaeological work in Iran began only between the First and Second World Wars. October 1927, which saw the inauguration of the Department of Antiquities, marked the beginning of national activity in this field. During the nineteen thirties foreign missions began to undertake systematic projects such as surveying for archaeological remains, aerial reconnaissance and the opening, chiefly by French and American teams, of major excavations like Tepe Giyan (1931), Tepe Hissar (1931), Tal-i Bakun (1932), Tepe Sialk (1933) and Tchashma Ali Tepe (1934).

From 1950 archaeological work in Iran has accelerated considerably. Besides the Iranian archaeologists working to rediscover the distant past of their country, many other teams crowded in with the permission of the Antiquities Service. The British, Germans, Belgians and Japanese joined the Americans and the French, who had been at Susa for many years. The sites chosen by these teams yielded a variety of information which all contributes vitally to the reconstruction of history. It is therefore very difficult to pick out any specific names from the whole, since an entire complex can be important. For example, during the second and first millennia BC Tchoga Zanbil, Hasanlu and Tchoga Mish were particularly noteworthy, as were also such Neolithic settlements as Tepe Asiab, Tepe Sarab, Ali Kosh, and other sites going back to the Paleolithic period such as the 'Belt' cavern or the cave of Hotu.

In other parts, the Luristan region has been studied. In fact, many finds from this area have reached museums or private collections by way

A detail of the frieze of archers of the royal guard at Susa. The frieze, which dates to the Achaemenid period, is carried out in coloured and glazed relief-moulded brick, like the much earlier Ishtar Gate of Babylon.

A view of the ruins of Babylon, the city which dated from 4000 BC. Destroyed and rebuilt again and again, Babylon was the greatest city of antiquity in the sixth and fifth centuries BC. Alexander died there, and his Seleucid successors built a rival city nearby on the Euphrates. Babylon's glory ended at this time (275 BC) and 2000 years passed before Robert Koldewey's great work on the site.

of the antiquities trade, which supplies no reliable information. Nowadays attempts are being made to assign these to their correct archaeological context once more. Finally, since 1960, several sites already investigated before the Second World War but not fully worked out, have attracted archaeologists who are examining them further. The results have justified the effort, although it is true to say that improvements in archaeological methods have sharpened observation.

These numerous projects are not uniformly dispersed over the land of Iran. Some regions are more favoured, or favourable (climate, access roads, etc.) and on the whole, in spite of continuous surveys, the archaeological sites of Iran are among the least well documented of Western Asia. It is to be hoped that the development of scientific research will gradually bring about far reaching assessments, and put an end to what is unfortunately typical of Iranian archaeology: the appearance of finds divorced from their archaeological context as a result of illegal excavations or chance finds. Among these were Luristan bronzes in 1928 to 1930, the Ziwiye treasure in 1947 and, more recently, the vessels in precious metals and a number of terracotta animals from the provinces south of the Caspian Sea which have appeared on the antiquities market.

By studying material collected over the years, experts have been able to reconstruct the broad outlines of the history of Iran. This has been largely possible for the Susa region, which early attained an advanced literate culture parallel to or identical with that of Mesopotamia, the history of which can be reconstructed by means of the texts and chronological parallels. The Iranian plateau, on the contrary, did not reach this stage of development and for a long time remained on the fringes of history, which it only effectually entered in the Achaemenid period. As a result, in the absence of written evidence scholars have had to rely on the purely archaeological techniques of dating and classification, stratigraphy and typology, which up till the present have been chiefly based on pottery, for the reconstruction of the development of this area.

Iraq

When the forerunners of modern archaeology first worked there, Iraq did not bear its present name. At that time it was only a province of the Ottoman empire which, after 1831, was subjected to new territorial divisions inaugurated by the reform movement started in Istanbul.

The First World War saw the unfolding of the Anglo-Turkish war in the Tigris and Euphrates plain. Then, in 1918, the Allied victory led to the break up of the empire and the country was placed under a British mandate (1919–20) before becoming the kingdom of Iraq in 1921, which was progressively emancipated until 1932, the date when the mandate ended. In 1958 the free republic of Iraq was defined as follows: to the west were the borders of Jordan and Syria, to the north, that of Turkey, to the east, Iran, to the south-east, the opening of the Persian Gulf and the Kuwait district, and to the south-south-east, the frontiers of Saudi Arabia.

The three-part physical division of this territory (to the north-east a crescent of mountain ranges and foothills, in the middle, plains watered by the great rivers, and to the west, desert rising towards the Arabian highlands)

Gudea, governor of Lagash during the troubled period immediately after the fall of the Akkadian kings, left many statues of himself. The illustrated example, though headless, is identified by an extensive cuneiform inscription. The plan of a building lies on his lap.

In the Early Dynastic period the people of Tell Asmar used to set up statuettes of worshippers to represent themselves in the temple. This male and female are the largest known examples. The base of the male bears vegetation symbols and the female once had the figure of a child beside her.

Personal goods in the ancient Near East were often marked with cylinder seals, which were rolled on damp clay to produce relief or intaglio impressions. Two very early examples from the Erech period (before 3000 BC) show processions of animals and human figures.

has influenced its historical destiny. So has the presence of the rivers Tigris and Euphrates, the waters of which, rising in Turkey and the Azerbaijan mountains, have been utilised in life-giving irrigation channels since the beginning of the ancient world.

In 1810 a young Englishman, Claudius James Rich, who had been appointed Baghdad representative of the East India Company (the company sent out officers from 1783 onwards) arrived in Iraq to find a country which was excessively hot and where there were sandstorms and occasional uncontrollable floods. Rich, who was rapidly enthralled by the buried past of Mesopotamia, conducted his methodical and scholarly researches first at Baghdad, then at Nineveh, and then in Persia, where he died of cholera in 1821. The archaeology of ancient Iraq, which came to be known as Mesopotamian archaeology after a few decades, originated with the posthumous publication of his work.

Great interest was aroused among the cultured public (Lord Byron referred to Rich's work in Canto V of his *Don Juan*) as well as in European academic circles. It stimulated the French expedition led by Paul Emile Botta, consul at Mosul, who conducted the first full-scale excavation of the site of Khorsabad in the neighbourhood of this town in 1843 and 1844. He found carved stone blocks decorating a structure which was later discovered to be an impressive neo-Assyrian palace. English excavations directed by Austen Henry Layard (1845) on the site of Nineveh and then on the ruins of other ancient Assyrian cities subsequently followed. These led to a veritable competition between the British and the French to unearth works worthy of a place in the British Museum and the Louvre. These were treasure hunts rather than scholarly research, and in their course, since no one had either time enough or any idea of horizontal excavating, tunnels were dug and shafts sunk to reach the interesting pieces more quickly.

This feverish activity in the north of the country as well as a few less important soundings in the south brought to light some inscriptions in cuneiform script engraved on stone monuments and clay tablets. To learn how these texts came to be deciphered it is necessary to revert temporarily to Iran, and to the decipherment which began with the Persepolis scripts a long time before. These studies had advanced considerably, largely due to the efforts of Henry

The cuneiform letter tablet from Alishar, now in Ankara, is still in the original but badly damaged case in which it was once dispatched.

Creswicke Rawlinson, who tackled another cuneiform inscription, also trilingual, on the cliffs of Behistun. In 1847 the first three lines were mastered: it was translated into 'Old Persian'.

It was later realised that the writing of the third part of the trilingual Persian texts was exactly the same as that which appeared, this time unaccompanied, on the monuments of the valley of the Two Rivers. As a result, several scholars were able to translate it; the same text, consigned to four experts (English, French and Irish), was transcribed and translated by them and the results, separately collected and placed under sealed cover, were strikingly similar.

Everything was ready for the reconquest. It was to be pursued and extended, Americans and Germans arriving at the end of the nineteenth century to join the British and French pioneers. Turkey under the rule of Hamdi Bey was beginning to regularise archaeological work at the same time. The Sumerian lowlands in the south became in their turn the centre of interest, with digs opening at Tello (1881), Fara (1900), and Bismaya (1904), and also at the capital Babylon in the middle of the country (1894–1914), Abu Hatab (1894), and Kish (1912). Meanwhile, in the north, in ancient Assyria which had always been studied, the dig at Assur was started (1903) and also that of Samarra (1911) where very early periods were reached but not recorded. Thus the ancient civilisation of Mesopotamia gradually revealed some of its facets, its pronounced regional characteristics and its successive historical eras.

The end of the nineteenth century and the beginning of the twentieth were particularly distinguished by great advances in scientific method; treasure hunting was no longer the primary aim. From this time forward archaeologists attempted to find walls and reconstruct the plans of architectural complexes built of baked or, more often, sun dried bricks, and the horizontal excavation technique gradually superseded tunnels and shafts. Victor Place thought of doing this at Khorsabad, and the German architects turned archaeologists mastered the technique at Babylon and Assur.

This first phase, ending in 1914, was also one of conflicts with the local authorities, who were not only incapable of keeping order, but often in revolt against the Turkish government. Real dangers – isolation, climate and even gunfire – awaited the archaeologist who strayed too far from the caravan routes.

After 1919, during the British mandate, archaeological research gathered momentum while being systematised and gradually coming under national control. A Department of Antiquities was created and a law concerning antiquities passed in 1924, and the Iraq museum established at Baghdad in 1926. When the mandate expired in 1932 eleven teams from five nations were working in Iraqi territory on sites whose names have become famous: Ubaid, Ur, Kish, Eshnuna, Tello, Nuzi, Tepe Gawra, etc. The advance of knowledge followed swiftly. In 1928 the seventeenth International Congress of Orientalists registered both the 'admission of Sumerology into the circle of the exact philological sciences', and the discovery of painted pottery in Mesopotamia. In 1931 the Eighteenth Congress agreed that the prehistoric periods should be named after the eponymous sites ('Ubaid, Uruk and Warka) and the first historic periods by the term Early Dynastic. In the period between the First and Second World Wars, too, prehistoric sites were identified for the first time at Zarzi and Hazer Merd (1928).

The basic chronological framework was now established, consisting of prehistory, protohistory and the successive historic eras from the early dynasties of the third millennium until the capture of Babylon by the Persians.

The unbroken advance in excavating methods and the enormous progress of knowledge proves that the professional archaeologist had arrived, and taken over from the enlightened amateurs and diplomats of the nineteenth century.

The Second World War, which enforced the progressive retreat of all foreign expeditions, saw increasing numbers of Iraqi archaeologists at work in the field, at Tell Uqair and Tell Hassuna (after 1940), and then at Tell Harmal (1945), Eridu (1946), Tell al Dhiba (1961) and Tell es Sawwan. Parallel to this field work, the Department of Antiquities, whose activities had not been impeded by the 1958 revolution, systematically continued to record the historic sites and monuments. Every year the number increased. By 1962, 6685 had been located. Since 1948 the foreign teams have been able to return. American, British, German and French scholars in turn went back to their excavations on the sites they were formerly investigating (Nippur, Nimrud, Warka and Larsa), or to new sites such as Tell al Rimah. Japanese and Italian teams have followed.

Elsewhere, more and more information was gathered about the early periods, such as the late prehistoric era on the sites of Karim Shahir, Jarmo and Tell es-Sawwan, and about more remote prehistory with the Middle Paleolithic site of Shanidar and the Lower Paleolithic site of Barda Balka.

This third phase was not distinguished by new methods in the field, but by the employment in chronological research of the informa

A statue dating to the reign of Gudea of Lagash in the twenty-second century BC is made of hard and costly diorite. Its sturdy self-confidence reflects nothing of the troubled times which were afflicting the rest of Mesopotamia.

A seated figure of King Idrimi in white magnesite bears an inscription telling the life story of this fifteenth-century BC ruler of Alalakh, and relates his difficulties and adventures in attaining the crown.

The goddess in a high headdress and heavy beaded necklace, from the palace of Zimrilim at *Larsa*, holds a libation vase. The jar has a channel drilled through it which was probably once connected to a water tank to enable the figure to function as a fountain. It belongs to the Isin-Larsa period.

tion furnished in the laboratory by carbon 14 dating.

A combination of all the work since Beauchamp and Rich, all the excavated monuments and objects and all the translated texts constitutes the present science of Assyriology, which shows us that after long formative periods it was in Mesopotamia at the end of the fourth millennium BC that the earliest known advanced civilisation evolved, later to become the ancestor, origin and model for all the other regions of Western Asia.

Turkey

When the Ottoman empire was broken up after the First World War its nucleus, Turkey, formed a national unity within new frontiers with the inauguration of the republic in 1923. Its borders lay between Turkey and Iran, the Soviet Union, Syria and Iraq. In 1938–39 this was slightly modified by the annexation to Turkey of the plain of Hattay (Sandjak of Alexandretta), which is quoted here as an example of the distribution of archaeological sites.

The peninsula of Turkey, or Anatolia, is so isolated from the rest of the continent that it is sometimes known as Asia Minor. As a result, *unlike its land locked neighbours Iraq and Iran*, it has a series of long coastal frontiers on the Black Sea, the Sea of Marmora, the Aegean and the Mediterranean. The heart of the peninsula, the Anatolian highlands, linked to the continent on the east by the vast mountain massif of Armenia, is enclosed by mountains on each side, the chief of which are the Pontic range to the north *along the Black Sea and the Taurus mountains on the other side*. The remarkably diverse relief of this country defines several distinctly articulated regions; the climate varies from the harshness of the mountains to the softness of the Aegean and the Mediterranean. Despite apparently impassable mountains, Anatolia was destined to establish contacts with neighbouring regions, especially those of the so called 'Fertile Crescent', by means of the plain of Cilicia, which largely opens into Syria, and the Tigris and Euphrates rivers which rise *in Anatolia* and flow through it before descending into Iraqi territory. The geography of Anatolia in the south and west also explains its very early contacts with Western countries, maintained and strengthened over the millennia, with a result that Turkey is remarkably rich in Greek and Roman remains, some of which are above *ground (Ephesus, Halicarnassus*, Magnesia on the Meander, Pergamon and Priene, to name a few). From 1764 onwards these inspired expeditions organised by the Society of Dilettanti, founded in London in 1732.

During the nineteenth century, while the British and the French pursued their investiga*tions and geographical* explorations in the interests of classical archaeology, they sometimes came in contact with impressive above ground remains which were still unidentified. The existence of a remote past was already undreamed of when Herodotus mentioned in his Book II the presence of 'royal figures carved on the rock, one on the road from Ephesus to Phocaea, the other on the road from Sardis to Smyrna'.

It gradually came to be realised, however, not only that these elements did not fit together into unbroken history or continuous culture, but that they were manifestations of separate stages *of development in different regions and nations*, speaking different languages and writing different scripts, although they often showed a striking technological continuity which was truly Anatolian. At random among the chaos of the discoveries there reappeared traces of the surviving Hittites of the first millennium, the Hittite empire of the second millennium, the *kingdom of Urartu, the Assyrian colonies, the mysterious tombs* of Alaca Hüyük, *Phrygian* Gordium and the Neolithic cultures.

The first step in the reconquest was not taken in Turkey itself, which made it even harder to understand, but at Hama in Syria in 1812, when the Swiss traveller *Johann* Ludwig *Burckhardt* found a stone covered with hieroglyphs built into the wall of a house. His attention was at once drawn to this unknown script, and other evidence of it was recognised or discovered in

A carved stele from Lagash (Tello) dating from the *rule of Gudea*, now in the Berlin Museum, shows two divinities. The figure on the left is leading *another by the hand*.

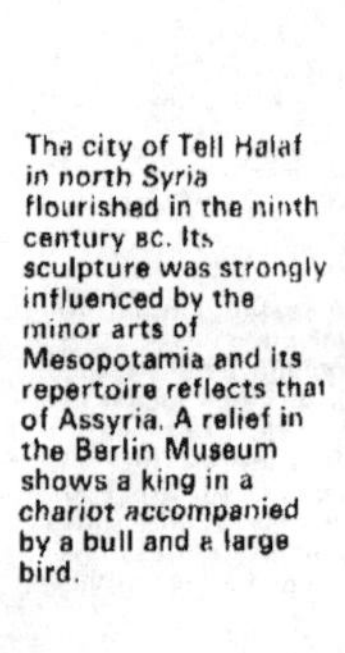

The city of Tell Halaf in north Syria flourished in the ninth century BC. Its sculpture was strongly influenced by the minor arts of Mesopotamia and its repertoire reflects that of Assyria. A relief in the Berlin Museum shows a king in a *chariot accompanied* by a bull and a large bird.

Syria, and then in Anatolia, until by 1900 the number of known hieroglyphs was large enough to warrant the compiling of a *Corpus*. In 1879 A. H. Sayce had already given these signs, which were identical in both countries, the name of Hittite, in spite of the doubts of all his colleagues. The references to Hittites in the Old Testament had not prepared scholars for the emergence of this people in Anatolia.

In 1820, in the distant region of Lake Van, the orientalist Jean Saint Martin noticed and pointed out rock reliefs in which the French became interested. They sent the German, Schultz, to the place, where he made copies of cuneiform inscriptions. Despite the tragic death of Schultz, who was murdered by Kurdish bandits in 1829, his notes reached Paris, where it was realised that one of the inscriptions was trilingual. Two of the scripts (Persepolitan and Mesopotamian Akkadian) made it possible to read the third, subsequently known as Vannic, and to recognise that a kingdom known to the Assyrians as Urartu had existed in the land of Van in the first millennium BC. Thus a new chapter in Turkish history was opened, to which the British, French, Americans, Germans and Russians were all to contribute.

It was in yet another region, Cappadocia, that the third and fourth steps of the reconquest took place. In 1834 the attention of Charles Texier was arrested at a place called Boghazköy by the ruins of a fortified town, the walls and carved gates of which were still visible, and by a group of rock-cut bas-reliefs in a defile in the rocks at nearby Yazilikaya. Georges Perrot and Karl Humann followed in his wake, but none of them dreamed of the protracted existence of a powerful empire, although they nevertheless contributed to its historical rediscovery.

About 1880 some clay tablets covered with cuneiform signs were offered for sale in the bazaar at Kayseri. Although their exact provenance was unknown, their meaning was clear. They formed the mercantile correspondence translated into Akkadian of the Assyrian colonist who had settled in the region in the nineteenth century BC. Then a second group of cuneiform tablets appeared, which roused a certain amount of astonishment in the academic world. Ernest Chantre described how 'During our first investigation of the ruins of Boghazköy in 1893, Mme Chantre, climbing up the debris on the slopes of the fortress ... collected ... several fragments of brick with cuneiform inscriptions. Our attention once drawn to this point, we soon found others, helped by the villagers of Boghazköy.' Chantre's discovery and his publication once more attracted the notice of the academic world. Then in 1905 Hugo Winckler and Macridi Bey undertook the first actual excavation, which resulted in the unearthing of more than ten thousand tablets. The Hittite empire of the second millennium thus became part of history, and gradually took its rightful place in relation to the minor Hittite revivals of the first millennium already indicated by the presence of hieroglyphic inscriptions in the Taurus region.

This picture of the origins of preclassical archaeology in Turkey would be incomplete without mentioning the saga of Heinrich Schliemann, who sought the site of Homeric Troy at Hissarlik from 1870 to 1890, of the awakening of national interest in archaeology, demonstrated by the establishment of museums in Turkey after 1850, the enlightened activities of Hamdi Bey after 1881 and the work of Macridi Bey at Alaca Hoyük in 1907; and of the arrival of foreign missions on the important sites, such

A basalt upright slab from Carchemish. The hieroglyphic inscription is Hittite and dates from the second half of the eighth century BC.

as the British at Carcemish (1878–81) and the Germans at Sinjerli (1888–92). These missions operated technically in much the same way as the contemporary Iraqi teams: completely unmethodically.

In comparison, the work of Schliemann's successor Wilhelm Dörpfeld at Hissarlik (1893–1894) was much more orderly and efficient.

Two parallel branches of study were continued between the First and Second World Wars, supporting and supplementing each other. On the other hand the epigraphists applied themselves to the decipherment of the rediscovered writings. In 1915 the Czech, Friedrich Hrozny, noticed affinities between Hittite and the Indo-European group of languages, while in 1919 M. Forrer demonstrated

The Scorpion-man, a strange figure with a human head, bird's feet and wings and a scorpion's sting, once formed part of the gatepost for the ninth-century palace of Kaparu (Tell Halaf).

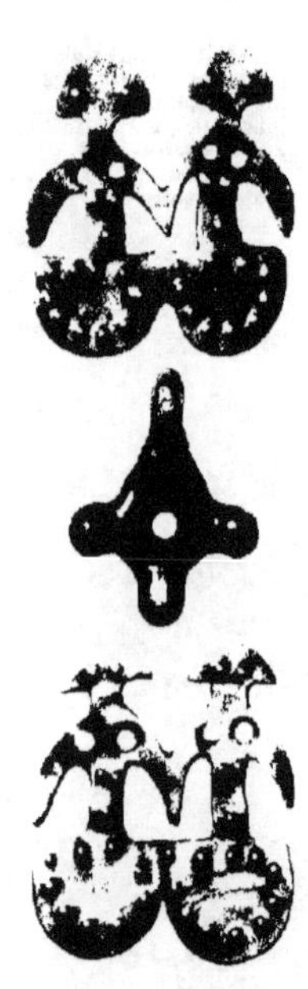

The Anatolian plateau *was one of the earliest* metal-working centres of the Near East. Highly stylised twin idols cut from sheet *gold and decorated* with punch marks and perforations come from Alaça Hüyük.

that the cuneiform texts of Boghazköy corresponded to eight different languages, one of which was Akkadian, another, Luvian, was subdivided into dialects, one of which was written in the hieroglyphs rediscovered in 1812.

Meanwhile the field archaeologists were at work. In 1925 the source of the tablets written by the Assyrian colonists at Kültepe, the ancient Kanesh, near Kayseri, was identified. With the organisation of the republic by Ataturk, the years 1930–35 saw the part played by Turkish archaeologists becoming more decisive. They took over the site of Alaca Hüyük, where the appearance of an abundance of gold, silver and bronze objects in the royal tombs proved the existence of a rich and brilliant culture, hitherto unknown, in Anatolia at the end of the third millennium. The years 1932, 1933 and 1935 saw the advent of scientific archaeology, i.e. methodical exploration and stratigraphical excavation with projects run by the Americans at Hissarlik and the British at Mersin. The year 1939 saw the already investigated sites in Sandjak of Alexandretta pass to direct Turkish control; these included Tell Atchana (ancient Alalakh) and the two still unnamed sites found by an American team in the plain of Antioch.

Archaeological research such as this, based on sound foundations, served by numerous skilled national teams including the American, German, British and Italian institutes, could not fail to advance and develop. Since 1945 many surveys have aided the accuracy of the excavation and chronological study of sites. Reopened sites and new digs have contributed valuable information on the main periods, which have already been outlined, of preclassical archaeology, whether it dealt with the tombs of Alaca Hüyük, the Assyrian colonists whose seals and ivories shortly appeared at Acem Hüyük, the Hittite empire, or Urartu.

The excavation begun at Gordium in 1949 has added important chapters to the history of ancient Phrygia. At Hacilar, Can Hasan and, most important, Çatal Hüyük, Neolithic cultures with astonishingly advanced technologies have been found since 1960, while the Upper Paleolithic can be studied at Belbasi and Beldibi.

These various results aroused intense interest in preclassical Anatolian archaeology and stimulated an exchange of ideas. It would be impossible nowadays to study the Neolithic period in Macedon, Thrace or Thessaly without a knowledge of the Neolithic farmers of southern Anatolia, or the plan of the *megaron* without a knowledge of Hissarlik and Beyce Sultan. The comparative data from Urartu can take the bronze expert as far as Etruscan Italy, and the architect as far as the royal Achaemenid buildings.

Finally, in Turkey as elsewhere, dating elements provided by carbon-14 play an important part in the improvement of chronology. It was also in these territories that a new method of investigation, underwater archaeology, was practised and perfected.

Syria

Apart from a few intermediate movements, Syria remained under the control of the Ottoman empire until its break-up after the First World War. At that time, however, this name was traditionally attached to a geographical area much larger than the land which bears it at present. The decisions of San Remo in 1920 placed the part of the country which still keeps the name of Syria under the temporary rule of a French mandate, but it achieved independence in 1945–46.

Except for a 'window' 175 kilometres long opening on the Mediterranean between the Turkish and Lebanese seaboards, the border between Syria and Turkey is now quite naturally marked by the line of the mountain ranges. To the east and south-east it is separated from Iraq by an arbitrary border. To the south and southwest it is separated from Lebanon by a frontier consisting of Mount Hermon and the Anti-Lebanon. Coastal and subcoastal zones defined by the parallel ranges of mountains crossing the country along its length enjoy a Mediterranean climate, in contrast to the regions with a continental climate – the highlands of Djezireh and central Syria.

A procession of massive divine figures was carved in the rock walls of the Hittite open-air sanctuary at Yazilikaya near Boghazköy.

The Hittite King Tudhaliya IV is embraced by the god Sarumma in a relief *carving from the wall* of the rock shrine at Yazilikaya.

These Syrian territories have always been much less isolated than the other Western Asiatic lands because of their geographical situation. They frequently played a strategic or cultural part inasmuch as they were way stations crossed by the great highways, not only those linking the various neighbouring countries but also those to Egypt and the West. Writings found in Iraq reflect this situation, since they relate the exploits of Sargon the Great of Akkad, who ruled Mesopotamia about 2300 BC and conquered Syria as far as the 'cedar forests' (*Amanus*) and the 'silver mountain' (*Taurus*).

Nearer to the Holy Land than Iran, Iraq or Turkey, Syria entered the arena of the West at an early date with the Crusades and the Frankish settlements. Greek and Roman remains could still sometimes be seen there above ground, like the ruins of Palmyra, the plans of which were drawn by Robert Wood and J. Dawkins in 1753. Many travellers and scholars, including Léon de Laborde (1927), M. de Vogüé (1861–1862) and others, made observations and interesting studies there, but they were not relevant to the early periods which are being considered. Those eras had vanished from human memory. Some observers, however, apparently foreseeing the future, established the first landmarks of the reconquest. Comte de [illegible] *for instance, travelling in Syria from* 178[illegible] to 1785, was struck by what we now know as *tells*: 'On the road from Aleppo to Hama [illegible] I particularly remarked a number of round [illegible] oval hillocks w[illegible] built-up appearance and steep slope on [illegible] lat plain indicated that they were man-made [illegible] rckhardt, too, was struck by the hieroglyph[illegible] Hama later found to be Hittite, as the Englis[illegible] Francis Rawdon Ch[illegible]ney was in 1850 by [illegible] colossal stone lions which give their name to the site where they can still be seen: Arslan Tash, 'the lion stone'. Porter in the eighteen eighties and R. Dussaud in 1903 were impressed by the basalt lion of Sheikh Saad.

When Hamdi Bey went to Arslan Tash and had the Assyrian sculptures, complete with their inscriptions removed to the Imperial Ottoman Museum at Istanbul in 1883–89, when Hogarth conducted his fruitful exploration of Tell Ahmar in 1908, and when scholars shed light on small kingdoms on various parts of the present Syrian frontier using 'Hittite' hieroglyphs, the discovery of the earlier periods of history in Syria could be said to have begun. Thus the reconstruction of this history began with the discovery of the existence of the middle Euphrates culture at the end of the second and the beginning of the first millennia and its frequent conflicts with the Assyrian empire.

From 1919 onwards the period of the French mandate saw the real beginning of preclassical archaeology in Syria, starting with the creation of an Antiquities Service. Aerial reconnaissance, *surveys at ground level* (that of the middle Euphrates tells by W. F. Albright in 1925 marks the start of an epoch: he used surface finds of pottery for chronological purposes), and excavations opened at Tell Nebi Mend, Mishrife and Neirab all accelerated the researches which were recognised by scholars who gathered together for an international congress of Syrian and Palestinian archaeology in 1926.

The reconquest of a further millennium followed when R. Dussaud discerned the existence of Syrian art of the second millennium, working from very few specimens. A group of clay tablets found at Neirab and the torso of a statuette from Sfire showed that the Akkadian

The sphinxes carved on the gateposts at Alaca Hüyük show affinities with Syrian prototypes. They are probably meant to be female, but their headdresses are a distorted version of the Egyptian pharaonic headcloth.

language and script were known in Syria at the same time. The eventful and chequered history of the most important kingdoms [illegible] this period, their contacts with Mesopotamian culture and their relations with the Egyptians, Hittites and Mycenaeans gradually became clearer during the course of the French excavations at Ras Shamra, ancient Ugarit, in 1929, and at Tell Hariri, ancient Mari, in 1933, and the British excavations started in 1939 at Tell Atchana (Alalakh) in Hatay.

Furthermore, the appearance in 1929 of the Ras Shamra archives and their decipherment in 1930 revealed that the Syrians not only used the Akkadian language and script in the second millennium, but that they perfected a new system for writing their own language, which was also Semitic. (Akkadian cuneiform, or Assyro-Babylonian, was syllabic, and included some 500 signs. Ugaritic cuneiform, which borrowed nothing but the wedge-shaped form from Akkadian, was a simplified and considerably advanced form, since it was alphabetic and included only thirty signs.)

As to the improvement in archaeological methods, it was as noticeable in Syria between the two wars as everywhere else in Western Asia, and the Danish stratigraphical studies at Hama on the Orontes (1931–38) with their detailed publication, could be quoted as exemplary.

The retreating boundaries of the past did not stop here. The work conducted by the Germans at Tell Halaf in 1911–13 and reopened in 1927, and some parts of the excavations at Mari and Ras Shamra, as well as the projects initiated by the British in 1935 and 1937 at Chagar Bazar and Tell Brak, made it possible to follow the main lines of Syria's development in the third and fourth millennia, and to discover its ancient contacts with neighbouring countries. By investigating the remains from the Middle Paleolithic horizons at Yabrud even earlier periods were reached.

The independence of Syria was accompanied by an upsurge of national interest in its antiquities, shown by a clause in the 1950 constitution providing for the protection of this heritage by the opening of numerous projects under the auspices of the Antiquities Department, and by the creation and expansion of museums. The necessity for surveying, either in the field or by aerial photography (more than 10,000 photographs were taken in Jezireh for archaeological purposes) was also recognised, since as complete as possible an inventory of sites is indispensable if excavation sites are to be selected according to scholarly principles.

After the interruption caused by the Second World War, the foreign teams, among whom were the Germans, Americans, British, French and Italians, collaborated with the Antiquities Service to reopen the chief sites (Ras Shamra in 1948, Tell Hariri in 1951). New sites like Tell Chuera, Tell Sukas, Tell Mardikh and Tell Rifa'at were begun and knowledge of the different phases of prehistory was advanced by the study of sites like Yabrud (1950), Latamne, Tell Ramad and Mureybat.

Lebanon

When the little country which now forms Lebanon was under the protection of the Ottoman empire, it was not differentiated from Syria, and their paths followed the same course. But it was only in Lebanon that a French expeditionary force intervened after cruel dissensions, and this was also to play a part in the country's archaeology. At the end of the First World War Lebanon, like Syria, found itself under French mandate, to become a republic in 1926 and gain complete independence in 1943.

This country, which was once known to the Greeks as Phoenicia, is flanked on the north and south by Syria and, for a short distance, by the state of Israel. Except for its long Mediterranean frontage, its contours are disposed in longitudinal strips, in which the lowland littoral of Beqaa alternates with the mountain ranges of Lebanon, the Anti-Lebanon and the Hermon. Its climate could best be described in the words of the Arab poets who sang of Djebel Sannin, wearing winter on its head, spring on its shoulders and autumn on its breast, while summer slept at its feet.

Its fate has always been dictated by its geographical situation. A place of passage, meeting and exchange, in touch, like Syria, with the interior of the continent, Lebanon is even more than this; from the earliest antiquity its inhabitants, dwelling in cities clustered on the shore, were sailors and merchants travelling to Egypt and the West, and they were later to scatter far afield, founding trading posts and colonies throughout the entire Mediterranean.

The memory of some Lebanese preclassical monuments had been preserved in the written traditions of the Greeks and Romans after Homer and in the Old Testament and the *Phoenician History* of Sanchoniathon. Direct contact with the antique land was continuous, thanks to the seaports which had remained in

Alaça Hüyük, a small tell or mound in the middle of the central Anatolian plateau, is an exceptionally rich site as it was continuously occupied from about 4000 BC onwards. The temple (shown here from the north-west) had an open loggia façade and a walled enclosure entered by a double gate.

touch with the rest of the world and were the points of arrival and departure for many of the travellers and scholars already met in other, less accessible countries. Poets such as Lamartine celebrated the charms of Lebanon. Gérard de Nerval restored a timeless significance to it: 'Is not this shore the very cradle of all the faiths in the world? Ask the first hillman you meet. He will tell you that it was in this land that the early scenes of the Bible took place, he will show you the place where the smoke of the first sacrifices rose. . . . If you come to it as a Greek antiquarian you will see coming down from these hills the whole smiling conclave whose worship was adopted and transformed by the Greeks and spread by the Phoenician migrations. These woods and mountains still echo to the cries of Venus mourning Adonis.'

It might be asked whether, in writing these lines, Nerval made the connection between the Lebanese coast and the script which the British scholar J. Swinton and the French Abbé Barthélemy succeeded in deciphering about 1750. It was a Phoenician alphabetic script dating to the second half of the first millennium BC, such as was used in the Phoenician Mediterranean colonies, its decipherment was easier because of its resemblance to Hebrew. Despite these similarities and this decipherment and despite the bas reliefs carved on the gorges of Nahr el Kalb by the Assyrian conquerors of the first millennium BC, it was the Roman ruins of Baalbek mapped out by Wood and Dawkins in 1757, which first attracted the notice of the 'antiquarians'. A century had to pass before the accidental discovery took place of the sarcophagus of Eshmunazar at Saida (presented to the Louvre by the Duc de Luynes) which bore an inscription of 990 words in Phoenician alphabetic characters of the fifth century BC, which changed the situation completely.

Lebanese preclassical archaeology began in 1860–61 with Ernest Renan's Phoenician expedition. He conducted a remarkable operation there, carrying out the methodical registration of the above ground Phoenician remains with the aid of the French expeditionary force (Aradus, Byblos, Sidon and Tyre), as well as embarking on the excavation of various sites. Then, about 1900, came the plunge into prehistory with the finds from Antelias.

In Lebanon, as in the other countries of Western Asia, the end of hostilities in 1918 and the subsequent political reorganisation brought a lively resurgence of archaeological work, which now took place within a national context. In 1920 a museum was founded at Beirut to house the national Lebanese antiquities, and in 1936 the administration was transformed into the Antiquities Service. From 1926 onwards the progress which had been achieved was clearly apparent. The archaeological congress mentioned above was able to establish a credit balance for Lebanon, too. The excavations at Jeball, ancient Byblos, started in 1920 and continued since then without a break, became one of the main centres of interest. Byblos, which had contacts with Egypt from the beginning of the third millennium onwards because of its timber trade, yielded some fine third- and second-century finds, which usually reflect the commercial character of the Lebanese coast by their mixture of styles (Egyptian, Mycenaean and oriental) as well as by the inscriptions which are sometimes added to them. One of these finds was extremely important. In 1924 the stone sarcophagus of Ahiram, king of Byblos, was discovered, and it bore an inscription, dating perhaps to the tenth century BC, in alphabetic Phoenician script which is the ancestor of Greek and therefore of our own.

Finally, the period between the wars saw the opening of a less spectacular but certainly important chapter of prehistory at Adlun.

Following the achievement of independence in 1945 the Antiquities Service was transformed

Alaça Hüyük was particularly rich in early metal work. Some curious staffs topped with ritual figures and known as standards were found in tombs dating to the Early Bronze Age.

Among the standards from the tombs at Alaça Hüyük were some curious examples with openwork tops to which wheel-like discs were loosely attached. They may have served as rattles for use in religious ceremonies.

Even when no buildings have survived, their appearance can sometimes be reconstructed from the minor arts. Part of a Hittite clay vase (fifteenth century BC) represents a city wall with battlements and windows.

The site of the ancient capital of the Hittites, Hattusas, is at present-day Boghazköy. The royal gate capitals had carved reliefs on either side, one of which can be seen here. These powerful figures have been variously called gods or guardians; the military dress and attitude suggest the latter and would be in keeping with the Hittites' martial character. Fourteenth century BC.

into a director general's department controlling all the museums, monuments and excavations.

For the last few decades Lebanese archaeology, following the same course as the rest of Western Asia, has been pushing back the boundaries of chronology. Nowadays many investigations, such as Ksar Akkil, Abu Halka and Rasdel Keb, are being pursued in the interests of prehistory, while the Neolithic levels of Byblos have revealed that this town had an unusually long life span, lasting from this remote period to the present day.

Israel and Jordan

In spite of various interludes, Palestine has shared the fate of Iraq, Syria and Lebanon since 1517. It was no more than a province of the Ottoman empire until it was placed under a British mandate in 1918. The Zionist movement, which has been asserting itself with growing strength since the eighteenth century, helped to bring about the partition of Palestine between the two states of Israel and Jordan. The division was accepted by the United Nations in 1947, and was implemented on 11 May 1948 with the proclamation of the independent state of Israel.

This political division does not apply to the progress of archaeology or the earlier events which it serves to reconstruct, since these take place without regard for frontiers and apply broadly to Palestine as a whole. On the other hand, the division has been obvious in both the spirit and the motivation of archaeology since 1948.

In Israel, where one of the first enactments of the civic and military powers was to provide for the protection of ancient sites and to organise further investigations, archaeology has become a leading preoccupation. Many skilled national teams are working on it, and even on a world-wide scale, their contribution is remarkable.

In Jordan, on the other hand, archaeology has not yet reached a national level from the scholastic point of view. Various foreign teams work there under the patronage of the Director General of Antiquities department, in an atmosphere nearer to that which used to predominate in Western Asia between the wars than to that which currently exists in the countries just reviewed.

The traditional boundaries of Palestine are the Lebanese mountains in the north, the Mediterranean in the west, the Arabian desert in the east and the Sinai peninsula in the south. Within these limits the relief stretches longitudinally from north to south, as in Syria and Lebanon, and defines four regions in succession (littoral, mountain massifs, the Jordan valley and the Dead Sea, and the Jordanian highland). After the end of the first Israeli-Arab war in 1949 this area was cut in two by an artificial frontier giving Israel a coastal fringe along the Mediterranean extending more or less, in the words of the scriptures, 'from Dan to Beersheba' (I Sam. III, 20), which was continued in the south by the Negev desert as far as the Gulf of Eilat on the Red Sea. Jordan, thus deprived of access to the sea, occupied the rest of the traditional land of Palestine until 1967.

The most frequently visited part of Western Asia since the earliest times is Palestine. The holy places there have always been a magnet for the three great monotheistic religions. From the first millennium AD onwards pilgrims and travellers have made their way there, and some of them have told the story of their travels. But in this country, as in all the others, there was no interest in archaeology before the eighteenth century. Here, too, only the visible classical remains which had never been lost (e.g. Jerash and Caesarea) roused an interest at first. Ernest Renan recognised only large-scale structures at Jerusalem and Hebron in 1860–61, and exclaimed 'Everywhere else there is a remarkable dearth of archaeology'. The American, Edward Robinson, undertook a survey of the area in 1838 with some new ideas in mind. He reported that a real investigation of Palestine would require the abandoning of the traditional routes followed by generations of monks and pilgrims, and that there was much to be learned from the clues furnished by placenames, which often preserved memories of towns mentioned in the Old Testament. These judicious methods enabled Robinson to find ancient habitation sites and to identify some of them correctly with Biblical places. Preclassical Palestinian archaeology was launched.

The value of the Bible as a historical document became obvious. It was confirmed by the discovery in 1868 of the stele of King Mesa of Moab (now in the Louvre) bearing a long inscription in a language similar to Hebrew and a script similar to the ancient Phoenician texts. It commemorates a passage of arms between Mesa and Israel which is described in the Old Testament (II Kings III, 4–27).

An expedition organised by the Palestine Exploration Fund, which was formed in England in 1865, continued the investigations so well begun by Robinson. A little later the excavation conducted by Flinders Petrie at Tall el Hesi played a preliminary part in establishing the principles which were to be the mainstay of archaeology in Western Asia: the importance of the stratigraphy of all the debris accumulated by human occupation, and the chronological importance of pottery sequences. These advances were unfortunately very isolated, since most of the results obtained by other projects before 1914 were not so scholastically reliable. Reference to the excavations at Gezer, Taanak, Megiddo, Jericho and Samaria, and the foundation of the American, German and Dominican School of Archaeology in Jerusalem by the Reverend Father Lagrange in 1900–01 is enough to indicate the weakness of the discipline, which could be classed as a renewal of

The Hittite sanctuary at Yazilikaya lies in a natural amphitheatre between two rocks. Figures in procession were carved on the walls and an extensive complex of temple buildings constructed in front of the shrine.

Kultepe in Anatolia was the home of a flourishing colony of merchants in the great days of the Assyrian empire, but its antecedents go back far earlier. A geometrically patterned vase in the Ankara Museum dates to the eighteenth century BC.

Biblical studies during this initial period.

At first the British mandate aroused a greater interest in archaeology in Palestine than in any of the neighbouring countries, with the establishment of a department of antiquities. From 1921 onwards the Americans, the British and then the French, who opened the École Française in Jerusalem in 1922, set to work on unexplored sites (Beisan, Bethel, Sichem, Lachish, Ay, Teleilat Ghassul, etc.) or reopened the big sites where work had been started before the war (Megiddo, Samaria and Jericho). Another immense project was started in 1932–33, and not completed until 1965; this was an archaeological survey of Transjordan, the Jordan valley and the Negev.

Except for Teleilat Ghassul, the names quoted above are those of ancient cities as they appear in the Old Testament, as identified by experts; the excavations conducted since 1920 therefore confirm the historical validity of the Bible. Buildings, objects and ancient Hebrew texts were also seen to be full of information on various periods of history as well as the contacts between Palestine and other countries. The excellent 1932–33 publication of the American campaign at Tell Beit Mirsim perfected the pottery sequences and established the basis of Palestinian chronology. At the same time the abundance of prehistoric remains in Palestine was recognised, and the study of prehistory took shape there more rapidly than anywhere else in Western Asia. After 1925, for instance, Galilee man and Mount Carmel man were found

The city of Susa was the capital of several successive empires. The brick relief (1170–1151 BC) of deities, now in the Louvre, is the work of the Elamites, who were overthrown late in the second millennium.

there; these are remains of human skeletons which can be attributed to the Middle Paleolithic. They were followed by the discovery of cultures dating to the Late Paleolithic (chiefly on Mount Carmel and in the caves of El Wad, El Tabun and Kebara), and to the Mesolithic, which was traced in the cave of Shukbah in the Wadi en Natuf.

The events of 1948, as we have seen, partitioned the archaeological sites on either side of a line of demarcation which has been much contested, and has neither geographical, cultural, nor historical significance.

The expansion of archaeology in Israel since 1948 is due to three leading bodies (the Antiquities Service, the Hebrew University and the Israel Exploration Society) and to a skilled team of national scholars. Little by little foreign teams have returned to collaborate with them.

The prolixity of archaeological work in Israel is such that it is even more difficult than before to pick out any specific sites for mention. The excavators are competing to find the location of towns known to the ancient world whose names appear in the Old Testament, like Hazor, Ashdod and Arad, and important complexes whose ancient names are still unknown, like Tell el Qassileh. They have reopened work on mounds which were already partially excavated like Megiddo and Lachish, and added new elements, and sometimes improvements and corrections to earlier results. They have made substantial contributions to the accuracy of Palestinian prehistory, while establishing new landmarks for the academic world: the Chalcolithic period (i.e. the time when stone tools were used side by side with copper) at Beersheba (Abur Matar and Safadi), the Neolithic at Sha'ar Hagolan and Munhatta, and the Mesolithic at Nallaha. One of the most remarkable of these discoveries took place several years ago to the south of Lake Tiberias at Ubeidiyeh. Some poor human remains there were associated with rudimentary tools made of flaked pebbles (pebble culture), which point to the Lower Paleolithic period. The Ubeidiyeh deposit is the earliest evidence known to date of the presence of man in Western Asia.

In Jordan, which was created in 1947, the organisation of archaeological research is not radically different, as mentioned above, from conditions under the British mandate. Foreign teams continue to work there. Towns mentioned in the Old Testament, like Gibeon, were brought to light, as well as complexes which are known only by their modern names, like Tell Far'ah and Tell es Saidiyeh. Some of the cities such as Ay, Bethel and Taanak, which had already been thoroughly investigated, have judiciously and advantageously been reopened here too. Projects like Jericho, Seyl Aqlat and Beidha have also improved our knowledge of the prehistoric period.

Results obtained in Jordan are thus broadly similar to those from Israel. Each complements the other, to the advancement of knowledge of the various eras of ancient Palestine. They join together in seeking to ensure that not a single

A white marble female head from Warka (ancient Uruk) dates from the late fourth millennium BC. The hair was once gilded and the eyes and brows emphasised by coloured inlays.

The stones of Byblos, the ancient Phoenician city and seaport, which was a centre of communication with Egypt.

The sarcophagus of Ahiram, king of Byblos in Phoenicia, bears the longest known inscription in early Phoenician script. Thirteenth century BC.

Opposite page: details of the elaborate sculpture which adorned the massive mortuary temple of Queen Hatshepshut at Deir el Bahri. In the illustration the figure of Osiris stands on the upper level; the woman's head which forms the capital of a column is not the queen's but a representation of the goddess Hathor.

Overleaf, left: one of the fine ivories found at Fort Shalmanezer and now in the Ashmolean Museum. A man before an altar, carrying a lotus plant. Right: the Assyrian relief carvings are among the finest works of art of the ancient world. They give a vivid impression of the likeness of the warrior kings and remarkable detail in their dress and arms. This detail is from the series from Nineveh, now in the British Museum, describing the progress of a royal hunt.

phase of Palestinian history remains in doubt, be it prehistory, the period of Egyptian rule, or the Israelite periods. In the field of chronology, now more accurate, attempts are currently being made to produce more points of reference derived from carbon 14, with the frequent additional advantage of links with Egyptian history.

Archaeological work today

The foregoing brief account of the development of research in Western Asia shows that each country has its own rhythm. It is important to bear this in mind in order to understand the environments in which field archaeology is evolving. The current position of these studies will now be described, but first a few lines must be devoted to the innumerable results obtained since 1818 when C. J. Rich arrived in Baghdad.

The monuments and objects which have been dug up and classified, the languages which have been deciphered and studied, and the historical periods which have been reconstructed are now united to form the enormous body of knowledge based on archaeological excavation. It is possible to give only two examples since there are so many: first there is slow reconstruction, with the help of 'silent evidence', of the cultural zones of painted pottery as they intersect and succeed each other in the chronology of Western Asia; then there is reconstruction, with the help of written evidence (usually on clay tablets), of the whole range of cuneiform, from the library of Assurbanipal from Nineveh, the palace archives from Mari on the middle Euphrates, the palace of Ugarit on the Mediterranean coast, Boghazköy in Anatolia, and even from Tell el Amarna in Egypt.

These major discoveries revealed that cuneiform script was used throughout Western Asia, that in the second millennium the Akkadian language became the international medium for the whole of Western Asia, and that great scholarly and literary Akkadian texts were read, studied and translated. 'The intellectual superiority of Mesopotamia', wrote René Labat, 'was universally acknowledged at that time. Its culture, to some extent, expanded spontaneously. As soon as neighbouring countries became civilised they naturally turned their eyes thither, towards the treasure house of knowledge they sought to share.' The intellectual reputation of Mesopotamia can be compared in many respects to that of Greece at a later date.

This enormous accumulation of knowledge in such a short span of years has not yet come to an end, since one of the typical characteristics of the third phase as it is currently developing is the perpetual increase in the number of

The excavation at Jericho, which was continuously occupied for many centuries, presented an unusual problem. The successive occupants had all left traces which had to be identified by archaeologists working in deep trenches.

excavations. Another consequence of this is the increasing tendency towards specialisation among linguists and archaeologists.

The latter's position with regard to this excess of material and specialisation [illegible] easily defined. One of the archaeologist's duti[illegible] keep abreast of all the results obtained by his colleagues as published in preliminary reports in increasingly numerous periodicals throughout the world, although he sometimes has to wait for a very long time for the definitive publication. Nor is it easier to work from records discovered at an earlier date and kept in the museums of the Old and New World, since these institutions have so far published only a fraction of their collections. He cannot always date his finds without difficulty since chronology and terminology have not been standardised everywhere, and he must weigh one against another before making his choice, or else add to the confusion by compiling new catalogues in a nomenclature of his own.

By and large, Western Asiatic archaeology has two conflicting aspects in this third phase of its development. The discipline is very young in years, and very vital because of its increasing and remarkable progress through the use of data furnished by the exact sciences, modern mechanisation, excavating techniques and methods of preserving finds. On the other hand, it has some unalterable conventions which

The ruins of Megiddo (Biblical Armageddon) are situated on the north side of the Carmel ridge in Israel. One of the most important hoards of ancient ivories was discovered upon this site.

Excavations proceeding at the site of Safadi in Israel, near Beersheba. The work was begun by the French Archaeological Mission in Israel in 1957, and it follows the methods of Sir Mortimer Wheeler.

Opposite page: the hypostyle hall of the temple of Amon at Karnak, skilfully lit for viewing at night. It gives a good impression of the atmosphere of this mighty structure as it might have been during nocturnal religious ceremonies.

One of the superb Assyrian reliefs from Nimrud. Mounted archers and war chariots express the apparently invincible armed might of the great empire which dominated Western Asia for so many centuries.

hamper the archaeologist and cause him a great many problems.

The speed with which the excavation campaigns produce new discoveries is not, therefore, matched by that of the definitive publication of the results. This backlog accumulates because the information can no longer be mastered by one person, as before, but requires the team work which is very rarely seen in practice. Western Asian archaeology is therefore suffering from suffocation. There will have to be reforms if knowledge is really to be advanced, chiefly in the matter of the publication of collections of both recently unearthed finds and of those already housed in museums.

A transformation would then be possible, so full use could be made of results acquired at a time like the present, when projects are properly planned and statistics, laws, and the natural and exact sciences can all be combined to help the archaeologist.

The conditions of archaeological work

Practical problems of the utmost importance face the field archaeologist, whatever his speciality may be. These may have been dealt with in the first part of this book. Three of them will be mentioned here concerning Western Asia, since they will supply specific examples and illustrate some of the current theories. How is an archaeological site selected for excavation? How is the site excavated? How are certain buildings and objects preserved after they are discovered?

In following these lines, however, we must never lose sight of the fact that these examples are only a selection. Indeed, they can only furnish an inevitably limited picture of results which are as numerous and varied as the land which is being discussed is vast and diverse.

During the nineteenth century and up to 1914 the sites which attracted most attention were, as we have already seen, those which were still visible above ground, like Persepolis, or invested with an aura of prestige by oral tradition, like the capitals of the Assyrian kings of the first millennium BC. As to the archaeological enterprises inspired by written tradition, the saga of Heinrich Schliemann in his quest for Homeric Troy is still the most romantic example.

Sometimes, too, interest was roused by the impressive nature of the remains. At Susa, the tell of the 'Citadel' dominated its neighbours, rising to a height of thirty eight metres above the plain; from the very first this imposing mound of earth caught the interest of W. K. Loftus in 1850, and later tempted Marcel Dieulafoy and Jacques de Morgan to start excavatio[illegible] Other artificial mounds [illegible]gh less imposing, none the less bore witne[illegible] former inhabitation. Soon after his arrival Ernest de Sarzec, who was appointed French vice-consul at Basrah in 1877, set out to survey a region of Lower Mesopotamia which was still regarded as dangerous because of the 'lawless nature of the Montefik Arabs'. His attention was attracted by tells recognisable by 'numerous remains strewn about the ground, potsherds, c[illegible]nes and inscribed bricks, and broken carvings – a completely intact field for excavation which could not fail to yield good results'. He added, 'On my first circuit on horseback I met with a

A relief from the palace of Assurnasirpal at Nimrud (ninth century BC) shows attackers swimming across a river to assault a walled city. The two swimmers in the foreground are clinging to inflated skins as the inhabitants of the region still do today.

magnificent fragment of a colossal statue with an inscription on its shoulder lying on the ground at the foot of the principal mound . . . The investigation of Tello followed. It was destined partially to restore a city inhabited by Sumerians during the third millennium. The inscribed fragment noticed by Sarzec at the beginning heralded the discovery of a whole series of statues of Gudea, ruler of Lagash (the ancient name of Tello). Several of these sculptures are now in the Louvre.

Some ancient cities lay hidden and were only discovered by accident. This is how two of these chance finds were described by the archaeologists who later excavated them. In March 1928, Claude F. A. Schaeffer related that 'a native working on his field not far from the sandy bank of the creek of Minet el Beida lifted a slab covering an underground passage leading to a corbel vaulted burial chamber'. The Antiquities service was alerted, and a year later an expedition accompanied by seven baggage camels settled in the district on the Mediterranean coast. It was to learn that Minet el Beida was once the port of Ugarit, an ancient city rich in wonders. Forty years later the same expedition was still at work there.

In August 1933, André Parrot's great work began when, on the middle Euphrates, 'some Arabs conducting a funeral unearthed on the tell of Hariri . . . a large stone which proved to be a mutilated statue. They informed Lieutenant F. Cabane, an officer of the special branch, who placed the object in safe-keeping and made his report'. An expedition dispatched to the place soon discovered that the earth they were digging had for centuries hidden Mari, an old town no less rich and celebrated in the ancient world.

Sometimes archaeologists find that it is not a matter of discovering a site, but of detecting the provenance of clandestine finds sold on the antiquities market. In Turkey ever since 1880 attempts had been made to trace the source near Kayseri of the tablets inscribed by the Assyrian colonists of the second millennium. A promising tell at Kültepe was investigated without result. Not a single tablet was found. The archaeologists' labours were rewarded in 1925 when the leader of a Czech expedition, who spoke fluent Turkish, was able to discover by means of local informers what had been concealed until then. In fact, the tablets came not from the tell but from the meadow at its foot.

With the reopening of earlier projects, the work on promising new tells, and chance finds, archaeologists sometimes settle on sites in what are to some extent isolated areas. If their selection was confined to this, the second phase of research would hardly differ materially from the first. But this is not the case, as can be seen by a new experiment which also began with some clandestine excavations.

In 1930 the Director of Antiquities was alerted by the appearance of a number of stone statuettes in the antique shops of Baghdad: these were attributed to Khafaja, a site in the Diyala valley. A team of American scholars, among whom Henri Frankfort must be singled out for mention, then acquired permission to excavate Khafaja and a group of neighbouring tells, Tell Asmar, Tell Agrab and Ischali. This was a good beginning. It was to show that investigations on an isolated mound were never completely satisfactory, since they separated ancient towns from their environment. Study of the Diyala sites demonstrates the importance of the regional context. Four cities were thus assigned to their overall setting, dating from the end of the fourth to the beginning of the second millennium BC. Houses, palaces, temples with many superimposed levels, stone statuettes of male and female votaries, terracotta articles and cylinder seals acquired a fuller meaning by virtue of this background. These results made it possible to establish such a sound chronological sequence that from then onwards it has been regularly used as a cross reference in scholastic work.

A survey of the plain of Amuq and the excavations subsequently conducted on some of the chief sites found there confirms the leading part which can be played by work on a whole regional complex. There, too, Robert J. Braidwood and the other experts of the Oriental Institute of Chicago have obtained a chrono

One of the massive winged man-bull figures which guarded the gateway of King Assurnasirpal's palace at Nimrud shows the curious 'double aspect' convention of perspective. A fifth leg has been added so that the figure can be viewed both frontally and from the side.

King Assurnasirpal II, who moved the capital of the Assyrian empire from Assur to Nimrud, is shown in a portrait statue found in the temple of Ninurta at Nimrud.

logical sequence divided into ten parts extending from 6000 to 2000 BC. Like the Diyala table, this has been extremely useful, and has *been in constant use, not only for Amuq, but* for the neighbouring regions.

When archaeological work started again in Western Asia after 1945, i.e. in the third phase, the time of the pursuit of large buildings and valuable objects was over. The importance of systematic clearing of the strata and detailed observation of the context was an accepted fact of prehistoric as well as historic archaeology.

The method of selecting a site has not, on the whole, changed very much since 1945. The various criteria enumerated above which dictate the choice are still valid. However, some new factors have appeared, and tendencies can be discerned, one of which at least opens up new approaches.

Thus accident plays an increasing part as a result of the modernisation going on throughout Western Asia. Whole regions threatened by the construction of dams (projects of this kind in Iraq have led to 'archaeological salvage operations' like that of Shemshara, which can sometimes be productive) and the building of roads, airfields, and other amenities often chance to bring to light finds which the experts have to examine in a great hurry. The new nations thus demonstrate the need for registration of their *cultural heritage,* and they *are now trying to* make a record of most of the ancient remains by means of surveying and aerial photography.

On the other hand, the modern surveying methods which other countries are applying so successfully to find ancient remains hidden *under the earth have hardly been used at all in* Western Asia. In this respect the experiment tried at Gordium in Turkey, the capital of ancient Phrygia, is still exceptional. In fact, a very specific problem faced the American team led by R. S. Young, who directed this project from 1956.

The Phrygians settled in Anatolia at the beginning of the first millennium BC following the great migrations which marked the end of the second millennium in Western Asia. They *settled (at Gordium and Boghazköy, for instance)* on sites previously occupied by the Hittites. These people brought with them burial customs requiring the raising of great tombs in the form of artificial mounds, or tumuli, many of which stand round Gordium. A tumulus higher than all the others aroused the curiosity of the archaeologists. Did it conceal a royal tomb? Given its impressive proportions (it was about fifty metres high), to find the answer by means of an ordinary excavation would have needed years and years of thankless, perhaps fruitless, digging, which they hesitated to undertake. An oil company came to the archaeologists' rescue by putting a sounder at their disposal. After taking many soundings, the instrument finally *located and defined a layer of stones lying about* thirty-nine metres below the summit of the hill. This was the upper part of a tomb.

The ceremonial entry to the city of Babylon the Ishtar Gate, was flanked by brick walls on which alternate rows of bulls and dragons were moulded in relief. The gate dates to the reign of Nebuchadnezzar (late seventh century BC).

Working this time from definite facts, the excavators were able to dig a trench for 140 metres along the base of the tumulus, and then a tunnel which led them to the interior of the wooden burial chamber, probably the last resting place of a Phrygian king. The mortal remains of this personage were surrounded by rich grave goods which were intact. Two inlaid wooden chests were filled with offerings. Nine three-legged tables, the fragments of which were collected, had been set up to hold nearly 100 bronze vessels, some of which showed strong Assyrian influences. Some large bronze vases, with handles decorated with bulls' heads and winged beings, stood on tripods of the same metal. They belonged to a style and technique very common in the kingdom of Urartu at this time (in the Lake Van region, also in Turkey), and favoured as far afield as Greece and Etruria.

However interesting this experiment at Gordium, it has scarcely ever been repeated in Western Asia where, now as at other times, in spite of the increase in the number of excavations, there is more work to do than experts to do it. It will be the same for many years to come. The enormous area of Western Asia is a veritable treasure house of archaeological remains, and the archaeologist is free to choose his sphere of activity either intuitively or by some system.

Systematic criteria have already inspired an experiment to which the name of the American scholar Robert J. Braidwood is attached; it is well worth describing. At the beginning of the proceedings a theory was formulated in the quiet of the research libraries rather than a tell chosen for any of the reasons which have been mentioned. A group of scholars discussed the state of knowledge of prehistory, protohistory and the early historic period. A little-known field which needed elucidation was selected, and in 1945 they were able to draw up a 'gap chart'. The Chicago Oriental Institute's project took shape. It was not a matter of concentrating on ancient towns, architecture or objects but a general problem touching on the evolution of human behaviour. A study theme was then selected: they were to set about the investigation of the transitional period separating the Pleistocene hunters from the first agriculturalists after their settlement in villages in Western Asia. In the field, the project called for the selection of small settlements close to the surface of the ground.

During their surveys the Department of Antiquities of Iraq had formerly located certain sites of this nature. Their work at Hassuna in 1942-44 showed the importance of Iraqi Kurdistan in prehistoric and protohistoric studies (this village, which dated to the Late Neolithic period, yielded two little 'farmhouses' consisting of mud-built rooms opening onto a court). When the project was implemented in 1947–48 advantage was taken of the results obtained by the Antiquities Service, and they therefore began with Matarrah and Jarmo on the flank of the Zagros foothills. They subsequently continued on other sites in the same geographical environment until 1955.

After 1958 the political situation in this part of Iraq brought the work to a halt, but it was possible to continue elsewhere, since it was confined to a geoclimatic area bounded on either side chiefly by the political frontiers between Iraq, Iran and Turkey. Having abandoned Iraqi Kurdistan for Iranian Kurdistan, the team investigated the valleys around Kermanshah in 1960, as well as Tepe Asiab, Tepe Sarab and

Among the Nimrud reliefs were the muscular figures of a prince or tributary and one of his followers, with two realistically portrayed monkeys, making submission to Assurnasirpal.

Ali Kosh in Khuzistan. In 1963 the scholars of the team departed for Turkey, where they began by conducting a survey of the provinces of Siirt, Diyarbakir and Urfa before settling on the small site of Çayönü.

We know that mankind was a hunter and food gatherer in the Paleolithic and most of the Mesolithic periods, and that after this he gradually learned to produce his own means of subsistence and to raise animals which lived close to him. The Australian scholar Gordon Childe called this transformation the 'Neolithic revolution', and claimed that it started in the oases and alluvial lowlands of the great Western Asian rivers.

The results of these new investigations confirmed that the original centre of the movement was Western Asia, but they also suggested other explanations, and led to spectacular advances of knowledge in several other branches of study, for the archaeologist does not work in a vacuum. Investigation of the physical environment in which the Neolithic revolution took place having been judged indispensable, experts from the natural sciences were called in (zoology, botany, meteorology, etc.), and they, in their turn, examined the beginnings of plant and animal domestication.

This collaboration was so successful that a wave of emulation in the academic world followed it. In 1950 a French enterprise, launched by the archaeologist Jean Parrot and based on the principle of scholarly co-operation inaugurated by the Oriental Institute of Chicago, obtained also excellent results in Palestine and Syria. Elsewhere excavations on Neolithic sites have increased in number, and extended from Turkestan and Western Pakistan to the Aegean and the Balkans.

The extensive investigation of what R. J. Braidwood calls the 'agricultural revolution' has

One of the relief-moulded lions from the processional way inside the Ishtar Gate of Babylon shows how the details were carried out in glazed brick. The colours are brown, white and yellow on a dark blue background.

contributed much new knowledge as well as revealing problems and hitherto unsuspected gaps. Apart from these limitations, we now have a good general idea of this extremely important period. The domestication of certain species of flora and fauna probably did not take place in the region of the oases and the great alluvial plains. It took place in the eventful lands bordering the arc formed by the mountain ranges dominating the 'fertile crescent' from Iran to Palestine and above them in Anatolia; i.e. in the regions which form the natural habitat of wild grain (wheat and barley) and domesticable animals. The range of stone tools (sickles, mortars, mills and pestles) found during these excavations shows that edible grain was probably used before mankind had reached the stage of practising deliberate sowing.

Villages like Jarmo in Iraq or Tepe Sarab in Iran illustrate the transition to agriculture and stockbreeding, which took place about 6500 BC. At Jarmo cultivated wheat and barley were found in a carbonised form in hearths as well as the impressions of whole ears in clay. The goat, the dog and the sheep were perhaps domesticated in the same period.

Everything from these proto-Neolithic and Neolithic horizons points to the approach of a hitherto unknown way of life, which is reflected by a considerable acceleration of material culture within the framework of domestic life. Although the details of this culture show different facets at times and places, it can be said that mankind, now partially or completely sedentary, lived in dwellings of various forms built of stone, mud or brick, equipped with hearths, ovens and silos for storing the harvest. He owned a quantity of tools and weapons, often made of obsidian from Anatolia; he used vessels of polished clay and wood as well as wicker work and he modelled clay figurines, chiefly animals and female figures. He also practised new techniques like weaving and pottery. Recent discoveries at Çayönü and Çatal Hüyük in Turkey and Tell es Sawwan in Iraq indicate, in addition, that metal was known earlier than used to be thought, and worked in the techniques of real metallurgy from that time onwards.

The complexes discovered by the archaeologists range from 'towns' to villages and hamlets. The oldest known urban complex was situated near the Dead Sea in an oasis 250 metres below sea level. This is Jericho, a Neolithic 'town' dating to about 7000 BC. It superseded an open settlement with round dwellings. Jericho, according to an estimate by its excavator Kathleen Kenyon, must have numbered some 2000 inhabitants, and was enclosed by a fortification consisting of enormous walls built of stone blocks, sometimes impressively large. This rampart was perhaps reinforced by several round towers, only one of which has been recovered. It had an internal staircase. The defensive wall itself was preceded by a deep man-made ditch. In the town, where some important buildings with rectangular ground-plans have been cleared, structures intended for storing water or food were built.

It is impossible to pass over this period at Jericho without mention of a very striking discovery. This was a group of skulls coated with clay moulded into a likeness of the deceased, the eyes rendered by inlaid shells. The appearance of similar skulls during the excavations now being conducted at Tell Ramad (Syria) indicates that this custom was not confined to Jericho.

The very advanced architecture and the often profound development of Jericho in this period sets a major problem for the archaeologist, who has to ask himself the reason for this remarkable expansion. In fact, the explanation is probably nothing to do with the agricultural revolution. It is true that they had reached the stage of utilising wild grain, since mortars, mills, etc. have been found, but probably not that of agriculture properly speaking. In addition, the Jericho enclosed within the protection of its walls in the period known as 'pre pottery B', did not yet practise pottery making. This early phase of the Neolithic period lasted a long time, since certain parts of the Jericho excavation showed twenty six superimposed occupation levels. This phenomenon is even more difficult to understand because this is the only example of it. However, it is to be hoped that excavations on other sites of the same period will result in a more thorough analysis of Neolithic development, so that Jericho can be assigned to a better known context.

In this respect the discoveries recently made or currently in progress in Turkey are extremely interesting. Çatal Hüyük, for instance, is a town of the latest Neolithic period partially cleared by James Mellaart. Although it was not sheltered by thick walls like Jericho, Çatal Hüyük was protected also from external dangers, since it looked like a sort of compact nucleus, presenting nothing but blind walls to new arrivals.

Its rectangular houses, set close together, had neither doors nor windows. Entrance was effected via the roof which was equipped with a trapdoor leading to a ladder for climbing down to the interior. The architecture, of both bricks and wood, as well as the arrangement of the rooms inside the dwelling was already stereotyped. These rooms were provided with hearths, ovens, and clay platforms for beds, under which the dead were buried. Some of these dwellings (or perhaps shrines) were decorated with an abundance of paintings, painted reliefs (female figures and animals' heads) and bulls' horns set in benches in an arrangement which still remains enigmatical. Fragments of textiles and figurines of clay or carved stone completed this architectural and decorative complex, and show a cultural level hitherto unsuspected in the Neolithic period. Several other details show that some of the artistic traditions of the Upper Paleolithic (paintings and reliefs) were not altogether forgotten at this time.

Twelve occupation levels have been found at Çatal Hüyük, dating, according to the excavator, from around 6700 BC (level XII) to 5700 BC (level I). Traces of copper and lead have been found after level IX.

Çatal Hüyük is a spectacular addition to the progress of archaeological research, and provides much information. Like Jericho, it is an isolated phenomenon. The picture of the type of village inhabited by the Neolithic farmer in the seventh to sixth millennium in Western Asia is better exemplified by Jarmo, a modest group of some twenty-five houses, and by the simple objects found there, which reflect the beginnings of an agricultural economy.

Future problems This brief review can only give a faint idea of the mass of results obtained since 1948. These sometimes throw the expert off the scent and set problems for him. To begin with, for instance, it used to be thought that the remarkable progress of Neolithic architecture was closely linked to the abandonment of nomadic life which necessarily accompanied the domestication of plants and animals. This statement is now suspect since it has been discovered, as Jean Parrot remarked, 'that the development of architecture is not linked to that of methods of food production, as it was believed'.

Questions are also being asked nowadays about the conditions in which the appearance of the new technique of ceramic working took place. It does not seem to be connected with the development of agriculture and stock breeding throughout the whole of Western Asia, nor with the changes in diet resulting from this, since according to the present state of knowledge, its origins are later than the transition

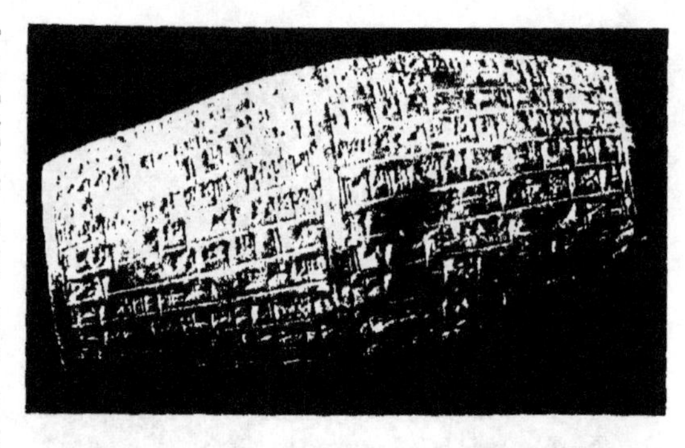

A small clay cylinder from Babylon, dating from the reign of King Nebuchadnezzar II, is inscribed with a cuneiform invocation in honour of the god Lugalmaradde.

A terracotta lioness from Dur Kurigalzh, near the site of ancient Susa which was the Biblical Elam. The workmanship is from the dynasty of Kassite kings who ruled there in the fourteenth century BC.

The famous law-code of Hammurabi was engraved on a stone stele found in his capital city of Susa. The top of this stele shows the king confronting the sun god.

Ideas about the colonisation of the Tigris and Euphrates plain and the development of the agricultural revolution in this region also [illegible] to have changed. It is questionable whether these ev[illegible] took place as slowly as has [illegible] thought since 1948. According to the p[illegible] botanist H. Helback, [illegible] development reached by cultivated [illegible] ains of which have been found at [illegible] implies the existence of important ea[illegible] cultures, as yet unde[illegible]d in this region. This gap would be due to the fact that any small complex which flourished here about 8000 BC would be buried under six or eight metres of alluvial deposits today.

Finally, the sum of the results obtained in Western Asia, and especially in Turkey, during the last few years, has enabled archaeologists to reach a better understanding of the routes by which the Neolithic revolution spread to Europe.

The general question 'How is a site chosen?' leads us to examine a new experiment which is i[illegible] many respects a model. It has led to spectacular advances, not only in archaeology, but in many other disciplines such as botany, zoology, etc. It has opened hitherto unsuspected avenues. It has brought about a transformation in methods which should play an essential part in the advance of Western Asian archaeology. This is proved by the fact that it has already inspired enterprises based on similar principles. In practice, archaeologists and research organisations now realise that the point of departure for an archaeological campaign can be academic reasoning. The problem confronting them is no longer simply the discovery of a site, but the preliminary choice between the empirical and the systematic method of selection, made within the framework of a project based on consideration of the gaps which need to be filled.

Selecting a site Whichever choice is made the selection of a tell can confront the archaeologist with many problems. These may be of a purely practical nature, but the manner in which they are solved can have repercussions, fortunate or unfortunate, on the advance of archaeological knowledge.

Experience has shown that in practice the complete excavation of an extensive ancient town is out of the question. Unlimited time, expert staff, numerous workmen and an uninterrupted supply of copious funds would be needed to achieve this. Since these ideal conditions are very rare the objectives to be attained are inevitably limited. Many sites are abandoned long before they are fully worked out. Moreover, after more than a hundred years of archaeological investigation in Western Asia, we know about religious and civic buildings, but practically nothing of town planning. Homes and streets are documented by a very few limited examples such as the residential quarters at Ur, the Diyala towns and Babylon. Nothing is known either about the contacts between the towns and the surrounding countryside (in this respect R. M. Adam's work in the Diyala valley is a recent happy exception).

If we think over the results obtained by the chief expeditions of the present time, which have been established on the main sites for many years, we must see that there is a grave problem here, and that it is highly relevant to the site and the responsibility of the archaeologist. In spite of exhaustive work for thirty, forty, fifty or more years, are Susa, Uruk, Ugarit and Mari any more than partially investigated?'

Excavating a large town The early second millennium palace at Mari has been excavated and its splendour revealed. It was already famous in its own time. Built of baked or sun-dried brick and measuring 200 metres long by 120 metres wide, 'the palace must have included a minimum of 300 rooms, passages and courtyards', wrote André Parrot, the archaeologist who excavated it. Its interior plan, its great courtyards, its workshops, school, mural paintings, statues, cylinder seals, terracottas and, above all, its abundant archives inscribed on clay tablets, have vividly restored to life the court of Zimrilim, king of Mari and contemporary of Hammurabi the Great, king of Babylon.

All these remarkable finds added to the excavation of this huge complex tell us nothing relevant to the town or the daily life of the subjects of this luxurious monarch. This is why the team, after finishing their work in Zimrilim's palace, turned their attention to some religious buildings to learn more about the town. By a curious paradox, the results of this excavation gradually led the archaeologists back to the palace. Then, during the fourth campaign at Mari in 1964, a vertical sounding taken for the purpose of establishing the stratigraphy was halted almost as soon as it began because it had lighted on another palace, very much earlier, hidden underneath King Zimrilim's residence. A combination of circumstances thus led to the excavation of a new royal residence dating to the Early Dynastic period in the third millennium.

This incident demonstrates the importance of the unknown, as well as the stratigraphical complexities lying in wait for the archaeologist in Western Asia, where successive buildings were erected on the same traditional site for thousands of years. If he starts out to dig a homogeneous urban complex by removing the layers horizontally, the excavator may be compelled, as at Mari, to stay in one place and make a vertical study of the successive occupation levels in one building, so as to retrace its history. It is not one town he has to dig in order to reconstruct the past, but several towns, sometimes covering an extensive area. The expedition then finds itself obliged to change its objectives and limit its views, since to this day, nobody has ever had enough time, experts and manpower to achieve this end. Is, then, the excavation of an ancient Western Asian city beyond the means of any one nation? Perhaps, and so in the future the need for more extensive results may lead to the adoption of a system of international collaboration in similar cases as well as much longer excavation projects.

An expedition's funds are sometimes so limited that the leader has to take drastic steps to reach the important part of the tell he is investigating as soon as possible. An experiment of this nature was made by an archaeologist who has spent very many years in the field in Western Asia. In 1951 Seton Lloyd, then Director of the British Institute of Archaeology at Ankara, carried out a dig on the impressive tell of Sultan Tépé on the Syria-Turkey border, with the co-operation of a Turkish archaeologist. This man-made hill rose about forty-five metres above ground level, and the modern villagers, not wanting to live on this eminence, had settled at the foot of the tell. The dig began, not at the summit, but about thirty metres above ground level, where two granite column bases were still visible. The slopes were attacked at this height in several places by means of vertical trenches, rather as cake is cut, and this revealed an occupation level of the neo-Assyrian period (first half of the first millennium BC). Then a trench starting at the summit revealed to the excavators that this neo-Assyrian level was covered with about fifteen metres of Hellenistic and Roman occupation levels.

The expedition's budget did not allow for the removal of this cap by archaeological methods, i.e. examining each layer in turn, from the summit downwards. The excavators therefore abandoned the tell when they had done no more than to follow the neo-Assyrian level all round it half way up, with its remains of several buildings: they made a miraculous discovery of more than six hundred tablets here. Unlike Nineveh, this was not a library, i.e. a systematic collection like that of the Sargonid kings, especially Assurbanipal, in that city) but was probably the result of the labours of a school of scribes attached to the temple. We know from other finds that students in Western Asia learned by copying other texts of all kinds which were thought to be important. One of those found at Sultan Tépé (it relates the conflicts of the 'poor man of Nippur' with the mayor of the town, on whom he plays a series of practical jokes) is well known because it is particularly entertaining, and has preserved all the flavour of a folk tale. Details of this kind make it all the more regrettable that the excavations of Sultan Tépé have never been continued.

A third example will help to demonstrate the fact that the difficulties described are not confined to the people who tackle the sites of the great ancient cities. They can also affect far more modest projects. When he began to excavate the site of Jarmo in 1948, R. J. Braidwood first carried out two deep cuttings which were continued throughout the two campaigns within the limits of two squares with fifteen-metre sides. At the beginning of the third campaign it was decided to change to horizontal excavation of the Neolithic village. Braidwood then made the following calculations: he would need a team of fifty workmen led by himself and his

Sir Leonard Woolley, one of the most famous men of modern archaeology. In 1924 he began the excavations at Ur which uncovered the oldest royal graves known to man.

The precinct of the temple of E-anna at Uruk was decorated in the Early Sumerian period with thousands of small polychrome cones driven into the damp clay to produce a colourful mosaic effect.

The ziggurat (temple tower) of the moon god Nanna was built by the Third Dynasty King Urnammu of Ur. Its outer face (now restored) is supported by buttresses, and sheets of matting were built into the original brickwork to strengthen it.

colleagues, and *twenty four* excavation campaigns to clear each of the successive occupation levels covering the whole surface of the village of Jarmo.

Within the established limits of the project, such a programme was out of the question. The team leader had to fall back on more modest plans. And then, hoping at least to 'sample' the site systematically, Braidwood devised the following procedure. The whole area was to be squared off into units of five metres wide by six metres long. In the middle of each unit a shaft two metres wide was to be dug. This was done, but the results were deceptive because the occupation levels were not evenly dispersed. As Braidwood wrote, the traditional comparison between 'the strata of a tell and the layers of a cake is an excessive simplification'. This much is obvious, in a town as well as a village, one quarter could be destroyed or abandoned while life went on without interruption in other parts of the same complex. In his 1955 campaign Braidwood succeeded in digging and investigating 151 holes to an average depth of 1.75 metres. Jarmo was then abandoned!

The excavation of a site

Some of the chief obstacles facing the archaeologist who plans to excavate the whole of a large ancient complex or sometimes even a small one have now been described. We are continually reminded that the artificial mounds the tell, tepe or huyuk covering this land consist of debris accumulated during unusually prolonged human occupation, and they are therefore a particularly complex archaeological zone. This leads us to the heart of another current problem, which was stated with admirable clarity by Seton Lloyd in 1963. 'Can these Western Asian mounds be excavated using principles which are applicable in other parts of the world where the terrain is different?' In his book *Mounds of the Near East*, conclusions based on a series of archaeological experiments conducted by himself in Iraq and Turkey, were discussed.

The 'anatomy' of tells and the eccentricities which occur in their formation demonstrate that the strategy of a dig cannot be other than empirical. 'It goes without saying that a talent

The Sumerian city of Uruk, looking south-west from the ziggurat.

A lyre found in one of the royal cemetery tombs at Ur was decorated with a bull's head terminal in gold leaf over a wooden core. The body of the lyre was lavishly inlaid.

for improvisation is an almost indispensable asset to an archaeologist in Western Asia.'

This being so, it may be asked what plan of campaign should be adopted when a dig has to be carried out. According to Seton Lloyd, the whole operation should be regulated from the start by the clearing of ancient walls. In Iraq these are mostly built of unfired brick, i.e. bricks which were simply dried in the sun. Elsewhere in Western Asia, in districts where stone is not so rare, the foundations of walls may be made of this material on which the bricks were then placed, while in Anatolia the brick-work was very often encased in wooden beams. In all these places, however, the dried brick wall, which has more or less disintegrated over the course of the years, is difficult to distinguish from the surrounding earth. The brief history of Western Asian archaeology cannot conceal the fact that there was a time when archaeologists did not recognise dried brick walls, which they demolished in the belief that they were clearing away the accumulated sand of the millennia.

As was shown earlier, it was the German excavators of Babylon and Assur at the beginning of the century who learned by trial and error how to distinguish the traces of brick walls. In their school they trained a team of workmen from the village of Sherqat near Assur, who became famous in the annals of archaeology under the name of the *Sherqati*. These first experts in the recognition and excavation of dried brick walls and the pupils they had taught became a select corps, and were recruited by other teams to whom they could transmit their skills. According to Seton Lloyd, to find the traces of a wall it is first necessary to scrape the surface of the ground vehemently, with a hoe, until either some structure actually made of brick or a line of contrasting colour, indicating the division between the wall and the accumulated earth forming the fill, is seen. The workman then makes a hole very close to the wall, in which he can crouch down. As he clears the wall the hole can be extended along its length. Since one wall often leads to another, the plan of the ancient building can be obtained in this way. The fill, which occupies almost all the space enclosed by the walls, can then be carefully removed, not without first taking care to cut a section to reconstruct the history of the excavated building. The fill can result from the action of the elements or the collapse of the building, but it can also have been formed in antiquity when successive occupants laid one floor on top of another in the same architectural complex. This method, which is known as *wall tracing*, has the great advantage in the eyes of many archaeologists of exposing the ground plan of the ancient buildings fairly rapidly.

A German expedition working at Warka in Lower Mesopotamia made use of a slightly different technique which is still being used today. This consists of proceeding, not by digging vertically down the discovered wall, but by clearing the wall horizontally at ground level on the surface. In this case the shape of every brick is examined with the point of a knife and a brush. All the experts must recall the wonderful plans published by the Warka team, on which not only the walls but the shape of every brick was shown. In this connection, Seton Lloyd has speculated as to whether the detail with which the architectural remains of Warka were repro-

The so-called royal standard of Ur, now in the British Museum, has tiny inlaid figures of shell on a lapis lazuli mosaic background representing peace on one side, war on the other. The 'war' side shows charioteers and foot soldiers in its three registers.

duced has not perhaps tended to 'obscure the broader conclusions to be drawn from them'.

Empirical and British methods As was shown in the chapter on excavation there is another method based on a different principle. It is frequently employed nowadays, and has been explained in several handbooks such as *Beginning in Archaeology* (1952) by Kathleen M. Kenyon who is a specialist on Western Asia, where she has led several expeditions, including one to Jericho. This system is based on a grid of squares covering the area to be investigated. Miss Kenyon wrote: 'The site is excavated in a series of squares separated by baulks which are left standing, and which thus provide keys to the stratification. The size of the squares or the width of the baulks will depend on the depth of the excavation and the type of soil.' When it is a question of exposing a buried building by means of this system, the excavator's first duty is not to separate the remains of walls from their stratigraphic context, that is the 'fill'. On the contrary it is essential to attack the walls at right angles so as to obtain a section establishing the connection between the wall and the afore-mentioned fill. This being so, 'it is a cardinal error', writes Miss Kenyon, 'to dig along the line of a wall. Such a trench cuts through the connection of walls and floors or other deposits, and destroys the evidence.'

The wall-tracing method described earlier is the direct contradiction of this. Seton Lloyd opposed it strongly and argued that the second method, which he called 'academic' and which is also known as the Wheeler method, was developed for use in Britain and is not suitable for Western Asia, where the difficulties presented by the tells can only be counteracted by an empirical approach, flexible and open to improvisation. One of the criticisms sometimes levelled at the Wheeler method is that the recommended procedure is so slow. The excavations of the huge complexes of buildings in Syria and Mesopotamia which were conducted during the first two phases of archaeological research could never have been achieved if they had been carried out as slowly and meticulously as this technique requires.

These are the elements of the obvious problems with which the Western Asian excavator has been confronted since 1963. He cannot escape them, and must therefore choose his method of working the site and then justify his choice so as to contribute to the establishment of a definite rule. He must decide whether the empirical method means chaos and the Wheeler method means discipline. It is up to him to decide if the latter, which has been successfully employed at several places in Western Asia such as Safadi and Ashdod in Palestine, Cayonü in Turkey and Hasanlu in Iran, is only suitable for a few sites. He must choose between quick results, which used to be valuable because they contributed to a new discipline the raw materials on which it grew so rapidly, and the infinitely steady advance of the expert who understands the weight of his responsibilities. 'In spite of everything, the best of excavators is a vandal who destroys his own evidence even while he studies it,' wrote André LeRoy Gourhan.

The preservation of buildings and objects

The destruction necessitated by the demands of archaeological research gives the excavator an

The life-sized hollow-cast bronze head of an Akkadian king from Nineveh has a stately dignity which is rarely seen in secular art of this period. The inlaid eyes, now missing, were probably made of some precious material.

A relief from the palace of Sargon II at Khorsabad shows a man in a sheepskin cape with upraised hands. The Khorsabad reliefs, unlike those of most of the Assyrian capitals, are chiefly representational rather than narrative.

overwhelming feeling of responsibility, and as a result makes him even more anxious to preserve the vestiges he has been able to rescue as far as possible. This raises innumerable problems.

We have already seen that learned travellers since the beginning of the nineteenth century have been struck by the rapidity with which the monuments are disintegrating. Sir Robert Ker Porter, inspecting Babylon in 1821, deplored the disappearance of much that Beauchamp and Rich had seen in 1782 and 1811: 'The appearance of the summit and the sides of the tell is constantly being spoiled by the incessant search for bricks ...' In our own times surveillance of the sites has put an end to this kind of looting, but it cannot guard against the no less formidable action of the elements. Torrential rains, floods, sandstorms and dust attack and cover newly dug walls cleared by archaeologists, particularly in Syria and Iraq.

In this respect a typical mischance happened to the excavators of Tell Abu Shahrain, a very early town in Eridu in southern Mesopotamia, where the climate is particularly extreme. During the first campaign the team built a house of 'Sumerian' bricks. 'When we returned for the second campaign', wrote Seton Lloyd, the house 'was buried to the eaves in dust, and had itself to be excavated before we could reoccupy it!' Even in less hostile regions ancient walls lose height quickly once they are cleared. Travellers going round the excavation sites in Iraq and Syria nowadays are surprised to find that they have difficulty in following the ground plan of the exposed buildings.

The restoration of buildings Anyone who wants to preserve a tangible reminder of excavated cities or buildings must adopt a salvage plan employing effective measures. An invidious choice must then be made. Given the number of buildings already exposed, it is impossible, for financial and practical reasons, to try to preserve them all. The archaeologist must therefore select the ruins which are both the most typical from the historical viewpoint and the most suitable for preservation. The subsequent task of restoration and consolidation is very difficult. It calls for scrupulous accuracy on the part of the archaeologist; it is not a matter of restoring or arranging, but of preserving. The Department of Antiquities of Iraq is currently conducting just such a salvage programme, and in this way parts of the ziggurats of Ur and Aqarquf have been preserved.

The process of reconstruction can also sometimes help the experts to a better understanding of the technical methods used by the ancients. This was what happened at Nimrud, the ancient name of Kalhu, one of the capitals of the Assyrian monarchs, which towers over the river Tigris and backs onto the hills of Iraqi Kurdistan, themselves dominated by more distant, eternally snow-capped peaks. The palace of Assurnasirpal (883–859 BC), facing north-west, stood on the acropolis of Nimrud. It was investigated by Layard (1845–51) in the heroic age of the archaeological reconquest. Many reliefs carved in soft stone were found there. These were once attached to the palace walls to decorate them and to tell the history of the king in pictures, showing his religious functions and giving pride of place to his military campaigns and victories. Layard had some of the reliefs taken to London where they can be seen in the galleries of the British Museum, while he had others buried again in the same place at Nimrud. A hundred years later (1951–52) a new British expedition

Opposite page: a massive relief figure, nearly fourteen feet tall, of the hero Gilgamesh effortlessly overpowering a lion while holding a serpent in his free hand. This relief was one of the guardians of the entrance to the throne room of King Sargon II at Khorsabad.

d by M. E. L. Mallowan unearthed these liberately buried reliefs once again, but their ethods were very different. While Layard had orked by means of tunnels and pits without the least regard for the architecture of the lace, the modern team cleared the building e imposing size of which then became apparent. The throne room and the great monial apartments opened on to a vast measuring twenty-seven metres wide and thirty-two metres long. The monumental gate as well as the corridors were flanked on pre by colossal carvings of winged monsters, benevolent spirits protecting the entrance.

Persia's ancient glory was only temporarily extinguished by Alexander. In the third century AD the Sassanids overthrew the Parthian overlords and founded a dynasty that shook the power of Rome herself. They carved vivid reliefs, like their predecessors, to commemorate their great deeds. Left: King Bahram II rides into battle. Below: Sassanid kings honouring their gods.

Details of the brilliant reliefs which line the stairway leading to the Apadana in the city of Persepolis. Above: a Syrian tribute-bearer. Right: a double line of spearmen and archers. Opposite page: a lion springs on to the back of a bull. The lion was a symbol of royalty.

Thanks to the work of the Mallowan expedition, the facade and the entrance to the throne room have been restored by the Iraqi Antiquities Service. The architectural decorations (the bas-reliefs and the colossi guarding the gates) have been replaced in their original position. These restorations have enabled archaeologists to understand a point which they never thought of until then. The colossi ornamenting the exterior facade could certainly never have stood in the open; if they had, the soft stone of which they are carved would not have survived the elements. To avoid such deterioration at present, the experts in charge of the reconstruction have designed a pent roof. But scholars would like to know what course was adopted by the Assyrian architects of the first millennium BC. Mallowan has suggested a colonnaded portico, but he points out that this notion is not entirely satisfactory since an arrangement like this would have obscured part of the carved decoration.

No less formidable problems are raised by the preservation of records and objects abruptly brought back into contact with the atmosphere after resting in the ground for thousands of years. Such problems frequently arise in archaeology. Some of them, however, are particularly acute in Western Asia. A typical case is that of the clay tablets, so many of which have been found, bearing the writings of the ancient world. Since they are often made of sun-dried clay there is a danger of their crumbling into dust, and they have to be treated very promptly by 'washing' to remove excess salts and then firing, sometimes on the spot and sometimes in the museum. There have been two spectacular successes recently, which are worth describing; one concerns the finds from Çatal Hüyük in Turkey, the other those from Nimrud in Iraq.

Çatal Hüyük The most complex of the preservation problems confronting James Mellaart, the excavator of Çatal Hüyük, were the inevitable consequence of the rarity and abundance of his finds, which included wooden objects, fragments of textiles and mural paintings dating back to the Neolithic period. These treasures, preserved by a miracle, had to be treated on the spot if they were to have any chance at all of surviving. The best course for the wooden objects, according to Mr and Mrs Mellaart, was to keep a thick crust of earth round them when they were dug out; this made it easier to transport them to the dig house, so long as they were then relieved of this protective coating as quickly as possible before it had time to harden. The wood was then kept moist but not wet by constant applications of suitable products. Great vigilance was necessary, however, since wood may rot if it becomes too damp.

The textiles from the tombs at Çatal Hüyük were often mixed with bones and earth. The recommended course of treatment was the opposite of that used for wood. The fragments were first left to dry out, which could take anything up to six weeks; then came the cleaning, followed by treatment and 'unfolding', which was extremely delicate when tissues were wrapped together.

In the case of the wall paintings, they were sometimes recovered from under thick coats of plaster which had to be removed a millimetre at a time using the finest tools. When they were uncovered they were protected from sun and

rain and photographed as they were before any action was attempted. Then several very thin layers of a reinforcing agent were spread over them with a soft brush. The painted area was covered up again with paper, then with several layers of padding applied like a surgical dressing and held in place by pieces of wood set in plaster of Paris. Prepared like this, the painting (on which a background no more than three centimetres thick had been left) could be first detached, then removed from the wall. The back was then treated in the same way as the face, and the piece was ready for removal.

The remarkable success crowning these endeavours emphasises the advances in technical processes used on the spot by expedition leaders. In view of the remarkable salvage of the Dura Europos murals, the time is now gone when wall paintings in Western Asia used to deteriorate and vanish as soon as they were exposed to bright light for lack of adequate treatment.

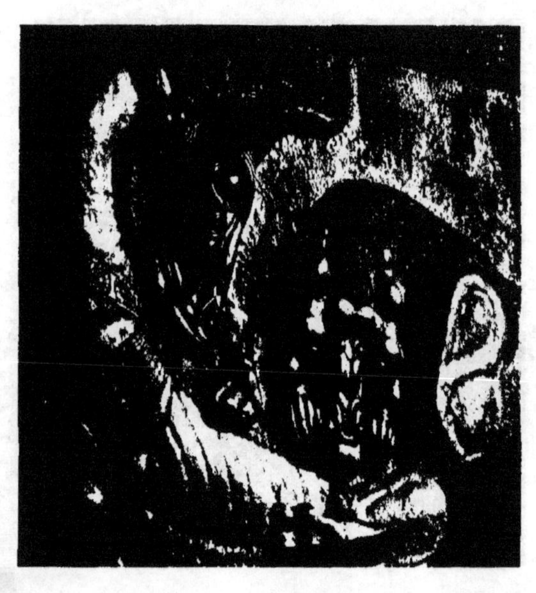

The preservation of ivories The preservation of the ivories found at Nimrud by the British expedition, which renewed work there in 1951, also called for great precautions. The objects, discovered early in the new excavations, owe their fine state of preservation to the fact that they were found at the bottom of a shaft in a damp muddy place. In fact, a deposit of clay had enclosed them and protected them for centuries from mechanical damage and excessive changes in the humidity and temperature. This coating also enabled them to survive intact the ordeal of being restored to the atmosphere. In spite of these favourable conditions, the ivories would not have survived disinterment if appropriate care had not been lavished on them. To begin with, the expedition had not the technical equipment for this kind of salvage operation, and its success was due to the perspicacity of Mrs. Mallowan, wife of the expedition leader, who is known to the world of letters by the name of Agatha Christie. Ivories, like wood, have to be dried out very slowly. Seeing what was needed, Mrs. Mallowan kept them for several weeks under some cloths, which were first wet, then only damp, watching them personally day and night until they were ready to be removed and entrusted to a laboratory.

The very considerable importance of the ivories was subsequently recognised, and the Nimrud expedition was accordingly organised. A specialist was trained in treating them. Since ivory cannot survive even a few minutes' exposure to the dry air of Iraq without the risk of breaking and cracking up, it was found best to treat them even before they were removed from the ground. The cleaning was carried out in such a way that each part was impregnated with the proper solution as soon as it was exposed. The number and splendour of the Nimrud ivories, now in various museums, bear witness to the complete success of this treatment.

Practical and theoretical archaeology

At the beginning of this chapter on Western Asia its aim was explained: to retrace the steps in the development of field research, the archaeologist's fundamental activity. The preservation

Ivory is a highly perishable substance which requires very careful treatment by the excavator. A mirror handle from Nimrud carved with female figures, now in the British Museum, was found in the south-east palace.

of groups of buildings, records and objects after their discovery is absolutely indispensable to this primary function. However, these practical operations are all equally designed to furnish the experts with material for their studies.

No review of archaeology would be complete if it omitted to stress the essential links between practical and theoretic archaeology. Thus the close connections between the new system of field archaeology adopted and put into practice by Braidwood, and knowledge of the 'agricultural revolution' has been already observed. Another example, involving the sophisticated use of a valuable material, will now be described. Like the previous example, it has been selected with a view to demonstrating that the study of certain tangible objects found in widely scattered sites in several different countries can only be valid on the scale of Western Asia as a whole, without regard for political frontiers. Ivory working was practised in Western Asia from a very early date. The inlaying of luxury articles with ivory or shell was also known in Palestine from an early period, as many remarkable pieces from the royal cemetery at Ur have proved. The use of ivory for inlaying was therefore no novelty in Western Asia when it returned to favour in Syria, Lebanon and Anatolia in the second millennium.

The use of this material in the fourteenth century BC at Ras Shamra, ancient Ugarit, on the Syrian coast, is attested by both a text written in the Akkadian language on a clay tablet and by the discoveries of the French team led by C. F. A. Schaeffer. The inventory of the trousseau presented to Queen Ahat-Milku on her marriage to the king of Ugarit mentions, among other treasures, 'three beds inlaid with ivory with a step . . . one ebony chair inlaid with ivory with a step'. Among the ivories found at Ras Shamra (plates, tusks, etc.) were panels carved in low relief which could once have been part of the decoration of these beds. They show the king and queen embracing, the king triumphing over his enemies, and the king suckled by a winged goddess crowned with the cow's horns which in Egypt are an attribute of the goddess Hathor. At first glance the similarity of these subjects with the Egyptian repertoire is striking. A closer examination reveals that these Egyptian motifs have been adapted to a different taste, probably Syrian or Phoenician.

An ivory box lid found in a tomb at Ugarit and attributed to the thirteenth century BC shows a completely different side of the picture. The scene so skilfully carved on it shows a seated woman with a fine profile in the centre, flanked by two goat-like animals standing on their hind legs and trying to nibble the branches she is holding up in her hands. The Mycenaean influence is as obvious here as the Egyptian style was in the motifs previously described. This is hardly surprising, since at that time there were close connections between the Mycenaean world and the Levantine coast. But this statement soon calls for qualification. Although the decoration on the lid is Mycenaean in form, it is not so in feeling. The motif of a central figure holding branches for upright animals to graze on belongs to an iconographic cycle known in the East from the beginning of the third millennium.

A third group of ivories supplements these impressions. It was found in 1938 at Megiddo in Palestine by an American team in a building thought to be the royal residence of the town, where about 300 ivories were heaped up in three rooms. They once decorated furniture, musical instruments, gaming boards, vases, combs etc. The example which concerns us is a rectangular plaque dating to the thirteenth century BC which probably decorated a box. It shows a crouching griffin with outspread wings in low relief. This mythical creature was part of the Minoan repertoire before it was adopted by the Mycenaeans.

Contrary to the two preceding examples, this admirable composition is purely Mycenaean in subject, composition, style and vigour of modelling. It is therefore a very important piece.

The decoration of these three fragments of ivory indicates a trade in luxury goods at the end of the second millennium in the Levantine countries which were exposed to Egyptian and Mycenaean influences. This trade was supplied either by imports (the Megiddo ivory) or by local workshops (the Egyptian-style plaque and the Mycenaean-style box-lid from Ras Shamra). These craftsmen, sedentary or itinerant and already masters of the art of interpreting foreign motifs in their own style, thus developed a composite one which was to appeal to an ever-increasing clientele.

What became of the techniques and motifs of these craftsmen during the events with which the second millennium closed? Population migrations beginning at the end of the thirteenth century brought chaos to many regions of Western Asia, and led to the disappearance of many capital cities which had flourished until then. This period is so poorly documented as to be known as the Dark Ages (1100–900 BC). It represents such a pronounced cultural break that when it was over the political face of Western Asia had completely changed. It is true that the second half of the second millennium had seen the struggle between the Egyptian and Hittite empires in Syria. But in spite of violent conflicts between these two great nations, other kingdoms (their vassals or foes) had managed to preserve their own autonomy. One after another they formed the diplomatic or cultural alliances which makes it justifiable to describe this as a truly international epoch.

In the first millennium everything surviving from all this was ground under the heel of the Assyrian military empire, which attained a height of power and extent hitherto unknown in the ancient world. Because of this, the culture of Western Asia became uniformly Assyrian in

Also from Nimrud, two fragments of an ivory plaque which seem to suggest that the whole was an elaborate piece of work. The remaining portion shows a man examining the fruits of a tree, and there was probably his counterpart facing him in the original plaque.

Nimrud and Fort Shalmanezer

Such a pronounced break and radical change could have put an end to the craft of ivory working. This was not the case. The ninth and eighth centuries BC saw this craft not only renewed but immeasurably enriched. Ivory was used not only to decorate furniture and objects, it probably also decorated luxurious residences. The chief sources of the raw material (Syria and Africa), which had supplied the needs of the ivory workers of the second millennium, were probably inadequate to meet this enormous demand. It seems that in the first millennium they had to call in India too, for tusks. Moreover, this taste was not confined to one region, as a succession of discoveries has shown. In order to trace the history of ivories it is necessary to cross a great many frontiers. There are ivories at Samaria in Palestine, at Arslan Tash in Syria, at Carcemish, Zinjirli, Gordium and the ancient kingdom of Urartu in the Lake Van region of Turkey; at Hasanlu and Ziwiyeh in Iran, and in

The most remarkable of the Nimrud ivories was a plaque depicting an Ethiopian being mauled by a lion. The plaque is inlaid with red and blue and the details picked out in gold leaf.

A carved alabaster vase from Early Sumerian Uruk shows processional scenes. It was found in the ruins of a temple dating to the Proto-literate period, and shows offerings being made to the fertility goddess Innana.

the former Assyrian territories of Khorsabad, Assur and Nimrud in Assyria. Finally, if we wish to follow the ivory trail far afield beyond Western Asia, we come as near to home as the Etruscan tombs.

Among so many finds, those recently made at Nimrud stand out even more remarkably because ivories had already been recovered from this site by Loftus in 1848 and Layard in 1852. These pioneers of the archaeological reconquest certainly never dreamed that the tell had yielded no more than a fraction of its treasures, and that several hundred more important ivories as well as tens of thousands of fragments still lay buried in the mud and sand.

The most numerous and perhaps the most spectacular groups of ivories were discovered in 1957, when a British team set to work on a mound outside the ramparts surrounding Nimrud and found a building flanked with towers, with sun-dried brick walls between 4.2 metres and 3.7 metres thick (the west wall was 290 metres long). These lines of massive towers and the whole impressive defensive array were typical of Assyrian military architecture. Nevertheless, the interior of the building had unusual features, since it was simultaneously reminiscent of a residence (it even included a throne room decorated with wall paintings), a warehouse, an arsenal and a fortress. The British excavators named it Fort Shalmaneser. It was mainly built by Shalmaneser III (859–824 BC). However, it showed traces of several successive occupations, among which were repairs by Adad Nirari III, probable use by Sargon the Great, one destruction in 614 and a final destruction in 612.

Luxurious beds inlaid with ivory panels were once stored in one of the rooms of the fort, along with bronze and iron weapons. The frames of the beds had disappeared, but the ivories survived. The number of panels indicated that there had been about fifteen beds, and very likely more, in view of the number of additional fragments. Several other rooms of the fort also yielded large groups of ivories. The assembling of this hoard of treasures within the protection of thick walls reflects the tragic end of the Assyrian empire under the attacks of the combined forces of the Medes and Persians. The excavators found evidence of several breaches which led them to believe that the fort was taken by storm once in 614, after which an attempt was made to restore order and repair the structure. The rearrangement of the valuable furniture dates from this period. Then in 612, the year of the fall of Nineveh, the final sack of Nimrud reduced Fort Shalmaneser to the condition in which the British team found it after 2500 years of oblivion.

The number and variety of the ivories found in the former Assyrian royal city made it possible to reconstruct many of the technical processes practised by their makers. They worked equally readily in high and low relief. Sculpture in the round is represented at Nimrud by several pieces which are as unusual for the employment of this hitherto rare technique as they are for the subject chosen by the ivory carver. These are statuettes which, set up separately on the same base, formed a procession of male figures with pronounced racial characteristics. Their faces are negroid and their bodies slender. These men move gracefully, carrying monkeys or small lions on one shoulder while they lead various animals such as antelopes, gazelles, etc., holding them either by the horns or on a leash.

Other ivory plaques have fretted decoration. In addition, whatever technique was used, the ivory might be picked out in delicate colours or with gold leaf. Other techniques such as cloisonné or champlevé work embellished the ivory with lapis lazuli, red and green stones or coloured glass paste. There seems to have been no end to the skill and inventiveness of the ivory-workers.

During the first centuries of the first millennium the role of ivory in the art of inlaying became even greater than it had been in the second. Panels of this material probably ornamented doors and walls, enriched beds, chairs, stools, steps, tables, fans, sceptres, gaming boards, combs, etc., and ivory was even used to make the harness of horses more extravagant. Ivory containers held cosmetics and boxes, displayed on tables as collector's pieces (an Assyrian relief probably shows one of these close by King Assurbanipal), held jewellery.

Evidence from Nimrud has shown that every scrap of ivory was scrupulously used up in the workshops. In spite of this economy, the period of the maximum use of ivory, dated by the experts to 850–700 BC, was followed by a period of decline attributed by M. A. L. Mallowan to shortage of the raw material caused by such intensive consumption.

Where were these workshops and who worked in them? The styles of the subjects supply some clues. As in the second millennium, Egyptian motifs were popular, but they were always treated with some modification which betrays the foreign hand; this is known as the Egyptianising style.

Some very different motifs (they belong to the 'composite' category already known in the second millennium, and attributed to the Mediterranean seaboard, which was exposed to a variety of influences), like those previously mentioned, naturally suggest the idea of workshops in transit and trading areas like northern Syria, Phoenicia and Palestine. The Nimrud ivories also help to prove the existence of a purely Assyrian style, faithfully reproducing as they do the subjects and style of the great stone sculptures and bronze groups dedicated to commemorate the monarch's achievements.

It is reasonable to assume the existence not only of well-known workshops supplying a lively luxury trade, but of groups of itinerant craftsmen offering their services to the most splendid courts, and of ivory workers captured by the Assyrian armies and taken to Assyria,

A Phoenician ivory inlay once decorated a piece of furniture in the north-west palace at Nimrud. Dating from the late eighth century BC, the formal lines of the hair show marked Egyptian influence.

where they worked at such centres as Nimrud. Raw materials were found at Fort Shalmaneser which could have been used for ivory ornaments. Sometimes the co-operation of Phoenician and Syrian craftsmen is beyond doubt, since the assembly instruction marked on the back of the ivories resemble Phoenician and Aramaic characters. In the case of the ivories from Arslan Tash, which are very like those of Nimrud (some of them are now in the Louvre), the inscription on one of them is even more explicit. It states that 'this' was made for 'our Lord Hazael'. Hazael was probably king of Damascus in the time of Shalmanezer III.

The prospect in Western Asia

Practical archaeology has both unearthed and preserved monuments and records which have subsequently been exploited by 'synthetic' academic archaeology. This in its turn supplies raw materials to other more specialist branches of study.

This process is illustrated by two examples chosen from the whole of Western Asia. The first, the 'agricultural revolution', goes back to the very remote past, and belongs to prehistory. The second, the manufacture of carved ivories, belongs to two historical periods of brilliant culture.

These examples are necessarily limited (they could have been multiplied, since a list of all the results so far obtained would be practically impossible to compile) but they are fair representatives of the extraordinarily swift development of Western Asiatic archaeology, as well as of some of its current tendencies. However, if questions are asked about this discipline, it will be seen that its future lies in neither the elucidation of previous successes, however tempting, nor in the repetition of previous methods. Nowadays the large number of teams from nations with a traditional interest in Western Asiatic archaeology, as well as teams from nations which are just beginning to be interested, can scarcely be described as methodically organised. In the broader sense, these numerous efforts, so pleasing because of their intrinsic merits, add to the spasmodic and scattered quality of the results while increasing the difficulties met by the student in search of information. It would be equally possible to envisage for Western Asiatic archaeology a future lying in an organisation encompassing the meticulous concerted definition of the innumerable areas still awaiting investigation and the many problems awaiting solution, a leading example of which is the lacunae in history, rather than any existing system. However, in order to succeed on these new lines, the archaeology of Western Asia would first have to do away with the stultifying effects of academic nationalism.

The ivory head of a woman from the hoard found at Nimrud, dating to the eighth or ninth century BC, was once decorated with inlays, now lost. Her half-smiling expression is slightly reminiscent of archaic Greek sculpture.

A beautifully stylised miniature of a crouching ox, now in the British Museum, was found in the ivory deposit at Nimrud.

The Nile valley

Egyptian archaeology

The beginnings of archaeological activity

On 27 September 1822 Jean François Champollion supplied his brilliant key to the decipherment of hieroglyphic script, publishing the meaning of the chief signs in his letter to M. Dacier. But it was not until Auguste Édouard Mariette's excavations at Saqqara in 1851 that large scale field research began and with it Egyptian archaeology.

From the Renaissance onwards innumerable Egyptian objects had been gathered to enrich collections of antiquities, but these came chiefly from excavations in Italy at Rome, Herculaneum and Pompeii. These stimulated an outburst of interest for all things Egyptian which, even before Napoleon's famous expedition, had inspired an Egyptianising style. The Rosetta Stone was found by Captain Bouchard while building fortifications on a tributary of the river Nile, while the main achievement of the archaeologists of this famous campaign was to compile a general inventory of antiquities throughout the valley, which was subsequently included in the memoirs and admirable collections of plates of the *Description de l'Égypte*. Vivant Denon described vividly this archaeological feat, conducted on the battlefield.

When Egypt was opened up to the West during the rule of Mehemet Ali there were many amateurs who participated in the hunt for objects and buildings. Some representatives of the great powers took advantage of their diplomatic privileges to organise veritable raids on the archaeological sites, and sometimes rival 'gangs' came to blows. The British employed the ingenious adventurer Jean Baptiste Belzoni, who discovered among other things the sarcophagus of Seti I and was one of the first to enter the great temple of Abu Simbel. The French consul, Drovetti, was also conspicuous for his cleverness. Thus the nuclei of the great European collections were formed, but, by a curious paradox, Drovetti's collection ended up in the Turin Museum and that of the British consul, Salt, in the Louvre.

Scholarly expeditions were conducted by Englishmen like Sir J. Gardner Wilkinson and Lord Prudhoe and the French architect J. N. Huyot. But it was Champollion once again who was the first to organise a thoroughly learned expedition. He toured Egypt and Nubia as far as the second cataract from August 1828 to October 1829 with an escort of draughtsmen and assistants, collecting notes and sketches for drawings which were not published until much later.

From the autumn of 1842 until the end of 1845 the celebrated Richard Lepsius pushed on further south as far as Meroe in the heart of ancient Ethiopia and made detailed studies of the Nile valley sites. The twelve huge folios of plates, executed with astonishing precision by the Prussian team, are still one of the foundations of Egyptology.

However, active archaeology, which sets out to find remains of the past hidden in the ground and tries to reconstruct the history of a site, the evolution of a sanctuary or the period when a building was occupied, was primarily the achievement of Auguste Mariette. He embarked on the first systematic excavation of the great complex of the Serapeum at Memphis. At the end of October 1850 he had been appointed to join an Egyptian expedition to the necropolis at Saqqara, a little to the south of Cairo. While drawing up some plans, he noticed the limestone head of a sphinx rising from the sand, and realised, after remembering a passage in Strabo, that he was on the avenue leading to the temple of Serapis. After a year of arduous toil Mariette entered the Serapeum on 12 November 1851. The Egyptians who had placed the last stone in the wall of the entrance to the tomb had left finger marks on the plaster facing. Their footprints were in the layer of sand in one corner of the burial chamber. Since then nothing had moved in this dedicated place where a mummified bull had lain for nearly 4000 years. Mariette found the tomb of Apis, the famous bull, and some invaluable stores of jewels.

A few years later, on 4 July 1858, by a decree of the Khedive he was appointed *mamour* (director) of Antiquities of Egypt. The Antiquities Service established by Said, the Viceroy of Egypt, and Mariette, rendered a real service to Egyptology. From this time onwards he worked through the valley, site by site, with tireless dedication. Tanis, Memphis, Dendera, Thebes and Edfu: he cleared and reinforced the ruins of sanctuaries and had inscriptions copied. Because he had so much to do his excavations were sometimes hasty, though always shrewd. At Abydos with a few strokes of his pick he revealed an unknown wall painting. An old *fellah* came and said to him, 'I have

Tauret, the Egyptian goddess of childbirth, was usually represented as a pregnant hippopotamus. A statuette of the Twenty-sixth Dynasty, from Karnak.

Opposite page: a slate palette of king Narmer, dating from the First Dynasty, illustrates one of the stages in the developing art of writing. The king, wearing the red crown of the north and followed by a servant carrying his sandals, is inspecting rows of decapitated enemies after the battle which united the two kingdoms of Egypt.

The reverse of the Narmer palette shows the commanding figure of the king, now wearing the white crown of the south, about to strike an enemy with his mace. The new hieroglyphic signs are used to identify the chief figures.

never left this village, but I never heard tell of a wall there.' 'How old are you, that you remember where it was?' 'I'm three thousand,' replied the imperturbable Mariette. 'Well, then,' answered the old man, 'if you're so old and look so young, you must be a great holy man. Let me look at you.'

Everywhere he went Mariette collected *stelae*, statues, amulets and other articles which he deposited in the small museum of Boulaq. (This was the origin of the wonderful collection of treasures of the Cairo museum.) His host in Cairo said of Mariette that he had 'a *profound understanding of life* as a whole except for the one facet which is so important in this perverse world, his own personal interest.' He could never manage his own money economically and was completely indifferent to fortune.' Worn out by travel and suffering from diabetes, Mariette Pasha died on 18 June 1881. He was succeeded by Gaston Maspero.

This period saw the beginning of modern archaeology, which gathered momentum throughout the whole of Egypt. The Antiquities Service was organised and its budget increased. Its staff of scholars included learned men of several nationalities. In 1902 a plan was drawn up for a huge general catalogue of the Cairo Museum, which today includes a hundred thick volumes, though this is only a fraction of all the objects awaiting publication. Learned societies, institutions such as the Institut Français d'Archéologie Orientale du Caire, government-supported bodies and rich patrons gave money which opened the way for new digs and extended research. The romantic age of the treasure seekers was soon brought to an end. Since 1924 improved methods for technical investigation have been scrupulously applied. German, American, British, Belgian, French and Italian teams and Egyptian scholars from the Antiquities Service and universities have been at work on innumerable digs. Every region and period has been investigated, but the sand and silt continue inexhaustibly to yield relics of the past. *In 1900 Maspero said, 'We have hardly* scratched the surface of Egypt'. This statement is still true today, so favourable are conditions there for the preservation of remains.

Under a cloudless sky Situated between the twenty-fourth and the thirty-first degrees north, Egypt is an extremely dry region. The *desert sands under cloudless skies promise* wonderful possibilities of preservation. There is some rainfall in the northern Delta, however, and at the bottom of the valley the earth is muddy and saturated with river water. The Nile, the life force of Egyptian civilisation, is a permanent threat to its antiquities, and modern science has made matters worse by seeking to raise the water level by building dams. The construction of the high dam at Aswan attracted world wide attention to the cultural drama of the final submerging of the whole of Nubia. As a result of the alluvial deposits carried by the river, the water level in the valley itself and the Delta is continually rising, a millimetre a year and a metre every thousand years. The foundations of the buildings in the valleys are being attacked, and the temples, although they are founded on beds of huge stones, are beginning to rock. This explains why part of the hypostyle hall at Karnak collapsed in *1899*. On a wider scale, the capillary action of saltpetre and other penetrating substances have spoilt a number of reliefs. This destructive process is particularly apparent when stone objects are dug up. Abrupt contact with the atmosphere, a different level of humidity and handling have caused crumbling and sometimes total disintegration of a find. This is why excavations must be followed *by publication as soon as possible.*

The geographic and climatic factors in Egypt combine to produce different excavating from any other region, even the neighbouring lands of Western Asia. Work therefore very often consists simply of clearing buildings of the stones and sand covering them. Research is obviously a more complicated matter when it involves dried brick buildings found in the alluvial mud of the valley. Only infinitely patient investigations have made it possible to separate the walls from the silt around them.

During the course of the centuries the remains of various civilisations have accumulated on top of each other. These heaps of superimposed evidence of the passing of time are known as *koms*, or tells and furnish a section of history, going backwards in time. In his quest for the past the archaeologist is helped by the fact that Egypt under the pharaohs was essentially a literate civilisation. Building stones are very often inscribed, and objects can in themselves be texts. Bearing in mind the internal and external criteria needed for the interpretation of any written document, from any culture, it must be admitted that Egyptology is favoured

The first stages of Egyptian monumental architecture are represented by the Step Pyramid and the mortuary temple of the Third Dynasty king Zoser. The name of the architect, Imhotep, was later revered as almost divine by the Egyptians.

in this respect. Most excavators in Egypt are Egyptologists who by reading the hieroglyphs make use of the invaluable information derived from 4000 years of history. The price of this great wealth of documents is that all too often minor remains are ignored in favour of the great historical inscriptions. Why have scholars been so slow in making thorough observations about the colour of floors or other architectural details, and why has so much care been devoted to collecting minute fragments when so many texts with fine hieroglyphics and fascinating scenes have already been recovered? Egyptian pottery, in particular, has been unjustly neglected and no one has devoted an impressive *corpus* to it, nor given it the kind of detailed study upon which chronology is based in so many other branches of archaeology.

Advantages and problems The wealth of Egyptology is boundless, and at present there is probably not a single Egyptologist who is an expert in every aspect of the subject. Indeed it covers many widely different subjects. The study of very early periods demands a knowledge of prehistoric tools and artifacts, essential for excavating cemeteries, studying tells, and for reproducing and interpreting rock engravings. If the study of historical periods is to be mastered, the Egyptologist must keep in touch with the varied stages of development in the language and script, in spite of the astonishing conservatism of the Egyptian civilisation and its fidelity to original prototypes. From the standpoint of practical archaeology, the Egyptologist has to know how to excavate tombs, how to draw architectural plans of temples and palaces, how to clean and preserve papyri, ivories, wood and a variety of different materials. He must be familiar with the records of succeeding historical periods, and sometimes he must pursue his studies up to the Alexandrian and Coptic eras. If he is capable of dealing with all these possibilities, particularly in the field where the unexpected is the rule, the archaeologist also finds that he needs to be specially well informed about one period, type of building or excavation technique. The originality of the scholar lies in this balance between broad general information and highly specialised technique.

The responsibilities of the field excavator are moreover very great. An archaeological record is, in practice, unique. It is impossible to reconstruct the setting of an object if it has not been detached with the utmost care, and if very detailed notes have not been taken to supplement the photographs of the successive stages of the operation. Meticulous preliminary studies are necessary of the type of finds likely to occur

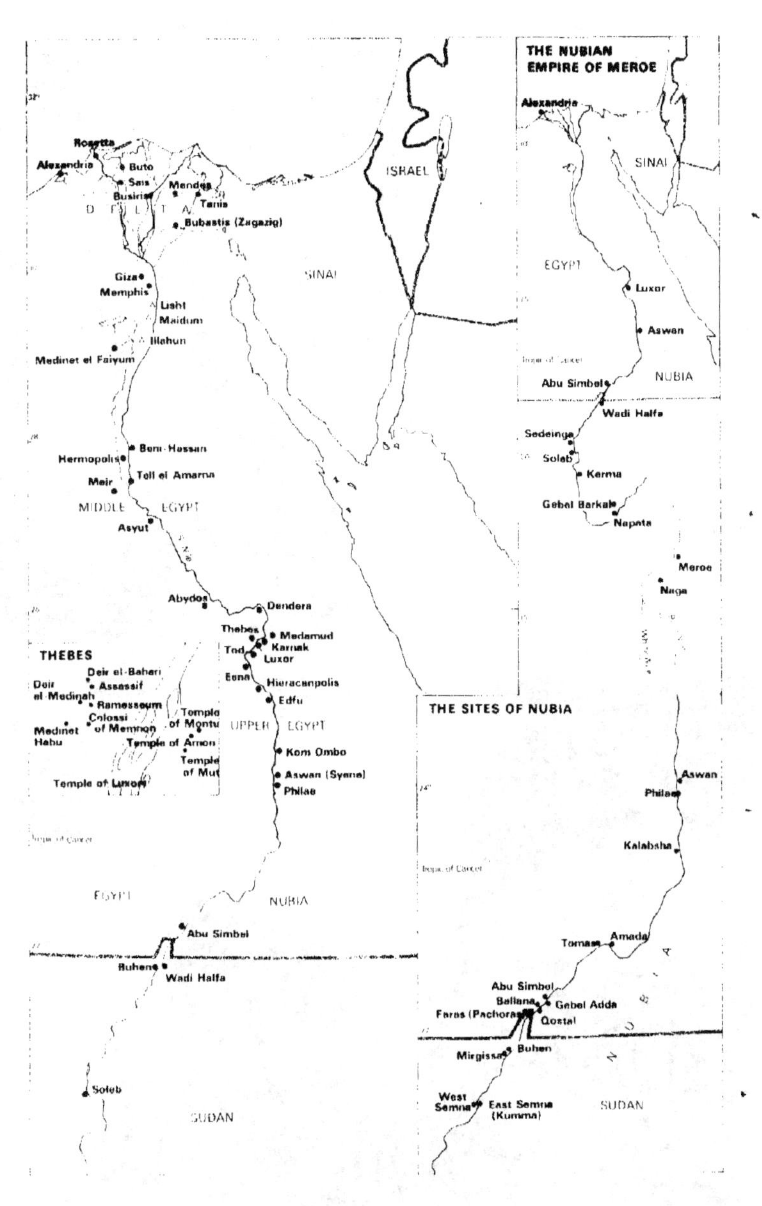

It is essential to be able to recognise immediately a bone, specify its original position and remove it without breakage. A typological study of material is also desirable and an overall plan of work. It is best not to hope for a major discovery, as this may lead to over hasty excavation outside the main plan of campaign. In addition there are the special problems of climate and humidity in Egypt which threaten the finds with rapid deterioration or even destruction once they are removed from the earth. Therefore an excavation, in Egypt perhaps more than anywhere else, must not be left in the hands of an ordinary overseer. It is, in fact, a source of the evidence which will later be the raw material of the process of interpretation.

An archaeological excursion

In view of its enormous extent, it would be impossible to present here a balanced account of Egyptian archaeology or even to describe the great variety of recent finds. Instead a few discoveries from different parts of the Nile valley over a period of 4000 years will be considered from which we can understand the history of the ancient Egyptians and penetrate to the heart of their daily life with a better appreciation of some of the great achievements of this fascinating civilisation.

Prehistory Egyptian prehistory was dominated for a long time by the British Egyptologist Flinders Petrie. At the end of the last century and the beginning of the twentieth he spent part of every year searching the soil of Egypt and unearthed pottery, tools, mirrors, weapons, combs and ivory figurines by the thousand. He regularly published the results of his excavations with amazing industry, and produced an impressive number of scholarly and popular books. Petrie established typological classifications to define more clearly the abundant material he had collected. Acting on the fundamental archaeological principle of distinguishing between earlier and later objects, he proposed a very useful grid of dates which permitted the comparative classification of various finds.

As knowledge advanced it soon became possible to distinguish several great cultures which were named after the chief sites. Although the Paleolithic period in the Nile valley is still badly documented, the principal stages of man's

The builder of the step pyramid, King Zoser. This is one of the earliest Egyptian royal statues and the painted limestone figure, which is life-size, was found walled up in a small chapel adjoining his pyramid. The eyes were removed in antiquity.

The great Fourth Dynasty king Khephren was responsible for many of the monumental buildings in the pyramid area at Giza. The Sphinx and the remains of the King's Valley Temple stand in front of his pyramid.

advance from the Neolithic period to the dawn of history have been precisely detailed. Egyptian civilisation seems gradually to have drawn apart from the African substructure common to Egypt and all the ancient cultures of the Nile valley, and finally to have assumed the original characteristics which persisted throughout its history.

The Tasian phase with its incised black pottery was succeeded by the Badarian, with coarse thick pottery. The two Naqada periods followed: first came the Amratian with two parallel series of pottery (red vases with black rims and red vases painted with white motifs in a geometric style), then the Gerzean with pottery covered with a pinkish slip painted with decoration in purplish ochre of very stylised plants and boats.

The Egyptians soon developed an extraordinary mastery in the carving of hard stone vases. It was in small objects, especially vases, that the Egyptian craftsman, who was a true artist, excelled. Throughout its history Egyptian art was distinguished by this remarkable skill in stone working. For thousands of years the Egyptians delighted in reproducing the same models, seeking to return to the original perfection which marked the earliest creations. It was perhaps for this reason that the stone-working civilisation of Egypt devoted such ardent worship to representations of its origins. Instead of looking to the future, like the metal cultures, it shut itself up in golden dreams of the past, and modelled its work on the traditional standard. The 'time of the gods' remained the undisputed model.

It is true that during these early periods the various parts of the valley seem to have differed considerably. The cultures of Upper Egypt show quite different characteristics from those of the Delta. It has been suggested that Upper Egypt was populated by nomadic stock breeders, in contrast with the sedentary agriculturalists of Lower Egypt. If this were true, the burial customs would be completely different, and the other finds dissimilar. In fact, this suggested contrast reveals more ingenuity than truth. It is supported chiefly by the fact that the nomads preferred schist palettes, which were eminently portable. In recent years the eastern Delta has also produced a few palettes. The discovery of these tiny fragments seems to have destroyed a theory propounded in impressive memoirs, but not soundly enough supported by evidence.

The establishment of Pharaonic rule

Egyptian history really begins with a particularly well known object, the Narmer palette from Hieraconpolis, the former capital of Upper Egypt. One side of this shield-shaped plaque shows the pharaoh wearing the White Crown of Upper Egypt sacrificing a foe. The symbolic swamp plant establishes the meaning of this historic scene. The hawk god Horus dominates the group, holding the head of another captive by a rope gripped in his beak. On the other side of the palette the upper register shows the king, Narmer, before two rows of

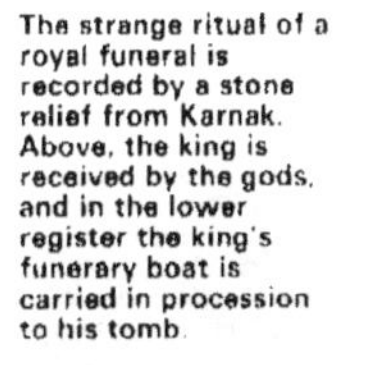

The strange ritual of a royal funeral is recorded by a stone relief from Karnak. Above, the king is received by the gods, and in the lower register the king's funerary boat is carried in procession to his tomb.

prisoners who have been tied together and beheaded, he wears the Red Crown of Lower Egypt. This indicates that he held this region, and that the two kingdoms had now been united. The south's victory over the north, the foundation of Egyptian history, was thenceforward to affect the entire political life of the country. In the 'Two Lands', the name by which Egypt was known ever afterwards, Upper Egypt always took precedence over Lower Egypt. In all the royal texts and titles the king is named first as Lord of the South and only afterwards as Ruler of the North.

The Narmer palette dates to about 3000 BC: the establishment of a unified kingdom, the organisation of an irrigation system, and the origin of writing are all attributed to the same period. The motivation for these three major achievements, which the evidence indicates to have been contemporary, is uncertain. According to the practical historians the inauguration of the irrigation system required Egyptian unity and led to the development of writing, which was indispensable to an overall plan. It may, however, have been the case (as suggested by the more theoretical standpoint) that the use of writing allowed the recording of orders and thence both the unification of the country and the establishment of the irrigation system [illegible]. It is possible that pharaonic rule was a necessary preliminary to the other developments.

This form of government endured for 3000 years, and was to shape the whole history of the land, thereby dominating Egyptian archaeology. From this time onwards Egypt made rapid advances, progressing by a series of clearly defined stages. After the archaic period of the two Thinite Dynasties one of the important stages was the transition from the Second to the Third Dynasty, with the achievements of Zoser and his brilliant architect Imhotep. The use of dried brick and wooden beams or poles for building was superseded by the regular use of stone. Forms originally designed for primitive techniques were suddenly translated into ageless material.

The Old Kingdom: the pyramid of Zoser

On the cliffs rising above Memphis, where the first dynasty pharaohs built their capital, Horus Neterikhet, usually known by his praenomen Zoser, erected the white stone wall with its projections and the recesses of his funerary complex. A six-stepped pyramid rose above the complex as a sort of majestic staircase for the king's soul to ascend to the sky and the gods to descend to the earth. Its size is colossal. The rectangular enclosure measures 544 metres long by 277 metres wide, or no less than fifteen hectares. The central structure includes six successive building stages, only the last of which could truly be classed as a step pyramid. Its base measures 121 metres long by 109 metres wide, and its original height was about sixty metres.

Zoser's funerary complex has been very thoroughly investigated. Jean-Philippe Lauer

A relief from the vast temple complex of Karnak shows crowned kneeling figures making offerings of food and flowers to the gods. The standing divinity in the centre is the god Horus, who is often shown with a hawk's head.

Cheops planned a large cemetery of flat-topped rectangular mastaba graves for the families of his high officials. These tombs are arranged in regular rows east and west of the Great Pyramid of Cheops at Giza.

devoting his entire career to it. The courts have been cleared, and the buildings of various types and functions identified. The architecture is deceptive, since all the buildings are only facades, the interior consisting of a filling of gravel. The doors or more often simulated doors are open or shut for all eternity. [illegible]all lime-stone facing blocks, about the s[illegible] the bricks for which they were substituted, [illegible] fitted and joined with remarkable accuracy. The excavators have also succeeded in penetrating the vast underground complex of the actual tomb, where a network of corridors and funerary chambers extended beneath the pyramid. Some of the walls were covered with magnificent decorations of small blue squares of lapis lazuli arranged in elaborate patterns in panels. Two of the great passages yielded an enormous hoard of schist and alabaster vessels annexed by Zoser from the royal stores where the property of his predecessors lay heaped up. More than 500 chests came from these two passages, or more than 30,000 vases, the examination and classification of which still remain to be done.

From this time onwards the pyramid was to dominate the civilisation of the Old Kingdom. All the pharaohs were determined to raise these mountains of stone blocks to cover and protect their tombs, thereby helping to ensure their eternal survival. The pyramid defied time and demonstrated the grandeur of the god-king. The Third Dynasty step pyramids of Zoser and the unfinished tomb of his successor Sekhemk[illegible]t were very distinctive, with their pattern of slabs which had been laid parallel to the sloping layers. The supreme simplicity of the great pyramids of Giza was attained from these origins, with several intermediate stages like the rhomboidal pyramid of Dahshur and the Maidum pyramid.

There is probably no other form of monument which has roused so much discussion, polemic and speculation as the group of great Fourth Dynasty pyramids. It must be confessed that these imposing masses of stone produce such a completely harmonious effect, so elegant in its strength, that description is totally beggared by the reality.

The great Herodotus was so moved by these pyramids that he propounded theories about the construction in which legend is combined with factual technical information. He says that it would have taken no less than ten years to lay the causeway by which stones were brought to the site. A few dimensions serve to increase our amazement. The Great Pyramid, known as 'Cheops of the Horizon', was originally more than 146 metres high, with sides more than 230 metres long. The second pyramid, 'Great is Khephren', had sides 215 metres long and was more than 140 metres high. The third pyramid, 'Divine is Mykerinos', is more modest, with sides 103 metres long and a height of 66 metres, but the stones of which it is constructed are very large. The area in which the pyramids (there are more than seventy) are located extends from Abu Rawash in the north to Lisht and Maidum in the south, and is far from having been completely studied. Every year new discoveries modify the data, but it is still inadequate. In 1954, for instance, eighty-three blocks each measuring more than a metre square were found during the construction of a new access road south of the pyramid of Cheops, no more than a few metres from an area which had been thoroughly investigated for many years by an Austrian team. The first forty-two covered a sort of dry dock about thirty metres long.

A cedarwood ship, nearly 5000 years old, was also found near Cheops. The prow and the poop were separate from the rest of the boat and arranged fore and aft, apart from it. The components, however (about 600 of them, mostly intact), had been stacked in a very orderly manner where they were due to be re-assembled. Both graceful and sturdy, the boat, shaped like a papyrus, was forty-three metres long, five metres high at the prow and seven metres high at the poop. This was boldly curved back towards the inside, and spread into the lovely floral shape of a papyrus head. The two longest cedarwood planks were twenty-three metres long, half a metre wide and seventy-two centimetres thick and weighed more than two tons. Placed down the middle of the boat near each side, they held together the transverse beams which passed from one side to another at regular intervals under the bridge. Each one of these ribs was attached to a ridged piece fastened laterally halfway up the side of the boat. It should be remembered that there was no main beam in the bottom of the boat, nor any keel. The ridged pieces had been pierced for ropes to go through, to hold them together. These ropes, except in a few cases, could not be seen from the outside. Only three copper nails have been found for the ship. It was therefore made of pieces of wood which were stitched together, and this explains how it could be dismantled so easily. The wood would swell in the water while the ropes would shrink, and the various pieces of the structure would therefore be certain to fit together perfectly, making any caulking of the boat unnecessary.

All the planks of cedarwood were in their proper places when the boat was dismantled, ropes, tackle, sails and twelve oars made from a single piece of wood, two of which were arranged behind as a rudder for steering. On the bridge was an elegant cabin, the side walls of which were doubled by a row of small columns holding a sort of dais, probably covered with sailcloth to procure a pleasant coolness by appropriate ventilation.

The conventions of Egyptian sculpture dictated that the size of a figure should be related to his importance. A colossal statue of Ramesses II at Karnak is accompanied by a knee-high figure of his wife.

This remarkable vessel has rashly been named a 'solar barque', but there are many reasons opposing this identification. For example, it may have been a ritual pilgrim ship bound for the holy cities of Abydos or Busiris or, more probably, since it appears to be perfectly functional, part of the funerary equipment of the king himself. Perhaps it was the catafalque used for his obsequies, to carry him to the shores of eternity in the west.

Within each pyramid the tombs of high officials are grouped around the pharaoh's, to reconstruct the society of the living world in the hereafter. A very rigid hierarchy prevailed in Egyptian life, in the form of a veritable social pyramid with the god-king pharaoh at the top. There is little information about the lower echelons of society whose members died without a tomb and were abandoned to the sands of the desert. By examining the impressive funerary goods of the nobles and the reliefs carved on their shrines, we can witness a wide variety of scenes of their daily life and thereby form a detailed picture of the Old

Kingdom civilisation, although the history of this period is still very obscure. The Egyptians did not attempt a systematic recording of their historical development. Absorbed in their deities and the preparation for afterlife, they never set any value on the practical evolution, so transitory, of deeds accomplished on earth. They took no more note of the specific events of their personal lives than of the history of their country, to which they accorded so little importance. Inscriptions carved on the tombs usually contain idealised biographies of the deceased. It would be a mistake, however, to accuse the Egyptians of commemorating only the victories and relating nothing but the virtues. For them it was not a question of transcribing facts for the information of future generations, but of appearing before their gods as embodiments of the essential principles of pharaonic civilisation.

Little by little the Egyptian economy collapsed under the fantastic weight of mountains of stone. The innumerable levies required for the construction of the pyramid complexes channelled off the best resources of the land. It was a curious form of capitalism in which the investment was unproductive—silver and gold, precious stones poured into the depths of tombs, efforts and achievements all devoted to the hereafter and deification.

The size of the pyramids gradually diminished, the pyramid complexes shrank and stone was replaced by writing, as the rulers no longer relied on the weight of the pyramids but on the magic of the written word to ensure their survival. During the reigns of Unas, the last king of the Fifth Dynasty, and of Teti and Pepy of the Sixth long columns were constructed, beautifully carved with hieroglyphics, and those pyramid texts formed the world's first books. It was about 1880 when Sir Gaston Maspero noticed a few fragments outside some of the smaller pyramids at Saqqara. After examining them he realised that, unlike the great preliterate pyramids of Giza, they were decorated with funerary texts. Sometimes written in sparkling verse but often obscure, these texts were part of the ritual incantations to guarantee the deceased a glorious burial and safe passage to the resurrection.

In the reign of Pepy II (about 2300–2200 BC), probably the longest reign in history, since the king succeeded to the throne at the age of six and was more than a hundred when he died, power gradually passed into the hands of the leaders of the *nomes*, or districts, and royal power no longer really existed. As a result there was no monumental art produced during this period, since pharaonic rule and Egyptian art are indissolubly linked. However, a very spirited local art developed at the bidding of the provincial governors; picturesque motifs replaced the great compositions formerly carved on the walls of tombs. The excavations at Asyut have yielded captivating women bringing offerings, made of wood painted with bright colours; standing upright under their burdens, these tall slender figures are models of spontaneous grace.

The Middle Kingdom: supremely balanced art About the year 2000 BC there was a return to centralised government, which the chiefs of the nome of Thebes imposed on the rest of Egypt. The Middle Kingdom dates from this period. The local gods, Mentu and more particularly Amon, attained the dignity of national gods, guarantors of pharaoh's victory

over the rest of the world. This was an era of classic harmony *par excellence*: the power of the Old Kingdom was tempered with an elegance which foreshadowed the art of the New Kingdom.

Few monuments have survived from this period. Shortly before the Second World War an Italian team excavated a sanctuary in a very good state of preservation in the sands of Faiyum at Medinet el Faiyum. Later on the Egyptian Antiquities Service cleared two temples in the Delta, one at Ezbet Rushdi north of Zagazig, the other at Bubastis. The importance of such discoveries cannot be exaggerated coming as they do from such a poorly documented period and region of Egypt.

The most remarkable building, however, is still the white limestone kiosk of Sesostris I which has been entirely reconstructed in all its supreme grace by Henri Chevrier at Karnak. There has been, and will continue to be, much controversy about the kind of reconstruction. Although the limestone blocks were treated with respect during their curious architectural odyssey, dismantling, storage, transport and rebuilding have damaged the reliefs. A thorough search within the pylons at Karnak and in the foundations of the colonnades and halls has yielded fragments of innumerable buildings which have abruptly ceased to stand in the temple, although their remains continued to live on, buried in the heart of the new monu-

The art of low relief in the Middle Kingdom is attested to in the funerary chapels of several provincial cemeteries. The realism of the scenes at Meir was rarely to be equalled; thus we see a remarkable, emaciated old man laboriously leading his flocks. As to the paintings, those of Beni Hassan are justly famous. Although they are badly damaged, they demonstrate the harmony of Egyptian graphic art with their warm colours and flowing lines. Their beautifully balanced composition makes skilful use of picturesque detail.

The gold work of the Middle Kingdom is fairly well documented. This is the period of sumptuous ornamental breastplates, enhanced by glittering jewel inlays. Freedom of invention informed the most traditional motifs. Although the medium was pre-eminently luxurious and decorative, the basic themes of pharaonic rule were still emphasised and illustrated: the glorification of the royal name, the union of the two lands, the slaughter of enemies and the triumph of Egypt over all subversive elements. There is, however, a pure grace in the gold and enamelled flowers which form the crowns of the princesses of Illahun and Dahshur. Among the jewels of the Princess Khnoumit is a beautiful, light and elegant crown consisting of openwork cruciform ornaments and a tracery of gold wire decorated with little flowers and blue beads, which represent grain.

However extensive the problems of Egyptian archaeology may be, the variety of techniques which can be employed is equal to the difficulties. For example, in studying the reign of Sesostris III (1887–1850 BC), the scholar has to rely on psychological analysis. An admirable portrait gallery, much of which was found during the French excavations of Medamud just north of Karnak, shows us the bitter features of this sovereign. His face, with its projecting cheekbones, twisted scornful mouth and sardonic pose is expressive of this period when psychological understanding of the human predicament became more personal. Pharaoh

The monumental stone architecture of Egypt preserves traces of the more perishable building materials used in earlier times. At Karnak columns derived from bundles of reeds are topped by capitals in the form of stylised lotus flowers.

The enormous enclosure at Karnak contains not only temples but a sacred lake and housing for the entire religious administration. Each pharaoh in turn added to the complex, and the copious inscriptions which cover the walls and columns record their contributions.

is still divine, but he is capable of feeling human joy and sorrow.

Sesostris III carried his military campaigns into the heart of Palestine, and Egyptian traders were active on the coast of Asia where the adventurous Sinuhe had an outstanding career. Several Middle Kingdom statues have been found at Byblos and Ras Shamra, while under the foundations of the temple of Tod in Upper Egypt four bronze chests with the name of Amenemhet II contained a hoard of Aegean and Asiatic gold work. Many craftsmen with Semitic names were employed in Egyptian workshops.

It was at this time that an enormous network of fortresses was built across the district of the second cataract on the African frontier in the south. The excavations of W. B. Emery and J. Vercoutter have revealed splendid military settlements at Buhen in an exceptional state of preservation. Thick accumulations of sand preserved these ruins unusually well, and the excavators had only to remove a colossal quantity of fill. The remains were visible only for a short time and the inexorable rise of the waters of the high dam at Aswan covered them again. The water has also destroyed the brick towers which the millennia have respected.

The massive brick walls of the Buhen fortress were twelve metres high and about five metres thick, and were strengthened by a series of rectangular bastions. The foot of the wall was protected by a rampart with a loop-holed parapet, overhanging the scarp of a rock-cut ditch 8.4 metres wide and 6.5 metres deep. The counterscarp on the far side was topped by a narrow brick wall from which a gently sloping glacis descended to the inside. Although this fortress is evidence of the power of the Middle Kingdom, it also betrays the importance of Egypt's enemies. These were Saharan stock-breeders, known by the simple name of 'Group C', living in the south of the kingdom of Kerma.

It was not however from the south that the greatest threat to Egypt lay. This came like a pestilence from Asia with the conquest by the Hyksos, the 'Shepherd Kings' who, from about 1730 BC onwards, controlled the northern Nile valley. Their invasion of Egypt has been thought to have been the result of the great movement of Indo-European horse-taming people throughout Western Asia at the beginning of the second millennium. It is also evidence of movement of Semitic people. Canaanite names such as Jacob Her and Anat Her have been found on some interlaced scarabs. The Hyksos period concluded a long period of decadence and regression in Egypt, a void for which the historians can find no better name than the Second Intermediate period. It was a time of religious anarchy when men began debating about god, the world and mankind. Some of the most interesting texts in Egyptian literature date from this period, but there are no works of art and few archaeological remains.

The New Kingdom Egypt suddenly attained its freedom, and once more political power was allied to an astonishing artistic flowering—that of the New Kingdom.

The expulsion of the Hyksos is briefly related on the Sallier papyrus. The excavations of the base of a colossus of Ramesses II north of the passage of the second pylon at Karnak in 1954 resulted in the discovery of reused blocks of a stele which told of the fall of the Hyksos in some detail, and of the Theban liberator Kamosis. One person, Nechi, who is shown on the bottom of the stele, is the narrator of these episodes. He tells of Homeric defiance under the walls of the Hittite capital, Avaris, and the waylaying of a messenger of the Hyksos king, Apophis, who had been dispatched with a letter to the ruler of Kush in Upper Nubia by way of a chain of oases. The messenger was captured in the oasis region.

The successful endeavours of the native dynasties to regain power did not stop with the deliverance and restoration of Egypt; the impetus once given, the Egyptians devoted themselves to conquest and created a widespread empire extending to the heart of the Sudan in Africa and to Syrian Palestine in Asia. The accumulation of wealth allowed the development of a brilliant and luxurious civilisation, the building of impressive temples and palaces, and the creation of works of art in all fields which are the glory of museums today. In a word, the second half of the second millennium (about 1580–1085 BC), the Eighteenth and Nineteenth Dynasties, is the high point of Egyptian history. This period, the most admired of all in ancient Egypt, has yielded the most numerous and the best documented remains to archaeological records.

The Theban Necropolis The capital of the empire, Thebes, was adorned with a wonderful architectural flowering, though very little has survived of the palaces which the kings built, each in his turn. The remains brought to light by the Americans of the Malkata Palace are, however, an indication of their incomparable splendour. It was to their gods and their dead that the Egyptians dedicated the magnificent creations which have defied time. The west bank of the Nile was reserved for dead pharaohs and the great nobles whose duty it was to rally round their kings in the next world as they had in this. Dominated by the magnificent natural pyramid of the Theban peak, the Western Mountain, the necropolis is divided into innumerable parts extending over an enormous area.

The tombs of the nobles are justly celebrated, and the reliefs and paintings decorating them have always been much admired, the best parts of them having been reproduced in art books to the unceasing delight of collectors. Since many of the tombs are unfortunately in a disastrous state of preservation, we have only details of the reliefs and paintings at which to look. For a long time the shrines and burial vaults were used as shelters and stables by the poor *fellahin* of the area, who were dislodged only with the utmost difficulty. Tomb robbers have also plied their trade, wrecking whole walls to get a few fragments of a relief. Nowhere have time and human agency been so fiercely destructive. If the words 'cultural preservation' have any meaning at all, it is here above all that they should be brought into play before more of these treasures perish.

What is more, the walls on which paintings were made often seem to be very mediocre. Though the Theban subsoil contained several seams of good limestone which enabled artists to cut amazingly fine reliefs, the rock was often marly. The flaky walls had to be covered with a coating on which the colours were laid. Egyptian painting, thenceforward so celebrated, was in fact a substitute, the best possible replacement for the painted bas relief. This explains the most familiar characteristics of the paintings: their popular spirit and spontaneity. The mediocrity of the stone thus contributed to the development of an unequalled expression of Egyptian genius, very different in nature from the solemnity of the official traditional art.

It would take too long to discuss in detail the work of the sculptors and painters, whose talents were deployed throughout the five centuries of the New Kingdom. The paintings in the tombs of Menna and Sennefer, the reliefs in those of Rahmose and Kheruef are among the artistic masterpieces of all time. Sennefer was curator of the royal gardens and parks; the 'grape ceiling' of his tomb is world famous.

With centuries of practice the stonemasons of Egypt acquired great skill in carving hieroglyphic inscriptions. The vast columns of the hypostyle hall at Karnak bear records of the history, titles and achievements of the kings who built it.

The god Amon, originally an undistinguished local deity, acquired great importance in the Egyptian pantheon when Thebes became the capital. Fine statues of his sacred animal, the ram, line the avenue leading to his temple at Karnak.

A detail of one of the columns in the great hypostyle hall at Karnak shows the counter-sunk (as opposed to relief) inscriptions favoured during the New Kingdom. Groups of signs, enclosed in the oval frame known as a cartouche, are the names of pharaohs.

The sarcophagus of the Twelfth Dynasty king Sesostris III is decorated with a low relief of the king wearing the Osiris beard and a diadem with two tall plumes. The three pharaohs of this name extended Egypt's political and commercial empire so widely that their conquests later assumed legendary proportions.

The tombs of the sovereigns of this period also merit a book in themselves. The burial vaults were usually cut in the cliffs of the Valley of the Kings, while the mortuary temples stood far away, on the plains of the west bank of the Nile before the cliffs at the edge of the Valley of the Kings. The only remains of the mortuary temple of Amenophis III are the giant statues which stood at its entrance, the famous Colossi of Memnon. The 'Ramesseum' at Thebes still stands witness to the glory of Ramesses II, while the battles of Ramesses III are recorded on the vast walls of the great temple of Medinet Habu.

The most extraordinary mortuary temple is that erected by Queen Hatshepsut in the great curve of Deir el Bahri at the very foot of the Theban peak. There, close to the temple of one of the kings of the Eleventh Dynasty, Hatshepsut and her architect Senenmut succeeded brilliantly in adapting their design to the site. A remarkable sense of proportion is apparent in the purity of the lines and the huge horizontals which stand out cleanly at the foot of the sheer cliff. Their guide lines seem to have been the architectural embodiment of unshakable faith in eternity. There is nothing else in the whole of the Egyptian art comparable to this shrine carved out of the rocky cliffs, flanked by terraces bordered with colonnades and famous reliefs. For over twenty years Queen Hatshepsut managed to exclude from power Tuthmosis III, the young son of her half brother and husband Tuthmosis II, and later her second husband. He came to loathe the memory of his aunt-wife, whose name he ordered to be obliterated from all monuments. It was behind Hatshepsut's temple that a Polish expedition found the remains of the funerary temple of Tuthmosis III, which had remained completely undiscovered in one of the most thoroughly excavated places in Egypt.

One of the most impressive monuments of Egyptian archaeology is still unpublished. This is the red shrine of the queen, the red quartzite fragments of which, with their extremely fine carving, were reused in various parts of the site of Karnak and reassembled after patient research.

Karnak It was the dynastic god Amon who chiefly benefited from the Egyptians' enthusiasm for building. There was not a single monarch who failed to dedicate sanctuaries and statues to him in the great dynastic temple of Karnak. The main east-west approach followed a row of six pylons, courtyards and halls before reaching the holy of holies, and continued at the rear via the great festival chamber of Tuthmosis III and a reverse succession of counterbalanced temple buildings. A second row of great pylons and buildings were arranged along another line running from north to south. A sacred lake, side chapels and priests' and workers' houses were all to be found inside a vast enclosure surrounded by a thick wall of dried brick. It enclosed more than twenty-five hectares for Amon's precinct alone. To this must be added the god Mentu's precinct to the north, and that of the goddess Mut, Amon's consort, to the south (the latter included a crescent-shaped lake). The distance from the north gate of Mentu to the south of the temple of Mut is 1300 metres.

In spite of 100 years of intensive work, many parts of Karnak have never been excavated. Vast piles of bricks and broken stone are still awaiting archaeological investigation. For several decades, however, hundreds of labourers have been at work under the direction of Henri Chevrier, and narrow gauge railways loaded with fill have been ceaselessly ploughing the avenues of the Great Temple. Whole chapters of Egyptian history remain forever buried in the heap of ruins of the Karnak sanctuaries.

Further south an avenue of sphinxes 2000 metres long leads to Luxor, the 'Southern Harem' of Amon. During the Feast of Opet the triad of Amon, Mut and Khonsu left the Karnak sanctuaries in solemn procession and proceeded by state barge to Luxor where a sumptuous welcome awaited them. The harmonious complex built by Amenophis III is one of the high points of Egyptian architectural art in its classic perfection. The great court is framed by a graceful peristyle with columns shaped like bunches of papyrus reeds.

The great temple of Soleb Papyrus-bundle columns are also found far away in the Sudan, in the great jubilee temple built by Amenophis III at Soleb about 1400 BC. Until 1957 this Nubian masterpiece was lost in the wastes of a region where the climate is extreme and the countryside infested with mosquitos. In that year permission for a full excavation was requested from the Sudan Antiquities Service by Signora M. S. Giorgini and placed under the patronage of Pisa University. The chaos of the ruins was cleared up and restored to order. In front of the temple are the ruins of its high pylon which, according to the inscriptions, were made 'of fine white sandstone'. This is certainly a true characteristic of the sandstone of Soleb which is particularly light and fine. Two courtyards have to be crossed to reach the hypostyle hall. Above the remains which time has laid low a tall column, miraculously preserved, lifts its palm-shaped capital, still crowned with its abacus and a fragment of the architrave. Behind the temple there is nothing to be seen but a picturesque mass of fragments scattered about the bottom of a sort of ravine. It is a curious paradox that in the heart of one of the most arid deserts in the world this temple was destroyed by a torrential flood. The holy of holies was equipped for the cult of the two divinities: Amon, the great imperial god who was, above all, the image of the divine king himself, and Neb-maat-ra, 'Ra (the sun) is the lord of truth', which was the second name of Amenophis III. Sun worship under his son and heir, the famous Akhenaten, culminated in monotheistic worship of the solar disc. 'The living form of the king', his divine features, appeared frequently in the temple, his face flanked by rams' horns, symbol of the god Amon, and crowned with a cylindrical cap with a crescent and a disc which were the symbolic attributes of his divine son Khonsu. The sacred representation of the god-king is also accompanied by the sign of Egypt's victory over the rest of the world, particularly Africa. 'Ra the lord of truth', if he was lord of the sky, was also lord of Nubia.

The column bases of the hypostyle hall were decorated with lozenges showing and naming the peoples whom pharaoh had conquered and those he hoped to conquer. An oval shape, usually crenellated, is surmounted by the torso, from which jut the arms tied behind the head. The picturesque features of Asiatics appear on the north columns and Africans on the south. The Mitanni, the Hittites, the Syrians and the Mesopotamians are shown in the same way, as are the people of Punt (the famous land of

View down the hypostyle hall at dawn

The seated statue of Sesostris III in the Cairo Museum expresses the strength and uncompromising realism of Middle Kingdom sculpture. The simplicity of the body and the headdress serve to focus attention on the vigorous and determined face of the conqueror.

incense) and the tribes of the upper Nile whose strange names cannot be identified. Biblical sites and the great Western Asian excavations like Byblos, Ugarit, Tyre and Ascalon are obviously easier to recognise. One tribe, surely from the land of Edom, bears the name of Yahweh - perhaps the earliest example of the famous tetragrammation.

The Amarna crisis The names of subject peoples inscribed in the lozenges on the Soleb temple comprise an inventory of the known world; but for the following reign, that of Amenophis IV, or Akhenaten, more precise details are available from the famous tablets found at Tell el Amarna in 1887. This discovery included the archives of the diplomatic correspondence between the king and his ministers and various Asian potentates. In the process of the excavations, which were first conducted by the Germans and then, after 1920, by the British, the capital city built by the heretic pharaoh was rediscovered.

This sick man (he suffered from adenoids and glandular imbalance) hated Amon and his priests, and wanted to devote himself to the sole worship of the sun disc. As 'God's drunkard', he composed immortal poems to the glory of the sun. Breaking the centuries old traditions of Egypt, he inaugurated a strange art style which fastened eagerly on his personal deformities and resulted in a stylisation applied indiscriminately to the queen, the extraordinarily lovely Nefertiti, the royal princesses and the whole court. Possibly the abundance of classic art produced during the reign of Amenophis III was bound to end in this explosion of affected mannerism.

However, the last word still remains to be said on the Amarna crisis. For tens of thousands of small stones, mostly sandstone and similar in size (57 × 27 × 22 centimetres) have been found forming the fill of the Karnak pylons. They have been discovered also in the founda

Dismantled monuments and eradicated inscriptions of discredited rulers and sects, rebuilt into the great pylons at Karnak, provide invaluable clues to changes in the religious and political climate of Egypt.

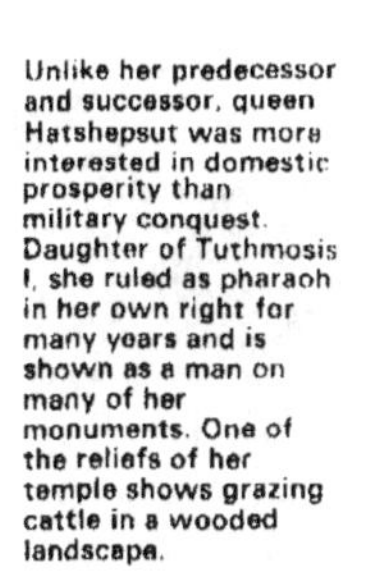

Unlike her predecessor and successor, queen Hatshepsut was more interested in domestic prosperity than military conquest. Daughter of Tuthmosis I, she ruled as pharaoh in her own right for many years and is shown as a man on many of her monuments. One of the reliefs of her temple shows grazing cattle in a wooded landscape.

tions of the hypostyle hall. Its substructure clearly consists of innumerable pieces belonging to a single construction, and these fragments came from Akhenaten's buildings. The planning of the hypostyle hall must be attributed to Horemheb.

The dismantling of the reused stones from the second pylon by H. Chevrier has shown that they were arranged in thirty-two layers of more than 700 blocks each inside the cases formed by the walls of the piers of the pylon, which were divided into compartments by internal partition walls. The outer casing of the north block of the pylon was made of several reused foundations of square pillars from a temple of Amenophis IV. On the whole, these reused pieces were not indiscriminately placed. Most of them were set with the decorated face upright. Parts of monuments, including several adjacent stones, had even been reconstructed and cemented together inside the pylon. Certain figures had been obliterated, but the blocks usually preserve their original relief and colour.

It is now more than a hundred years since the reused stones in the ninth and tenth pylons at Karnak first made such an impression on visitors, particularly Pris d'Avennes. 'They seemed to have belonged to an edifice which had been deliberately dismantled, since the paintings were well preserved and the corners of the stones very sharp', Comte Louis de Saint-Ferriol also noted in his diary in 1842.

In order to understand the part played by Karnak in the Amarna period it is also necessary to take into account the [illegible] statues found east of the great temple in the most eminently solar cardinal direction. They show the ruler holding the Osirid sceptre and flail. He is almost nude and shows the symptoms of sexual bimorphism, but perhaps this should be interpreted as a reminder of the basic principle of creation, which is both male and female at once. The change in historical viewpoint which these discoveries entail is clearly apparent.

The head of the colossal seated statue of Ramesses II outside the great temple at Abu Simbel. This extraordinary expression of serene omnipotence has been saved from destruction and now rests above the waters of Lake Nasser.

The greatest achievement of Hatshepsut's reign was the expedition to the unidentified African kingdom of Punt to establish trade relations. Much can be learnt about ancient transport from the details of tackle and rigging shown in the reliefs of the queen's temple at Deir el Bahri.

The temple of the Eighteenth Dynasty queen Hatshepsut at Deir el Bahri.

Opposite page, above: the Acropolis, seen from the hill of Philopappos. Though this, the most famous site in history, has never been lost to view it has suffered appalling damage. When Greece was within sight of regaining her independence from the Turks, in the revolt of 1821–26, both the Erechtheum and Parthenon were damaged during a bombardment. Below: one of the lesser-known palace sites in Crete is at Mallia. Rich finds there and at other Cretan sites confirm the splendour of the Minoan culture. The picture shows the stairway to an open paved floor which was probably the scene of religious ceremonies.

There are many other uncertainties in our knowledge of this period, although numerous records have survived. For instance, we know nothing of the origins of many of the leading figures. Who was Queen Tiy, wife of Amenophis III? Was she the daughter of Syrian immigrants, or perhaps a Nubian? And was Nefertiti the sister-wife of Amenophis IV, or a princess of the Mitanni? Nor do we know anything about the relationships between the rulers who preceded the celebrated Tutankhamun.

Tutankhamun These questions were not answered by the discovery of the possessions of this monarch in November 1922. The achievement of Lord Carnarvon and his excavator Howard Carter was one of the most brilliant successes in the history of archaeology. The American Theodore Davis broke off his investigations in the Valley of the Kings in 1924, believing that there was not a single undiscovered tomb left in the site. Tutankhamun was the only remaining Eighteenth Dynasty pharaoh whose tomb had not yet been discovered. Minor clues had been found, even in the area close to the tomb of Ramesses VI; a faience cup hidden under a rock bore the name of Tutankhamun. Nearby was a pit grave where a box was found containing fragments of gold leaf with the name of this pharaoh and his consort. Finally, some large sealed jars were discovered in a fissure in the rock, and these held linen marked with seals bearing the name of Tutankhamun and the priests of the Theban necropolis. It seemed feasible that this was part of the material used at the time of the king's burial.

Carter then began to dig in this area, but without results. Though he unearthed some workers' huts, probably belonging to the men who had built the tomb of Ramesses VI, he did not pursue his investigations any further so as to avoid obstructing tourist access to a much-visited tomb, and continued his search close to the tomb of Tuthmosis III. The excavators grew discouraged. Nevertheless Lord Carnarvon provided funds for a last campaign which began at the end of October 1922. Carter returned to the area of the tomb of Ramesses VI and after taking plans of them decided to remove the pharaonic workmen's huts. Under the first cabin he found the steps of a staircase filled with stones which took a few days to clear. This ended in a walled-up door bearing the seals of the Theban necropolis and those of Tutankhamun. The opening, which had been reclosed later, showed that the tomb had been visited after the burial, but it could not have been touched since the reign of Ramesses VI, since the entrance had been covered by the huts of the labourers who worked on the tomb. Behind it there was a passage way, also full of stones. Finally, on 26 November, a second walled up and sealed door appeared, which also bore traces of a break-in. When an opening was made in this wall, the archaeologists saw wonders by the light of their torches. The room in which the king's fabulous tomb furnishings were piled was no more than an antechamber. An adjoining room opened off it, in which the excavators found the same heaps of furniture and goods. On the right was another walled-up door guarded by two black and gold wooden statues standing like sentinels on either side.

This was not opened until February 1923, when it gave access to the burial chamber in which four coffins, one inside another, one quartzite chest and three mummiform sarcophagi, one of solid gold, held the mummy of the king. In a last small room, known as the treasure room, the beautiful canopic chests containing the king's viscera were found as well as other items. This was only the tomb of a modest pharaoh, one of the last of the Eighteenth Dynasty. What incredible wealth must have been buried in the tombs of the great New Kingdom conquerors! However, they had been looted in antiquity.

Nefertari was the wife of the long-lived Nineteenth Dynasty pharaoh Rameses II. Her tomb in the Valley of the Kings at Thebes is covered with paintings of the presiding deities of the after-life and the dark blue ceiling is studded with stars.

The tombs of the Pharaohs The astonishingly well preserved remains of most of the famous Egyptian monarchs have been recovered along with some of the funerary goods meant to equip them for the next world. After his appointment as head of the Antiquities Service, Maspero was surprised to discover several objects bearing the names of various royal and princely figures dating from the end of the New Kingdom appearing on the black market. An efficient police enquiry in March and April 1881 quickly revealed the source and made it possible to identify the ringleader and arrest him. The witnesses, however, remained obstinately silent, though one of them did reveal an irretrievable hoard. In the south flank of the escarpment of Deir el Bahri the ancient Egyptians had dug a pit two metres wide and more than ten metres deep. A narrow corridor seventy metres long at the bottom of it led to an oblong room. The passage was blocked with objects, coffins, boxes of shabtis, and canopic chests. Clambering onwards, the bemused archaeologists recognised in the light of their candles the names of the greatest pharaohs and their consorts: Amenophis I, Tuthmosis III, Queen Nefertari, and in the chamber, Ramesses II. There was no possibility of making sketches, or even taking notes. The investigators only just had time to get the objects out by working forty eight hours at a stretch in the height of summer, and to move them to the foot of the hill. The booty was transported across the Theban plain to a barge which carried off this astonishing cargo which took no less [illegible] four years of study to catalogue. All the [illegible] down the Nile the joyful reverence of the pe[illegible] greeted the passage of the barge, the shrill cries of women and the salvoes of gunfire made by the men.

Sixteen years later Victor Loret discovered another cache of royal mummies in the tomb of Amenophis II, right in the middle of the Valley of the Kings. Tuthmosis IV and his illustrious son Amenophis III as well as the later rulers Ramesses IV, V and VI were added to the

Opposite page: the restored palace of Knossos looks out on the hills of Crete. Schliemann did some exploratory work on this site in 1886 but he died the following year. Arthur Evans acquired much of the site in 1894 but political disturbances prevented him from full scale work until 1900 - it was to go on for twenty-five years. Evans has been criticised for his wholesale restoration but his triumph in archaeology cannot be challenged.

The royal tombs in the Valley of the Kings were an irresistible lure to numbers of archaeologists, and their intensive work almost prevented the most famous discovery of all. The large doorway on the lower right leads to the tomb of Ramesses IV - work upon which obliterated all visible traces of the more modest dimensions of the tomb of Tutankhamun.

When Howard Carter opened the coffin of Tutankhamun he found a solid gold mummy-case containing the young king's body. He is shown examining the wrappings, which had become extremely brittle as a result of the unguents poured into the coffin during the burial rites.

splendid collection of royal mummies.

Before they were placed in the tombs, the mummies and funerary goods had usually an eventful history. As a result of the incessant robberies which disturbed the necropolis during the later reigns of the New Kingdom, they had to be moved many times. Legal records relate the manoeuvres involved. The mummies had to be rewrapped, and in the end the priests succeeded in thwarting the thieves. Incomparable treasures must have been lost in the course of all these refittings and removals.

Among the finds from the cache were several papyri which identified the personages of the Nineteenth Dynasty. For a long time little had been known about it, and the monarchs themselves, Solomon's contemporaries, had remained very indistinct figures. It was not until 1939, on the eve of the Second World War, that their remains were discovered in the eastern Delta, a long way from Thebes. In 1928 Pierre Montet, who was then a professor at the University of Strasbourg, decided to reopen work on the huge tell of Tanis, which had been abandoned since the excavations of Mariette and Sir Flinders Petrie. From 1929 onwards he discovered some fine statues and remains of sacrifices, illustrating the original culture of this town which was more exposed than the rest of the country to Asiatic influences. It was only with his eleventh campaign that he reached the necropolis of Psusennes and Osorkon of the Twenty-first and Twenty-second Dynasties. This enterprise was continued until 1940, and after the Second World War in 1945.

In contrast with the long corridors and rock-cut chambers of the Valley of the Kings, the tombs of Tanis are small rooms built into a layer of alluvial silt. Large stone sarcophagi annexed from earlier owners were placed there before the heavy roofing slabs were set in position. The equipment, too, is different from the finds from the Upper Egyptian tombs. Apparently perishable goods were avoided as much as possible. They would have no chance of surviving in the humidity of the Delta. However, some brilliant silver sarcophagi, gold masks, precious jewels, vases and fine goldsmith's work have been preserved.

The pharaohs who ruled in the Delta during this period adhered to the custom of having themselves buried in the temple precincts of their gods.

The Ethiopian renaissance It is to be hoped that one day excavations will be carried out at Sais, since it was there, so the Greek

The contents of the crowded antechamber of Tutankhamun's tomb were found in confusion but apparently intact. Seeing a tomb-robber's hole through the inner wall, the excavators feared that the mummy had been plundered, but the thieves were evidently surprised before they were able to penetrate as far as the royal sarcophagus.

historian Herodotus tells us, that the sovereigns of the Twenty-sixth Dynasty were buried.

However in the Theban district, at the end of the eighth century BC, while the Assyrians were striving to overcome the north of Egypt, the Sudanese princes of the fourth cataract formed a powerful empire which took control of Egypt. This was the period of the Ethiopian dynasty, of Shabaka and Taharka, whose formidable names re-echo through the books of the Old Testament. Many Theban monuments show the features of these kings, so typical of the upper Nile valley, with their strongly marked cheek-bones, powerful jaws and greedy lips. Rather low on their foreheads are two uraei, symbols of the double monarchy of Egypt and the Sudan. As worshippers of the ram-god Amon, they conformed to the religious sentiment of their times by building innumerable small shrines to Osiris, 'Lord of life and eternity', the god of resurrection. One type of building seems to have originated in this period. This is the colonnaded propylaeum in which elegant lotus papyrus columns alternate with round-topped walls decorated with scenes of royal jubilees and, on the exterior facade, the roll of the names. Shabaka endowed one at Luxor and perhaps at Medamud, and Taharka built them at the four cardinal points at Karnak, at each of the main entrances to the great precinct of Amon and Maat. When they were making war on the northern borders or retiring to their distant residences in the land of Kush, the rulers entrusted the state of Thebes to some of their relatives, the divine votaries. These were sacred virgins dedicated to the service of the god Amon of Thebes. A true artistic resurgence, which should be known as the Ethiopian renaissance rather than the Saït as it is generally called, dates from this period.

The mortuary palace of Mentuemhat

Convincing proof of this renaissance is furnished by a tour of the wonderful mortuary palace which Mentuemhat, the leading figure of this period, caused to be carved in the Theban Assassif. He was 'fourth prophet of Amon, lord of the city'. Excavations in the nineteen-fifties have revealed still more about him. Ancient travellers, like modern scholars, never paid much attention to the Assissif tombs. They are buried under a thick layer of assorted debris, and their long labyrinthine passages are difficult to penetrate. They are haunted by flocks of humming and twittering bats. The hot and heavy air permeated with the nauseating smell of bats' excrement, the dust from limestone chips, the decomposing mummies and funerary goods have scarcely helped learned investigation. On the other hand, illegal excavators have had a free hand, and have taken full advantage of it. Apart from a few fragments here and there in various museums, our knowledge of the tomb of Mentuemhat is confined to a single chamber, although in fact, the dimensions of the entire complex were extraordinary. The enclosure, made of dried brick, consisted of a great arch and a series of bays and bastions. The interior of the tomb was approached by a long ramp and an anteroom carved in the rock. The approach continued from there into a vast chamber, also carved in the limestone, and then emerged in an easterly direction into two courts on to which a series of rock-cut shrines opened. Their doors, decorated by splendid motifs of two conjoined papyrus stems, were framed with columns decorated with the names of royalty and greetings dominating the seated statue of the dead man. Here is a translation into monumental terms of a theme borrowed from the decorative repertoire of the Old Kingdom. In the west wall of the second court was the true entrance to the subterranean part of the tomb, a series of vast rooms, shrines, passages and stairways with walls covered with inscriptions and scenes in a mature artistic style, the elegance of which does not detract from its strength. There are no less than four storeys of rooms joined by three stairways leading to the shaft room, the walls of which are carved with niches for powerful statues. Underneath the shaft room a fresh suite of chambers leads to the sarcophagus room.

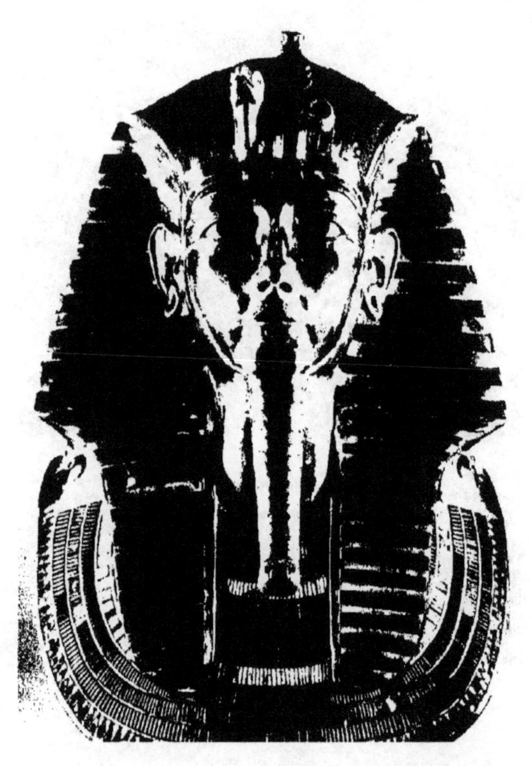

The funerary mask of the young pharaoh, Tutankhamun. This is the finest funerary mask ever found and is made of beaten gold, which was inlaid with coloured glass-paste and semi-precious stones

It is to be hoped that a thorough excavation will be undertaken to restore all these elements to their proper place. The publication of a mortuary palace like this would add one of the most splendid monuments of the ancient world to the credit of Egyptology, and greatly extend our knowledge of the period.

Hellenistic Egypt

In spite of an unquestionable state of decadence, Egypt was to produce monuments for hundreds of years, even though these masterpieces have not been of great interest to students of Egyptology. There are few monuments more impressive or better preserved than the great temples of Upper Egypt built by the Ptolemies

One of the best-preserved monuments of ancient Egypt is the temple of Horus at Edfu. The excavators even found some of the roofing intact, and though the colour which once enhanced the reliefs is gone, the complex has changed very little from its ancient condition.

A gold mount from a New Kingdom ostrich-feather fan shows the king hunting ostriches from his chariot. In this case the problem of reconstruction is simplified by the fact that an exactly similar fan is shown on the left of the mount, just behind the chariot.

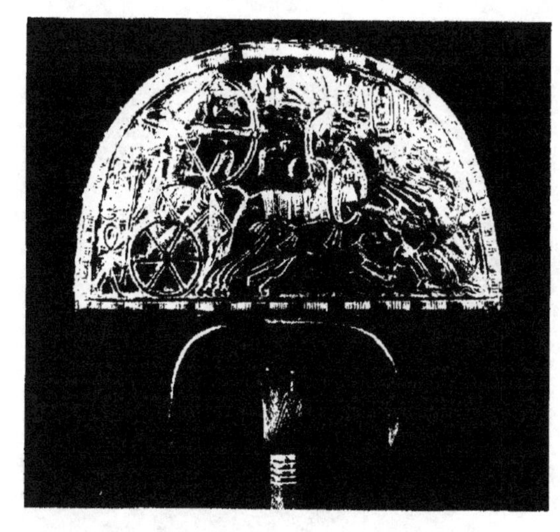

and the Romans. We should not forget that the Macedonian kings who succeeded Alexander were regarded as pharaohs by the Egyptians, and the same was true of the later Roman Caesars. Their cartouches appear along the lofty walls of Edfu and Denderah and contain the traditional formulae of the royal title. Priests of the later period managed to obtain funds from the successive rulers of the land for the glory of the dynastic god Horus and for Hathor one of the most gracious of all goddesses. Innumerable paintings show Pharaoh celebrating the rites of the cult and the deities granting their favours to the part of Egypt he represented. Inscriptions endlessly repeat the traditional epithets of the gods and hymns of praise. These monuments are remarkably well preserved.

The temple of Edfu, excavated by Mariette, which had been buried in a mass of assorted debris for centuries, displayed the simple arrangement of its component parts on an enormous scale. Beyond a great court, closed off on either side by two porticoes, the actual building, which is surrounded by a corridor, began with a pronaos with impressive columns. This was followed by a second hypostyle hall known as the 'hall of audience'. Before the sanctuary and the adjacent rooms the 'chamber of offerings' gave access to stairs leading to the roof. The terrace was encircled with a high wall shielding it from profane eyes. Some ceremonies were conducted there, particularly in a small kiosk similar to those on the roof of the temple at Denderah.

If we leave the grandiose piles of these temples we come to one of the loveliest jewels of ancient art: the entire island of Philae. It is only 400 metres long by less than 160 metres wide, but it contains masterpieces. Nectanebo built a wonderful portico there. The temple of Isis was reached by a double portico carved with the name of Augustus; the nature goddess was worshipped until the reign of Justinian in the sixth century, when the empire had already been Christian for a long time. The little birth temple, or *mammisi*, dedicated by Ptolemy III (Ptolemy Euergetes), was completed in the reign of Tiberius. Nearby is a small shrine of Osiris probably built under the Antonines, a little sanctuary of Imhotep, the charming pavilion of Trajan, and finally the temple of Hathor begun by the Ptolemies and completed by Augustus.

During the first half of the twentieth century these sanctuaries have been hidden by the flood. Paradoxically the new Aswan Dam constructed upstream has allowed Philae to escape, for the present water level downstream is lower. The island has thus been restored to the bright sky of Egypt all the year round.

Nubia and the Sudan

The Aswan High Dam and its consequences The high dam (Sadd el Ali) which was built south of the present Aswan reservoir regulates the flow of water and has, with its immense generators, doubled the electric current available to Egypt and reclaimed 80,000 hectares of arable land. Thus it has increased the country's farming land by twenty per cent. On the other hand, it has raised the water level from 120 to 180 metres by damming up a huge artificial lake which has spread upstream above the second cataract, in the very heart of Sudanese Nubia. The valley has been completely submerged over a vast area 500 kilometres long by fifteen to thirty kilometres wide, and the whole Nubian sector of Nilotic archaeology above the dam has been lost. The Nubians were compelled to evacuate the flooded area and settle in regions some distance away. In addition a veritable cultural disaster threatened the innumerable unexcavated temples, fortresses, cemeteries and remains belonging to the Christian period which lay buried in the sand. In comparison with Egypt Nubia has been rather neglected by archaeologists because of its extremes of climate, harsh working conditions and its comparative indifference (or so it used to be thought) to its own artistic heritage.

In 1955 strong feelings were roused in archaeological circles by the news of the plans for building the high dam at Aswan. From that time onwards, archaeologists had to compete with the rapidly rising waters, which were due to reach their maximum level in 1968. Thirteen years was considered very little time, in view of the task to be accomplished. In 1959 Egypt and the Sudan asked UNESCO for help, offering fifty per cent of the finds to any team which agreed to come and work in the condemned area, and teams from twenty-two countries soon converged on threatened Nubia. They had to conduct an exhaustive study of an immense

The god Horus, shown as a crowned hawk, watches the entry to his temple at Edfu in Upper Egypt. The complex, begun in 237 BC and finished 180 years later, is the work of the Ptolemies.

The quartzite sarcophagus of King Tutankhamun, one of the last of the Eighteenth Dynasty pharaohs, is protected by the outspread wings of four guardian goddesses who watch each of the corners.

area, in a region which was difficult to reach, very short of manpower, and had a climate of tropical violence. They surveyed numerous hitherto neglected regions and analysed the architecture and epigraphy of existing monuments. Hundreds of new sites were identified, graffiti and inscriptions were reproduced in detail. With the aid of photogrammetry employed by the French national geographic institute, a series of conjoined photographs made it possible to record the details of the monuments and their decoration, and even to reproduce the buildings themselves in relief, with all their carved walls in exact detail. As a result of these patient investigations world archaeology and particularly African history, which was hitherto so obscure, have been enriched.

The projects for removing the Nubian temples, which would have been drowned when the Aswan Dam was built, were spectacular — more spectacular than profitable to scholarship. It was essential not only to study the monuments but to attempt their rescue, and as a result, several sanctuaries were completely dismantled and rebuilt on the banks of the future artificial lake, or at Khartoum. The labour needed for this gigantic operation can easily be imagined. Every stone had to be listed and numbered, and every detail recorded. In the case of Amada, the temple was removed from its old site after lengthy preliminary preparations carried out with attentive care by a French team. Between the middle of December 1964 and the end of February 1965 the building was shifted a distance of 215 kilometres, with a difference

The island of Philae at the First Cataract contains an architectural gem of the Roman period but can only be visited for one month of the year as it has been submerged by the new dams ever since the first hydraulic works at Aswan. It is hoped that a coffer dam may be built to save the island.

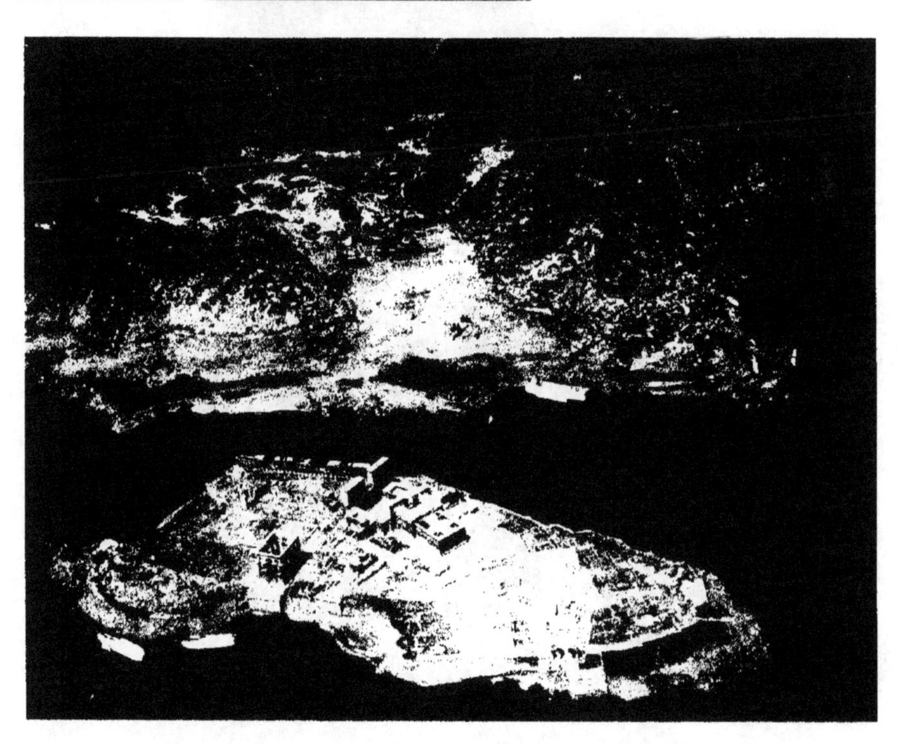

The dismantling of the temple of Kalabsha and its reconstruction near the high dam was a titanic operation. The adjoining structures, the quay, the Nilometer, the *mammisi* and even the enclosing wall were also moved. It had been necessary to prepare in advance a new site forty kilometres away in a specially planned area, and to make roads and docks for the transport of 13,000 blocks. This exploit was carried out by German scholars and engineers with the aid of hundreds of Egyptian labourers, working in shifts day and night for months on end in very trying climatic conditions.

The most difficult operation of all, however, was the project for saving the two temples of Ramesses II and his wife Queen Nefertari at Abu Simbel. The scheme for protecting the temples by a surrounding coffer dam was not adopted because of the danger of leakage and the high cost. The proposal to lift the temples in one piece by means of giant steel jacks was also rejected for the same reason. Finally, a simpler solution, thought to be less difficult, was adopted. The colossi of Ramesses II were cut into sections and removed. The statues, pillars and reliefs on the ceiling and walls were cut away with saws, and these detached pieces were then attached by iron staples to a concrete frame built on top of the cliff in the shape of the two temples. Unfortunately, this has destroyed the remarkable impression that the temple gave of being rooted in the earth, perhaps one of the most attractive aspects of these splendid rock-cut monuments.

Technical developments The impressive scale of these projects for saving the Nubian temples made international headline news. But the technical operations involved should not overshadow the new archaeological methods used throughout these investigations and excavations as a result of the collaboration of scholars from different disciplines who had scarcely ever worked in the Nile valley until this time. Egyptologists were not the only ones to be involved. Many teams of prehistorians, geologists and anthropologists appeared in Nubia, mostly sent by American universities and museums. Scandinavian and Viennese scholars employed the expert knowledge they had acquired in Nordic and central European prehistoric studies. Hundreds of sites ranging from the Acheulian period to the beginning of the Neolithic were found. Human remains were brought to light and carefully examined and analysed to identify their ethnic affinities. International experts, with their different backgrounds and training, brought with them methods of working and procedures from which the Egyptologists could learn, to their own advantage.

Whole regions were 'raked' with the most advanced equipment; soundings were taken wherever a site was found, and these areas distributed among the interested teams for further excavation. All the results were pooled and marked on maps and aerial photographs. In Nubia, where dated inscriptions and major works of art were rare, unlike Egypt where they are so numerous, it was essential to undertake an exhaustive examination of all collected evidence, however insignificant. Pottery and tools, so often relegated to the sidelines by Egyptologists, were studied and classified. The meticulous excavations of some very ordinary settlements shed considerable light on the life and work of ancient Nubian craftsmen and peasants. Scholastically speaking, the leading part was played by the humblest material remains, excavated when a wide range of research techniques were placed at the disposal of a relatively limited branch of archaeology.

The culture of ancient Nubia The excavation of the Sudan revealed that its past was not just that of a simple follower of Egypt, but an original culture with genuine African elements, much more important than had been previously realised. To begin with, the study of a large number of rock engravings of widely different dates, some very early, made it possible to add another chapter to our knowledge of prehistoric Africa. For instance, at Tomas, one of the sites of Egyptian Nubia, along the cliffs there are remarkably forceful and strong representations of great beasts of subtropical Africa. Several schools of rock artists are represented here. Elephants are shown singly or in pairs and have been faithfully rendered in their different attitudes by means of varying techniques, using simple outlines, dots of different sizes, sunken relief or cut-away backgrounds which leave the designs in reverse. In one case the hunters, two for each beast, are shooting at them with bows at point-blank range, probably a magical scene rather than a picture of a real hunt. Herds of giraffes and ostriches are also represented; native to a dryer climate, they are possibly evidence of the fauna of their period. They seem to have been kept in some kind of domestication, as a few of the ostriches have ropes knotted round their necks and elsewhere a large lasso is aimed at a group of giraffes. Whole herds of gazelles, antelopes,

Innumerable photographs and countless exhaustive records are all that now remains of the original rock-cut temple of Ramesses II at Abu Simbel, the site of which is now covered by the waters of the new Aswan dam.

Nefertari, wife of Ramesses II, shown in appropriately small scale beside the seated colossus of her husband, had her own temple on the doomed site of Abu Simbel. The sculptures and reliefs had to be cut away from the rock face and removed to safety.

goats and cattle are also shown. The rock artist has succeeded with rare skill in showing the tremor that runs through the herd when it registers a dangerous noise or scent. Several animals are shown near a grill, some beside linear or broken marks which should surely be interpreted as a real or magical trap.

The goats and cattle, also drawn on the cliffs, must have belonged to a group of settled people, rather than a race of nomadic hunters. These animals were probably already domesticated, and they furnish some equally curious details of life at that time. Some have pendants hanging round their necks, sometimes their horns show signs of deformities, and finally their hides are decorated with various chequered patterns. Many of these same characteristics can be seen in the rock art in parts of the Sahara, notably at Ennedi and Tibesti. Similar details have also been noted on vases of the culture known as Group C. This being so, many engravings can be attributed to this basically pastoral civilisation, which is known to have dominated Nubia for a thousand years (2600–1580 BC).

Studies of the burial objects found in the numerous Group C cemeteries, mostly by the Scandinavians, supplied more facts about this warrior race and enabled us to define the limits of its expansion, the extent of its power and the circumstances under which it became integrated with Egyptian culture. In the present day when so many contacts, sometimes violent and too often cruel, take place between widely differing cultures, it is interesting to examine a similar process in the distant past.

In addition to the periods contemporary with the great ages of Egyptian history (which are, for this very reason, less obscure), Nubia had others which have been named by simple letters of the alphabet, a fact which shows how little we know about them. After the kingdoms of Kush and Meroe (800 BC–AD 350), for which recent excavations have yielded some important texts in a highly unpleasing cursive script, another culture appears: the X Group.

There are many tombs dating from the X Group era, often poorly furnished, mostly in the southern part of Nubia. A remarkable discovery concerning these obscure years was made at Ballana and Qostal a few miles south of Abu Simbel. The Nile lowland there is scattered with large mounds which Johann Ludwig Burckhardt, the indefatigable Nile explorer, had already noticed at the beginning of the nineteenth century. It was not until 1929, however, that the excavations of the British team of W. B. Emery and L. P. Kirwan discovered that these tells contained the tombs of powerful local chieftains. Buried with their servants and their richly caparisoned horses, these Nubian kings took their treasures with them to the grave. Jewels consisting of heavy silver crowns and bracelets set with coloured cabochon stones are curiously reminiscent of the gold work of their western contemporaries, the barbarian kings.

The glories of Christian Nubia Some very recent excavations in a cultural setting so different from our own have brought to light the touching remains of Christian Nubia, perhaps one of the most brilliant periods in the country's history. In 1960 the Warsaw University expedition led by Professor K. Michalowski set to work on the citadel of Faras on the

present border of Egypt and the Sudan. Beneath the Arab fortress on the top of the *kom* the archaeologists found the cathedral of Pachoras buried in the sand. This was a vast basilica with magnificent mural decorations flanked by adjoining structures which were the tombs of pious bishops. The ruins of a second church were also unearthed, with many stelae and inscriptions concerning the history of the buildings and the eparchs of Nubia.

The cathedral of Pachoras has three naves extended by aisles and annexes, the central nave leading to the semicircular apse and a platform with seven steps. All the walls are covered with paintings in a splendid state of preservation. In certain places the partial collapse of the top layer of the structure has exposed layers belonging to earlier periods. Very careful clearing has made it possible to recover various panels corresponding to the successive stages of the decoration. Among the 150 masterpieces which constitute the best known source to date of information on medieval Nubia are representations of the archangels Michael and Gabriel in the eighth century narthex, an eighth- to ninth-century portrait of Archbishop Ignatios of Antioch, and portraits of Bishops Petros and Marianos dating from the late tenth to early eleventh centuries. The pictures of the three Israelites in the fiery furnace can be dated to this same period. Against the orange-red background of flames, the angel's robe and his two great wings stand out with graceful asymmetrical movement, accentuated further by the green, brown and yellow embroidered garments of the three figures in traditional Parthian dress. The paintings of the Madonna of the Nubian ruler and the great composition of the Nativity on the wall of the north transept date from a later period.

Christian frescoes on the wall of an early church at Faras in the Sudan presented a difficult problem. Several superimposed layers had to be studied without damage to the brittle plaster and ancient pigments.

A twelfth-century fresco of an angel guarding the three Israelites in the fiery furnace, from the Christian church at Faras, is examined by the excavator, Professor Michalowski.

The capital of the southern kingdom was transfered from Napata to Meroe towards the end of the sixth century. The Sudanese rulers gradually lost interest in Egypt, but their buildings, especially a group of pyramids, show that they had been profoundly influenced by Egyptian culture.

The Virgin, a large figure, is shown resting on a bed watched over by archangels. Behind her Christ's cradle is flanked by an ass and some sort of humped cattle simply drawn. Leaning on their sticks, the two shepherds are standing on one foot, the other leg crossed over the first like the present day shilluks of the upper Nile. The three kings, dressed in the Parthian style, ride curvetting horses. This is a new aspect of Nubian art. The painting of Christian Nubia is very different from Coptic art in both style and subject matter. It is a separate but parallel branch of the first eastern Christian cultures, the origins of which should perhaps be sought in other parts of Western Asia. Paintings from this period also shed new light on the origins of the decorations of the earliest Ethiopian churches.

Dedicatory stelae found in the wall of a building south of the church at Faras bear the date of AD 171, or the year 11 during the reign of King Merkurios 'the new Constantine'. In all, more than 200 inscriptions in Coptic, old Nubian and Greek, as well as more than 200 graffiti have been found on the plaster of the church. This was a windfall for the experts, since texts written in old Nubian are extremely rare. The most remarkable document from the historical point of view is undoubtedly the list of the twenty seven bishops who held the See of Pachoras (Faras) between the seventh century and AD 1169, i.e. a few years before this part of Nubia was conquered by Islam.

The discoveries in Nubia are therefore more varied and more important than might have been supposed, and Nubia has shown itself to be rich and diverse beyond expectation. Its prehistory reveals that it was one of the most outstanding civilisations of early Africa. It was from similar origins that Egypt drew apart after a series of very swift mutations and consolidated its own characteristics and its individual density. Throughout the course of history Nubia has remained the permanent relay station between Egypt and the rest of Africa. One of the most essential tasks of historical deduction is tracing the direction of influences between the two areas as they shifted with the passing of time, and interpreting their meaning, their fluctuations and reverses. If Meroe seems to play an important part in the development of the civilisation of Africa, it was via Nubia that the splendid culture of the pharaohs was transmitted to that continent.

The empire of Meroe The culture of Meroe, known as Ethiopia to the ancients, the most southerly of the Nile civilisations, was for a long time plunged into obscurity and recalled only by some fragmentary descriptions in a few classical authors like Herodotus and Diodorus Siculus. Only recently has this civilisation been rediscovered by archaeology.

The excavations conducted by the American George Reisner in the present Sudan from 1916 to 1923 on the sites of Napata and Meroe were the first to shed light on the history and origins of the ancient Meroitic empire. Its forerunner, the kingdom of Kush, which was a vassal of Egypt at first, gradually attained a *de facto* independence towards the end of the New Kingdom. The kingdom of Kush reached the peak of its power towards the end of the eighth century when the Kushite Twenty first Dynasty (the Ethiopian kings) established its hegemony over Egypt. Expelled in the middle of the seventh century by the Assyrian armies, the Kushites returned to their traditional territories. Egypt soon became hostile, cursing their memory and eradicating every trace of their domination. After the break in commercial and diplomatic relations with Egypt, the first capital, Napata, below the fourth cataract, was found to be too close to pharaoh's frontier and too economically isolated in a region without resources. It was probably this unfavourable situation which led to gradual desertion in favour of a new capital much further south, a little downstream from modern Khartoum – this was Meroe.

Meroitic archaeology is a new discipline, and it has to answer a great many questions. Did the sovereigns buried in the imposing cemeteries of Napata and Meroe belong to two or more parallel dynasties, Napata being the residence of the minor provincial branch? What exactly were their relationships of kinship and succession? Future archaeological discoveries and the constant progress of Meroitic studies may supply the answer to some of these riddles. Although F. L. Griffith's decipherment has explained the signs of the Meroitic script, the language has so far eluded translation.

Our knowledge of Meroe is limited, in fact, to the evidence of the finds made by the pioneers, John Garstang, George Reisner and Griffith, and by the recent methodical explorations. The resulting evidence makes it possible to recognise some of the chief characteristics of the Meroitic civilisation. Although it was a hybrid in many respects, the persistent vitality of the primordial Kushite stock allowed the assimilation of foreign elements as well as the development of a homogeneous culture.

The Meroitic sovereigns, who were god kings like the pharaohs, borrowed many customs from Egypt. On the walls of the temples at Meroe, Nagaa and Mousawarat es Sofra, they are shown in traditional royal poses, worshipping a god or slaughtering prisoners. The designs of their buildings and the architectural elements indicate Egyptian inspiration, but the proportions, style and the details of the reliefs are far removed from pharaonic canons. The splendour of brute force is demonstrated in Meroitic art, combining the elements of massive strength and personal luxury, in the richly embroidered garments and the heavy gold bracelets of the kings and 'Candaces'. There is also an element of cruelty and a taste for horror, shown by kings transfixing their prisoners with spears or vultures digging their talons into the backs of bound enemies. The Meroitic pantheon, as it appears in the portraits and inscriptions decorating the walls of the temples, also combined the Egyptian gods with foreign deities like Debyumker and Apedemak. The latter is a lion god, sometimes three headed. On another wall the crowned head of a lion rises above the coils of a snake, which itself emerges from the calyx of a flower. All this points towards some influence from the direction of India; and recent excavations have also brought to light a curious elephant headed statue reminiscent of the god Ganesha of the Hindu pantheon.

Other elements of Egyptian religion seem to have penetrated the Meroitic empire and evolved into a local variation there. The same is probably true of burial customs. The pyramids standing in the desert wastes of Napata or in the wadis of Meroe, or in the cemeteries of the capitals of Sedenga or Gebel Adda were modelled on the pharaonic form, but their construction, dimensions, and angles are different, as is the skylight left in the top, perhaps for the mysterious *ba* bird. The Meroitic mortuary complex is still little known. A lintel, a stela and an offering table, with the pyramid and the burial vault, seem to have been the chief components. At the top of the texts carved on lintels, stelae and offering tables it is interesting to see on many of them an invocation to Isis and Osiris. The pre-eminence of the goddess over the god can be compared with the increasing popularity of the Isaic cults which spread across the Mediterranean world from Alexandria and throughout the Roman empire. Rather surprisingly, it is generally Anubis who is associated with Isis in the pouring of libations on the offering table. The ancient *psychopompos* god, conductor of the dead to the next world, thus seems to have enjoyed special popularity at Meroe, and his jackal head often accompanies the processions of the Isaic ritual.

Egyptian influences never completely submerge the basic African foundations at Meroe. If the Meroitic empire based its commercial prosperity on its privileged position as the intermediary between Egypt and the rest of Africa, it also played this part in the field of culture. These links and relationships with the rest of Africa are still poorly defined, but certain evidence like the pottery or elements of the language seem to prove beyond question some kind of relationship. Furthermore the god Amon, whose sanctuary at Napata was famous, appears in the form of a ram, an animal found all around the Sahara and associated with storms and rainfall as well as in his aspect as dynastic god. Should this be regarded as pharaonic tradition transmitted through Meroe or rather, as the survival in Egypt of the remains of an ancient African cult, more specifically Saharan, which evolved before the civilisation of the Nile developed?

The most important contribution of Meroe to the history of Africa was the diffusion of iron working. The Libyan Berbers could also have played an important part in the spread of iron technology across the Sahara, but the huge heaps once more show the extent of the role of industrial activity at Meroe, so much so that British scholars have even called it 'the Birmingham of Africa'. A technological role like this could have made the Kushite capital a centre of attraction for all the countries of the south and west.

It was from the east, however, that the final blow came, which laid low the empire which had held its own against Augustus and troubled the *Pax Romana* in Egypt. Meroe was gradually weakened by the repeated attacks of the Black Nuba, the Bobat, the X Group peoples and the *Blemmyes*. Its glory was no more than a memory towards the beginning of the fourth century after an attack by the raiders descending from Axum, the capital of Ethiopia, the new kingdom which had arisen on the Abyssinian highlands.

The great political centre of the Sudan was located at Napata from about 750 until the end of the sixth century BC. Sudanese kings even conquered Egypt and ruled there for many years.

The shrines of the Meroe pyramids are decorated with low relief carvings of processions of small figures bearing palm branches. Excavations have confirmed the importance of Meroe as a centre from which Mediterranean arts and techniques were distributed to the rest of Africa.

The archaeology of Ethiopia

A new study Ethiopia, a vast empire with a history covering 3000 years, provides an immense field of study which has only just been opened to the archaeologist.

Until recently its past was chiefly known to us from the descriptions of ancient travellers. One of the earliest was the Alexandrian Cosmas Indicopleustes who, in his *Christian Geography*, furnishes some information on ancient Axum of the sixth century. After this it was not until the sixteenth century and the beginning of the seventeenth that missionaries, mostly Portuguese, wrote accounts which include some encyclopedic descriptions of the Ethiopian empire. As to the somewhat contrived journal of the Scot James Bruce, who spent three years (1769–72) at the court of Gondar, it is an unrivalled source of information. Some nineteenth-century explorers carefully described the remains they encountered on their travels: Lefèbvre and his companions, the Abbadia brothers, Rochet d'Héricourt, Combes and Tamisier between 1840 and 1850, the British troops of General Napier's expedition in 1868, Heuglin, Raffray and Theodore Bent. It was at the beginning of the twentieth century, however, that the first scholarly excavations took place. In 1906 a campaign was conducted in the Axum and Yeha regions by the German scholars E. Littman and K. Krencker. The work of Italian scholars should also be mentioned. Nevertheless, the rich history of Ethiopia, illustrated by the glorious memories of Solomon and the Queen of Sheba, the monarchs of Axum and finally, a church which clung to its faith in spite of so many attacks, was still awaiting excavation.

Accordingly, Emperor Haile Selassie, wishing to 'protect the monuments and other remains of the past which illustrate the history of the empire and its inheritance', called in French experts in 1952 to establish an archaeology department in the Ethiopian Institute of Study and Research. The tasks awaiting the new department were many. They had to draw up a catalogue of known antiquities and ensure their protection, compile evidence about the past from ancient books and writings, and finally carry out a programme of excavation. The department had the advantage of the enlightened support of the emperor, and could benefit from the most advanced techniques. The air force, for instance, made several flights in search of ruins, and systematic plans were drawn of the Axum region, which is rich in ancient remains.

The archaeological work of recent years has been very diverse, and has thrown a new light on the past of Ethiopia. The archaeology of the Red Sea countries, east Africa and the Indian Ocean has also made significant advances, as has knowledge of the links between the Mediterranean world and the Far East.

Prehistory The prehistory of this country, which was a meeting point between the highlands of the great lakes and the Nile Valley, is still very obscure. However, a study of the engravings and paintings on the rocks at Harar and the Diredawa region, Tegre and Eritrea and the specimens discovered at Sidamo suggests that Ethiopia is one of the most important sources of African rock art. The discovery and examination of a large group of dressed flints at Melka Kontourea represents a significant step forward in the study of east African prehistory. Ethiopia also belongs to the great band of megalithic cultures, as shown by the abundance of worked stones sometimes shaped into phallic symbols. However, caution is necessary with Ethiopia as with the rest of Africa in interpreting and dating, since prehistoric techniques can occur well into the historic period.

In any assessment of Ethiopian culture, the salient fact to remember is that in the area where the pastoral people of the great Wiltonian culture lived an original civilisation developed, which is distinguished by its similarities to that of southern Arabia. This area covered most of the north east part of Africa, and many of its characteristics have survived in certain Kushite and Nilotic elements still to be found in these regions. Perhaps we should conclude that Sabaean people of Semitic origin crossed the Red Sea and overran Ethiopia, to impose their rule on the local peoples. In fact, according to the most recent theories suggested by the latest discoveries, Ethiopia does not seem to have actually been a colony of southern Arabia. It is true that there are many resemblances between the two sides of the Red Sea, but there are also some basic differences. In a word, the archaeology of Ethiopia is unquestionably original.

The stages of its development have not yet been defined. A period which might be called Sabaeo Ethiopian (fifth to fourth century BC)

In the fifth and fourth centuries BC a culture showing strong similarities to the south Arabian Sabaean civilisation flourished in Ethiopia. A curious carved frieze adorned the temple in the city of Yeha in the Adowa mountains.

corresponds to the oldest levels so far known, in which Ethiopian reaction against south Arabian influence illustrates the dynamic side of its character. Several hundred years of obscurity then followed, about which we know very little indeed. Finally came the Axum period (towards the end of the first century AD or the beginning of the second) which is marked by the flowering of the most characteristic art and culture of ancient Ethiopia.

The Sabaeo-Ethiopian period Recent excavations have revealed the Sabaeo-Ethiopian period. Ten kilometres south-east of Axum in the Melazo region an investigation of the little mound of Gobochela yielded a sanctuary which seems to be the oldest known building in Ethiopia. The style of the inscriptions suggests the fifth century BC. Some votive tablets show that it was dedicated to the south Arabian moon god Almaqah. Diverse cult objects have been collected at Gobochela. There are two statuettes of bulls, one of which, made of schist, is small but remarkably forceful and has inscribed on its right side a partially readable inscription to Almaqah in south Arabian characters. The other bull is made of alabaster, simply yet vigorously designed. Both statuettes demonstrate the exceptional mastery of animal art attained by the ancient Ethiopian. Several altars have been found decorated with the symbol of the conjoined disc and crescent as well as some inscriptions in the magnificent monumental script, so regular and fine, of the 'land of the Queen of Sheba'.

H. de Contenson's researches in the Melazo

The Meroitic lion-god is shown with the body of a snake emerging from a flower on a pylon in his temple at Naqua. The relief on the adjoining wall shows king Natekamani, queen Amanitare and their son at worship.

Among the most striking monuments of the Ethiopian kingdom of Axum are the enormous stone obelisks, the largest of which (now fallen) would be 33·5 metres high. Decorated with architectural features, they may have been tomb markers.

The Ethiopian kingdom of Axum was one of the earliest realms to adopt Christianity as its official religion. The change dated to the early fourth century. Fine churches in the classic Axumite style have added a new dimension to Ethiopian studies.

region have brought to light the remains of two buildings, the exterior walls of which were surrounded by a bench. On it were a number of terracotta votive objects, mostly figurines of cattle, but also miniature jugs, models of houses and rather crude figures of pregnant women. Parts of two important monuments which are now the pride of the Addis Ababa museum were also found here.

The first is a tall baldachin 1·4 metres high decorated with carvings in very low relief, the style of which is slightly reminiscent of Persepolis. The front of the steps and the edge of the dais are decorated with crouching ibexes, their long horns intertwined to produce a fine decorative effect. On each of the side walls two figures are advancing. One is tall and holds what appears to be a flail before him. The other, smaller, figure precedes him, leaning on a stick with both hands.

The other masterpiece is a statuette in white limestone, eighty centimetres high, of a seated woman with a face more striking than beautiful and expressing great vitality. She is dressed in a full robe with long pleats and wears a heavy collar with a counterweight at the back, signifying her undoubted high position in society.

Discoveries made at the foot of the hill of Haoulti have produced some blue faience amulets—a fragment of a Hathor headdress and a figure of Ptah Soter. These charms raise questions about the relationship between ancient Ethiopia and the Nile valley. Other imports from Egypt or the kingdom of Meroe have been found on various Ethiopian sites: at Axum a scarabaeoid in whitish stone, a statuette of a hermaphrodite and perhaps a cippus of Horus on his crocodiles (or so the traveller James Bruce believed); at Matara a statuette adorned with a Horus lock and two urei was found; at Yeha some large vases similar in shape to the *hes* vases found along the Nile banks. At Hawilea Assaraw some metal vases were found which suggested Meroitic influence, especially a cup ornamented with nineteen lotus stems separated by knobs under a frieze of nineteen frogs, symbol of eternity in Egyptian iconography.

The hoard at Hawilea-Assaraw produced other objects of great importance. Firstly there was a delicately executed limestone statuette, fifty centimetres high, of a woman seated on a stool which stands on a dais. She clenches goblets in the hands, and wears a robe decorated with rosettes, probably once inlaid with precious stones to give the effect of embroidery. The expression on her face is cruel, her thin-lipped mouth dominates a very small chin, her eyes are big, and the pupils were once marked by inlays which have been lost. Her hair is dressed in little curls and cut off cleanly on the forehead to form a sort of pad in front of her ears, which are curiously represented as stylised lunettes. Is this a deity, or a votary who dedicated her own likeness to secure the blessing of motherhood? The dais, on which appear the words 'In order to grant a child to Yemenet', may not belong to this statuette. There is also a fine carved altar which bears a dedication to the god Almaqah made by a 'ruler (*mukarrib*) of Daamat and Saba' to commemorate a victory. Probably the most important find, however, is a carved bronze sceptre or a model of a knife, bearing a commemoration of two victories of 'Geder, King of Axum', written in a script which is Ethiopian (the earliest-known characters in this script). Geder, whose name is similar to one on the 'king list', seems to be the king previously mentioned in the Sabaean texts.

The work of F. Anfray at Yehma, a city in the heart of the terrible mountains of Aduwa, has revealed a necropolis opening on a slope dominated by a south Arabian temple. Many objects were found here, among which are some small metal pendants shaped like an ibex, a lion and a bird, decorated with written characters. The richness of the design and the facility of execution bear witness to the mastery of the craftsmen of ancient Ethiopia in the Sabaeo-Ethiopian period and a little later, before the existence of the kingdom of Axum.

The cross-shaped roof of Beita Georgis, the most important of the early Ethiopian churches, rises above the slope of the hillside in which it is built. Finds from the area attest unbroken contacts with Roman civilisation as well as providing valuable details of daily life.

The kingdom of Axum How did the kingdom of Axum arise? We have not yet enough information to be sure. Excavations at Axum allow us to assign the beginnings of the kingdom to the first or early second century AD. The history of the first kings is still obscure. The first ruler whose name is known was Zoscales. He was covetous but high-minded and captivated by Greek civilisation. In his reign and those of his successors Axum was a prosperous city where trade was active.

Among the things discovered at Axum was a monumental stepped base on which seven 'obelisks' or perhaps giant stelae had stood. The biggest of these stelae, now broken into pieces, was a monolith 33.5 metres high. These stelae are decorated with carvings of tall houses with several storeys like those still to be seen in southern Arabia. The houses have doors with lattices and above them several rows of grilled windows between which the ends of the second-storey floor beams can be seen. Did these stelae perhaps mark the site of a tomb? Stone thrones were also found in Axum. Some had a votive appearance and were dedicated to deities whose likeness was carved on the back or the sides, while others were perhaps memorial thrones dedicated to the memory of deceased kings. Also found at Axum, and as charming as they are puzzling, are red terracotta heads with traces of grey slip. The faces, generally fine featured, are framed by a large wig which juts out at the back. The heads sometimes have a hole in the top and a rather long neck forming a collar and were perhaps used as jar lids.

It is not only at Axum, however, that traces of the Axumite civilisation are to be found. Archaeological sites are awaiting the excavator's spade everywhere in this vast kingdom, particularly among the coastal cities, since the kingdom of Axum derived its political and economic power from the cities along the Red Sea which traded with India and the Far East. The caravan trails of Adulis were the trading routes to the middle Nile and Meroe. The question of free passage to the Red Sea ports could have been one of the causes of the quarrel between Axum and Meroe which ended in the victory of Axum in the fourth century AD. But even before that time Axum's oppressive treatment of Meroe was a major factor in the decadence of this once powerful kingdom. Investigations of the ancient Red

The ruins of the Axumite city at Matara Tigré south of Asmara are dated by their earliest contents to the fourth and fifth centuries BC. Intensive excavations of Ethiopian archaeology are shedding new light on African history as a whole.

Sea ports will no doubt tell us more about the connections between east Africa and the shores of the Indian Ocean, and the relations of the Mediterranean countries with the Far East. In any case, the first excavations at Adoulis have shown that the Axumite kingdom had close contacts with the Mediterranean basin. This explains the discovery near Axum of a statuette of a hermaphrodite and a small bronze bull, possibly imported from Alexandria.

Christian Ethiopia Soon, however, a new factor appeared and completely transformed the Axumite civilisation. At the beginning of the fourth century Frumence, who had been shipwrecked as a child and picked up by the Ethiopians, returned to convert the country. As a result the sign of the cross supersedes the crescent and disc on the coins of King Ezana, the sovereign at this time, while a stela from Axum mentioning a victory over the Nubians and Kushites at the confluence of the Nile and the Atbara is dedicated to the 'Lord of the Sky', surely the Christian god. Ethiopia therefore became Christian at the very time when Constantine was making Constantinople the capital of his empire.

These successive periods can still be discerned in sites such as Matara near Senafea in Chimezanea. From 1959 F. Anfray conducted important excavations there, which revealed splendid buildings in the classic Axumite style, with exterior walls in tiers with alternating bastions and recesses. Objects such as cooking stoves, bowls, pans and small lamps from a large number of finds reveal what everyday life was like. There was also the discovery of a magnificent treasure in the ruins of a church built in the middle of one of the complexes being investigated. In the heart of the Ethiopian highlands gold crosses were found, as well as rich chains and a statire with fourteen gold leaves which exhibit portraits of the emperors from Nerva to Septimius Severus. These provide fresh evidence of the prolonged contacts between this powerful African kingdom and the civilisations of the West.

The work carried out by the Ethiopian Institute of Archaeology has therefore encompassed the excavations of a number of important sites in a few years, as well as the illumination of many obscure points of Sabaeo-Ethiopian pre-Axumite and Axumite history; but there is still a lot to be done in these fields. At a time when the African continent is beginning to take an active interest in its past, the findings from the part of Nubia now submerged by the high dam have brought about a better understanding of the relations between the Mediterranean, the Nile valley and the rest of Africa. It is vital to be able to integrate the highly original culture of ancient Ethiopia into the scheme of all the civilisations, from the most classic to the least known.

Archaeologists are often presented with problems when later generations adapt buildings and shrines to their own requirements. Here an existing shrine of a pharaoh has been changed for Christian worship. St Peter is seen above an altar surrounded by pharaonic reliefs which have been painted over.

Opposite page: a view of Delos from Mount Cynthus. The birthplace of Apollo and his sister Artemis, Delos was a sacred island to the Greeks and is mentioned in the *Odyssey*. The archaeological exploration of Delos was carried out by the French School at Athens: work began in 1873 and Théophile Homolle uncovered Apollo's temple in 1877.

The Aegean world

Many Cretan motifs and processes were adopted by the Mycenaeans, but after modifications to mainland taste they are often barely recognisable. However, they supply important evidence of close and influential contacts between Crete and Mycenae during the later Bronze Age. A much corroded silver rhyton in the form of a bull's head picked out with gold details (the horns are reconstructed) from the shaft graves offers an interesting comparison with the similar rhyton found at Knossos.

The History of Aegean archaeology

Aegean archaeology is a comparatively new discipline which did not start until the end of the nineteenth century. Until then our knowledge of the Bronze Age in Crete and Greece was necessarily drawn from the *Iliad* and the *Odyssey*, distorted as they were by the annotations of twenty centuries [illegible] scholarship. [illegible] the existence of Minos and Agamemnon was confirmed or invalidated [illegible] by philological and artistic criteria. The '[illegible]gians' (a name given by later authors to the first inhabitants of Greece) were known only [illegible] few sparse and fragmentary remains.

The decisive step from myth to history and from literature to archaeology was taken by Heinrich Schliemann. According to his autobiography, which is not devoid of self-congratulation, but on the whole sincere, Schliemann's belief in Homer began in childhood and inspired him with the idea of publicly vindicating the poet's veracity. Obliged to earn his own living at a very early age, his commercial talents and a remarkable gift for picking up languages enabled him rapidly to accumulate a considerable fortune. About the age of forty he resumed the study of Greek, which he had hitherto refused to do because of his belief that this fascinating language would completely enthrall him and interfere with his business. Having read and reread the *Iliad* and consulted the accounts of travellers, he departed for the Troad. Among the tells which broke the monotony of the Trojan plain he chose Hissarlik (*Hissar* is Turkish for castle) which had long been known as the site of Hellenistic and Roman Troy. During the years 1870 to 1890 Schliemann excavated nine successive Bronze Age levels and thus definitively established the site of Homeric Troy. It was more than a personal triumph.

Many successors have corrected or verified Schliemann's theories, the first to do so being the architect and archaeologist Wilhelm Dörpfeld, who was his assistant during his later years and carried on Schliemann's work in 1893 and 1894. Then there was the American team led by C. W. Blegen, which conducted seven meticulous campaigns from 1932 to 1938. Nobody, however, can deprive Schliemann of his claim to be the founder of Aegean archaeology, having discovered the kingdoms of Priam and Agamemnon.

Though the site of Mycenae had been known for a long time, it had never been investigated. In 1876 Schliemann brought to light a circle of royal tombs (now known as Grave Circle A), the contents of which far surpassed all the finds of mainland Greece in richness and amply justified the Homeric phrase 'Mycenae, rich in gold'. After the discovery of the famous gold burial masks he was able to send a cable to the German emperor in which he said he had 'looked on the face of Agamemnon'. Schliemann's chronology was later shown to be inaccurate by the Greek archaeologist Christos Tsountas, who excavated the palace at Mycenae at the end of the last century, and J. Papadimitriou, who found a second circle of royal tombs known as Grave Circle B in 1953. The British archaeologist A. J. Wace, who made many discoveries at Mycenae, and the American C. Blegen, who excavated the palace of Nestor at Pylos, have also corrected Schliemann's chronology. Nevertheless his claim to be the originator is beyond question. Apart from the site of Mycenae, Schliemann also dug that of Tiryns. Shortly before his death he had planned to go to Crete to look for the kingdom of Minos, and only the diplomatic difficulties of the time (Crete was still under Turkish rule) prevented him from being the excavator of Knossos.

Thus was to be a privilege reserved for the Englishman Arthur Evans. There was a great contrast between Schliemann, the inspired, self educated man, and Evans, the son of a wealthy archaeologist, with a traditional English classical education. After several expeditions to the Balkans, he settled down in Crete where he formed a collection of gems engraved with hieroglyphic signs, and as soon as the island was released from Turkish rule, began to excavate the site of Knossos. From 1900 to 1905 he virtually cleared the 'Palace of Minos'. Evans' record of the work was to be the Bible of Cretan archaeology for many years and is still a valuable source of information. It described the palace (which Greek imagination subsequently identified with the labyrinth, the lair of the Minotaur), and proposed a detailed chronology, established on the evidence of accurately dated Egyptian objects among the Minoan finds. Present-day archaeologists accept the fact that Evans' theories often need modification, because he came to regard Crete as the centre of Aegean civilisation, and could not bear to be contradicted on this point.

Opposite page: a figurine from Crete, dating from about the sixteenth century BC and now in the Fitzwilliam Museum, Cambridge. The figure is different from the others found at Knossos; it is marble, not terracotta, and the familiar serpents are absent from the arms. It may be simply a statuette of a worshipper, or possibly of a priestess of a lesser rank.

Dr Heinrich Schliemann (1822–90), a self-educated linguist and scholar, was the first to reveal that the epic poems of Homer were confirmed by sound archaeological fact. His investigations of the Bronze Age cultures of Greece and Troy, which added a whole new dimension to classical scholarship, were based on his unwavering faith in the accuracy of the Greek literary sources in the face of universal academic disbelief.

In his middle years Schliemann married a distinguished Athenian girl named Sophia who subsequently accompanied him on all his excavations. During his expedition to Troy he discovered a golden treasure in the early Bronze Age levels which he firmly believed to be a relic of Priam's city. Although he was mistaken in this theory the jewels, in which his wife was photographed, shed a great deal of light on the advanced metal-working processes of ancient Anatolia.

A. J. Wace was virtually banished from Mycenae for several years for having been the first to suggest the theory, soon to be confirmed, that it was the kings of Mycenae who conquered Crete, and not the other way round. Evans' memory has been somewhat unjustly condemned by the 'Mycenologists'. The fact remains that neither his faults nor his prejudices can obscure the greatness of his work.

After Evans a number of excavations, some of which produced spectacular results, were opened in Crete. Among many others, the names of Luigi Pernier (the palace of Phaistos), Fernand Chapouthier (the palace of Mallia) and Joseph Hatzidakis (the Minoan villa of Tylissos) should be mentioned. However, the island is far from being thoroughly explored; in 1963, for instance, N. Platon found a new palace at Zakro in east Crete.

Although his work was more philological than archaeological, it seems right to end with a reference to Michael Ventris, to whom we owe the decipherment of the Mycenaean language. Evans and Blegen had found hundreds of inscribed tablets at Knossos and Pylos, and the work of two generations of scholars had established that this was a syllabic language using a decimal system of numbers. An architect by training, Ventris, like Schliemann, had a remarkable flair for languages. After several years' work, this dedicated amateur succeeded in proving that the Mycenaean language, which had formerly been thought to be either Hittite or related to Etruscan, was actually very early Greek, much more primitive than Homer's, but with similar names for some gods and familiar words for vase, tripod, ass and horse. Announced in 1952 and published in 1953 with the collaboration of the philologist John Chadwick, this decipherment was accepted by the great majority of linguists and archaeologists.

At the moment Aegean archaeology is a thriving discipline in which any preconceived notion may be called into question from one day to another. In the following pages more hypotheses than certainties will be found. Indeed, there is still much to be done, and archaeological investigation is far from complete, either in Crete or in mainland Greece, while in Asia Minor the surface has hardly been scratched. The oldest Cretan texts (the hieroglyphic script and syllabic writing known as Linear A as opposed to the Mycenaean form known as Linear B) have not yet been deciphered. This being so, there is no question of presenting an account based on a series of categorical facts. We more often find ourselves confronted by problems which are still awaiting solution.

The Paleolithic and Neolithic periods

Though scattered and fragmentary, incontrovertible evidence indicates that Greece was inhabited at a very early date. Indeed, a series of Middle Paleolithic (Mousterian) sites have been discovered in Thessaly, the Argolid and Elis. They show a much less advanced culture than that of Western Europe, with neither sculpture in the round nor wall painting, and rather poor stone tools.

The Neolithic period, on the other hand, is much better represented. The richest sites are found in the Argolid (Lerna, Corinth and Prosymna), Boeotia (Eutresis and Orchomenos), Thrace (Dikili Tash and Galepsos) and especially Thessaly, where many tells have yielded quantities of finds (Sesklo, Dimini, Gremnos Magoula, Argissa Magoula, etc.). Although chronology is still very doubtful, prehistorians generally agree that the Neolithic period began at the end of the seventh millennium and lasted until 3000 BC.

During the Early Neolithic, towards the *middle of the sixth millennium, pottery made its* appearance. There can hardly be any doubt that this skill was imported from Western Asia, where it had already existed for a long time. Vases, made by hand, are mostly in the form of large bowls, with or without a low foot. Monochrome at first, from about 5000 BC they began to be decorated with some relief ornaments (lozenges and bosses), but more often with paint alone. The motifs were exclusively geometric (broken lines and dog tooth patterns) and were first carried out in red ochre on a light clay ground. Then technique improved, the shapes of vases became more varied and the decorative repertoire extended: they made goblets, cups and amphorae decorated in white on a red ground or conversely, in red on a white ground with lines, triangles and combinations of dots and crosses.

The next stage (the Middle Neolithic) is marked by the appearance of primitive glazed ware known by its German name as *Urfirnis*, the vase being covered with a thick slip with a shining warm coloured surface which was painted with geometric motifs in red. Finally, in the Late Neolithic, the so called Dimini culture is typified by pottery with incised decoration. The incisions were filled in with white to make the design stand out better. The motifs became increasingly varied and complex, including spirals, volutes, concentric circles and the earliest rough meanders.

As well as pottery, there are other examples of Neolithic craftsmanship to be found in a whole range of polished stone tools (many fashioned from flint and obsidian), axes,

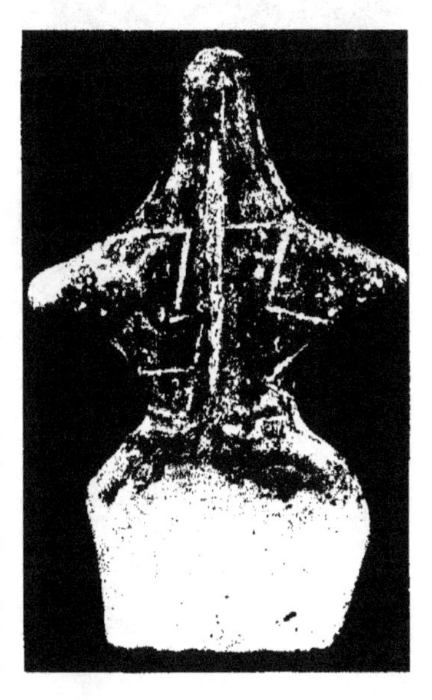

polishers, blades, chisels, grinders, mortars and pestles. There are vases carved in white or grey marble, and limestone. Bone also served as raw material for making needles, awls, gouges, scrapers and daggers.

Early Neolithic houses, gathered together in fortified villages, were round or rectangular huts with wattle walls. Then walls of sun-dried brick appeared, with or without stone foundations, while the rooms were arranged in a plan which consisted of a porch, its roof supported by one or two posts, opening into a large room with a hearth in the middle, the ceiling of which was also supported by posts and had a hole to let out the smoke. The adjoining rooms, bedrooms or storerooms, were arranged behind the main room on either side. The floor was often covered with moist beaten clay.

The Neolithic cultures, as well as the Paleolithic, worshipped a mother goddess who is often represented by a figurine holding a child on her lap. Her shape, which at first was exaggerated (steatopygous goddesses), soon developed normal proportions. During this period the dead were never cremated but always buried either in the open or in a natural cave, and sometimes (as at Soufli Magoula) in an urn. Children were buried in jars. The body was placed on its side in a contracted or foetal position, the hands and knees under the chin. It is possible that the custom of restoring the dead to the position they assumed before birth indicates a hope of survival, as if the cycle of life and death could be repeated for the same person.

Archaeology shows that the men of the Neolithic period were settled farmers endowed with some social organisation and religion. This organisation favoured progress towards a more comfortable and sophisticated existence. There are, however, still many doubtful points about this period and absolute chronology, despite the aid of carbon 14, is often vague. Were there invasions, or perhaps just ordinary trading exchanges among people? What part did Asia Minor and the Danube countries play in these influences? The recent discoveries of K. Kenyon at Jericho and of J. Mellaart at Çatal Hüyük in Anatolia have shown that these countries had a very advanced civilisation at a very early date. It is reasonable to assume that their influence spread throughout the whole Aegean area.

Crete in the Bronze Age: the Minoan civilisation

The Bronze Age in the Aegean area covers more than 1000 years. From 3000 to 1100 BC. we

Aegean archaeology began in 1876 with Heinrich Schliemann's discovery of the royal grave circle inside the citadel at Mycenae. The fabulous golden treasure from the graves was the first indication of a major Greek civilisation predating the classical era by a thousand years.

Retracing trade contacts and movements of peoples is an important part of the archaeology of pre-literate periods. Emanating from the Cyclades and widely distributed throughout the Aegean, a series of austerely stylised stone and clay idols (far left) demonstrates the unity of the east Mediterranean Neolithic culture.

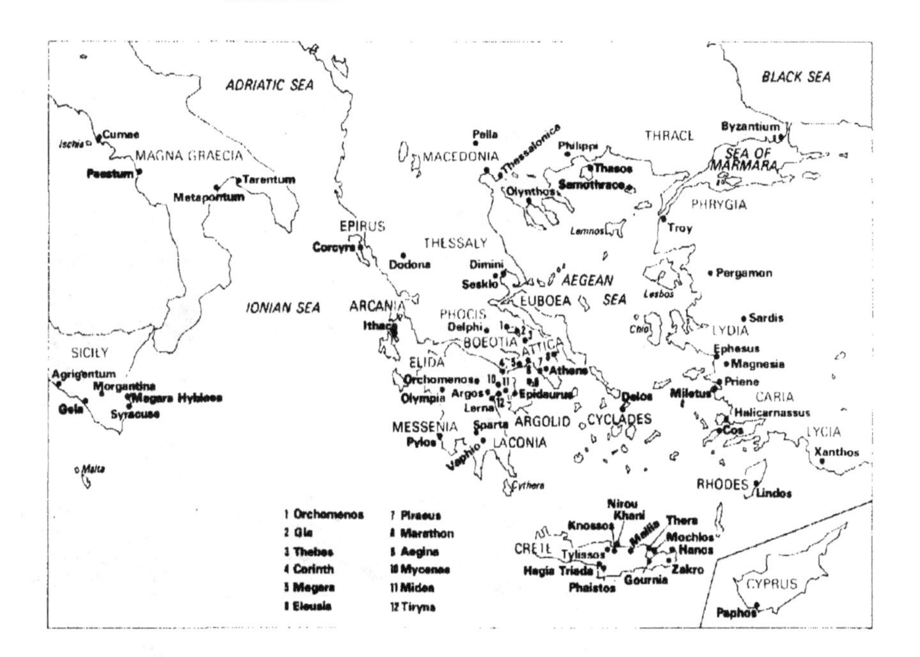

The Cycladic culture was remarkable for its production at a very early period of stone idols of outstanding distinction and elegance. Reduced to the barest essentials and stylised with striking simplicity, these figures vary in size from a few centimetres to well over a metre in height and are found widely distributed around the eastern Mediterranean.

see, first in Crete and the Cyclades, then in mainland Greece, the gradual rise and the subsequent fall of a civilisation which was sometimes warlike, sometimes peaceful, sometimes crude and sometimes sophisticated, but always amazingly rich and luxurious. Geographically this civilisation covered three regions: Minoan in Crete, Helladic in the southern part of mainland Greece, and Cycladic in the Aegean islands. Evans introduced a series of three-part divisions, founded on pottery styles, in the chronological evolution of these three cultures: thus we speak of Early, Middle and Late Minoan for Crete, and Early, Middle and Late Helladic for mainland Greece. Each of these sections in turn has three divisions and subdivisions (e.g. Late Helladic III C). Despite the artificiality of this system it is still used, with certain modifications, for Helladic studies, although a more flexible system now tends to be used for studying Crete. Minoan civilisation is characterised by a number of palaces, and its chronology is linked to the construction and destruction of these residences. Thus the pre-palace period, the first palace period (proto-palatial), the new palace period and the post-palace period can be classified.

It is difficult to distinguish between the pre-palace period (about 2600–2000 BC) and the Neolithic, so that this period is sometimes known as the sub-Neolithic. The light-on-dark decoration on pottery was still purely geometric, but increasingly dynamic, curved lines entwined in braids with waves and spirals, and straight lines divided the pot into horizontal registers. The virtuosity of craftsmen during this period is demonstrated in the carving of soft stone like schist, as shown by the figure of a dog, both stylised and lively, sitting on a pyxis lid found at Mochlos. The shapes of vases, dating from the third millennium, show influences from Asia Minor (the long-spouted sauceboats have parallels in Iran), but they also betray a certain endemic imagination. For instance it is impossible to guess the function of one particular vase shaped like a gourd with a large flat mouth and curved handle, which could neither have been suspended nor stood upright.

Apart from vases, the archaeological finds from the pre-palace period include copper tools (double axes and daggers), terracotta figurines of the Earth Goddess, jewellery of gold (diadems, pendants and leaves) and rock crystal. There is also a particularly fine collection of seals, some carved with geometric motifs based on spirals, some with heavily stylised animals or human beings associated with magical subjects such as the tree of life. Here, too, the influence of Asia Minor is obvious.

The pre-palace period, then, seems to have been an age of technical progress and advance, influenced by the more developed civilisations of the east. The first palaces, built soon after 2000 BC, were the natural result of this evolution. Little is known about them, because after their sudden destruction (about 1700) their ruins were almost entirely obliterated by the new palaces built on the same sites. The plan of the first structures can be seen best at Phaistos on Crete, and has been studied there in detail. Though three successive palaces preceded the new palaces, their general construction does not seem to have differed materially from that of their successors. There were no fortifications, the main entrance was reached by crossing a large paved court, and the buildings of the palace were grouped around a rectangular interior court. The first course of the façade consisted of orthostats, or large slabs set on edge. The façade itself was broken up by indentations. The rooms inside, separated by thick walls were arranged regardless of alignment or perspective, around a small hypostyle hall similar to those found in Egyptian temples. These same characteristics reappeared in the palace at Mallia which, according to P. Demargne, was only partially rebuilt in the new palace period. This clearly indicates that there was no fundamental difference between the buildings of the two periods.

Vases, jewellery and seals furnish a more complete picture than architecture of the early glories of Crete. At this time pottery technique was improved by the introduction of the fast wheel, which made it possible to execute tall vessels with thin sides. With the style known as Kamares (the name of the cave in which the first examples were found, the best, however, come from Phaistos) pottery reached an exceptional technical and artistic level. While the walls were thin and the shapes free, jugs, cups, craters and pithoi were treated with amazing polychrome decoration in red and white on a blue ground, in which natural motifs were so stylised as to be almost geometric, and geometric motifs in their turn were enlivened with intense vitality. Fish and spirals, open flowers and shining stars play over the surface of the vases without overcrowding. Applied motifs were sometimes added to the painted decoration, which resulted in an almost baroque appearance. Kamares vases have been found in Egyptian tombs, a discovery which has provided valuable clues on dating.

A similar mastery of technique can be seen in the metal work and seals which dates from this period. A gold pendant found at Mallia shows two bees holding a cake of honey between their feet and is distinguished by both the authority of the composition, forming concentric circles, and the fineness of the granulation. A ceremonial sword with a hilt ornamented with a gold roundel was also found at Mallia. The figure of an acrobat was worked in repoussé on the hilt, the curve of his body following the edge of the disc. Engraved seals, apart from their real artistic merit, furnish much information about Minoan civilisation, showing for example, the type of ship which enabled the rulers of Crete to become a maritime power and dominate the Aegean.

We would know very much more about this if we could read the hieroglyphic script which was in use during this period, but it has not yet been deciphered. One of the most puzzling objects is the Phaistos Disc, the front and back of which are covered with hieroglyphic signs arranged in a spiral. It is not even certain that this object is actually Cretan, though it was found in that island.

The archaeological problems of Troy were far from being solved when Schliemann located the site at Hissarlik on the Dardanelles by following up the clues he assembled from the geographical descriptions in the *Iliad*. His unscientific approach destroyed much valuable evidence, but it has since been possible to establish that many successive cities were superimposed on each other. The excavations, begun in 1870, have continued with a few interruptions since Schliemann's first campaigns and are still in progress.

This culture met with a violent end. About 1700 BC a catastrophe destroyed the palaces, but its exact nature is not known. Some scholars believe that it was an earthquake. Whatever the truth may be, Minoan prosperity was not permanently affected. New palaces were built without any apparent break in continuity on the same sites, and technological progress continued unchecked. These second palaces lasted until about 1450 BC. This was the peak period of Cretan civilisation.

The high point of Cretan civilisation

The Cretan palaces of Knossos, Mallia and Phaistos have a number of details in common and are larger and better preserved than the early palaces. First of all, they each have a rectangular central court, the size of which (roughly fifty metres long by twenty metres wide) varies

The early stages of human activity in Crete are still obscure and for many years very little interest was taken in this aspect of the island civilisation. However, recent research at Knossos is helping to provide details of the flourishing Neolithic settlement, which produced such artifacts as this sturdy vessel with its incised decoration.

Once Sir Arthur Evans had established the outlines of the Minoan culture, discoveries of other Cretan centres followed in rapid succession, each contributing fresh information on the life and thought of the islanders. An Italian team is still investigating the palace at Phaistos which is the only one of the palaces to occupy a strategic position. Its architecture, however, with wide unprotected stairways and no fortifications, betrays no concern with defence.

After the collapse of the Minoan palaces about 1475 BC a brief revival took place at Knossos. The pottery produced during this period, which is decorated with stiff formal patterns bearing little resemblance to the exuberance of the earlier styles, suggests that an alien element had entered Cretan life imported by new overlords from mainland Greece.

very little from one palace to another. It could be said that its area was as rigidly prescribed as that of a tennis court or football pitch. On one or more sides this court was bordered by a portico, the supports of which (pillars or wooden columns) might be joined together by a balustrade forming an enclosure. In the middle of the court at Mallia was an altar which does not occur elsewhere. The exact purpose of the court is still uncertain. It has been suggested that it was an enclosure where religious ceremonies, in which acrobats somersaulted around and over a bull which had been let loose, were conducted. That such festivals actually took place is indisputable, since there are large quantities of objects showing acrobats and cavorting bulls. It is by no means certain, however, that these festivals actually took place in the central court, which hardly seems suitable for this form of exercise. It is very likely, though, that it was the scene of some religious ceremonies. At Knossos and Mallia a group of rooms of unquestionably cult character opened on the west side of the court.

The original appearance of a shrine at Knossos, with a double entrance framing a slightly raised central section, has been preserved by a miniature fresco from the same palace, which probably shows a ceremony taking place in the presence of a great crowd of people. The bull is symbolised here by the 'horns of consecration', a motif to be seen all over Cretan palaces and sanctuaries. Inside the shrine, among other things, a small room was found furnished with one or two thick pillars in the middle. These pillars are not explained by any architectural function, and do more to obstruct the room than to hold up the ceiling. In fact, they probably had a ritual significance, representing the tree of life. The religious complex at Knossos was completed by the throne room, in which a chair of state made of gypsum was found, flanked by benches backing on a wall decorated with frescoes of griffins heraldically arranged on either side of the throne. In the same room but on a slightly lower level (there are a few steps down to it) was a small paved area surrounded by a balustrade known as the 'lustral area'. The presence of a ritual area in the political environment of the throne room indicates how closely the two functions were connected for the Minoans. The king could only have been a priest-king.

The residential quarter of Knossos is situated on the east side of the central court. Several original features typical of Cretan architecture occur here; these include firstly the 'light well', a small internal courtyard with several living rooms opening on to it. At Knossos one of them is bordered on three sides by a monumental staircase with its upper flights resting on columns. Secondly there is the type of room sometimes inaccurately known as the Cretan megaron, or just the megaron. This is a large oblong room, one whole side of which ends in a colonnade opening on to a terrace or light well. The interior of the room is also divided into compartments by a series of easily closed bays, thus the size of the room, its lighting and its exposure to the sun could have been regulated at will, according to the weather and the circumstances. This highly flexible arrangement, particularly suitable to the Cretan climate, seems to have been a Minoan invention. It turns up, with a few variations, at Mallia, Phaistos, Hagia Triada, the Little Palace at Knossos and several villas at Knossos and Tylissos. It is indisputable evidence of a certain feeling for comfort, and this impression is confirmed by the presence in the residential quarter of some bathrooms, many of which have an advanced drainage system. Apart from these two typical features, the Minoan palace also included numerous magazines. These look like a long row of compartments connected by a corridor. Those of Mallia are particularly well preserved. Each compartment was fitted with two benches on either side of the central gangway, on which jars and pithoi were placed or inset. A system of drains was provided to deal with overflows and spills. Finally, at the end of the main corridor, a seat was placed for a watchman.

Two other features, one constant and the other variable, should be mentioned at the end of this description of the architecture of the Minoan palace. The unchanging characteristic is the peristyle hall which is medium sized and usually situated to the north of the palace near a door; perhaps it should be regarded as a guard house. As to the entrances, they vary considerably from one palace to another in both appearance and construction. Sometimes there is a wide flight of steps (Phaistos), sometimes a simple porch with a roof supported by a single column (the west entrance at Knossos), sometimes a propylaeum (north entry at Knossos), and sometimes a paved corridor (south entry at [illegible]). These entrances have no features in common, nor was one of them planned to present the visitor with a monumental view worthy of the palace. The south entrance at Mallia probably gave directly on to the central court, and it must be observed once more that, placed as it was in the south-west corner of the court, it would have been impossible to see the whole court before emerging from the corridor. *The monumental stair at Phaistos opens after* two anterooms on to a broad hall with a little door, almost invisible, through one corner. The west porch at Knossos leads to a long corridor which turns twice at right angles before it reaches the propylaeum.

This constant rejection of straight lines and direct approaches is not due to any idea of defence (the palaces were not fortified), but to an architectural predilection, the reason for which is as obscure to us as it was to the Greeks. Indeed, it was to explain the mysterious nature of these arrangements that the Greeks devised the legend of the Labyrinth. A network of corridors and rooms designed to confuse the unwary and unescorted visitor, a snare from which only a hero could escape, was what became of the memory of the palace at Knossos 1000 years later. Nothing could actually be more unjust. The function of many of the rooms is recognisable, and the overall plan is comparatively clear. But the basic inspiration of the architectural style is very foreign to the Greek spirit. It is distinguished by a more flexible and less rigid logic, which betrays that the Cretan art of living was incomprehensible to the Hellenes.

Apart from the actual palaces, the investigation of Minoan towns still has a long way to go. The only one to have been extensively excavated is the village of Gournia on the Gulf of Mirabello. *The houses with their dry stone walls are* small and poor, and usually include only two rooms. A larger building, still only modest in comparison with the great Minoan palaces, may have been the governor's residence. On the other hand, the villas surrounding the palace of Knossos or standing on the north coast at Tylissos and Nirou Khani are much richer, even luxurious. Here, too, the Cretan megaron which

Sir Arthur Evans (1851-1941) was one of the first Bronze Age archaeologists to approach excavation in a scientific and scholarly fashion. At the beginning of this century he secured the concession to excavate the mound of Knossos in Crete for which Schliemann had been negotiating just before his death. The mound proved to be a huge labyrinthine palace, the heart of the brilliant Minoan civilisation which produced such outstanding naturalistic masterpieces as the stone bull's head ritual vessel with which Evans is shown.

An essential item in every Late Minoan household was a small figure of a household goddess. The upraised hands and poppy-head diadem give a touch of individuality to the mass-produced wheel thrown base.

A clay disc from Phaistos is covered on both sides with impressed signs which look like some form of writing. Many explanations, some more fanciful than logical, have been offered but no satisfactory solution has ever been produced, and no other examples of this so-called 'script' have ever been found. It is possible that the signs are not a script at all, but some form of gaming-board.

has been described in connection with the palaces occurs, as well as internal courts and light wells. It should finally be mentioned that near the palace of Mallia a complex has recently been discovered which includes a large court and a peristyle hall, and this complex indicates a much higher level of urban living near Mallia than at Gournia. It is still too early to interpret this complex with any certainty. It is enough simply to state that we are still far from being familiar with all the aspects of a Minoan town.

The decoration and interior arrangements of Minoan palaces and houses show the same concern for comfort as the general plan. The walls of the living and reception rooms were often covered with frescoes, but these have survived only in pieces. Working from these few fragments, the younger E. Gilliéron painted many reconstructions under Evans' directions, and because these are easier to reproduce, they are used as illustration more frequently than the originals. The reconstructions conform closely to the Minoan frescoes. Gilliéron restored nothing that was neither certain nor highly probable. Nevertheless it is impossible not to suspect that they are slightly out of key with the Minoan genius. The warm, stately colours of the originals have been replaced by flat, almost gaudy shades. The vitality of the Minoan painter has given place to a studious but cold technique.

The genuine fragments, however, are enough to enable us to appreciate the quality of Minoan painting. One of the processions is a little too stiff for modern taste, but its freshness and virtuosity can be readily admired. Animals are sketched in a lively jungle where they disport themselves freely. At Hagia Triada the painter who reproduced the cat, with its fluid and controlled movement, noiseless, ready to spring, lying below a bush in wait for a heedless pheasant, achieved the same felicity of expression as La Fontaine did in describing Raminagrobis. There is, however, no strict representationalism; the colours are conventional and indicate that the chief preoccupation was with decorative effect.

The same brilliant spontaneity can be seen in the famous head of a woman nicknamed 'the Parisienne'. She is probably a priestess; in fact, she wears on the back of her neck the sacral knot which indicates her office. The eye is shown frontally in a profile face, in the Egyptian convention; but the lock of hair (which may also have had a ritual function) on her forehead, the flirtatious tilt of her nose and the vivacity of her parted lips lend this face an alluring charm.

Creative freedom is to be found on a different level in the painted pottery. The polychrome technique of the Kamares style seems to have become impoverished, since from this time onwards the décor was almost entirely executed in dark paint on a light ground. But geometric stylisation gave place to free and light hearted naturalism. The most popular motifs were borrowed from plant life and the marine world. Enormous octopuses sprawl their coiling tentacles over the whole surface of round bodied flasks. On a rhyton dolphins leap out of the water; on another, argonauts surround a star fish. On an amphora from Knossos the stalks of lily flowers which spread above the body are modelled in relief like so many ribs to break the monotony of the flat surface.

The Minoan craftsman, therefore, used his technical skill in the service of lightheartedness, and his art is devoid of suffering. This is not to say that it shows no religious feeling. Several bronze figurines show a man holding his hand to his forehead in token of worship. The main events of agricultural life were the occasion for thanking and honouring the goddess of fertility. A steatite vase from Hagia Triada shows in low relief a procession of harvesters in a row, with their pitchforks on their shoulders, singing at the tops of their voices, following a choir master who is shaking a sistrum. The same site yielded the most revealing record of Cretan religion, but also the most difficult to interpret. This is a painted sarcophagus. On one of the long sides are scenes of sacrifice and offering which are an invaluable source of ritual history. The same is true of the left hand part of the other long side. But on the right hand part of the same panel is a scene which may well be

The jewellery of the Middle Minoan period is outstanding in the ancient world for its artistic excellence and technical skill. A pair of winged insects from the Mallia cemetery, decorated with the advanced and difficult processes of granulation and appliqué, indicate contacts with the metal-working centres of Anatolia, but the inspiration of the design is purely Cretan.

taking place in the next world. The dead man is shown standing in front of his tomb. He is wrapped in a long tunic and looks as if he has neither arms nor legs, which accentuates his unreal, supernatural character. A procession of offering-bearers is arriving before him, two of whom carry whole calves or bulls in their arms, and the third, what seems to be a huge basket. The fantastic character of the scene is the result of the lack of proportion between the human figures and the animals. In the sacrifice scene the slaughtered bull is the normal size, but in the supernatural scene the animals are far too small. It appears, therefore, that the artist was trying to show what happened in the next world to offerings made to the dead here on earth. If this interpretation is correct, it indicates a conception of the hereafter fairly close to what we know of Egyptian beliefs. The individual lived on in a particular form. He needed the servants and riches he had enjoyed on earth in the next world. It was the duty of his survivors to enable him to prolong this new life by their offerings. There is nothing despairing in such beliefs, and nothing to suggest a lack of interest in seeking happiness on earth.

The destruction of the Minoan empire

Nowadays it is accepted that the destruction of the Minoan empire (or the various Cretan principalities) was the result of the eruption of the volcanic island of Thera. The problem is complicated by the fact that traces of two successive destructions have been found at Knossos. According to the British archaeologist M. F. S. Hood, who carried out a series of stratigraphic soundings in the central court of the palace in 1960 and 1961, Knossos was damaged and the neighbouring townships of Tylissos and Nirou Khani were destroyed between 1500 and 1450 BC (the period known as Late Minoan I B in the Evans chronology). After this disaster Knossos was conquered by the Achaeans and was finally destroyed at the end of Late Minoan II (about 1400), or shortly after. From this point of view the Linear B tables (i.e. Mycenaean Greek) found at Knossos date to the end of Late Minoan II, the date proposed by Evans, rather than Late Minoan III B (about 1300-1200 BC) as recently suggested by the philologist L. R. Palmer.

At the highest point of their civilisation, the Cretans exerted more influence on the Greeks than the Greeks did on them. However, a tendency to stylisation and symmetry can be observed in the painted pottery of Late Minoan II which is highly typical of the Achaeans. This is nothing like the exuberant and spirited stylisation of the Kamares period, but more of a geometric distortion. The motifs are largely unchanged (apart from the introduction of martial elements such as helmets), but they are treated differently and are broken up to form a pattern. Thus elements which had previously been shown in outline are now fretted, speckled or striped with bands of different types. The decorative elements seem to dissolve and disintegrate as we look at them. This is the invasion of formal logic into a hitherto free and instinctive art form.

It cannot definitely be stated whether this change of style is the effect of an unnatural alliance, or simply that of the destruction of Knossos. Nevertheless, Cretan art declined, as inspiration and technique grew poorer. Deserted by their inhabitants, the Cretan palaces gradually collapsed at the end of the Bronze Age.

The Bronze Age in the Cyclades

The Aegean islands saw the development of a brilliant and original civilisation in the Bronze Age. The origins of this wealth are to be found beneath the ground in these islands. Melos had veins of obsidian, there were copper mines on Naxos and Paros, and the gold mines of Siphnos had perhaps already been exploited.

It is on Melos that the most important discoveries of this period have been made. Among the most productive sites, those of Syros (Chalandriani) and Kea should also be mentioned. Nothing is so vulnerable to attack by pirates as a small island, and the inhabitants of these Aegean islands tended to gather together in the interior, in small fortified towns. Phylakopi on Melos is defended by a double enclosure, both walls of which are two metres thick, and at Chalandriani the inner rampart is surmounted with round towers. The Cycladic houses behind these walls were usually rectangular, with several small rooms in a row connected by doors. It was unusual for the complex to take the form of a megaron (a comparatively late development, occurring in the 'palace' of the third town of Phylakopi in the Mycenaean period). The houses often had an upper storey, and the walls, made of mud brick at first and later of stone, were often covered with frescoes. At Phylakopi a painted decoration of flying fish

The Early Minoan culture was strongest in the east of the island where contacts with the more developed Levantine cultures were more frequent. The site of Vasiliki yielded attractive pottery decorated with a mottled effect produced by holding a burning branch against the vase immediately after firing (below).

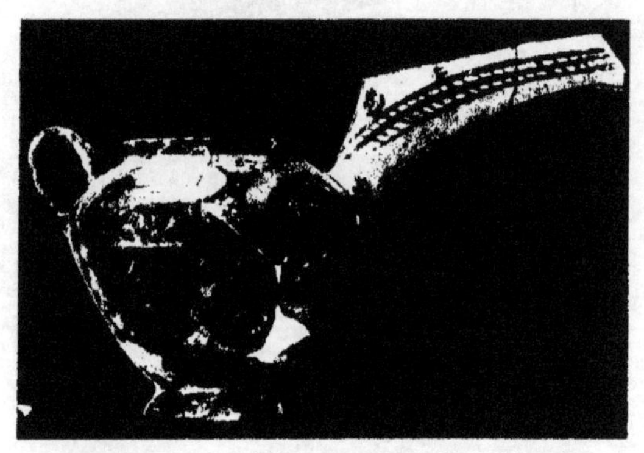

Male and female worshippers flank a larger divinity with raised hands, as they were found in the Late Minoan shrine of the palace of Knossos (bottom). The palace was the centre of both religion and bureaucracy.

To the east of Knossos the land drops away to a shallow ravine. The suites of rooms on this side where the royal family had their private apartments are terraced into the slope, and open bastions and loggias provide a view over the opposite hills.

in the purest Minoan style has been preserved.

It was not in the houses, however, that the most original productions of Cycladic civilisation were found, but in the tombs. These are usually small, of the type known as cists. The four sides of a pit dug in the ground were lined with slabs forming small drystone walls. The grave goods were often very rich—ornaments of gold, silver, bronze, bone and stone have been found, and especially marble idols, vases and pottery. The latter have a variety of shapes. Particularly notable are the *askoi* (jars with an off centre mouth and a handle on top), animal shaped vases, *kernoi* (cruet like objects with several vessels mounted on one base, which perhaps contained the first fruits of the earth to be offered to the fertility goddess), and finally, some curious vessels shaped like frying pans with straight edges and a very short handle, which may also have played part in ritual. Cycladic vases were inci stamped or painted. The incision method produced some

Knossos, like the other Cretan palaces, lay hospitably open to all comers. The north entrance is a shallow ramp which passes a gallery frescoed with a charging bull along the approach to the central court. Considerable controversy was roused by the extensive restorations of the upper storeys which Evans undertook, replacing long-perished timbers with concrete and setting modern reproductions in the positions originally occupied by the fragmentary paintings.

very fine decorative results, in which geometric motifs (especially spirals) and natural elements (*fish, boats, the sun*) set off and counterbalance each other with the utmost felicity. Painted decoration, although it was perhaps influenced by mainland art, still shows a degree of independence. Thus, instead of seeking to fill the whole field, the artists liked to make a small number of linear or floral motifs stand out in isolation on the light surface of the vase.

There is hardly a single tomb which has failed to furnish at least one example of the marble figurines known as Cycladic idols. They vary in size, the largest being up to 1.5 metres high, but they are usually small and all very much alike. The face is reduced to a slanting or trapezoidal expanse on which only the long thin nose stands out. Neither eyes nor mouth are represented plastically but were probably indicated in paint. The highly stylised arms are folded at right angles across the abdomen; the female genitalia are clearly marked. The legs are straight and flat. All these figures are simplified to the point of abstraction and demonstrate a wish to reduce the human body to its distinctive features and basic elements. Similar shapes recur many centuries later in some forms of African art.

The same qualities can be seen in a series of broken terracotta statues recently found on the island of Kea off Attica by an American team. When these statues were intact they must have been about one metre high; they may show the earth goddess and her attendants or votaries. They date from the fifteenth century BC, and are strongly influenced, particularly as regards their clothes, by Crete. The actual heads, however, are typically Cycladic. One of these statues met with a curious fate. After the destruction of the building in which it stood, the body and the limbs lay buried in debris. But in the eighth century the head, subsequently recovered and probably handed down from generation to generation, was set on a base specially made for the purpose and placed in a small shrine as a sacred object. Nothing illustrates the Greek feeling for the continuity of their culture better than these honours paid centuries later to a survival from the Bronze Age.

Over the course of several centuries the palace at Knossos was improved. Earthquake damage about 1600 BC gave the Cretans a welcome opportunity to add a magnificent staircase, built round an open light well, to link the several storeys of royal apartments.

One of the most spectacular activities of ancient Crete was the bull-leaping game in which young men and women participated. Modern rodeo experts, when consulted on the subject, have declared it to be impossible, but the evidence of *Cretan* art is incontrovertible.

The oldest throne in Europe is the gypsum chair which still stands in its original position in the ceremonial apartments at Knossos. Very little Cretan furniture has survived.

The Early and Middle Bronze Ages in mainland Greece

Our knowledge of the early Bronze Age in mainland Greece has been perceptibly advanced by the very recent excavation of the site of Lerna by an American team led by J. L. Caskey. The name of this mound was traditionally associated with one of the labours of Heracles. There is still a marsh on the site of the lair of the many-headed Hydra and the entrance to Hades. On the actual mound an important settlement of the Early Helladic period has been discovered above the Neolithic levels. A strong wall with towers and casements protected a veritable palace. Its heart consisted of three large rectangular rooms surrounded by passages with doors opening into them. This was an original arrangement with no relation to the megaron. The walls were of mud brick on a stone base. In the course of the dig a large number of flat rectangular tiles were found, which probably covered a flat roof. The remains of this palace, which was destroyed at the end of Early Helladic II, continued to be an object of veneration for the inhabitants of Lerna. Its approximate site was enclosed by a circle of stones, and no new buildings were constructed within it. The rest of the site was occupied without a break until the end of the Bronze Age, but it was never again as important as it had been in the Early Helladic period.

The violent and sudden destruction of Lerna was probably the work of the Indo-Europeans. Nowadays it is agreed that their arrival took place about 2200 BC, perhaps in successive waves spread over several generations; they were probably the first Greeks. They brought with them a wheel-made grey pottery of excellent quality, freely imitating the shape and appearance of metal vases, known as Minyan ware. The contemporary pottery, known as matt-painted ware, bore very simple, purely linear motifs, in dark paint on the light clay background. During the first centuries of its existence (Middle Helladic, 1900–1580 BC) this Indo-European civilisation developed somewhat slowly. In the cemeteries of Korakou, Zygouries and Prosymna the slow advance of a community trying to find its feet can be traced. It was not until the next period that it achieved its full growth.

The peasant classes of Crete, although less privileged than the lords of Knossos, obviously had their consolations. A stone vase from Hagia Triada, showing a procession of shouting, singing farmers, radiates vitality and high spirits. Objects which throw light on the social life of the ordinary people of the ancient world are, in many ways, more valuable to archaeology than the art of the great palaces.

The Late Bronze Age (the Mycenaean period)

The Mycenaean age was the first to unite the necessary conditions for a true historical study. After the decipherment of the Linear B script by Michael Ventris, the monuments could actually be studied in the light of these documents. The texts are, however, far from being completely deciphered. There is not a trace of literature in them. They are the archives and inventories, incised on tablets made of sun-dried clay, of the palaces where they were found. In general this kind of object would only have a very short life, since inventories were frequently renewed. Paradoxically it was to the catastrophe which destroyed the palace that we owe their preservation. The fire had the effect of baking and hardening the clay tablets so that they stayed intact underground. It is no use expecting more from these records, lists of offerings and inventories of stores and lands than they can give. However, they furnish the main outlines of the political and religious hierarchy.

At the head was the king (*wanax*), a title also applied to the gods, which seems to indicate that his person was sacrosanct. Beside him there was a sort of grand vizier (*lawagetas*). He and the king each had their own province, or *temenos*. The rest of the land was divided according to a somewhat complex system of tenures. The *hequetai* (literally, 'followers', cf. the French word *comte*) probably formed the military aristocracy. The ruling classes lived chiefly on the population of farmers and craftsmen, and a large number of slaves. This community was centred, as in Crete and the Homeric poems, on the person of the king. This is indicated archaeologically by the presence of palaces in the chief cities of the principalities. The best preserved are those of Tiryns, Mycenae and particularly that of Pylos. Their construction was radically different from the Cretan palaces. The central element, which they all have in common, is the megaron inherited from the early Bronze Age houses, but enlarged and improved. Now it included a porch with a ceiling supported on two columns *in antis*, i.e. enclosed between the projections of the side walls. The wall at the back of the porch had a door through it leading to an oblong anteroom which opened in its turn on to the principal room, also sometimes known as the megaron in the strict meaning of the word. This was a rectangular room with a large round hearth in the centre. Around the hearth were four columns supporting the ceiling, which must have had a lantern in it to let the smoke out. Against the right-hand wall a slab which probably served as the dais for a chair of state was found at Pylos and Tiryns. This room was therefore the throne room or state apartment of the palace. The megaron usually had a courtyard in front of it which might have a colonnade on one or more sides. This courtyard was entered by a propylaeum, a gateway with two columns *in antis* before and behind it. Around the courtyard and the megaron were clustered the service quarters (magazines and archives), the domestic quarters, which often included a smaller megaron known as the 'Queen's megaron', and the bathroom. It was in the bathroom of the palace of the Atreidae at Mycenae that Agamemnon was murdered. At Pylos a terracotta bath was also found, whilst at Tiryns the floor of the bathroom was made of a single huge slab of gneiss, set at a slight angle

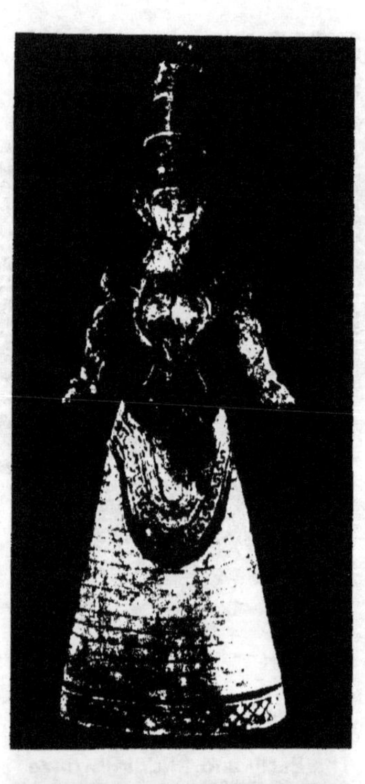

Snakes were a prominent feature of Minoan religious practice. An elegant little faience figurine of a woman dressed in the height of fashion, who may be either a goddess or a priestess, has serpents entwined round both arms.

The sophistication of Minoan high society is preserved in a religious figurine of a woman grasping a snake in each hand. Her many-flounced skirt, tight-waisted bodice and low neckline produce a distinctly worldly effect which suggests that to the Cretans religion, like so many other aspects of their life, was a social and pleasurable activity rather than a matter of superstitious gloom and terror.

At the height of the Minoan civilisation the palace bureaucracy employed a form of script known as Linear A (right). It has not yet been deciphered and even the language it transcribes is unknown. New statistical methods of decipherment have proved inapplicable in this case because they need a large number of specimens for successful operation and too few tablets in this script have been recovered. In the last years of the Cretan culture a new script known as Linear B (far right) was adopted. It has now been established that these tablets are written in a primitive but nonetheless recognisable form of Greek, confirming the evidence of the other artifacts which suggest the presence of mainland conquerors on the island during the final years of the Minoan civilisation.

The Minoans were familiar with the art of writing from an early date, but this skill may have been confined to a limited class and the majority preferred to mark their possessions with seals, in the carving of which their craftsmen excelled.

to enable the water to run off into a drain. The interior decoration of the megaron was very fine. Not only were the walls covered with frescoes, several beautiful fragments of which have survived, but even the floor was plastered, divided into squares, and each square painted with a different decoration.

The Mycenaean palaces are very much smaller than those of Crete. As far as their location was concerned, the difference between Mycenae and Tiryns on the one hand and Pylos on the other should be noted. The latter stands on a low hill some distance from the coast, on an open, unfortified site. Tiryns on the contrary, perched on an outcrop dominating the plain of Nauplia, is defended by extremely strong walls, and Mycenae is protected both by natural ravines on the south and east and by a rampart. While Pylos, an open city, does not seem to have been aware of any threat, Tiryns and Mycenae are strongholds, the lairs of warrior kings like the medieval 'burgs'. This is true of the majority of Mycenaean sites. The same details are found at Gla in Boeotia, Midea in the Argolid and the Acropolis of Athens itself. At Tiryns, while the palace had its own defences, an extension of the wall enclosed the lower town where the inhabitants of the open country round about could take refuge in time of danger, just as in the Middle Ages the enclosure of the castle sheltered the peasantry while the keep was reserved for the lord of the manor. The walls were thick, either in the style known as Cyclopean (huge rough stones piled on one another) or in large dressed blocks. Many precautions were taken in case of siege. The two gates of Mycenae stand at the end of passages formed by bastions in the wall which made it possible to attack the enemy on his unprotected side. The entrance to Tiryns is reached via three successive gates, also set between very strong walls. The problem of food and water was not neglected. At Mycenae a secret passage runs under the Cyclopean wall and continues underground outside the fortress to give access to a subterranean cistern where the water from a neighbouring spring collected. A similar arrangement was found at Tiryns and the Acropolis of Athens. For storing food and arms, a series of casemates were built into the thickness of the wall at Tiryns, approached by a long corbel-vaulted corridor which is one of the most spectacular achievements of Mycenaean architecture.

The same warlike spirit is to be seen in the productions of Mycenaean art. In Crete it was unusual to depict violence, but it was common enough on the mainland. One of the best-known Mycenaean vases shows a row of soldiers. A silver rhyton worked in repoussé shows a besieged city; a very badly damaged fresco from Mycenae shows a warrior falling off a roof; some of the grave stelae from Mycenae have reliefs of a warrior in a chariot pursuing a foot soldier; an ivory from Delos shows an armed warrior, and so on. Finally, the oldest known Greek cuirass was recently discovered in a tomb at Dendra (Midea).

It is not the palace and the spacious houses around it which furnish the most accurate picture of the wealth of Mycenae. Destroyed and pillaged in antiquity, these have yielded only a few valuable finds to the archaeologists. On the contrary, the splendour of the Achaean court was revealed by the excavation of the tombs. The practice of grouping tombs together within a circle marked by a ring of stones was adopted at the end of the Middle Helladic period. The stone circle round the ruins of the palace of Lerna may be another example of the same rite. Two circles of royal graves have been found at Mycenae. The oldest of those is also the most recently discovered (1953). It lies outside the ramparts. It was in use at the end of the Middle Helladic period and the beginning of the Late Helladic (about 1600–1500 BC) and its existence seems to have been forgotten shortly after this. Indeed, one of the beehive tombs of the fourteenth century BC, inaccurately known as the Tomb of Clytaemnestra, overlapped part of the circle. Later on the Tomb of Clytaemnestra in its turn was partially covered by the seats of a Hellenistic theatre.

Grave Circle B is thirty metres in diameter and includes twenty-four tombs. More than half of

Unlike the Egyptians, the Cretans do not seem to have been plagued by an overmastering obsession with the after-life. However, a painted plaster sarcophagus shows that some sacrifices to the dead man (shown as a diminutive figure outside his tomb) were customary. The sarcophagus, which dates from the period after the fall of the palace centres, is one of the very rare identifiable scenes of Cretan funerary practice.

these are shaft graves. At the bottom of a shaft of varying depth (not more than 3·5 metres) a grave was dug, its sides lined with dry stone walls and the floor covered with a layer of gravel. The corpse was placed in the extended position with his weapons and funerary gifts. The grave was covered with a wooden roof, clay and brushwood. The shaft was then filled in to ground level and the tomb was marked with a stela. The same grave was often used for several burials in succession. The earlier skeleton was pushed aside to make room for the newcomer. One of the tombs is an unusual construction, it takes the form of a rock cut

When Cretan culture began to fade and the island no longer presented a serious challenge to the inhabitants of Greece, the city of Mycenae came to prominence. Huge new fortification walls were built, enclosing an extended area of ground. The citizens showed their reverence for their predecessors by throwing out a wide curve in the wall so as to include the hallowed ground of the royal grave circle. The mouths of the shaft graves, buried under several metres of fill when the new enclosure was made, are now uncovered.

At the peak of the Mycenaean civilisation royal graves known as *tholos* tombs were built in hillsides. These tombs are astonishing feats of engineering and craftsmanship, and were used only for the burials of the ruling classes. The so-called 'Treasury of Atreus' at Mycenae is faced with skilfully dressed masonry and has a lintel weighing well over a hundred tons.

chamber with walls lined with slabs and a roof shaped like an upside down V, approached by a corridor built in the same way. This type of tomb, unique at Mycenae, may perhaps be an imitation of certain eastern tombs. It has parallels, admittedly later in date, at Ras Shamra.

Grave Circle A, excavated by Schliemann, is slightly later (the second half of the sixteenth century BC). It measures twenty-seven metres in diameter, and included only six shaft graves, five of which contained more than one skeleton. They were marked by stelae at surface level. Unlike Circle B, Circle A was venerated until the end of the Bronze Age. About 1250 BC, when the fortifications described earlier were built round the citadel, Circle A was enclosed within the ramparts which were extended at that point to honour it. The ground was levelled and the whole was enclosed in a double ring of stone slabs set on edge with a third slab laid flat across the top. The grave goods of these royal tombs were exceptionally rich. It was an epoch-making discovery in the history of archaeology. The most spectacular finds were probably the gold and electrum masks (electrum is a natural alloy of gold and silver) which covered the faces of the men in Grave Circle A (only one was found in Circle B). Worked in repoussé, they possibly portray the individual features of their owners. It is startling to notice the occurrence of the 'Greek profile' on some of them.

The Mycenaean princes and princesses were buried in state robes with all their armaments and jewellery. Their clothes were covered with small gold discs, glued or stitched on, decorated with animal, plant or linear designs. More than 700 of them were found in Shaft Grave III of Grave Circle A, which contained the bones of three women and two children. There were also many diadems and necklaces, usually decorated with spirals and rosettes. The vases lying near the bodies, usually by the head, were of pottery, gold or silver.

The weapons included swords with gold hilts, bronze daggers inlaid with gold and silver, and high helmets of leather covered with boar's tusks. Faced with such abundance, in contrast with the poverty of the Middle Bronze Age tombs, the obvious question arises as to the source of this wealth. The problem has never been satisfactorily solved. It is very tempting to believe that these are spoils from Crete. But the definitive occupation of Crete by the Achaeans cannot have taken place before the middle of the fifteenth century BC, and Circle A and definitely Circle B are certainly earlier. Perhaps the conquest was preceded by profitable raids. It is still an open question.

However that may be, after the final conquest of Crete in the fifteenth century BC the wealth of the Mycenaean Dynasty continued to accumulate. Indeed, enormous resources of money and manpower must have been needed to build the *tholos* tombs. These appeared at the end of the Middle Helladic period. From the fifteenth to the thirteenth century the building technique visibly improved, but the type never varied from the beginning. A long entrance passage or *dromos* was dug into the side of a hill, and at the end of this was the actual vault. The latter was a circular beehive-shaped structure with a high corbelled dome closed at the top. In only two cases (the 'Treasury of Atreus' at Mycenae and the 'Treasury of Minyas' at Orchomenos in Boeotia) was a small room with a flat ceiling constructed on the right of the dome. This may have been the burial chamber while the vaulted room served for offerings and sacrifices. The corbel vaulting technique was fairly primitive at first. In the earliest beehive tombs at Mycenae, at the beginning of the fifteenth century BC, small rough flat stones were used. Then in the second half of the fifteenth century dressed stones appeared. The dromos and the façade were lined, and the main door was provided with a heavy lintel surmounted by an empty triangular space (the relieving triangle) to take the weight off it. In the Treasury of Atreus, one of the latest and most advanced of these tombs (thirteenth century BC), the vault is nearly fifteen metres in diameter, and almost the same height. The façade is ten metres high, and the door measures 5·4 metres to the underside of

The so-called 'Mask of Agamemnon', found by Schliemann in a shaft grave at Mycenae, actually dates to a period some centuries earlier than the epic siege of Troy. Slightly primitive in treatment, it is nevertheless important because it is one of the very rare attempts at individual portraiture from the pre-classical world.

The ruins of Agamemnon's palace at Mycenae. A natural rock citadel and the centre of the great Helladic civilisation, it lies half-hidden in a mountain glen. Mycenae was also Agamemnon's capital and Homer described it as 'rich in gold'. It was destroyed by the inhabitants of Argos in 468 BC.

the lintel, which is a vast block 8 metres × 5 metres × 1.2 metres, weighing several hundred tons. It was probably set in place by *sliding it down the slope of the hill in which the* tomb was hollowed, but even this was a great engineering feat. The thirty-three courses of bonding stones forming the inside of the vault must have been decorated with metal ornaments, the nail-holes for which are still visible. The facade was faced with green marble with an engaged column on either side of the door, surmounted by two smaller columns at the relieving triangle, which was covered with a slab of the same stone carved in low relief with geometric motifs of spirals and broken lines. These tombs were too splendid and too eye-catching to escape looting. Apart from one near Pylos, every known example was plundered in antiquity. We therefore know nothing of their occupants (the traditional names like the Treasury of Atreus and the Tomb of Clytaemnestra are not to be taken seriously) *and no more has been preserved of their* funerary equipment.

The question of Cretan influence

Is this type of tomb a native Mycenaean invention? This has been suggested, but it seems more likely that it was Cretan in origin. Several very primitive circular tombs which were probably roofed with branches have been found in the Mesara plain just below the palace of Phaistos. Moreover a tomb with a stone vault like the mainland beehive tombs has recently been found on the site of Lebena in central

The shaft graves at Mycenae contained a great quantity of gold shaped into personal ornaments, ritual objects and weapons. Some of them are *obviously imported* from Crete, others show the undeveloped forms and crude workmanship typical of current Mycenaean taste and others again were clearly made by Cretan craftsmen although the choice of themes indicates Mycenaean patronage. In the latter category is a bronze dagger inlaid with a lively hunting scene executed in gold, silver and niello.

The approach to the Lion Gate at Mycenae runs between two huge bastions of dressed stone so that the entry to the citadel is both strong and impressive. Elsewhere the walls are made of rough-dressed masonry, so massive that the later Greeks believed that they must have been the work of giants.

In the Bronze Age most religious rites were accompanied by the pouring of libations through funnel-shaped vessels known as rhytons which were often made in the form of animals. An angular gold lion's head rhyton from Shaft Grave IV at Mycenae is typical of local work with its unsubtle planes and harsh lines.

Crete. The Mycenaeans can perhaps claim credit for the improvement but not for the invention of beehive tombs.

The problem of Cretan influences affects nearly all Mycenaean art. Indeed, it seems to be beyond doubt that the Achaeans imported many Cretan artifacts or (which comes to the same thing) that Cretan craftsmen came either voluntarily or as conquered slaves to work on the mainland. For example, two gold cups found in a tholos tomb at Vaphio near Sparta are typical imports. The pastoral theme, the human figures, the farmers wearing a double pointed loin cloth, and especially the manner in which the two scenes are treated, are all Minoan. The motifs decorating the inlaid daggers from Circle A, too, owe nothing to the mainland; they show, for instance, the octopuses and argonauts so familiar from L.M. II pottery. Some gold cups inlaid with silver found in the royal tombs of Dendra (ancient Midea) betray the same source of inspiration. Where another branch of thought is concerned, the objects bearing religious motifs are so similar to their Cretan counterparts that there was good reason to assume a common Minoan-Mycenaean religion until the Pylos texts introduced a few different shades of meaning. The ritual objects are the same, and the representations show identical religious scenes: the mother goddess, the fertility rites and the tree and pillar cults all appear. On seals and impressions the goddess and her attendants are shown dressed in the Cretan style with a long flounced skirt and a short sleeved jacket leaving the breasts exposed. The relief above the Lion Gate at Mycenae which leads to the approach to the citadel shows even more obvious Cretan inspiration. The column set on a pedestal in the middle might have come from the palace at Knossos. As to the two confronted lions guarding it, their identical counterparts appear on a Cretan seal which shows them guarding the great goddess, who is standing on a mountain peak. Whatever Evans may have believed, it is by no means certain that the Mycenaean column is an aniconic symbol of the great goddess. Others have suggested that it is a symbol of the palace and town itself, which is under the protection of the beasts. The treatment is totally Cretan, even if the symbol stands for a Mycenaean subject.

In many cases it is possible to distinguish the original Achaean elements in the archaeological material from the mainland. First of all, the numerous representations of warriors and scenes of combat discussed above can be accepted as exclusively Mycenaean, since they seem to be largely absent from Cretan art. In addition, the Achaeans had a taste for symmetry, almost rigidity, which is completely foreign to the Minoan spirit. We need only compare the 'Parisienne' from Knossos with her counterpart from the palace at Tiryns; the dress and hair are almost identical, but the difference is unmistakable. While the Cretan lady is vivacious and spontaneous, the Achaean is stiff and stilted. In the same way, when a Mycenaean craftsman borrowed the octopus motif from Crete, he did not let it sprawl freely all over the sides of the vase; the creature, as if tamed, coils his tentacles tidily on either side of a symmetrical axis. The surface to be decorated is not treated as a unity, but divided into metopes, or zones. In plastic art, this liking for order and geometric break-down is betrayed by a taste for flat planes and acute angles. Thus the gold lion's head from Mycenae looks as if it was carved out with an axe (it is actually made of beaten gold). On the other hand, a silver bull's head, also from Mycenae but certainly of Minoan workmanship, is entirely carried out in curves. These are two fundamentally opposite artistic concepts.

Perhaps these qualities of order and discipline which can be discerned among the Achaeans explain their success over the Cretans. The remarkable expansion of their trading empire can also be attributed to it. In Tarentum in southern Italy and in Sicily there are numerous imports. They are very common in the eastern Mediterranean. Rhodes and Miletus both had a flourishing Mycenaean trading post. Mycenaean vases have been found on the whole of the Syrian and Lebanese seaboard in the Levant. The remains of some 800 Mycenaean vases were found at Tell el Amarna, the short-lived capital of the pharaoh Akhenaton in Egypt. Aegean painting probably also influenced the Amarna frescoes. This expansion was obviously based on a powerful army and fleet. The former included notably a strong squadron of chariots, the shape of which has been preserved by vase paintings and clay tablets. The latter were pirate ships equipped with a beak and powered by oars. Trade was essentially barter. A series of copper ingots have been found, especially in the royal villa at Hagia Triada, and in the Mycenaean wreck off Cape Gelydonia in southern Turkey, which seem to be the Mycenaean equivalent of money. Each ingot have a predetermined value in spite of appreciable variations in weight, for which no explanation has yet been found.

This being so, it is not surprising that imported eastern objects have been found in Greece. The most important find was made in 1963 at Thebes in Boeotia. In the ruins of the Mycenaean palace, attributed by legend to Cadmos, several lapis lazuli cylinders and seals with cuneiform inscriptions were brought to light, as well as some tablets inscribed in Linear B. The seals mostly belong to the Babylonian Kassite dynasty, and therefore date to the fourteenth century BC. Although the precise reason for their presence in Thebes is by no means clear, they are certainly evidence of the extensive contacts between the Mycenaean world and Western Asia.

The end of the Mycenaean civilisation

The high point of Mycenaean civilisation was reached in the second half of the fourteenth century BC. Towards the middle of the thirteenth century signs of decadence were beginning to appear.

We now enter on an extremely obscure period in Mycenaean history. Nothing, or almost nothing, is certain. We do not know if the Trojan war actually took place nor if it did what its date was within fifty years or so. As to the final destruction of the Mycenaean sites we do not know whose work it was. Even the precise date of the Dorians' arrival in Greece, which marks the beginning of the Iron Age, is equally uncertain. There is, therefore, no question of definitive statements now, we must confine ourselves to setting out a series of theories. It has already been shown that Mycenaean expansion reached its maximum at the end of the fourteenth century (the beginning of L.H. III B). There are 143 known Mycenaean sites belonging to this period compared with ninety for Late Helladic III A. On the other hand, the number drops to sixty-four for the subsequent period (L.H. III C) which began at the end of the thirteenth century. The decline, therefore, was spectacular, and must have begun in the preceding period. There are, in fact, several indications that the Mycenaean kingdoms were anticipating attack towards the middle of the thirteenth century BC. The ramparts at Mycenae were extended to include the entrance to the underground spring. An inscribed tablet from Pylos reveals that bronze from a sanctuary deposit was requisitioned for making arrow heads. Finally, a wall was built to bar the Isthmus of Corinth.

These precautions did not halt the invasion which broke over the Achaean kingdoms. There were probably several raids in succession. Thus there were at least two destructions at Mycenae: the first, at the end of L.H. III B, only affected the houses outside the citadel, the second, in L.H. III C, reached the granary in the citadel itself. The same thing probably happened at Tiryns. On the other hand, the palace of Nestor at Pylos, the citadel of Gla in Boeotia, and the village of Zygouries in the Argolid were finally

The details of Cretan religion are obscure although many ritual scenes have been preserved, particularly on the bezels of large gold seal-rings. A common theme is the elaborately clothed goddess attended by female votaries and flanked by tree shrines. One of the most difficult problems in dealing with an essentially pre-literate society is the reconstruction of religious practices. An enormous number of cult objects and scenes are known, but their real significance remains a mystery.

The largest vault in the ancient world constructed before the Roman Pantheon was the Treasury of Atreus at Mycenae. The dome is not a true arch but a corbel vault, which is made by laying converging courses of masonry and smoothing off the angles of the stones on the inside. Perforations in the stonework and some fragments of metal in the fill of the tomb indicate that the inside was embellished with bronze rosettes.

Recent excavations near Pylos, the legendary home of Nestor, on the west coast of the Peloponnese have revealed a *tholos* tomb, the vault of which has now been restored. An important palatial centre near this site once more attests the accuracy of the Homeric poems as an archaeological source.

destroyed at the end of L.H. III B, or towards the end of the thirteenth century BC.

We do not know who the enemy was. Literary tradition mentions many internal dissensions both before and after the Trojan War. At Mycenae itself, to quote only one example, the strife between the descendants of Perseus and those of Pelops and the subsequent deeds of the Atreidae indicate the instability of the political administration. But it seems doubtful that the total disintegration of a flourishing empire could be attributed solely to civic strife. This is even more unlikely in view of the fact that the period from 1250 to 1150 BC was one of unrest and destruction throughout the whole of the eastern Mediterranean. Egypt was attacked twice (in the reign of the pharaoh Merneptah about 1200 and again under Ramesses III about 1190) by a curious coalition known as the Sea Peoples, who have been identified, among other things, as the Lycians and the future Philistines before their settlement in Palestine. It is not improbable that one of these peoples ravaged mainland Greece, though it is difficult to imagine what route they followed. This hypothesis, however, is far from being substantiated.

The problem is further complicated by the Trojan War and the question of the Homeric poems. According to the written evidence and also, no doubt, that of other lost sagas, the historical truth of the event was never questioned by the classical Greeks. Ought we to have as much faith as they? Homer's partisans never fail to emphasise the number of typically Mycenaean features mentioned in the epics. Some of the Homeric epithets ('Mycenae rich in gold' and the 'well-greaved Achaeans') have been verified by excavation. The boar's tusk helmets found in the royal tombs of Mycenae are accurately described in the *Iliad*. In the *Odyssey* Nestor uses a cup with two doves with outspread wings on the rim as if they were drinking from it. A vessel answering this description was also found at Mycenae. The Catalogue of Ships in Book II of the *Iliad* paints a picture of Mycenaean Greece which is fully confirmed by archaeology. Comparisons can also be drawn between the language of Homer and that of the tablets. All this implies that there was a continuous tradition from the end of the Bronze Age (about 1100 BC) to the Homeric period (about 800 BC). Studies of the oral tradition of the Yugoslav popular epics conducted between the wars have shown that this is by no means impossible. Relying on a turn of phrase based on set formulae to fill half or a third of the line, the bard was able to repeat the majority of both the substance and the letter, while keeping his freedom to improvise and contributing many modifications.

The excavations at Troy

Finally we come to the archaeological evidence furnished by the actual excavation of Troy. At the end of the Bronze Age the situation was as follows: about 1300 BC the site was occupied by the sixth city, which was itself in its eighth stage (Troy VI-H). It was a strongly fortified citadel, parts of the wall of which are still standing. The walls were up to nine metres high and four metres thick, had battlements and were flanked with towers. The houses inside were large and spacious. Many were built on the megaron plan. This city was destroyed about 1300 BC by an earthquake which caused several fires, but there was no trace of violence. It was succeeded by a city (Troy VII-A) which occupied the same site but was certainly much poorer than the preceding city. The ramparts were repaired in much rougher masonry, the houses were smaller and there were fewer finds of pottery. It has been observed that huts were crowded against the ramparts as if there had been a great influx of population. Moreover, C. Blegen interpreted (perhaps mistakenly) the fact that provisions were stored in pithoi sunk in the ground as a precaution against a siege. However this may be, this city too was destroyed, but this time by war. There was a fire and the human remains indicate battle and sudden death. C. Blegen is convinced that this was the work of the Achaean expedition, and that Troy VII-A is Priam's Troy. The Homeric tradition is thus amply verified.

Several more or less valid objections have been urged against this theory. One argument is that Troy VII-A is too poor a city to answer to Homer's magnificent description. Troy VI-H would make a much better candidate. In fact, Troy VII-A occupies the same ground as Troy VI, which has not yielded any very valuable objects. Above all, allowance must be made for poetic licence on Homer's part. The objection to the dates proposed by Blegen is more serious. He believes that Troy VII-A was destroyed before Pylos at the end of L.H. III-B, or about 1260 BC. Other experts, however, place a different interpretation on Trojan pottery, which leads them to assign the destruction to L.H. III C, or around 1200 BC. This being so, either the destruction of the Mycenaean palaces must be brought down by two generations, which is hardly possible, or it must be allowed that the destruction of Troy is contemporary with, if not later than, that of the Mycenaean settlements. If this is true, it could not be the work of the Achaeans, whose kingdoms collapsed at the end of the thirteenth century. In any case, if the date of 1260 BC is accepted for the fall of Troy, it might be asked why the Mycenaeans, already in danger and acutely aware of it, went and wasted their forces on distant expeditions.

This brings us back to the question of the validity of the Homeric texts. It should first be observed that various errors and anachronisms are to be found there as well as the Mycenaean characteristics confirmed by archaeology. For instance, Homeric warriors used their chariots as 'taxis' to take them to the battlefield, while Mycenaean warriors actually employed them as assault vehicles in the Hittite manner. The dead were sometimes cremated, sometimes inhumed, while in fact cremation was seldom practised before 100 BC. Furthermore, the accuracy of a few details proves nothing about the validity of the poem as a whole. On the contrary, in an epic of a relatively recent date like the Middle Ages, a comparison between historical events and the sagas to which they gave rise shows that very little reliance can safely be placed on the epic tradition. For instance, the poet of the *Chanson de Roland* makes an enormous battle out of a minor skirmish and reverses the outcome to suit his story. In the *Niebelungenlied* the attacker becomes the victim and vice versa. Where the *Iliad* is concerned, the existence of the city of Troy is as undeniable as that of the pass of Roncevaux, Agamemnon can be accepted as being as much of a historical character as Charlemagne. It is equally certain that Troy was destroyed in the second half of the thirteenth century BC, and this was surely the historical event which gave rise to the epic. Beyond this point we enter a realm where we are not yet in a position to distinguish between legend and history. It is not even at all certain that the assailants were the Achaeans, and at present this does not even seem to be the likeliest solution. Could this be the Sea Peoples again? Once more, everything is uncertain, and in the present state of archaeological studies, the question remains open.

The Dorian invasion

The fall of the Mycenaean kingdoms in L.H. III C which can be seen perfectly clearly on the sites, is perhaps reflected in the traditions about the homecomings of the Homeric heroes and the palace revolutions which followed, for instance at Mycenae. It was probably at the close of this period that the event took place which finally put an end to the Bronze Age in Greece: the Dorian invasion. The historical reality of this occurrence, which figures in the histories as the Return of the Heracleidae, admits of no doubt. A map of the distribution of Greek dialects in the classical period clearly shows the route followed by the newcomers, who took first Epirus, Aetolia, Acarnania and the north coast of the Corinthian Gulf before seizing the Peloponnese and spreading through the Aegean to Crete, the Dodecanese, Melos and Thera. Attica, on the contrary, was spared, and became a refuge for many dispossessed Achaeans.

The Dorians, we must remember, were not really foreigners; they spoke a Greek dialect. The Return of the Heracleidae means to the Greeks the return of distant cousins to claim their heritage. They probably did not come from far away, and the most likely theory is that they were the inhabitants of northern Epirus and Macedonia who moved into the south for unknown reasons. Their standard of culture was lower than that of the Achaeans, and they do not seem to have brought an original style with them. It is remarkable, indeed, that the pottery known as Geometric, which developed from the tenth century onwards, originated in Attica, the very place where Dorian influence was least felt.

Soon afterwards, the division of Greece between Dorians and non-Dorians (also known as Ionians) was to become a decisive factor in its history: the Greeks themselves explained most of the strife which divided them until the Hellenistic period as conflicts between those two branches of the same people. By the end of the eleventh century BC the framework of Aegean civilisation had been destroyed and that of classical Greece was already established.

The huge fortification walls of the citadel-palace of Tiryns in the Argolid are pierced by corbel-vaulted casemates once used for storage of food and armaments in time of siege. These massive defences are a clear indication of the unsettled conditions of Mycenaean Greece. The opposition to the returning heroes of the Trojan campaign as told in classical poetry was plainly a historical fact.

Classical Greece

Early archaeologists

The earliest classical archaeologists are to be found among the ancient Greeks themselves. As soon as they began to speculate about their own history, they paid special attention to the evidence of objects and works of art. Thucydides followed a rigorous system which no one nowadays would reject. In the introduction to his *Peloponnesian War*, which is known as the archaeological section [illegible] the investigation of the origins of the Greek peoples, he analyses the power of the Greeks before the rise of the Minoan thalassocracy, [illegible] 'The pirates were the Carians and the [illegible] islanders; these were, in fact, the people who inhabited most of the islands, and this is the proof of it: when Delos was purified by the Athenians during the war and all the tombs on the island were cleared away, it became clear that more than half of them were Carian, as could be seen from the panoply of arms buried with the dead and the form of burial, which they still practise today.' A little further on, Thucydides showed the problems and dangers which are encountered when archaeology deduces historical facts from excavations alone.

'If the city of the Lacedaemonians were to be abandoned and only the sanctuaries and the foundations of the buildings remained, I do not think that future generations would believe that it was as powerful as they had been told, and yet the Lacedaemonians occupied two fifths of the Peloponnese, ruled the rest and had many allies further afield. But as their city was not adorned with sanctuaries and fine buildings, and as it was composed of hamlets in the manner of ancient Greece, it would appear very humble. If the same fate befell the Athenians, to be judged by the appearance of their city, their power would seem to be twice what it really is.'

The facts are discovered by discriminating use of the records and the establishment of parallels.

The respect enjoyed by Greek history and Greek artists during the Hellenistic period and still more under the Roman empire was such that authors extolled their works of art, which were eagerly sought by Roman collectors. The most renowned sites had innumerable visitors, their monuments were described in books, and their history, sacred and profane, was recorded. The most useful traveller to Greece was Pausanias, who toured the country in the second century AD in the company of local guides, and wrote up his notes in a book, the *Description of Greece*. He was painstaking but credulous, and is often accused of stupidity. Pausanias was trying to produce a work of art, a [illegible] and mythography, but he was interested almost exclusively in the monuments of the great periods of Greek history, the archaic and classical eras. When confronted with a work of art he sometimes indulged in digressions which occasionally make his record rather confused. Nevertheless, until the great nineteenth-century excavations, Greek archaeology rested on his evidence and that of the other literary sources.

The temple of Apollo at Delphi

The Riddle of Delphi is the title of a study attempting to solve several obscure and controversial questions raised by the excavations of the sanctuary of Apollo. This could, however, be the subtitle for the whole published record, so completely did the freaks of nature and the savagery of mankind succeed in wrecking the god's domain between its abandonment at the end of pagan times and the first systematic investigation, carried out from 1892 onwards by the École Française, which was directed at that time by T. Homolle. Photographs of the site are too common and familiar for extensive description to be necessary here. The sanctuary covers the rising ground [illegible] above the narrow gorge of the Pleistos, a few kilometres from the sea, at the foot of the rocky cliff of the Phaedriades. Earthquakes and rock falls are part of the history of the sanctuary. [illegible] the barbarians at the height of their attack and prevented them from reaching Delphi during the Persian Wars. During the Galatian invasion of 279 BC a similar incident occurred. In 1905 great rocks broke off from the Phaedriades and smashed the remains of the old temple of Athena which has only just been uncovered. Was it an earthquake and landslide, or only a fire, which destroyed the temple of Apollo in 373 BC? Whatever it was, a terror certainly brought about the total destruction of the temple.

Then came the desecration by human hands: the melting down of all the valuable metal offerings by the Phocians in order to pay mercenaries, the methodical and more dangerous pillage of the Roman proconsuls and emperors who enriched Rome with the spoils of Apollo

In 1896 the French excavators of Delphi were investigating a water channel dating to a late period when they lighted on one of the finest examples of early classical sculpture, a bronze figure of a charioteer which had once been part of a group. The statue is wonderfully preserved, even the inlaid eyes and wire eyelashes surviving to give a disturbing realism to its austere gaze. The accompanying inscription on the charioteer's plinth indicates that the group was dedicated as a thanks offering by the Sicilian tyrant Polyzalos.

Opposite page: the *temenos* (sacred precinct) of Apollo, built on a steep mountainside at Delphi, is an unusually complex archaeological problem. Laid out in the sixth century BC and only abandoned in early Christian times, the history of the sanctuary is one of centuries of human depredations and natural disasters. Several ancient authorities have left descriptions of the site before its destruction, but even this first-hand information can seldom be easily reconciled with the archaeological findings. A view of the Phaedriades and the terrace of the temple of Apollo seen from above gives some idea of the precipitous fall of the ground.

Building techniques and architectural ornaments can help in dating and reconstructing both individual buildings and large sites. At Delphi the polygonal masonry of the retaining wall to the right dates it to the early construction works of the sixth century, while the Athenian Treasury to the left shows a number of refinements although it was already finished before 480 BC. Variations in the height of the courses of stonework break the visual monotony of the blank sides, and the upward-tapering section of the walls lends an illusion of height.

However, the Romans did leave the architectural setting intact. After its final desertion at the beginning of the Byzantine era, the sanctuary was used as a stone quarry and a source of metal. In a period when metal was scarce it could be got by collecting the staples and pins made of bronze and coated with lead, which joined the blocks together. The effect of this pilfering can be clearly seen in the ruins of the great temple. The columns have been systematically dismantled and the slabs of the peristyle prised up in order to get at the fastenings of the course below. The first two downward courses of the foundations were not completely removed and only the third course was found in place. The site had not, however, been deserted, and a Byzantine village had been established above and around the temple. The cellars of the houses of this village rested on the terrace walls, notably the polygonal wall, and naturally enough the ruins provided the necessary building materials. Blocks of marble, very little reworked, formed the walls of houses and ancient paving slabs were reused in the main street, which roughly followed the course of the Sacred Way. This site was recognised at a very early date. As early as the fifteenth century Cyriacus of Ancona, the first of a long line of Western scholars, visited it and copied inscriptions.

Even a brief survey demonstrates the difficulties facing archaeologists. As long as the village of Kastri was standing topographical and epigraphical researches had to be conducted under extraordinarily trying conditions. Before excavations could be started in 1892, the French government financed the rebuilding of the village on nearby land. Tons of stones on a very steep slope were shifted as well as accumulated debris several metres thick, all mixed up with stray ancient blocks, but the harvest of works of art was extremely rich, as was that of architectural fragments and inscriptions. The enclosing wall of the *temenos*, or domain, of Apollo was excavated as well, and in its present state it can now be seen as a huge enclosed rectangle. The excavators cleared also the temple terrace, which is retained on the south side by the great polygonal wall and closed to the north by the *Ischegaon* (retaining wall). The complex was crossed by the Sacred Way which ascended in zigzags from the main entrance at the south-east corner of the enclosure to the temple terrace. A large number of foundations of all sizes were found which had to be identified and examined in order to discover the relationships of various buildings. At this stage of academic interpretation difficulties redoubled. The stones could have been moved any distance or direction over the surface of the sanctuary, some falling all the way down the slope, others having been carried up again and replaced. There is no need to stress the damage done to stones by their having been moved or the eroding action of rain and frost. Reliefs were found which had been hammered to pieces to make roofing slabs, nothing remaining but a mutilated wreck of the original stone. It was thus very difficult to find which monument it once belonged to and to establish its original function.

The problems of using an ancient guide

A study of the first section of the Sacred Way clearly illustrates the complexity of the clues from which the main elements of the ancient sanctuary have been reconstructed. Just inside the entrance the Sacred Way was flanked by offerings made by pilgrims from the cities, and these offerings were regarded as the sign of their acknowledgment of Apollo and his recognition of them. Pausanias described these monuments in great detail. He said that at the entrance there was a bronze bull dedicated by the Corcyrans after they had obtained a miraculous draught of fish. Then there was an *ex-voto* made by the Tegeans in gratitude for the spoils taken from the Lacedaemonians. This was a plinth on which stood statues of Apollo, Nike and the Arcadian heroes (this was the region in the central Peloponnese where Tegea was situated).

Opposite was the monument erected by Sparta after the victory over Athens which ended the Peloponnesian War. In the foreground the Lacedaemonian admiral Lysander was being crowned by Poseidon, who was surrounded by the chief deities. Behind Lysander were the other Spartan leaders and their armies; behind them stood the gigantic Trojan horse, a gift from the Argives, and in front of the horse a plinth with an inscription stating that it bore statues from the spoils of Marathon: these represented Athena, Apollo, the victorious *strategos* Miltiades and a series of Attic heroes. According to Pausanias 'The said statues are the work of Phidias'. Beside them were other Argive offerings, a group of the Seven Against Thebes led by Polynices, then a group of the Epigones, i.e. descendants, who again laid siege to the city a generation later. These groups were set up in the middle of the fifth century BC. Pausanias continued, 'Facing these are other statues. The Argives dedicated them for their share in the foundation of Messenia, with the Thebans and Epaminondas', or a century later. These are statues of heroes and Argive kings.

Pausanias' *Periegesis of Greece* seems to be perfectly clear about this. He first describes two monuments on one side of the Sacred Way, then five more on the other side and finally comes back to the first side. The only surprising thing is the attribution to Phidias of the statues set up with the booty from Marathon. If the monument was erected soon after the battle, according to custom, the career of the sculptor would have to be predated a long way. If not, it must be accepted that there were a good twenty years between the victory and the offering to Apollo. Perhaps Pausanias was mistaken. In fact, it has recently been proved that this delayed dedication was very well explained by internal political considerations. Cimon, son of Miltiades, the leading statesman of Athens, wished to remind the citizens of his family's merits and simultaneously to demonstrate his devotion to the gods.

Excavation, however, has shown how difficult it is to identify the various unearthed remains, including a number of foundations and stray stones, some of which are inscribed. On the north side, just beyond the entrance, a base was found which measured roughly 6.3 × 2.6 metres. This would have remained a mystery if it had not been realised that a stone from much higher up, near the polygonal wall, fitted perfectly with it. This stone bore a dedicatory inscription and the signature of the sculptor Theopropos of Aegina, to whom Pausanias attributed the Corcyran bull. There then remained only one mystery. The staples which joined together the blocks of the base were from the fourth century BC or even later, while Theopropos worked in the first third of the fifth. It was shortly deduced that the Corcyran dedication had been removed and set up on a new base soon after the middle of the fourth century. Not long afterwards a place described by Pausanias was accurately identified.

Immediately behind the Corcyran bull, also on the north side, the excavators discovered a base more than ten metres long, which could be very fully restored. Two courses of grey limestone on a red brecchia plinth bore a series of black marble slabs. Most of these slabs were recovered. Holes were found in them for attaching the feet of bronze statues. This was surely the Tegean dedication. The right hand end, which was wider, bore Apollo. The identification was confirmed by a verse inscription engraved on the steps, commemorating the origins of the Argive heroes, just as Pausanias records. Without a doubt he read the same text, or had had it read to him by the guides.

After the Tegean base and in line with it came the remains of three small Hellenistic bases, as shown by their architecture. Pausanias does not mention these. An equestrian statue of Philopoemen, an Achaean general of the end of the third century BC, probably stood on the first. The evidence of Plutarch and that from another

One of the rare circular buildings of classical Greece the Delphi *tholos* lies in the precinct of Athena Pronaia below that of Apollo, and is therefore particularly vulnerable to damage from the earthquakes and landslides which are so common in the area.

In a great classic complex such as that of Delphi, the work of reconstruction is not to be lightly undertaken. Almost every stone in the *temenos* was individually tailored for one specific place and no other, which means that substitution is virtually impossible. The tiny Athenian Treasury beside the Sacred Way is one of the cases where sufficient masonry has been identified to make successful anastylosis possible. The treasury is a reproduction in miniature of contemporary temples.

When the Persians destroyed the old city of Athens in 480 BC they presented the citizens with an ideal opportunity for the building programme which resulted in the construction of the Parthenon, the Erechtheum (left of the Parthenon), the monumental gateway to the Acropolis and the exquisite little Nike temple on the west bastion beside the entrance. Inscriptions state that the majority of the work on the Parthenon was begun in 447 and completed by 432 BC, an amazing feat of creative achievement.

stone bearing the Achaean dedication justify this assumption. The remains of the other two bases are still unidentified. The most eye-catching monument on this side of the Sacred Way, however, is an immense niche behind and above these bases. It is basically a rectangular platform more than twenty metres long. Its north west side is partially carved in the rock, and its south east corner stands on a brecchia [illegible] wall. Walls on the other three sides framed a splendid dedication set on a ledge that widened into two massive plinths at each end. There is no surviving superstructure to identify either the dedicator or the offering, and Pausanias never says a word about the most important monument of this part of the Sacred Way. As it happened, two semicircular monuments immediately behind the niche, one on each side of the Sacred Way, were clearly proved by their inscriptions to be the two Argive offerings, and Pausanias distinctly states that they stood opposite each other.

There are only a few poor remnants left of the other side, where Pausanias locates the Aegospotami dedication, the Marathon base and various Argive offerings one after another, or sixty three statues and a colossal horse. Some limestone blocks just opposite the Epigones

The Acropolis in 1819. In this view from the west the Turkish fortifications can be clearly seen. They join the Propylaia to the theatre of Herodes Atticus.

monument are the remnants of a massive base; some more blocks a few metres away from the corner are from a second base. The monuments they once bore cannot however be deduced. Immediately beside the entrance the excavators had to dig fairly deeply, only to find nothing in its original position but an ancient wall resting on virgin soil a long way below the fifth-century level. The ground here slopes rather steeply, and a large amount of fill had to be removed to put this part of the Sacred Way in order and reduce the slope to a reasonable degree. This side, too, was badly damaged. Nothing remains of one of the most important offerings but a few statue bases and stray stones found near the entrance to the sanctuary and identified by the names and inscriptions on them. Such total destruction of the bases is strange, as is the silence of Pausanias on this subject as well as that of the great niche on the north side. This coincidence led to the conclusion that either the written tradition was wrong, or that the Greek text had been misunderstood, or more simply, that Pausanias did not remember the arrangement of the monuments very clearly at the time he wrote his book, having confused his notes and therefore placed the Lacedaemonian admirals on the niche. The pedestals of the statues [illegible] very [illegible]

The two temples on the Acropolis at Athens, the Parthenon and the Erechtheum, form a subtly contrasting counterbalance of the Doric and Ionic orders of architecture. Skilfully adapted to the uneven contours of the ground, the Erechtheum is particularly rich in beautifully carved architectural ornament but its chief glory is the porch of caryatids whose serene poise and rising lines brilliantly avoid any feeling of weight and pressure and constitute one of the most successful examples of this form ever to be created.

The Erechtheum in 1750, long before the Acropolis was cleared. This view from the south-west is by James Stuart, co-author of *The Antiquities of Athens Measured and Delineated*, which was published in 1762.

for the monument, and it has been possible to *reconstruct a sort of room opening via a portico* on to the Sacred Way, with the help of the ancient authorities and the archaeological remains.

Most archaeologists accept this solution and the majority of plans include it but there are still a number of problems. If Pausanias is convicted of one error, he would have to be charged with many others. Thus someone who knew Delphi *very well wrote: 'Pausanias says that the twenty-*eight naval captains were lined up behind the principal group. He does not seem to have expressed himself accurately here . . .', and corrects him by arranging the main group on the ledge in the middle, dividing the admirals into two equal groups on the right and left. What is more, the architecture of the niche does not agree with the very early fourth-century date which was soon assigned to it. Apart from the fact that it seems to have been built after the Tegean base when it ought to have been fifty years before, some details such as the form of the staples seem inconsistent with so early a date. After noticing these points, some authors have been led to revise the traditional interpretation, *accept the silence of the texts about this* great niche, and faithfully follow the descriptions of Pausanias.

In view of the controversies raised by the topographical problems in this small section of the Sacred Way, it can be imagined what difficulties surround a description of the whole sanctuary. Archaeological *findings and ancient* evidence are two contradicting sources which *gainsay each other and seem, at first, to be in*compatible. If the sanctuary is to be properly understood and restored as far as possible to its ancient appearance, however, they must be re*conciled. Indeed, it is not enough* to uncover the different archaeological levels; the finds must then be explained. The best preserved monuments are still on the site waiting to be revealed, but as one scholar observed:

'It is not a question of making an artificial stone landscape for tourists, or of restoration. Actual reconstruction, or *anastylosis*, has a very special character in Greece, which distinguishes it from *the restorations, discreet or otherwise, conducted* elsewhere. In a Greek monument of the best period nothing is interchangeable, every stone has its place, and only one place. We are confining ourselves, therefore, to placing those stones which can be definitely identified in their original unique and mathematically determined position.'

A town-planner's excavation: Priene

When Priene was excavated, the German director Theodor Wiegand gradually acquired a perfect understanding of the characteristics of *the site. This scholar, who worked at Priene* from 1895 to 1898, explained that he was not trying to find masterpieces, nor to study an isolated monument, however remarkable, but to establish the plan of the city so as to be able to reconstruct both its history and the daily life of its citizens, with the help of inscriptions. Priene was the ideal *site for such a study. As it hap*pened, the whole city was rebuilt in the middle of the *fourth century* BC on a hitherto unoccupied hill above the plain of the river Maeander. It is thought that the first city, founded at an

An unfinished temple at Segesta in Sicily provides many clues to late fifth-century building methods. The inner cella was not even begun and the structure was never roofed, but the colonnade is complete and the shafts of the Doric columns are ready for fluting. The steps up to the temple base still bear the stone bosses which were used for securing ropes during transport and positioning, but the top surface has already been smoothed into the subtle rising curve which helped to give a feeling of lightness to the classic Doric temple.

unknown date during the Ionian colonisation of the coast of Asia Minor, occupied the lowland by the sea. Alluvial deposits from the Maeander gradually made the port unusable and then raised the level of the water table until the city was liable to continual flooding. The same fate was to befall Miletus several hundred years later. There was nothing for the inhabitants of Priene to do but abandon their city, which was then completely buried in silt. They looked for a refuge from these troubles and chose the south slope of a hill at the foot of a cliff, which could serve as an acropolis for defence in the last resort. Their first care was therefore to build ramparts while they laid out the plan of the new city. This enclosure did not entirely surround the city but followed the natural contours of the ground in the usual Greek style. It formed a triangle on the top of the hill, descended the sharpest slope, breaking off wherever the cliff was sheer enough to prevent attack, and widened out to enclose all the land on which the city was built. Wherever possible, attempts were made to make the roads and the gates coincide, but discrepancies between the two show plainly that the city and the wall were planned separately. There were always large empty spaces inside the wall.

It soon became clear that this was a novel plan. The architects had divided the habitable ground into a number of rectangular lots measuring about forty seven metres long by thirty five metres wide with streets between them. They then shared out these lots between the sacred ground dev[illegible]d to the sanctuaries of the patron deities of [illegible] city (chiefly Athena Polias and Zeus), the [illegible]atre, public squares, civic buildings and othe[illegible] like gymnasia, and private property on whi[illegible] individual houses could be built. The complex was provided with a system of intersecting streets, five main roads running from east to west cut by side streets running from north to south. The application of so rigid a system to irregular ground, in contrast with the free adaptation of the wall to the same ground, raised several problems. In order to provide a more or less horizontal area for each lot, the architects had to compensate for the natural slope of the ground by digging here and there, or by dumping fill and supporting the plot on strong retaining walls and terrace[illegible] The[illegible] considerations resulted in some [illegible] the str[illegible] being made into flights of steps. M[illegible] [illegible]ng opinions of this plan. S[illegible] appr[illegible] [illegible]f the terracin[illegible] and the lightness of the complex [illegible]ublic centr[illegible], such as the theatre, the colonnades of the sanctuaries and the agora mostly faced south, giving strollers a magnificent view over the Maeander plain and plenty of light. The sturdy steps and terraces also broke the monotony of the plan and gave a very pleasing harmony to the whole. Other scholars, however, believe that the road system, so ill adapted to the contours of the land, must have impeded internal traffic considerably, prevented the use of chariots (but surely most transport, particularly in so small a city, was by means of donkeys) and are less impressed by the ingenuity of this way of arranging the details than by the basic inconvenience of the whole idea.

If the Priene excavations have demonstrated the architects' working methods so clearly, it is because the plan of the city has survived virtually intact. There were, admittedly, various repairs carried out during the Hellenistic and Roman periods, but these were always very restricted, and can easily be recognised. The agora was extended, a new gymnasium built, new sanctuaries were consecrated and private houses remodelled. These alterations were always carried out within the framework of the oblong lots and had no appreciable effect on the plan of the city. The late Empire and the Byzantine period did not bring any major changes. The baths were enlarged at the expense of the gymnasium, the Christians built small churches, and the townspeople turned the sanctuary of Zeus into a fortified citadel because of the impossibility of defending a town which was forever sliding downwards, but these alterations were minor and can be easily identified. The builders abandoned the fine classical ashlar work for a much less elegant material, a heterogeneous mixture with a lot of rubble in it; they also used mortar, which is never found in classical Greek architecture.

The temple of Athena Polias

During the empire the population of Priene, which must have been between four and five thousand at the height of the Hellenistic period, continually diminished. This was due, no doubt, to the Maeander, which continued to silt up the gulf and transformed Priene into an inland city, badly situated above a marshy and pestilential plain. To the south, Miletus was gradually abandoned at the same time and for the same reasons. Although Christian Priene survived a little longer, the site was completely abandoned after the Seljuk invasion of the thirteenth century. In the eighteenth century the first Western visitors arrived, one of whom was Jacques Spon of Lyon, who identified the site and copied some inscriptions.

When Athenian ships sailed away from Attica their last sight of their homeland was the Doric temple of the sea-god Poseidon set high on the sheer cliff of Cape Sunium. Built about 440 BC, its details are so similar to those of the Temple of Hephaestus in the Athenian Agora that many scholars are inclined to ascribe both buildings to the same architect.

Another view of the massive bronze figure of a god from the sea off Artemisium, already shown on page 50. It might represent Poseidon himself, but detailed studies of the hand have failed to establish decisively whether it held a thunderbolt or the sea-god's trident. Whoever the figure may be, its austere magnificence and virile power place it in the front rank of Greek sculpture.

The first excavations were begun in 1765 by Chandler and Revett on behalf of a group of British scholars and antiquarians, the Society of Dilettanti. Their methods, like those of their successors who were led first by Gell in 1812 and then by Pullan in 1869–70, show clearly the aims of the first archaeologists who wanted to rediscover the temple of Athena Polias which had been celebrated in the ancient world. Indeed, it was regarded as the prototype of the Ionic temple. The architect was Pythius, who also built the Mausoleum, the tomb of King Mausolus at Halicarnassus. He recorded the details of his work in a b[illegible] which was one the classic works on Greek architecture in the ancient world and was used by Vitruvius.

The British archaeologists set about studying the structure of the temple of Athena Polias, drawing the architectural fragments and finding the decorative carvings which were eventually taken to London, by which means they escaped being burnt for lime, the destined fate of all the pieces of marble which the excavators dug up and abandoned on the site. The archaeologists wasted no time, they traced the outlines of the temple and published a description of its simple design, which was adorned with elegant decorations. Examination of the temple, which had been destroyed by a severe earthquake in the Middle Ages, enabled them to reconstruct the history of this monument and give valuable evidence of the difficulties which had faced the citizens of Priene.

The ornamental work of the temple is not strictly consistent as can be seen in the design of the palmettes, lotus flowers, acanthus scrolls, and the shape of the mouldings which follow a traditional style. By comparing these elements with those on securely dated buildings and other evidence such as pottery, it has been possible to trace the evolution of styles throughout the course of Greek history. This, in its turn, makes it possible to date a whole building or part of it fairly accurately by its decoration. Parts of the temple of Athena Polias clearly belong to the fourth century BC, while others are obviously later. Furthermore, according to the custom in Greek cities and sanctuaries, the stones of the walls were covered with inscriptions, votive dedications, edicts of rulers and important civic decrees. Some of these texts have also helped to illustrate the evolution of the building. One of the most important is the dedication of Alexander, made after the Battle of Granicus (334 BC), which is carved on the top of the south *anta* (the front extension of the *cella* wall): 'King Alexander dedicated this temple to Athena Polias.'

Building operations were probably started at the time of the foundation of the city about 350 BC, but they were delayed for a while, or actually stopped, by lack of funds, which was a fairly common contingency. Alexander then made a generous contribution which enabled the work to be continued, and received in return the privilege of dedicating the temple in his own name. Macedonian subsidies, however, were inadequate at Priene. Although the capitals on the front *antae* are clearly fourth century BC and the rear part of the walls is covered with third-century texts, it is plain that the back was left unfinished for nearly 200 years. The surviving parts of the entablature belong at the earliest to the second century. It was at this time that a donation from a Cappadocian prince paid for the erection of a cult statue, a copy of the Athena Parthenos as shown on some coins of Priene, as well as for the great monumental altar on the sanctuary terrace, which was modelled on the Great Altar of Pergamum. The completion of the temple to Athena Polias should probably be credited to this ruler. Still more work was carried out later. The *opisthodomos* was walled off and the temple rededicated to Athena and Augustus.

The British and German excavators have re-created two aspects of Priene; the former sought to find the work of a famous artist, Pythius, the latter reconstructed the archetype of the Hellenistic *polis* by excavating its squares and public buildings.

Delos: a great religious and commercial centre

The history of this tiny island in the heart of the Cyclades is very curious. In Mycenaean times it was probably already a sacred place, and in the seventh and sixth centuries BC it was a religious centre for the Ionians, who worshipped Apollo with processions, games and sacrifices. Temples and *oikoi* (sacred buildings for the offerings of the faithful in which some ritual gatherings were held), votive statues and other dedications already made it one of the most important Greek sanctuaries.

All around Apollo's domain, however, there extended a much later city which developed very rapidly – so rapidly that it could be called a mushroom town. Whole areas were built up in

A complex of Greek temples on the site of Paestum in southern Italy is, perhaps, in a better state of preservation than many in Greece itself. Most experts regard the temple of Poseidon, which was probably begun soon after 480 BC, as the best preserved surviving Greek temple. It is built of a matt golden local stone which is completely unlike the polished white marble usually employed in Greece. In the lush green Italian landscape the massive columns, nearly nine metres high, are remarkably impressive. Although it is traditionally called the temple of Poseidon little is known of its original dedication

the middle of the second century, but the prosperity of the port was to endure for less than a hundred years. In the second century BC the holy island was the biggest trading port of the eastern Mediterranean, and one of the leading slave markets of its day. This sudden upsurge was the result of Roman intervention. Wishing to be revenged on the Rhodians, who owed their power to their maritime activity, the Senate had decided to give Delos back to Athens, although it had been independent for nearly two hundred years, on condition that it remained a free port. The island's convenient location contributed also to its prosperity. 'The port is admirably situated for anyone sailing from Italy or Greece to Asia, ... and the Athenians, accepting the island, administered the sanctuaries and ports skilfully', wrote the Greek geographer Strabo. From 1872 onwards archaeological research was conducted first by A. Lebègue and then by T. Homolle, but it was particularly active between 1904 and 1914 thanks to the generosity of the duc de Loubat. With a few exceptions it has continued since then without interruption.

The port of Delos, excavated by the École Française, was extremely large. The group of harbours which form the port extend along the more sheltered west coast. They are flanked by quays for nearly two kilometres and linked to the public squares and private warehouses. The chief of these harbours is known as the Sacred Port because it was once thought that pilgrims disembarked there. From the archaic period onwards a large mole of granite blocks protected it from the *meltem*, or north wind, and for several hundred years improvements were added to the harbour. The oldest quay of gneissic stone lies parallel with the sanctuary wall and allows access to it. This enabled it to be dated to the last half of the seventh century BC or a little later. At the end of the third century the part immediately south of it was filled in to make a larger area of ground available between the sanctuary and the sea, and a new square called the Agora of the C[illegible] built on the newly reclaimed ground. It was at the [illegible] that Philip V of Macedon['s] portico was installed, flanking the avenue leading to the [illegible] with a parallel colo[nna]de opening [illegible] quays. It was then [illegible] closed on the south si[de] by a small mole at the point where a large n[illegible] k of land consisting of excavation fill new stretches out. At the same time or shortly afterwards, about 125 BC, Theophrastus carried out construction work in the north sector of the harbour and built the agora there, which bore his name. Various inscriptions containing accounts of the Delian magistrates and the dedication of the statue in the agora, set up in honour of Theophrastus, have made it possible to fix these dates accurately.

In a row on the south side of the Sacred Port were the four other harbours, with their quays, stores and warehouses, like the 'Shop of the columns' which included a central court with a peristyle and a court, on each side of which the

The oldest Hellenistic city to be investigated as a unified effort at town planning rather than as a series of unrelated buildings was Priene in Asia Minor. It is located on a steep slope below an acropolis overlooking the mouth of the river Maeander, and no attempt was made to adapt the uncompromising block-system of the city's layout to the irregularity of the ground. The result is a pleasing variation of streets and stairways, with open terraces giving a magnificent outlook, but the obvious inconveniences to wheeled transport must have constituted a serious drawback.

The island of Delos was a religious centre from the earliest times but as a commercial port its great days of prosperity lasted only from 136 to 88 BC. The average house followed the traditional fifth-century plan of rooms grouped round an inner courtyard but the building methods and materials were greatly improved. One courtyard has a splendid mosaic floor with a central roundel filled with running spirals and meanders and a magnificent pair of harnessed dolphins ridden by a tiny winged figure in each corner.

ground floor stalls and first floor residences opened. Harbours and shops formed a complex which was built all at the same time by the merchants and their guilds. The line of quays was not continuous but formed a series of irregularities, each part serving one or two establishments. It has been remarked that these installations were highly suitable for transit trade. 'Communications between the sea and the warehouses are as good as they are poor between the warehouses and the town. These arrangements seem to indicate that the goods, brought by sea and unloaded into the stores, were not going to be transported to the town, and that, on the contrary, they were going to stay in the warehouse, ultimately to be re-embarked and dispatched by sea.' The docks were served by roads which, because of the slope of the ground, were on a level with the first floor. As at Priene, a large part of the transportation must have been done by porters or donkeys.

Hellenistic Delos Delos, which Pausanias called the great emporium of Greece, became the centre of the mercantile population for the Hellenistic world, which included Phoenicia, Syria and Egypt as well as an increasingly important colony of Italians. The mercantile population was comprised of people from very different backgrounds and social structures – there were freedmen and citizens, enjoying varying degrees of wealth. A variety of merchants' guilds arose to unite the merchants of any one nationality and enable them to work together in their businesses and honour their native gods. These groups are often known only from some sparse inscriptions in honour of their patrons, but the name of one of them, the Society of Posidoniasts of Berytus (i.e. Beyrouth) of Delos, which included businessmen, armourers and warehousemen, has been preserved because

The figures of the gods from the eastern Parthenon pediment are arranged in attitudes carefully selected to fill the difficult triangular space to the best advantage. The apex of the pediment shows the birth of Athena, at which the reclining figure of Dionysus (or perhaps Theseus) was a spectator. One of the most remarkable features of the temple is the complete unity of style of the innumerable decorative sculptures.

the meeting place of the society, has been identified and studied. It looks rather like a large private house with a peristyle court with a cistern in the middle, a common feature in Delos. Running downwards on the south side is a series of rooms connecting with the road but not with the establishment, and these must have been stores. Although the rooms above, on the same level as the court, are now lost they were probably intended for the use of itinerants. In the west wing was the Posidoniasts' sanctuary consisting of four shrines preceded by a small anteroom which opened on to a little court with four altars.

It has been established that there were only three shrines at first, the largest of which was dedicated, according to the inscription, to Poseidon of Berytus, patron of the society. The other two had no distinguishing marks and have been attributed to Astarte and Eshmun-Asklepios, the three gods forming the chief triad of Berytus. The last shrine, the result of later remodelling, housed the statue of the goddess Roma, which was a highly academic piece by a sculptor working at the end of the first century BC. It was probably about 110 that the increasing influence of Rome and the development of exchanges with Italy caused the Posidoniasts to introduce Roma among their patron divinities. The rest of the building was occupied by an open air courtyard paved with a crude mosaic. Its purpose is uncertain; it may have been a sort of commercial exchange, or perhaps just a meeting place. The plan clearly shows the composite nature of the establishment, which was a shrine,

A fine statue of a youth was found in the sea off Anticythera. Identification of the statue is uncertain, but some scholars believe that the curved fingers of its upraised hand indicate a round object, from which they deduce that it is Paris presenting the Apple of Discord to Aphrodite.

The east pediment of the temple of Zeus at Olympia shows the preparations for the chariot race between Pelops and Oenomaus, the results of which brought generations of disaster to the family until Agamemnon's son Orestes was able to halt its tragic course. It is typical of fifth-century sculpture that even when scenes of incipient or actual violence are shown, an air of stillness and serenity pervades the figures. Contorted faces and limbs were not introduced until the fourth century.

a meeting place, a commercial exchange and a travellers' hotel at one and the same time'.

The most important of the foreign colonies was that of the Italians, who left much archaeological and epigraphical evidence on the island. They too built themselves an establishment, much larger than that of the Berytans, just beside the Sacred Lake. This had required a lot of work. To protect the area from flooding they had to build up the ground on which the various buildings of the Italian agora were erected after closing off the lake with a dam. This complex was, in fact, a huge trapezoidal square about seventy metres in length and fifty metres along the shortest side. All round it was a colonnade with two storeys, with 112 Doric columns on ground level and an Ionic storey above. It more or less followed the plan of the enclosed agora which was a common feature in the Greek world at that time. This, however, was a private house and was used, not as a market, but as a meeting place. It is true that commercial transactions were not completely banned. The south and east wings were flanked on the outside by a row of shops which must have been rented out for the society's profit. These shops opened on to the road and had no direct communication with the interior.

The surface of the large court, which was not even paved, was free from buildings and thus provided the wide empty space needed for the performance of religious ceremonies and the holding of assemblies. The interests of the Italian community were discussed at these, and honours voted to benefactors and patrons, who were often famous men like the consul *Quintus* Pompeius Rufus. The wall at the end of the portico was pulled down in several places in order to build ceremonial rooms, niches, loggias, and exedras (rooms open to the public) which housed statues and other commemorative monuments. The Italians also devoted great care to their own comfort. They had small baths set up in the north-west corner. The entrance to these was splendid, ornamented with blue marble pilasters and columns. After crossing a large hall, one reached the hot room, which was a small circular room with a basin, also round, in the middle, to which steam was conducted through a terracotta pipe. The drain ran towards the Sacred Lake.

The building expenses were borne by the members of the community, who donated here a section of the portico, there a single column, as shown by the dedications on the blocks they presented. By studying these texts it is possible

A fragment of one of the sculptured metopes of the fifth-century temple of Zeus at Olympia. Atlas supports the vault of heaven; Athene will relieve him of the burden while he fetches the Golden Apples of the Sun for Heracles, who waits with arms outstretched on the right.

to date the building of the agora and, with the help of the architecture, to follow the various additions. The texts also provide invaluable evidence about the Italian community—e.g. the north portico was donated by the banker Philostratos, a native of Ascalon who became a naturalised citizen of Naples.

The Italian agora, however, is not the only building to attest the increasing importance of this colony at Delos. Archaeologists have named an agora after one of the Italian societies; this is the agora between the first and second harbours known as the 'Agora of the Competaliasts'. According to an inscription the Competaliasts were a religious body consisting of from five to twelve members, nearly all slaves, who took care of the Roman rites of the gods of the crossroads, the Lares Compitales. This agora was clearly not comparable with the Italian agora. Built, it seems, during the course of the second century BC, it is a wide walk left almost empty to facilitate trading. In this place, where several roads met, some small monuments dedicated by the Competaliasts have been excavated, including altars, little marble cylinders decorated with garlands and bucrania, and the bases of commemorative statues. One of the most unusual of these monuments is an Ionic facade with four columns, backing on to the Stoa of Philip. This has been called a *naiskos*, with an offertory box in front for the gifts of passers-by, set up by Caius Varius, a freedman of Caius. In the middle of the agora were two small sanctuaries, a square shrine in the Doric style and an Ionic rotunda, which were donated by the Hermaists, an Italian religious college, who were devotees of the cult of Hermes, patron god of merchants. These colleges were made up of much humbler folk, slaves and freedmen, but their connection with the rest of the Italian community is not known.

The Italians were not alone in introducing new cults into Delos. Egyptian and Syrian gods arrived on the island with eastern sailors and merchants, not to be worshipped by a mono-national society, but to receive public veneration in a sanctuary built for them. Excavators have been able to identify three Serapeia dedicated to Serapis and other Egyptian gods, the huge sanctuary of the Syrian gods and, on the north slopes of Mount Cynthus, a great many sacred buildings, mostly dedicated to unidentified gods. These buildings have a very distinctive plan, with an open air court in the middle of which is an altar. Semitic sanctuaries have also been found on Delos, as confirmed by the few surviving inscriptions like that of the god Sin from the Arabian peninsula. A square building near

Opposite page: the temple of Olympian Zeus, sixteen huge Corinthian columns of which survive, was the largest temple ever dedicated to the supreme deity of the Greeks. It was begun by the tyrant Pisistratus but after the fall of his sons the temple went through many vicissitudes and was not completed until a benefaction by the emperor Hadrian enabled the Athenians to finish the work nearly 400 years later. It stands to the south east of the Acropolis near the bank of the river Ilyssus.

The west porch of the Erechtheum on the Athenian Acropolis bears the famous caryatid figures, with one exception. The third figure from the left is a cast, the original being in the British Museum. Widespread international depredations of the Greek cultural heritage *took place while the* country was under Turkish domination, and at the time they were partially justified by the necessity for safeguarding the unprotected art treasures.

the stadium used as a meeting place must have *been a synagogue, as its discoverer has proved* from an inscription invoking the Most High God.

The three Serapeia on Delos show most *clearly the installation of foreign gods on the* island. The first is very simple. The various structures—two porticoes and a room where ritual feasts were held, are arranged round a court closed off by a wall along which ran a bench. There were three altars and an offertory box in the middle. According to the inscription which tells its history, this construction was a private shrine, built about 220 BC. Although the second Serapeum has been very badly damaged, the same general arrangement can be made out, with a portico, a large hall and a small temple. The third one which is very much larger was the official sanctuary built by the people of Delos after the beginning of the second century. In 167 BC, when the Athenians took possession of the island, they recognised the nature of the sanctuary and *undertook extensive building* operations, constructing the chief temple, while private individuals dedicated shrines and various small monuments.

The downfall of the island in the first century must have been as cruel as its success had been brilliant. During the first war between Rome and King Mithridates of Pontus Delos was captured and sacked by the royal fleet in 88 BC. Scarcely twenty years later it was again attacked by pirates. Many of the buildings, like the establishment of the Posidoniasts, bear signs of these destructions and show that they were abandoned. The clearest evidence of this unrest in the Aegean is the wall which was hurriedly built by the Roman legate Triarius. Most of it has been traced and parts are fairly well preserved. It *does not seem to have protected all of the town* or much of the port, probably because a defensive rampart had to be built as quickly as possible. The wall, which was constructed of ruins, shows how much the town had suffered. Reused architectural elements, statues, stelae and inscriptions were tumbled pell-mell into the masonry. The port at Delos never rose again. The reconstructions dating to the Empire and the Byzantine period are insignificant, and material from the Christian era is rare among the considerable amount of everyday objects, such as lamps and coins, found in the houses. This decline was not confined to Delos. There was widespread insecurity in the Aegean in the first century and, as Delian archaeology shows, the cities of the islands and mainland Greece went through some extremely troubled times.

The excavation of the agora of Athens

The agora excavations conducted by the American School at Athens are very different in nature from those at Delphi, Priene or Delos. The undertaking involved a great many problems: the site had been continually altered in the ancient world, but the chief difficulty was that, apart from a brief period of desertion in the early medieval period, it had been in continuous occupation. The agora lies in a hollow between the Acropolis to the south, the Areopagus (the hill of Ares, where the city's first council met) to the south west and the Kolonos Agoraios (where Sophocles placed the death of Oedipus) to the west. When Athens was reduced by the Slavic invasion of the sixth century AD to a wretched village huddled on the slopes of the Acropolis, the drainage system rapidly disintegrated. The agora became a reservoir for all the water *running off the hills, its buildings were subjected* to severe water erosion, and the whole place was silted up. Later on, under the Byzantine *empire, when the city began to recover, then* during the Crusaders' invasions and finally under Turkish domination, this flat ground at the foot of the fortress was an inhabited area. New houses were built, and this naturally contributed to the disappearance of the remaining ancient structures, the materials of which were reused. These houses suffered very much during the two sieges of the war of independence (1821 and 1828), and when Athens was chosen as the capital of the new kingdom, the architects suggested building the new town north of the old, so as to leave a clear field for excavation. Unfortunately their advice was not followed and more houses were built on the agora.

Nevertheless, the site was identified, partially from the ancient authorities, and the Greek *archaeological society* was *able to carry out* limited investigations and to identify a few monuments, mostly fairly late buildings like the Stoa of Attalus. At the same time German archaeologists took some soundings on the west side of the agora, while the construction of the Athens to Piraeus underground railway on the north side uncovered various ancient artifacts and parts of a number of ancient buildings. Before these were destroyed, records were made of the remains. A picture of the site, no matter how vague, was beginning to emerge by 1922 when Greek refugees from Asia Minor threatened to occupy all the open space and prevent access to any work. It was then that the American School was able to make use of its considerable financial resources, the gift of John D. *Rockefeller Junior, and*, with the approval of the Greek Parliament, they bought out all the occupants of houses on the Agora and started systematic excavations which lasted from 1931 to 1940, and from 1946 to 1960. The whole of the surface they owned was cleared except for a few sections left as a control for the archaeologists to check their findings. At the same time the public was kept informed by reports, which were published immediately after the excavation campaigns in a review, and then by a definitive publication which was both rapidly and meticulously produced.

An outline of the history of the site was compiled from ancient sources. The agora had become the centre of political life in Athens from the sixth century BC onwards, when the framework of aristocratic government was disintegrating and an increasing number of citizens were acquiring political rights. They convened in assembly, or *ecclesia*, at the place which was to become the agora, or meeting place. From that time forward the Acropolis had only a religious function, while all round the new square a whole series of buildings were erected for the patron gods of civic life and the various magistrates. The main axis was the Sacred Way from Eleusis to the Acropolis, along which moved the great Panathenaic processions organised by Pisistratus about the middle of the sixth century. It was here, too, that Cimon laid the bones of Theseus, the chief Attic hero. But the agora was not just the scene of religious ceremonies and political activities, it was also an increasingly active market place, so that the assemblies ultimately had to abandon it for the quieter area of the Pnyx. The square was then occupied by a whole series of flimsy shops which Demosthenes ordered to be burned to warn the citizens that Philip had just taken Elataea. The swarming bustle of one time can be reconstructed from the works of orators and poets. During the next centuries the agora continued to play a leading part in Athenian life. The construction of new buildings indicates its importance. It enjoyed great prestige in the middle of the second century, when Pausanias visited it and wrote a detailed description.

The American excavators who were trying to reconstruct the details of Athenian democracy had to begin by uncovering the level of ground which had belonged to the fourth and fifth centuries AD from beneath a mass of medieval and modern debris from two to twelve metres thick. These layers of later civilisations were not without interest and were carefully studied. Then the excavators had to proceed downwards from the late Empire to the classical and archaic levels, analysing alterations in the buildings and their sequences, especially on the west side, where the main buildings stood. The complex had been altered very little by the Romans. In the south west corner, at the entrance to a road leading to the agora, was a marble boundary stone bearing in early fifth century lettering the words, 'I am the boundary of the agora'. Remains of other boundary stones show that the agora was a sacred area which criminals were forbidden by law to enter. Close beside it but clearly later, the excavators noticed a marble base of the first century AD, on which they replaced a *perirrhanterion*, or lustral basin used for purification.

Immediately in front of the boundary stone was the channel which drained off the water from the Areopagus and the Kolonos Agoraios into the Eridanos, a little stream [illegible] of the agora. The channel was made of polygonal blocks a metre square, very carefully fitted together and covered [illegible] big slabs

A Trojan bowman, carved from marble. *One of the figures from the east façade* of the temple of Aphaia, Aegina.

The Doric temple on the island of Aegina was once thought to have been dedicated to Jupiter Panhellenius but is now known to be the sanctuary of the local goddess Aphaia. When C. R. Cockerell drew it in 1860 it was in a state of great disrepair, but work had already been started on putting the building in order and studying the details of its construction and *decoration. It had* previously been regarded as little more than a source of pieces for sale to the many collectors who habitually toured the *Mediterranean*. Several of the carvings had previously been removed to the studio of the sculptor Thorwaldsen, whose 'restorations' altered *them almost beyond* recognition

It is not easy to form a true estimate of free-standing Greek sculpture as hardly any attested masterworks have survived. The great gold-and-ivory figure of Athena, made by Phidias for the Parthenon, and his Olympian Zeus, one of the Seven Wonders of the ancient world, are known only from coin-types or uninspired and pedestrian Roman copies. There is one fortunate exception to this rule: the fourth-century statue of Hermes holding the infant Dionysus, made by Praxiteles for dedication at Olympia. The delicacy and subtlety of the surface treatment and the refined, pensive beauty of the face make it plain that this statue could only be the work of a great artist.

building method and the potsherds found underneath it enabled the excavators to date this drain to the beginning of the fifth century BC. It had been well maintained, and had survived until the end of the ancient world. The drainage system was so efficient that the Americans pressed it back into service to protect the excavations from possible floods.

Some remnants of Athenian democracy

The excavators also uncovered a small round building less than twenty metres in diameter in the south west corner up against the Kolonos Agoraios. It had six columns inside, a monumental porch with four columns on the east side and a small annex on the north. The complex showed traces of several modifications. The interior colonnade had been rebuilt in the second century AD, very probably in the reign of Hadrian, a pavement of marble set in mortar had been added, and the walls lined with marble. The big porch and the earlier marble paving, which can still be seen in places, probably date back to the Augustan period. In the first stage there was a beaten earth floor, forty five metres below the present level. This building has been identified by its shape, which has been carefully preserved and embellished. This is the tholos, the function of which was described by Aristotle in his work on the Athenian constitution. The *prytanes* or members of the tribunal who presided over the *Boule* or Council of Athens took their meals together there. A kind of kitchen was set up in a little room on the north side. The tholos itself dates to the second quarter of the fifth century BC but according to the pottery found in the foundations it superseded *an earlier building* dating back to the middle of the sixth century. The plan of the earlier building was more complicated. Around a triangular court flanked by two porticos stood a whole series of rooms of different shapes and sizes. There is nothing to denote their function.

The Agora at Athens was the civic heart of Attica as the Acropolis was its religious centre. As time went on it was increasingly invaded by commercial establishments and the government offices were moved elsewhere, but it was always the chief meeting place of the citizens. The American School, who conducted the excavation of this demanding site, reconstructed the Stoa of Attalus which now houses the finds from the excavation and the School's workrooms. Every stone has been reproduced from an authentic source and even the original marble quarries were pressed back into service, but the new Stoa presents a rather harsh appearance on this ancient site.

The building has been compared to some Ionian palaces, and has been assigned to the period of the Pisistratid tyranny, where it may already have fulfilled the same function as the tholus.

A short distance to the south at a slightly lower level the excavators uncovered a cemetery with tombs, closely dated by deposits of Geometric and Orientalising pottery, ranging from the eighth to the beginning of the sixth centuries. This strict chronological series clearly illustrates the turning point which occurred at *the beginning of the sixth century* BC. A *burial* place until then, although the remains of a few houses have also been found there, it was abruptly called on to play an entirely different part. No further ancient tombs have been found there.

According to ancient tradition, the Bouleuterion, the meeting place of the famous Council of Five Hundred, should have been immediately to the north of the tombs, where there was a group of ruins of various dates. The latest complex, dating to the second century AD, included *three adjoining rooms and a small court with* rooms round it, the whole opening eastwards via an Ionic portico. There was virtually nothing left of this but the foundations. The Americans interpreted it to have been the Metroon, or the sanctuary of the mother of the gods, which also served as the city archives. Some terracotta tiles with a stamp indicating that they belonged to the Metroon were found round about it. Their shape is typical of this period. The sanctuary of the goddess would have been in the centre, the two rooms on either side being reserved for the archives while the north part would have been the official residence. The Bouleuterion must be identified with a huge rectangular foundation immediately opposite the Metroon on the west, on a site partly dug into the hillside. It was a large hall with a ceiling supported on four columns. It seems clear that when it was built at the end of the fifth century, wooden benches were arranged round three walls, forming a U-shaped auditorium which faced east. In the course of the next century a portico was added to the south façade and a monumental propylaeum opening on to the agora.

According to the experts, however, the Bouleuterion replaced an earlier building on the site of the present Metroon. There is nothing left of this but a few metres of limestone foundation, the major part of it having been destroyed at the time of the second century building operations. Nevertheless, a plan can be reconstructed of a square with sides twenty two metres long and four pillars in the centre. Its resemblance to the *Initiation Hall at Eleusis has been stressed*. Is this simply because they were both meeting halls? However, it has been pointed out that once a year the Council sat in the *Eleusinion* at Athens. Its construction probably dates back to the last years of the sixth century, at the time of the reforms of Cleisthenes who, it has been noted, assigned a very important part to the Council of Five Hundred. The hall would also have housed the famous statue of the mother of the gods attributed by the ancients to Phidias or one of his school, and would have been used as *an archive deposit*. Because of a lack of space, the two functions would have been separated and a new Bouleuterion built. A small archaic temple contemporary with the old Bouleuterion was also found under the Hellenistic Metroon; this would have been the early sanctuary dedicated to the mother of the gods.

These structures, however, were not the earliest. At an even lower level more remains were uncovered. A few rough walls were found from which the plan of two uneven buildings, constructed separately but both attributable to *the beginning of the sixth century* BC, could be reconstructed. Perhaps these were the first of a long series of public buildings. If so, they may have been connected with the reforms of Solon, which were carried out at this time. Although these identifications are not conclusive and the theory of the close relationship between modifications made to the city of Athens and archaeological levels is sometimes suspect, the modest size of the buildings and the brevity of

Towards the end of the sixth century BC the Athenian tyrant Pisistratus built the Telesterium to house the Eleusinian Mysteries at the ancient holy place of Eleusis. It was superseded by several successive buildings for the same purpose, most of which embodied the same basic elements. A colonnaded porch led into the hall round which were banks of several steps. They are too narrow for seating, so the audience must have stood. Rows of cross columns in the hall supported the roof in the Egyptian style and in one corner was a miniature representation of the palace in which the Mysteries were enacted. The Greek literary sources have largely kept the secret of the Mysteries, and what little we know is chiefly derived from the buildings associated with them.

The age-old struggle between the original chthonic religion and the new Olympian gods was symbolised for the Greeks by the battle between Apollo, god of light and rationalism, and the serpent Python for possession of the Delphic tripod. The serpent was defeated, but retained its importance in the sanctuary where it was represented by the priestess who gave the famous cryptic oracles under direct inspiration from Apollo.

Opposite page: the theatre at Leptis Magna seen from the stage. The birthplace of the Emperor Septimius Severus, the city was favoured by him and much enlarged; its ruins are the most extensive and imposing in Roman Africa. The city flourished during the late empire and Byzantium before being destroyed by the Berbers.

Overleaf, left above: Hadrian's defensive wall across the north of Britain was begun in AD 122 and completed in 128. It stretched for eighty miles and ran from Wallsend-on-Tyne to Bowness on Solway. It was destroyed and rebuilt several times while the Romans ruled Britain, and eventually abandoned in the fourth century. Left below: the temple and fortress of Kumma, the modern Semna, stood on the right bank of the Nile and would have been engulfed when the waters rose on the completion of the High Dam. The ruins were dismantled by the Sudanese authorities and reassembled in the museum at Khartoum. Right: the mosaic flooring in the Baths of Caracalla, the greatest building of its kind in Rome and now used for spectacular performances of opera. Begun in AD 211, the whole covered an area of 750×380 feet and stood 20 feet above the ground.

their period of use clearly illustrate the upheavals and crises which Attic historians prefer to consider as an inevitable result of the glorious birth of democracy.

The temple of Ares, nearly in the middle of the agora, is another example of the many transformations of the site. Visited by Pausanias and identified from his description, it is a small Doric temple dating to the middle of the fifth century BC. The few pieces of marble which have been preserved have made it possible to date the temple and compare it with the pseudo-Theseum, so that both buildings have been attributed to the same architect. However, not a single classical or Hellenistic author mentions the temple of Ares, and on closer examination each piece of the fragmentary remains was seen to bear a mason's mark indicating its exact position in the structure. The lettering and sherds from the foundation date the temple in the reign of Augustus. Excavators concluded that the temple had been completely dismantled and brought from its original site, probably in the small mountain village of Achaenes, to be carefully rebuilt on the agora at the very end of the first century BC. A contemporary inscription honours Gaius, the adopted son of Augustus, as the 'new Ares'.

The agora excavations have also yielded a great many objects which shed light on numerous details of public life. The museum houses a collection of 1200 *ostraca*, the sherds on which the Athenians scratched the name of the man they thought most likely to endanger democracy by aspiring to tyranny. If there were more than 6000 votes against any citizen, he was ostracised, and condemned to ten years in exile. Some of the most illustrious political names of the fifth century such as Aristides, Cimon and Pericles have been found on these ostraca. A pit was discovered which contained 190 sherds, all bearing the name of Themistocles. The form of the carefully inscribed lettering indicates that many of them came from the same hand. They must have been prepared in advance by his enemies, led by Cimon, for distribution on voting day.

Other finds are connected with the working of the tribunal. They clearly explain the passages of Aristotle on the selection of jurors and the procedure for debates. Each citizen had a small bronze plaque called a *pinakion*, which bore his name, that of his father, and that of his deme. It was a sort of identity card, authorised by an official stamp of an owl or a gorgon's head. These plaques, large numbers of which have been found, also served for drawing lots for jury service by means of a machine called the *kleroterion*. Its somewhat complicated mechanism was not understood until excavations produced a specimen dating to the Hellenistic period. Another pit yielded a *klepsydra* or water clock which measured the time allotted to each speaker. This clock looked like a flower pot with two handles, with a small spigot at the bottom. The water poured into it could be very accurately measured because of its wide mouth. Finally, on the estimated site of the tribunal a box was found which contained six voting discs. These were made of bronze, some of which were solid and some hollow. The 'public vote' was engraved on them, which enabled the jurors to cast their votes—acquittal for solid discs and condemnation for hollow ones.

These objects and others, such as dedication by victors of the Panathenaic games and the tools of craftsmen who set up their booths in the wide square, are splendidly displayed in the Agora Museum, the Stoa of Attalus which the American School decided to rebuild to house the results of the excavation. Its display cases show how much care has been taken to reconstruct every detail of the activities of the Agora. The archaeological material, however, only assumes its full significance when compared with the ancient writings. Delphi, Priene, Delos and Athens have all borne a part in the great investigation which began with the ancients themselves. The original condition of each site demanded specialised archaeological methods. But from one site to another and from one campaign to another the common heritage of earlier studies has contributed to the enrichment of the endless dialogue between the site and the excavator.

The future of Greek archaeology

The study of Greek archaeology came first in the general development of archaeology and held pride of place for a long time. It was in this field and that of Egyptology that the discipline scored its first successes, showed its ability to produce absolute truth, and triumphed, not without a struggle, over other more literary methods of studying the ancient world.

Greek scholars have long favoured the most senior branch of archaeology, and showed a distaste for research which was not connected with the field of aesthetics. Right up to the present day the chair of Greek archaeology in all the major universities was actually the chair of the history of sculpture. This subject has been exhausted by its own perfections. Unless there is some miraculous discovery, we cannot hope to improve our knowledge of Phidias, Polyclitus or Lysippus. These masters will henceforth be the province of the philosopher rather than the historian. But architecture, pottery and the so-called minor arts can still teach us much. History, which is chiefly based on the work of the epigraphists and numismatists, can hope to profit from stratigraphical excavations accompanied by exhaustive study of the finds like those at the Agora. Finally, it should not be forgotten that our knowledge of Greece is drawn from a few large towns and important sanctuaries, and that it is very detailed for the great periods, but very fragmentary immediately before and after them. Only archaeological research can fill the gaps, and the information is far from complete even in the case of this favoured civilisation.

The Etruscans

A nearly life-sized clay head dating to the early fifth century came from a figure of a divinity (perhaps Jupiter) associated with the temple of Mater Matuta at Conca. It still has the stylised treatment, slanting eyes and the hint of a smile typical of the archaic period in Greece and Etruria. Like most statues of the archaic and classical eras, the figure was once painted and the addition of colour to the eyes and lips must have given it a disturbing vitality and realism.

Etruscan archaeology has been popular for a long time, particularly since the Second World War, and it is easy to see why. It is centred in a region in the heart of the Italian peninsula which is justly famed for its natural and artistic wealth, and is now attracting an increasing number of tourists. Access roads are better and transport is faster, so that Tuscany has become familiar to crowds of Italians and foreigners who knew nothing of it before. This admirable region was the seat of an important culture which flourished before Rome and prospered for more than half a millennium. This alone would be enough to attract public attention to the Etruscan excavations but the fact that the Etruscans occupied a special place among the peoples of the ancient world has aroused amazement and curiosity.

They were not like any of the other people who lived around them, and the ancients believed that they had come from the shores of Asia Minor. Indeed, their language has defied all attempts at decipherment, even though it can be read perfectly well, since the Etruscans used the Greek alphabet. Their religion was similar to that practised by eastern peoples, in which divination and sacrifices played an important part. Finally, unlike the Italian peoples who were their neighbours, they developed an impressive artistic culture from the seventh century BC onwards. This was certainly inspired by Greece, but it has no equal in non Greek Italy. Etruscan art attracted attention even before archaeology became a scientific discipline. This can be explained by the situation of Etruscan cities and cemeteries in the heart of a continually inhabited province, often right at the gates of towns like Florence and Rome. The Etruscans' interest in burial rites and the cult of the dead has resulted in a collection of well preserved remnants of their civilisation.

Etruscan archaeology

The encroachment of mechanised society

Etruscan tombs consisted of a burial chamber or group of chambers, built of stone in northern Tuscany and carved in the tufa of the volcanic hills in the south. The dead man and his grave goods, which included some of the finest treasures from the home of his survivors, were placed in these underground houses, built to withstand the assaults of the elements and the years. As in Egypt, the opening of an Etrurian tomb often yields perfectly preserved vases, jewels and other objects which were laid beside the dead man to provide him in the hereafter with everything he had enjoyed on earth. Since ancient times tomb robbers and treasure seekers have found much of value in the Etruscan territories and a number of great families have founded their fortunes on the riches the Etruscan nobility laid beside their dead.

Although at present Etruria is one of those areas where archaeologists can be certain of finding important remains, it seems advisable here to mention the dangers currently threatening Etruscan archaeology. Immense damage has already been done to the area which is fundamentally the result of the construction of highways and dams, and of machinery used for these operations and for farming. This threat to the historical and archaeological wealth of Italy began in the early nineteen fifties, which was when other countries like France began to experience the destructive effects of modernisation. The Abu Simbel affair has demonstrated that the same ravages have taken place elsewhere.

The damage in Tuscany is particularly noteworthy because of the great archaeological wealth of the countryside. Perfectly equitable considerations of social justice have caused the government to take over huge properties, mostly in the Maremma and southern Italy, which belonged to leading families and had been used exclusively as pasture land. These tracts of land were distributed among farmers for cultivation and improvement with all the chemical and mechanical resources known to modern agriculture. This kind of agrarian reform, which has been widely applied, has resulted in the cultivation for the first time of districts of great archaeological wealth. Although the state has protected a few of the most important areas, damage has been widespread and continues.

Of course there is a brighter side to this problem, and a number of accidental discoveries have been reported to the authorities instead of being destroyed or appropriated, so that everything has by no means been lost to knowledge. But the co operation of a few well disposed farmers and the praiseworthy activities of the official bodies have been offset by the schemes of well organised teams of clandestine excavators who have efficient methods and sometimes ample means. They are familiar with the ground they are pillaging and are often in radio contact

Opposite page: a satyr's head, Etruscan style, from the necropolis at Cerveteri. The Etruscans left many arresting examples of their art – and a mystery about themselves since their writing has never been deciphered. They are in fact a people who loomed large in history but left no history of their own; even their origins cannot be ascertained. The satyr's head dates from the sixth century BC and is now in the Ny Carlsberg Glyptotek, Copenhagen.

The main port of the Etruscan territory of Caere was Pyrgi (*Santa Severa*), which was famous in the classical world for its riches. The excavations started in 1957 brought to light a building known as Temple A, dating to about 480–470 BC and dedicated to the Etruscan goddess Uni. It had a colonnaded porch and three inner chambers, and yielded fragments of a fine terracotta representation of the battle between the Olympian gods and the giants.

with the people involved in smuggling their precious finds out of the country. One may get an idea of the importance of the material that has vanished or been lost by reading *Italia Sepolta* by the well-known technician and engineer C. M. Lerici. He has collected in this book articles from Italian newspapers and periodicals, all of which are warnings provoked by the situation. Their titles are eloquent: 'The losing battle against the tomb robbers', 'The end of Etruria?', 'The Mafia of the Ruins' and so on.

The cliffs and ravines formed by the river Marta to the east of Tarquinii were chosen by the Etruscans as the site of veritable cities of rock-carved tombs. These chiefly take the form of cubes or gabled structures. Cube tombs at San Giuliano have a false door carved in the façade, which represents the front of a house and thus supplies valuable evidence about Etruscan domestic architecture.

The threat to Etruscan archaeology is great. It would be better if we were equally aware that, though the danger here is less pressing, it is just as real and mechanised farming and civil engineering can be the death of our buried heritage.

Aerial photography Photographs taken from the air have proved to be immensely valuable in Etruria. The first aerial photographs of an ancient site were those taken of the Roman forum in 1897 and 1908. Because of the interest in aerial photography during the First and Second World Wars, an immense collection of photographs was accumulated and an archive was established in Italy where a great variety of useful shots was made available to archaeologists. In 1958 a Library of Air Photographs was organised by the Italian government, and to this collection the British and American schools of archaeology in Rome added the photographs they had acquired from their respective air forces.

In Etruria such photographs have furnished excellent overall views of great mortuary districts such as Tarquinia and Cerveteri. Many vanished tumuli, flattened by centuries of farming, have reappeared on prints in the form of whitish patches caused by the mixing together of the broken fabric of the tumuli and the earth. The aerial survey carried out in the Po delta established the exact location of the Graeco-Etruscan town of Spina and archaeologists can now understand the arrangement and distribution of its *insulae* and streets. For a long time now, excavations in the huge Etruscan cemeteries in the Pega and Trebba valleys have been producing Etruscan bronzes and Greek painted pottery from the tombs in the watery regions of the Po estuary. Archaeologists have discovered more than three thousand tombs, and the archaeological museum at Ferrara has been continually enriched by masterpieces of Greek

The vast Etruscan necropolis attached to the ancient city of Tarquinii presented a serious problem for *archaeology*. So many tombs were known that neither finds nor personnel could be provided to excavate them all. Many are empty and blank but as they belonged to the peak period of Etruscan art, others might contain fine paintings such as the banquet scene from the Tomb of the Shields. New processes of periscope photography were developed to examine these graves from the surface and determine which were worth further investigation.

pottery from this poor and arid land. Until recent years, however, the habitation site to which these tombs belonged was unknown. It was an aerial photograph of the district which led to the discovery of the site, and revealed the plan of the town.

Physical and chemical surveying methods

It would have been surprising if the headlong advance of the so called 'exact sciences' since the beginning of the twentieth century had not led to some archaeological innovations and some new surveying techniques involving geochemistry, electromagnetics and electricity. From the point of view of Tuscan studies, the work of C. M. Lerici, an engineer who introduced new procedures, has resulted in many years of particularly spectacular and numerous discoveries. A single example illustrates the extraordinary fruitfulness of his campaigns. In 1960 Lerici wrote:

'In the Monterozzi necropolis at Tarquinia 516 tombs were discovered in an area measuring only five hectares. The necropolis of Monterozzi covers more than a hundred hectares, so it can be estimated from this that the total number of tombs may approach 10,000.

'Among the 516 investigated tombs, two were exceptionally important painted examples (the Tomb of the Olympiad and the *Tomba della Nave*) and twelve others had fragmentary frescoes. It can be assumed, therefore, that there are still more tombs worth saving.

'Our knowledge of the reasons for the destruction of the paintings enables us to state that in a few years there will be no tombs left worth rescuing, and the area will literally become a cemetery of painted tombs.'

This last sentence is notably pessimistic. But in the present discussion of new methods employed to locate and investigate a number of hypogea in a short period of time in a limited area statistical evaluation of the tombs already excavated enabled Lerici to state that ninety five per cent of them had already been opened and sixty eight per cent were empty.

Soil conducts electricity, but its degree of conductivity varies according to the nature of the soil. The presence in a given sector of archaeological remains, such as buildings, walls, roads, tombs or ditches causes localised alterations in the conductivity, and these variations can be plotted by a simply operated measuring apparatus. Electrodes are planted at

The settlement of Populonia, which lies on a promontory opposite the island of Elba, was a metal-working centre *producing first bronze* and later iron. Industrial activity helped to preserve the tombs in the neighbouring cemetery of San Cerbone, which was largely buried under iron slag. To make the local circle grave, a circular wall of flat stones was built round a square chamber, over which a grass-grown tumulus was heaped.

regular intervals, a reading on a dial shows the variations in conductivity which are plotted on a graph, and the curves of these graphs reveal *the presence of the remains, if there is nothing else to account for them.* This method was systematically applied by Lerici and his team on the sites of the Tuscan cemeteries, and this is how the spectacular results mentioned above were achieved. *Other geophysical and geo-chemical methods were also applied.* But a problem arose here because of the enormous number of discoveries. In areas as rich as the cemeteries of Cerveteri and Tarquinia, discoveries run into hundreds or even thousands. What is more, locating a tomb is not the end of *the matter: it has to be excavated,* and this calls for personnel, funds and time. It proved impossible for the existing teams to investigate all the tombs which were located, and, furthermore, it would be pointless to excavate most of them, because of the statistical evaluation of their probable condition. It was therefore vital to be able to find out the condition of the tomb and its contents before starting the excavation.

The Etruscans devoted a great deal of time and thought to the after-life, but the tombs painted during the zenith of their culture indicate that their attitude to death was far from gloomy. The Tomb of the Leopards at Tarquinii is decorated with a light-hearted drinking scene, carried out with considerable verve in gay colours and brilliant brushwork. Fair-skinned girls are shown reclining on equal terms with young men, waited on by nude slaves.

The most spectacular form of burial in the necropolis of Caere was the massive tumulus, which could reach a diameter of up to thirty yards. The stonework of the surrounding wall was neatly dressed with a *moulded cornice and* the contents of these tombs often included several successive burials, access to which was provided by an antechamber. So large was the cemetery that it came to be arranged like a city with a main road furnishing the route from Caere.

Photographic sounding devices and the Nistri Periscope Some ingenious methods have been invented for obtaining an overall interior view of a tomb which has been located but not yet excavated. As the Etruscan tomb takes the form of a chamber, a periscope or a probe equipped with a camera can be introduced before beginning to excavate. In order to do this it is necessary first to make a hole by piercing through the ground or the ceiling of the tomb. A portable drill powered by an electric motor then quickly makes a small hole, and the damage to the structure is minimal. Once this preliminary operation has been carried out, a cylinder with a camera attached to it is introduced into the cavity and operated from outside. By rotating the probe thirty degrees after each flash bulb shot the perimeter of the tomb can be photographed.

The results obtained by photographic probes in the survey campaigns of 1956–60 at Cerveteri, Vulci and Tarquinia made it possible to investigate hundreds of previously located tombs and provide information about their present state of

Like the Greeks and Romans, the Etruscans took part in banquets reclining on a couch known as a triclinium, one of which gives its name to a painted chamber tomb at *Tarquinii. These tombs* usually consisted of a single chamber with a pitched roof cut into a sloping hillside. Everyday life is the favourite subject and they therefore shed a great deal of light on *Etruscan customs and* preoccupations.

Little survives of the ancient city of Caere (modern Cerveteri) which is estimated to have had about 25,000 inhabitants at the height of its power. However, the two huge cemeteries of Banditaccia and Monte Abatone have provided a picture of the city's development through its burial *practices.*

One of the most successful Etruscan art-forms was that of engraving on bronze. Incised drawings of masterly assurance and skill were used to decorate mirrors and lidded cylindrical caskets, to which free-standing figures were added as handles and feet. In the late fourth century the city of Praeneste (modern Palestrina) was a flourishing centre for this type of industry.

preservation, the location of their entrances, the presence of water and mud, and the type of remains within the tombs. Sometimes these photographs serve as records of remains which fail to survive the actual excavation because of *disintegration on contact with the air. It is easy* to see how valuable this ingenious application of photography has been to archaeology.

The Nistri Periscope, named after the engineer who developed it, was first used in September 1957 when an Etruscan tomb in the Monte Abbatone necropolis was opened in the presence of His Majesty King Gustavus VI of Sweden. The Nistri Periscope is adjustable and can be extended from 3 to 4.5 metres. An electric light on the end of it illuminates the tomb during the inspection.

The human factor in research Thanks to the ingenuity of some fine technicians and to the characteristic shape of the chamber tomb, Etruscan archaeologists have been able to take *advantage of the valuable surveying methods.* But it is true that luck and the archaeologist's flair still play a large part in successful excavations. When it is no longer a matter of investigating a district which is already known for the wealth of its archaeological remains but of finding towns and cemeteries which have not yet been located, the new methods of research described above are certainly useful, but the actual discovery often depends on the knowledge and good judgment of the archaeologist himself. He must, for example, establish good relations with the inhabitants of the region where he is working. In archaeology as in war, this kind of intelligence is vital. In places like Tuscany, where an ancient culture once flourished, the local inhabitants have an intimate *knowledge of their land and its history. A large* number of them have discovered remains of one sort or another in the course of farming their land or have witnessed an accidental find. In a given area an archaeologist can find out much that has happened during the past hundred

The main street of the necropolis at Caere, like most busy centres of the ancient world, still bears the ruts worn by cartwheels in the soft tufa stone of the roadway. On either side of the track are ranged the circle graves which became popular in the sixth and seventh centuries BC.

The exact nature and extent of the Etruscans' debt to Greece was hotly debated by scholars for many years. A small bronze statue of Ajax falling on his sword shows the strength of Greek influence in the fields of both art and mythology.

years or so if the farmers are willing to pass on what they know or have heard. It is true that he must know how to interpret the stories that he is told, but this is comparatively easy, and one archaeologist found that many of the discoveries he succeeded in making resulted from the close co-operation of his foreman. This sort of collaboration with the people of the area where the archaeologist is working can make his labour much richer and more attractive. It is not enough to be in touch with the people of the distant past; their territory and the problems facing them can best be understood by someone living on good terms with the people who work the same fields today and have the same landscape and elements before their eyes.

Recent Etruscan discoveries

Although it is just as useful and indispensable as in other branches of archaeology, stratigraphy has perhaps been used less in Etruscology than elsewhere. Earth does not accumulate on the volcanic hills which form a large part of the country as it does on flat ground, and the present level is often the same as it was in the ancient world. The importance of stratigraphy in sites like the Roman forum, where the different occupation levels following the earliest occupation of the site have been traced in detail by the excavators, is widely known.

At the present time excavations are being conducted in Etruscan cemeteries and cities. Urban sites have hitherto been neglected because they do not promise the inexhaustible, sometimes fabulous wealth of tombs. However the archaeologist's aim is not confined to the discovery of beautiful or precious objects. His ultimate purpose is the advancement of historical knowledge of the people and cultures in question. From this point of view, the excavation of the sites where the Etruscan cities once stood is very important. Although these sites produce very sparse remains in comparison

The discovery in 1836 of the Regolini-Galassi tomb at Caere was the beginning of the formal study of the Etruscans. Its fabulously rich contents contributed a great deal of valuable evidence to the new discipline, but, not unnaturally, they also attracted a large number of treasure-hunters whose depredations have done untold damage. The main part of the tomb was a long chamber, partly carved in the rock and partly corbel-vaulted.

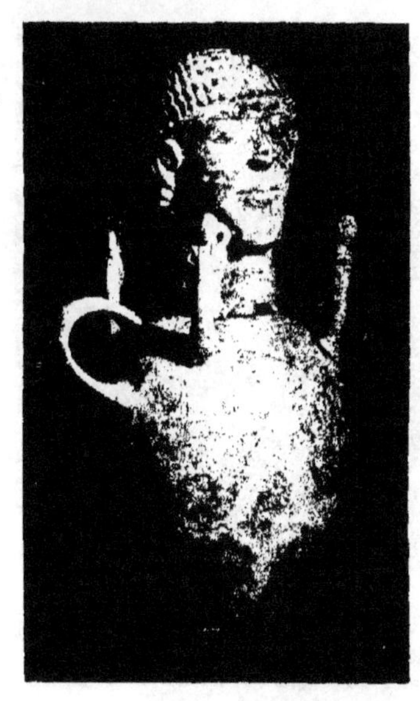

A sixth-century clay urn from Cetona was designed to hold the ashes after a cremation. Four curious winged goddesses on the shoulders of the urn probably functioned as guardians of the dead.

with those found in the tombs, the knowledge gained about urban life and religion can be invaluable. Two examples have shown how productive such excavations can be: these have been at Marzabotto in the Apennines and Pyrgi on the coast to the north of Rome.

The excavation of Marzabotto The excavations of Marzabotto have provided a solution to the controversy about the design of the Etruscans' towns. Their sacred books, large portions of which were translated by the Romans, give detailed descriptions of the ceremonies for founding a city which, according to tradition, were later adopted by the Romans. The *augur*, having consulted the auspices and finding them to be favourable, had to mark out the main axes of the city, the *cardo and decumanus*, which were established by preliminary sightings. After laying out the main axes, which crossed at right angles, the *pomoerium* and the space inside it were marked out in a chequerboard pattern, each square of which was to contain one *insula*. In practice, however, such geometrical regularity as this was nowhere to be seen in the Etruscan sites of Tuscany. The ground was highly irregular, and since the sites for towns were chosen for their defensive possibilities and their natural protective advantages, it would have been impossible to make the geometrical arrangements demanded by the sacred rites.

However, systematic excavations on the plateau of Misano near the present small town of Marzabotto thirty kilometres south of Bologna which have been proceeding for several generations now, have revealed a city with its streets, houses and sewers arranged according to a geometric plan. On the acropolis, which dominates the town, the foundations of five religious buildings were revealed with exactly the same orientation as the city itself. The plan of the complex was perfectly right angular, based on a main north–south road, the *cardo*, which was intersected by three perpendicular east–west roads, the *decumani*. It is clear, therefore, that the plan, which corresponds to the Etruscan ritual, was carried out in practice, and that the rules in question were actually applied. Furthermore, it can be seen why excavation of the Etruscan city of Pian di Misano revealed so much about Etruscan urban and religious principles, signs of which had been sought in vain in Tuscany. This city was organised according to an integrated design and was laid out on a huge plateau at a fairly late date (the very end of the sixth century), as shown by the pottery from the site. At that time the Etruscans launched great expeditions into the Po lowlands and on the way, before emerging into the plain, took advantage of a favourable site to build a regular and strictly oriented city. Their own cities in Tuscany were, on the contrary, established by degrees on irregular ground which was highly unfavourable for rigid town planning.

This is not an isolated example. Rome was similar. In spite of the principles of planning they inherited from the Etruscans, which suited their own taste for order and regularity so well, the Romans were completely unable to endow their own city with the precision so dear to their hearts. Over the course of centuries Rome was constructed on the irregular banks of the Tiber, and in spite of the wishes of the magistrates and then the emperors, the network of its streets was always confused, and writers never failed to mock the inconveniences of Rome, with its tortuous streets. Even Nero tried to *build an urbs nova*, a new city after the famous fire of Rome, but was unable to remodel the defective plan of the city after the example of the great orthogonal Hellenistic towns. Only the colonial cities, built on huge virgin sites, enabled the Romans to create military and urban complexes which foreshadow by nearly 2000 years the implacable regularity of the modern American ones. The ruins of Timgad and Lambesa in the south Algerian deserts show the Romans' enthusiasm for precisely planned cities.

Pyrgi The second example of a highly productive urban excavation is that at Pyrgi, which was one of the two ports of the Tuscan metropolis of Cerveteri. Its site has been known for a long time and was identified as the modern Santa Severa, a picturesque little place about sixty kilometres from Rome, dominated by the austere mass of its castello. Cerveteri, like many other large Etruscan sites, had developed a flourishing maritime trade and used two ports, Pyrgi and Punicum. A systematic excavation at Pyrgi was begun in 1957 under the learned leadership of M. Pallottino, Professor of Etruscology of the University of Rome. This was the result of a chance find along the coast of some architectural terracottas which indicated that there might be a religious building in that vicinity. Several ancient sources had mentioned a famous sanctuary at Pyrgi itself which, according to the Greeks, whose descriptions have survived, was dedicated [illegible] to the Greek goddess Eleithyia, as some say, [illegible] to Leukothea according to others. The we[illegible] of this sanctuary had been celebrated [illegible] the time of Dionysus, tyrant of Syracuse, [illegible] pillaged it in 384 BC when he attacked the [illegible]scan coast. It was this famous sanctuary [illegible] the Italian excavations were to reveal.

All along the shore appeared the foundations of two temples, one measuring twenty four metres wide by thirty four metres long, with three cellae, or one cella with two wings, the other was twenty metres wide by thirty metres long. They have been named Temple A and Temple B respectively. Work on the first has produced some extremely important sculptures and inscriptions. A high relief in painted terracotta, which must have adorned the pediment of the buildings, shows Athena in combat with the Giants. This has been largely restored and is worthy of a place among the finest Etruscan works of the early fifth century BC. Directly inspired by Greek work, it betrays the strength of Hellenistic influence in the Caere district, which was regarded even in the ancient world as a partly Greek town. Two fragmentary plaques bear the word *Unial*, genitive case of *Uni*, the name of a great Etruscan goddess who was counterpart of Hera and Juno.

The third-century Tomb of the Alcove at Caere represents a different trend in burial practices. Tumuli were abandoned towards the end of the fifth century, and were superseded by one-roomed tombs with benches round the walls for use in a number of successive burials. The dominant features of domestic architecture, such as the ceiling rafters, thick fluted pilasters and the recessed niche for the bed with its stone head-rest, are reproduced in the burial chamber.

The latest Etruscan tombs at Volterra are undistinguished either by painting or by any interest in architectural features and decoration. One of them, the Inghirami tomb, has been completely dismantled and reconstructed in the grounds of the Archaeological Museum at Florence. It is a simple round chamber with roughly dressed walls and the ceiling supported by a stone central pillar. A shelf round the walls held the sarcophagi of the Atria family, the owners of the tomb.

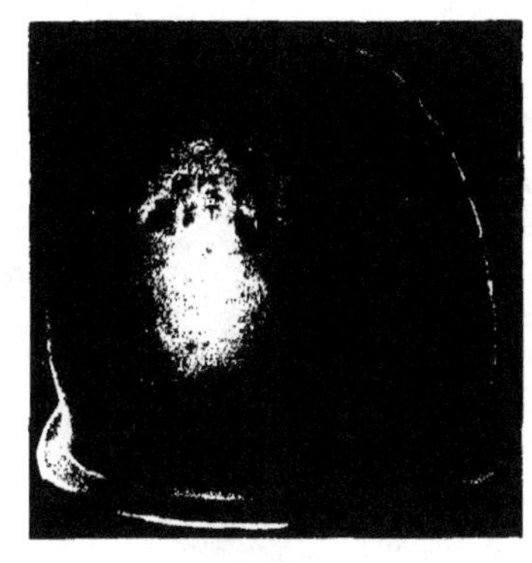

In 474 the Etruscans were already beginning to lose part of their empire in central Italy and in their need for a commercial outlet in Campania they raised a fleet to attack the city of Cumae. The citizens called on the aid of Hiero, tyrant of Syracuse, who inflicted an overwhelming defeat on the aggressors. A bronze helmet bearing an inscription commemorating the victory, part of the spoils from the battle, was sent to be dedicated at Olympia.

It was in the area between these two temples, however, that a discovery of the utmost importance for Etrurian history took place in the summer of 1964. A stone basin was found which was full of material, including many architectural terracottas apparently from Temple B. Three gold plates were also found, which had been originally nailed to the temple wall, one of these bore an inscription in Punic and the other two, Etruscan inscriptions. Hopes were raised that here at last was one of those bilingual inscriptions which, in the present state of knowledge, could decisively advance the decipherment of the Etruscan language. This proved not to be the case – even though the texts of the inscriptions were very similar, they were the parallel texts of dedications all made by the King of Caere, an Etruscan called Thefarie Velianas, and addressed to the same goddess, who was called Astarte in the Punic texts and Uni Astarte in the Etruscan. They probably date to the beginning of the fifth century BC. Therefore at the time of the rule of the Tarquins in Rome, the king of a powerful Etruscan city was giving homage to the goddess, who was Uni, i.e. Juno. Apparently he thought it best to have the dedication accompanying the offerings translated into two languages, one of which was Etruscan and the other Punic, and he also worshipped the goddess under the name of the powerful Phoenician deity Astarte.

It can easily be imagined how much information such inscriptions provide, by both their presence and their general meaning, even though the details of the Etruscan texts are still very obscure. They have shed some light on the relations between Carthage and the Etruscan cities, who joined forces against their common enemy, the Greeks. Polybius says that Rome and Carthage concluded several successive treaties of alliance, the earliest dating to the year 509 BC when Rome was still ruled by the Tarquins. This treaty has been the object of considerable speculation among modern scholars but the latest finds from Pyrgi have given strong confirmation of its existence. It is in the matter of religion that the inscriptions provide such significant information. Sixty kilometres north of Rome an Etruscan king, who ruled about 500 BC, showed total devotion to the great Semitic goddess Astarte. As Dupont-Sommer has written, this is the first sign of the fascination which the eastern cults exerted on the people of Italy for so many centuries. As to the assimilation of Astarte with the Etrusco-Latin Juno, it is an example of the situation clearly described by Virgil five centuries later at the beginning of the *Aeneid*. Queen Dido, the patroness of Carthage to whom Aeneas wished to raise a splendid sanctuary on his arrival in the city, must be the Semitic Lady of Tyre and Carthage, Astarte Tanit. Thus the close connection shown in the archaic period by the Pyrgi inscriptions between Uni-Juno and Astarte was still clearly felt by Virgil. In the material remains of a much earlier period and in the legend he used we can see the religious situation with which the Roman poet was familiar. It is a striking proof of the profound truths embodied in the Virgilian epic, the histories of Livy, and the legends about the origins of Rome in ancient literature in general.

Etruscan archaeology, perhaps more than any other discipline, has suffered from the activities of numerous prolific and often highly skilled forgers. The quaint stylisation of archaic Etruscan pieces is not difficult to reproduce, and the organisation of the illegal excavators supplies a ready-made channel for distributing and marketing such objects without the reliable excavation 'pedigree' which helps to safeguard against this type of fraud. The picture on the facing page shows a genuine Etruscan sarcophagus which shows a married couple reclining at a banquet.

Analysing Etruscan remains

Only the most important discoveries of recent years have been described, but there are many excavations in Etruscan territories which produce new facts every year, chance finds which end up in museums, as well as numerous clandestine digs which distribute objects with no known provenance. The rate of expansion of the archaeological collections is causing many problems. Sculptures, frescoes, vases and jewels need to be meticulously studied, restored and preserved, and this involves many operations demanding care, patience and precision.

Methods of analysis and dating used in the laboratories have advanced considerably during the last thirty years, and here again the archaeologist has the advantage of assistance from chemistry and physics. All forms of photography including spectrographs and X-rays have been used also by the archaeologist. Metal analysis has been used to examine the filigree and granulation employed by the Etruscan goldsmiths to decorate their work. The technique of granulation called for exact measurements and assured skill – the gold was shaped by the craftsman into tiny spheres, all exactly the same size, and these spheres, or granules, were artistically soldered on the surface to be decorated. (Archaeologists are still unsure about how such soldering was done.) In the technique of filigree the gold was worked into very fine wires which were placed according to the artist's design, on the jewel or the piece of gold work to be decorated and then soldered to the surface.

A comparison of Greek and Etruscan Art

In order to be certain of the exact date of a piece of Etruscan jewellery of the fifth, sixth and seventh centuries we need more information about the composition of Etruscan and Greek work of this period. Stylistic judgments are always unreliable. There is no method based on artistic judgment by which Greek and Tuscan pieces can readily be distinguished. Nevertheless, slightly more richness, exuberant decoration and more skilful metalworking techniques can sometimes be discerned in Etruscan jewels. Are there no criteria by which Greek creations can be distinguished from Etruscan? This is a question which often vexes the minds of both archaeologists and art collectors. We need first to understand the differences between various periods and various art forms. In the archaic and Orientalising periods the workmanship of certain workshops and goldsmiths in Greece and Etruria was so similar that there is not a single stylistic difference between their productions. The plastic arts of the same period, or the jewellery and gold work of earlier times are another matter entirely. The difficulty we have emphasised is perfectly genuine: the art critic or the scholar is confronted by very serious difficulties in attributing a jewel to a workshop in Asia Minor or Greece or Etruria. In the absence of stylistic criteria, there is a great need for detailed chemical analysis.

The same situation applies to Etruscan figurines and bronzes. The Etruscans were always regarded as highly skilled bronze craftsmen and in the archaic period small Tuscan bronzes were exported to various countries, where their style, strongly influenced by Ionian work, was popular. Their Hellenic appearance is very pronounced, and it is not easy to tell a sixth century Etrurian bronze from one fashioned in Magna Graecia, particularly in Tarentum. The greatest experts sometimes hesitate when faced with a fine piece. It is true that an occasional elongation of the hands and feet of figures and a certain disregard for anatomical proportion can betray Tuscan origin, but in the archaic period these characteristics diminished and became almost imperceptible. The archaeologist must turn to the specialised laboratory technician for an answer to this dilemma.

Etruscan forgeries

The present popularity of all sorts of antiquities and the consequent rise in prices all over the world has led to an increase in the number of forgeries created by men who have no scruples but are not without knowledge and skill, and who succeed in deceiving the public, and sometimes even the experts. An interesting discovery in archaeology or one of the arts made by an official body or an authorised individual often provokes an outburst of simultaneous activity among the forgers. The creation of the famous tiara of Saitapharnes, which was displayed for a while in the Louvre, was the result of some great discoveries at ancient Crimean sites. The public interest in Etruscan art of, let us say, the last fifty years has led forgers to increase their output of 'Etruscan' fakes. This has been made easier by the characteristics of Etruscan art, which did not acquire the rigid principles of evolution applied to Greek art. A workshop in Orvieto which specialised in forging works of art made the colossal warriors which adorned one of the rooms of the Metropolitan Museum in New York until the curator of the museum in collaboration with its laboratories conducted a thorough enquiry and succeeded in exposing the fraud. In recent years there has been a fresh outbreak of forged Etruscan archaic bronzes and terracottas. There are many of them and their numbers are continually increasing. This disturbing increase has arisen chiefly from the credulity of buyers who conclude dubious purchases without first consulting the opinion of an expert. Although such an authority is not infallible, he can usually distinguish ancient work from modern. Chemical and physical analyses can also help in evaluating doubtful objects. Finally, there is the problem of the day-dream which leads buyers to believe that artistic works of great rarity and value can be bought very cheaply. This is not impossible. But the masterpiece so cheaply bought much more often turns out to be a more or less crude facsimile of an ancient work.

A small sixth-century bronze figure of a youth from Pirizzimonte near Prato wears a toga with a heavily embroidered border and elaborate footwear. The pose, with hand outstretched as if in greeting, is a purely Etruscan concept which does not occur in contemporary Greek art. The proportions are broader in the shoulder and larger in the head than the Greek canon demanded, but the statuette, one of the best preserved of its type, has an undeniable vitality and charm.

Frescoes

Conservation, restoration and exhibition are problems which affect Etruscan antiquities as much as any others, but they are probably more pressing because of the increase in the number of finds in Tuscany. The most seriously threatened class of objects are also the most valuable: these are the frescoes which adorn the walls of the tombs in cemeteries, especially those in Tarquinia, Chiusi, Orvieto, Vulci and Cerveteri. The frescoes form a group of exceptional artistic and historical interest. They serve as a record of the Etruscans' religious and personal lives, and help us to understand ancient painting, which has otherwise vanished in an almost universal disaster.

Paintings done on walls and panels were highly prized by the Greeks and Romans, and enabled great painters to give free expression to their genius. Unfortunately these early paintings have almost all been lost because of the climate of the west which is too damp and changeable for the survival of such perishable materials.

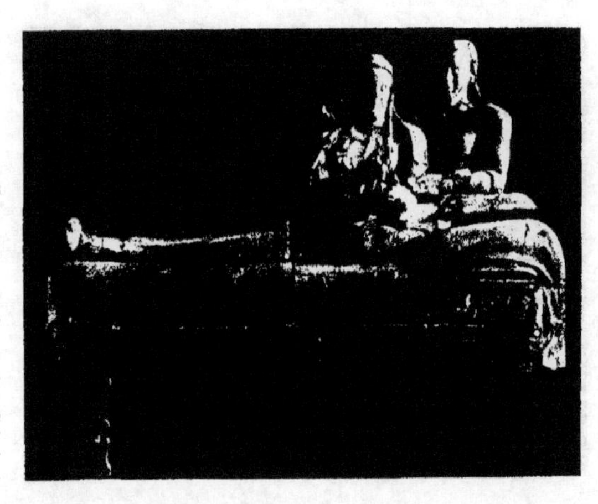

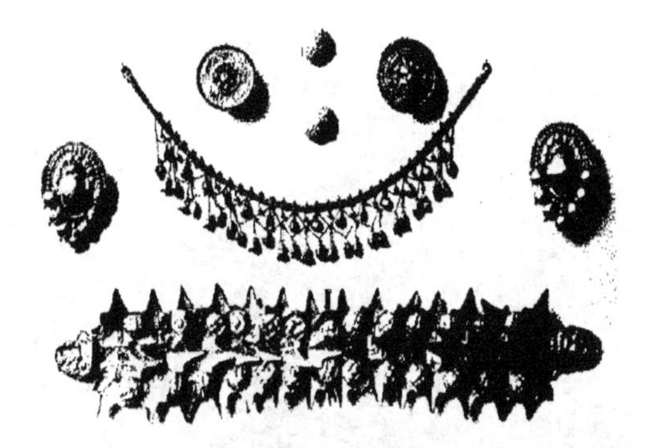

At the zenith of the Etruscan culture the wealth of the cities encouraged a high degree of development in the luxury arts. Their jewellery, above, worked in beaten gold and frequently decorated with minute beads of the same metal, is particularly fine. The use of thin wire-work, tiny cut-out figures attached to the surface of the piece and patterns carried out in beading (granulation) are typical of Etruscan work.

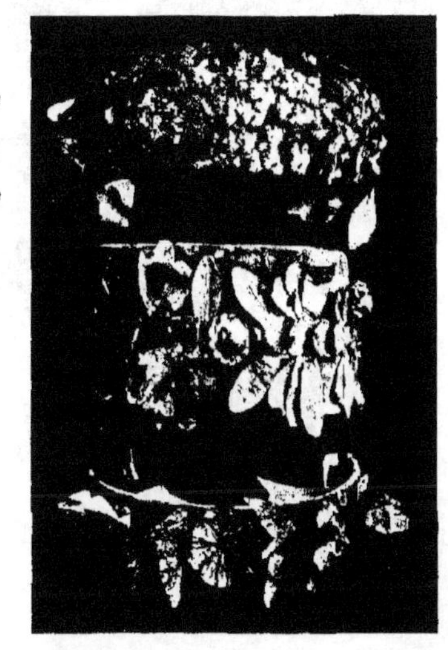

During the festivities and banquets which played such an important part in Etruscan daily life, wreaths and diadems were worn. For funerary purposes these floral garlands were reproduced in gold, usually very thin and much too fragile for frequent use but amply demonstrating the skill of the goldsmiths who made them.

Etruscan burial chests were made of terracotta and surmounted with a figure of the dead man reclining at the funeral banquet. The earlier examples are characterised by an attractive light-heartedness, but in later days the faces often wear an expression of haunted sadness. The sides of a sarcophagus from Volterra show the hunt for the Calydonian boar.

as wood and cloth when they are buried for several centuries. Only in hot dry countries like Egypt have a few paintings survived. Almost all Greek painting has vanished over the course of the years, and this is an irreparable loss. The disaster of AD 79 in Italy protected the frescoes of Pompeii and Herculaneum by burying them beneath a thick protecting layer of ash and lava. Etruscan artists painted huge frescoes on the walls of the hypogaea, and in many cases these frescoes have not yet deteriorated, though this situation is at present threatened by the chemical fertilisers used on the farmland which sometimes penetrate into the painted tombs and do serious damage to the decorations. Lenci has emphasised the effects of chemical fertilisers in connection with the Monterozzi necropolis at Tarquinia. He stressed the fact that the distribution of farming land, which includes the site of this, the most famous of all cemeteries, has led to a striking alteration in cultivation methods over the last few years. The owners use new products to increase the fertility of the lands and to improve the yield of their fields, as they naturally have the right to do. Surveys by the Milan Polytechnic have shown that rain water mixed with nitrogen products has seeped into painted tombs and caused the formation of extensive deposits of saltpetre on the tomb walls, which damages and destroys the pigments of the frescoes. Moreover, the roots of the plants, strengthened by the application of new fertilisers, often force their way into the tombs through the ceilings and walls, and contribute their share to the destruction of the paintings. All these factors call for more research into ways of halting the ultimate destruction of this precious artistic heritage.

The movement of tourists through archaeological monuments can also bring dangers like those which enforced the temporary closing of the famous Lascaux caves in France. Unless effective steps are taken to protect them, Etruscan frescoes deteriorate rapidly when left in their original position. All the examples found at Veii, Chiusi, Orvieto and Cerveteri have virtually disappeared. Although the situation is better at Tarquinia, the danger is not over, and is even greater nowadays because of the activities of illegal excavators. It is difficult to keep an effective watch over the huge area of the Tarquinian cemeteries, in view of the inadequate number of guards available. Clandestine excavators, like the scholars and technicians, grow more skilled with practice, and an appalling number of thefts of fresco fragments have been carried out at Tarquinia in recent years. They have been cut away from the walls and spirited away to unknown destinations.

All these threats point towards the necessity for saving everything possible. The Institute of Restoration at Rome, which has expert staff and advanced equipment, has begun to remove the frescoes from their rock background, mount them on canvas in frames and remove them to museums where no further harm can befall them. This operation is difficult because the frescoes are painted on a very thin layer of plaster no more than a centimetre thick, or sometimes, as at Chiusi, directly on to the rock. Once the transfer has been successfully effected the faded and nearly invisible colours can be discreetly restored. The removal of frescoes from the tomb of the Biga, the tomb of the Triclinium, and several others at Tarquinia has been completed.

The Tuscan school of painting can be studied by examining the frescoes. Though directly inspired by Greek painting, it expresses a completely different genius. These were the people who inaugurated the vogue for covering the interiors of homes and the walls of tombs with glistening brilliant colours. There is a feeling for colour in all early Italian art, which is typical of Etruscan and Roman work. The frescoes, which cover several walls in each tomb and are sometimes closely integrated, also show a pronounced sense of composition. The area to be filled has been worked into the design, and the artist has arranged the different elements of his complex work. To understand the desired effect the spectator must consider both the artistry of each individual element and the part it plays in the design as a whole. This feeling for space and composition has been another significant characteristic of Italian art.

The Etruscan frescoes, particularly the archaic examples of the sixth and fifth centuries BC, marked as they are first by Ionian and then by Attic influence, display a boldness of treatment, which freely simplifies and stylises the forms, and a spontaneity which never excludes occasional sophistication and refinement. It is easy to understand why crowds of tourists push into the doors of the hypogaea at Tarquinia, and why protection of the frescoes is more necessary than ever.

Displaying Etruscan remains

The preceding remarks will have demonstrated something of the problem facing the Roman and Tuscan museums into which these archaeological finds are pouring in ever increasing numbers. The way museums display this material is being reconsidered in many countries. Much thought has been devoted to it and many new ideas have been put into practice over recent years. Museums were once places where ancient objects were assembled and preserved, and the whole collection displayed in galleries and glass cases full of pieces which were often very much alike. Some museums are still housed in the venerable setting of ancient palaces. The Etruscan Museo Gregoriano of the Vatican, one of the richest in the world, is a case in point. The Etruscan museum of Florence has also preserved its old fashioned atmosphere, and an endless variety of material is displayed there. This material was badly damaged when Florence was flooded in the winter of 1966–67. The Villa Giulia Museum in Rome is a fine example of what is known as the modern museum, which displays only a selection of choice pieces to the public while the rest of the collection is stored in reserves where the scholar can study it at his leisure. In spite of the heated disputes and polemics which marked the modernisation of various museums, it is difficult to appreciate the artistry and ingenuity of their glowing rooms and light galleries without a knowledge also of the Renaissance architecture of the former villa of Pope Julius II.

Etruscan art

An *aristocratic* style Etruscan art seems faithfully to reflect the aristocratic society which produced it. The scenes represented are drawn from the life of the ruling classes — we see them enjoying a banquet, reclining on richly draped couches, and surrounded by attentive servants. Music and dancing were part of their revels, and here too we can glimpse the luxurious atmo-

[illegible] which [illegible] the [illegible] oligarchy and perhaps contributed to [illegible] downfall. The lower classes of society [illegible] seldom depicted in the frescoes and reliefs, except as servants and slaves, all busy in their master's service adding to the pleasure and festivity of the banquet. The craftsman is never shown at his daily task, unlike the later Roman reliefs, especially those representing life in the imperial provinces such as Roman Gaul. The lives of noblemen were what interested the artists who decorated the frescoes and reliefs. The tombs where the frescoes and sculptures were found were designed to imitate the houses which belonged to the great families. The spaciousness of the tombs reflects the care lavished on the Etruscans' houses and palaces.

The number of beautiful little bronzes, jewels and all kinds of feminine toilet articles now displayed in museums also helps to recreate the luxuriousness of Etruscan culture, and the variations of styles from various periods never obscure this opulence. In the tombs known to the Italians as *tombe degli ori*, or golden tombs, dating to the very beginnings of Etruscan civilisations in the early seventh century BC there was found an enormous collection of sumptuous jewels of unequalled richness and artistry. We

Aerial surveying has been successfully used in the case of the town of Marzabotto on the banks of the river Reno near Bologna. Founded in the sixth century, the city was laid out in well-planned blocks with wide streets, the centre reserved for wheeled traffic and the sides raised for pedestrians. Much of the city has been destroyed by the shifting of the river-bed, but the remainder provides valuable *information* on Etruscan town planning.

Etruscan scholars are faced with the problems of shortage of funds and personnel which seem to be almost universal. In Italy, however, highly advanced use is made of modern technological short cuts which keep wastage of time and effort to a minimum. The site of the Graeco-Etruscan commercial port of Spina on the Adriatic was surveyed by air, which revealed a complex of small canals and ditches with a larger channel to make a sea approach in the lagoon at the mouth of the river.

[illegible]d mention only the finds from the Regolini-[illegible]lassi tomb, discovered more than a hundred years ago by Archbishop Regolini and General Galassi. The remains from this tomb are now methodically exhibited in a room specially reserved for them in the Museo Gregorio of the Vatican. The beautiful gold work dates to about 650 BC, when the Etruscans were at the beginning of their period of greatest power and influence. The princely tombs found in the little Latin town of Praeneste, forty kilometres north of Rome, also date to the seventh century BC. Here we have yet another rich store [illegible] ivories, finely wrought silver-gilt vases and jewellery with unbelievably rich and elaborate decoration. The influence of eastern Mediterranean styles is very obvious and can be explained by the maritime trade which united the eastern and western Mediterranean.

As explained earlier, Etruscan art, from its earliest development, was the expression of a community which was avid for wealth and luxury, drawing the sources of this wealth from its extensive maritime activity – trade and piracy – which exposed it to a wide variety of foreign influences. For a long time the Etruscan fleets fought with the Greeks for control of the western Mediterranean. Consequently Etruscan art was exposed to influences from overseas, particularly from the Aegean.

The houses at Marzabotto are all rigidly confined to the oblong block system of the city's layout but within this framework considerable variation was found. They are separated from each other by party walls and have a courtyard with a well instead of a garden. These courts and the foundations of most of the walls were made of pebbles from the river, the upper courses being presumably of sun-dried brick which has not survived.

The great wealth of Etruscan ivory and gold and silver remains thus betrays the tastes of a rather barbarous society, dazzled by the sheen and glamour of precious metals. No other classical civilisation succeeded in acquiring such an ingenuous and childish taste for excessive luxury. The most opulent golden jewellery of the ancient world was found in the Scythian barrows near the Black Sea and in the Celtic tombs of Switzerland and Burgundy dating to the early Iron Age - for example the diadem from the famous Vix burial which is unrivalled in its size and decoration. Admittedly, this jewellery from the still barbarian lands of the Scythians and Celts came from workshops partly belonging to the Greek world. But these somewhat excessively large remains give the impression that they were commissioned by chieftains who appreciated Greek art without really understanding its restraint. Similarly, jewellery fashioned in Tuscany by Etruscan goldsmiths often surpasses in richness and complexity of ornament the Hellenic creations which inspired it.

The importance of feminine finery The Etruscans never lost their taste for luxury which accounts for the great wealth found in some tombs and the evidence of an insatiable desire which drove the local people to seek out any object among the remains of their ancestors which could impart a touch of long vanished splendour to their daily lives. The demand for feminine ornaments stimulated the production of elaborate goods until the time when Tuscan workshops disappeared. There are few objects as pleasing as the Etruscan bronze mirrors and chests delicately ornamented with incised scenes from Greek mythology or decorated with fine arabesques which were to inspire such artists as Picasso.

Women in Etruria were not confined in the women's quarters as they were in Greece but made their presence felt in the private and public affairs of the country. The jewels which the great ladies of the courts were proud to wear illustrate graphically some famous episodes of Roman history, for example, the Rape of Lucretia, the virtuous Roman matron, by Sextus

An air photograph of the necropolis area at Tarquinii gives some idea of the task facing the archaeologists at work on this site. Underground workings produce a thinner growth of vegetation which shows on the air view as a light patch, and it can readily be seen that the entire district is stippled all over with unexplored tombs and with the domed humps of tumulus graves. Thanks to modern methods, it is now possible to scan the interior of these tombs without a full-scale excavation, and gain some knowledge of what they contain.

The huge gold plate-fibula (dress pin) from the Regolini-Galassi tomb, which dates to the first half of the seventh century, is so massive and elaborate that it shows a far greater preoccupation with ostentation than with true artistry of design. The relief lions, the free-standing ducks on the lower plate, the palmette border on the upper plate and the chevrons on the hinged centre pieces are all carried out in the finest possible granulation, worked with beads so minute that they can hardly be seen with the naked eye.

Tarquinius, which led to the expulsion of the Tuscan dynasty from Rome. Although the details of the legend are apocryphal, Livy's description of the contrast between the luxurious life of the young Etruscan princesses at the court of Rome and the hard life of the Roman matron is not just a literary device, as used to be thought, but a statement of genuine fact.

The importance of the after-life Etruscan art also illustrates their religious beliefs and their cult of the dead. The vast dimensions of the cemeteries, the architecture of the tombs and the richness of the relics found beside the dead are explained by the importance of the next world to the Etruscans. Their religion is distinguished by its faith in the omnipotence of the gods and of fate, which led to a feeling of the insignificance of the individual faced as he was by superior powers and concerned for the well-being of the dead. The man who listened all his life to talk of divine powers is made aware of the fate which is to be his after death. Although the cult of the dead was important in the Greek and Roman worlds, in Etruria life after death meant more than life on earth. To find a comparable attitude we must turn to the civilisation of ancient Egypt.

One of the most important facets of Etruscan art and archaeology is the result of this obsession with death. Tuscan architecture is far better *known from its tombs than from the rare traces* of its houses and temples. The huge Etruscan cemeteries in which tumuli were succeeded by chamber tombs, surely deserve the name of *necropolis*, since they were indeed cities of the dead, with their innumerable tombs ranged neatly along the funerary way, as at Cerveteri. Sculpture had its origins in this same preoccupation with death. A portrait of the dead man was needed to ensure a sort of magical survival for him. As for frescoes, they were intended to enliven the chill gloom of the sepulchre, to give it colour and movement, and to enable the dead man to participate for ever in the games and feasts shown on the walls of his last home.

Etruscan archaeology therefore leads us to a world which was far removed from the Greeks in its basic preoccupations and values. However, Greece is always present because of its powerful artistic influence in Etruria. This influence is particularly apparent and strong in the refined and aristocratic products of Etruria known in German as the *Hochkunst*, and much less obvious in the popular products which were especially widespread during the Hellenistic era.

Etruscan art: a peripheral style At the end of this brief study it seems useful to establish

The Tomb of the Monkey at Chiusi (ancient Clusium) was painted in the early fifth century with scenes of dancers, wrestlers and riders. The work is not so well drawn as the frescoes of contemporary tombs at Tarquinii, but it shows the vitality and animation which characterises Etruscan draughtsmen at the height of their culture. After this period, tomb-painting was abandoned for some time and when it was renewed the scenes were pervaded by a gloom and solemnity completely unknown in the fifth century.

some parallels between Etruscan and Greek art. The appearance of some objects found in Etruscan tombs will thus be less surprising. Although Greek and Etruscan art developed more or less simultaneously, the function, purpose and nature of art were disparate. Despite the eastern influences to which it was exposed, and which are particularly apparent in its early stages, Greek art followed its own individual evolution and advanced in a straight line, every step of which was based on earlier traditions and was explained by the preceding stage. Compared with this natural and straightforward advance, Etruscan art seems much less indigenous. Its development was accomplished by means of a sort of compromise between its own aspirations and foreign influences, particularly those from Greece, which were unbroken and sometimes very powerful. Etruscan art is sometimes irregular and chaotic in character, and its successive stages, far from arising one from another, are inexplicable without reference to models adopted from the Greek world.

In this sense Etruscan art is peripheral to Greek art, and this must be borne in mind if we are not to be disconcerted by underdeveloped or conflicting styles and dissimilarities between works of the same date. As to originality, the Etruscans expressed this by means of their choice from among the available models and their transformation and adaptation of these models to a spirit totally unlike that of the Greeks. It is not harmony of form and perfection of detail which count here so much as total vision, which is often embodied in spirited movement. The stylised and sometimes audacious personal vision of the Etruscan artist enchants the people of today, who recognise their own preoccupations in it.

In the archaic period the richest Etruscan tombs show that luxury goods were sought and obtained all over the ancient world. Ivory from Syria, faience from Egypt and bronzes from Anatolia were found, and the Bernardini tomb at Praeneste yielded a particularly fine silver-gilt Phoenician bowl decorated with concentric bands of hunting, battle and religious scenes executed in splendid repoussé and chasing.

The Romans

One of the last and most determined defenders of Republicanism in Rome was Cato of Utica, whose name came to represent resistance to tyranny for many years after his suicide in 46 BC, after the defeat of Pompey and Julius Caesar's rise to power. A bronze bust found in the Roman colony of Volubilis has a number of individual features, notably the beaky nose and large ears, but it is more likely to be an ideal portrait than a strictly representational piece.

The study of the Roman world

Rome boasted of being eternal, and history has, in the last analysis, borne her out, since no fixed date has yet been assigned for her death. In the Middle Ages the majority of Europeans – Italians, Germans in the Holy Roman Empire, Byzantine Greeks – doubtless still regarded themselves as Romans. Latin was the official language everywhere, and the works of its literature were piously copied and read. However, for the very reason that they felt themselves to be Romans, it never struck the men of this period that Cicero's contemporaries could have lived in a manner different from theirs: this is made clear enough by a glance at the miniatures illustrating the copies of ancient manuscripts. People had also forgotten the significance of Roman monuments, although they were far more numerous and better preserved than they are today: popular legends, which were often manufactured by imagination working upon the little comprehended remains, attributed their construction to Biblical personages or to heroes of *chansons de geste*.

The first attempt to recover the true face of Rome was made in the Renaissance. Princes and scholars vied in their eagerness to retrieve and understand the monuments of antiquity – rarely, alas, to preserve them. Some discoveries, like that of the 'grottoes' of Oppius – actually rooms of Nero's Golden House – stirred the imagination and provided inspiration for contemporary artists. But it rarely occurred to the humanists to apply archaeological evidence to enrich and correct the knowledge they drew from literary records; at most they sought a few examples suitable for illustrating their work, and still proved themselves far less scrupulous *in this quest than in textual criticism. The art of* dating was, however, in its infancy, and few scholars suspected that the framework of life could have altered radically in the twelve centuries separating Romulus from Theodosius. Thus the naive picture formed in the Middle Ages gave way to one of a conventional Rome – a synthesis of widely different elements which would long inspire artists working in the classical style.

This attitude did not change significantly in the seventeenth and eighteenth centuries, in spite of the admirable work of great scholars like Peiresc (1580–1637) and Montfaucon (1655–1741). Historians such as Lenain de Tillemont, who began to publish in 1690 his *Histoire des empereurs*, a model of erudition and order, used inscriptions and coins wherever they could, but never dreamed of making use of monuments. Even the discovery of Herculaneum and Pompeii, excavated from 1719 onwards, brought about an artistic treasure hunt woefully destructive to the buildings. Doubtless the great 'antiquaries' like Caylus (1692–1765) added lustre to the Age of Enlightenment. But although they proved themselves sagacious and *methodical in their study of objects, they were* not in a position to reconstruct the framework of ancient life in its entirety, nor, consequently, to make a positive contribution to the study of civilisations. As for the pioneer of the history of ancient art, the German Johann von Winckelmann (1717–68), who made his home in Rome, did much to promote the idea, still frequently entertained, that Roman art is no more than a degenerate form of Greek.

It was during the nineteenth century that scientific methods of investigation were first *applied to Roman archaeology, which attained* maturity only after Egyptology, Assyriology and Greek archaeology had reached theirs. Until recently historians relied less on its authority than on epigraphy or numismatics, and even today philologists tend to regard Roman archaeology as a somewhat distant auxiliary. To make an impact, it had to demonstrate that *it could solve, or help solve, problems before* which the other disciplines were powerless. It is these problems, which we shall review and which will give us the opportunity to examine the working methods used in Roman archaeology.

The origins of Rome

In the time of Augustus Roman scholars taught that their city had been founded on 21 April in a year corresponding to 753 BC; this date, definitively established by one of them, the learned Varro, differed little from that calculated about 200 BC by the historian Fabius Pictor. But in earlier times the event was freely set back three centuries, some time after the Trojan War which had driven Aeneas west in his escape from the flames of Priam's city. Fabius Pictor had applied a simple but very logical line of reasoning. He found that tradition mentioned

Opposite page: the Roman city of Augusta Emerita (modern Merida, near Badajoz) was founded in 25 BC and quickly became the capital of the province of Lusitania and one of the leading towns in Iberia. It was so large that it easily housed a garrison of 90,000 men and was well supplied with majestic civic buildings. Apart from a fine granite bridge built by Trajan and kept in repair by the Visigoths, little Roman *architecture survives but the remains of the* great city wall, a triumphal arch and some beautifully decorated temples in the Corinthian order.

the names of seven kings who had ruled Rome before the revolution which set up the Republic; the date of this latter event was well established, since it was only a year before the consecration of the temple of the Capitol, in the wall of which a nail was driven to mark each year: the date corresponded to the year 509 BC. Fabius calculated that each king must have reigned thirty to thirty five years on average; this permitted the foundation to be set at about 750 BC.

From the beginning of the nineteenth century, modern scholars have recognised the uncertainty of the first centuries of Roman history. In imperial times guides used to point out to tourists a hut of branches piously preserved on the Palatine hill, which dominated the Forum to the south and was almost entirely covered by the residences of the Caesars; there, so it was said, Romulus had lived. In 1903 Boni, the director of excavations on the Palatine, noticed beneath the several superimposed archaeological levels which constituted the ruined palace of the Flavians some post holes cut into virgin rock and fragments of clay plaster which had served to fill in wattle screens. He immediately recognised the traces of a primitive settlement. Subsequently excavations, the latest of which took place in 1948, have fully supported his theory and have revealed hut foundations cut in the tufa on the other summit of the hill, the Germalus. Boni also showed that the inhabitants of this village buried their dead in the valley of the Forum, which was once marshland.

These tombs contained an archaeological haul, notably characterised by urns holding ashes of the dead which reproduced the wattle huts in miniature and permitted their reconstruction with great fidelity. Such urns are found in the cemeteries of the Alban Hills, where tradition placed Rome's mother city, the famous Alba Longa of tragic destiny. Lastly, these finds can be dated and were in use in the eighth century BC. Thus the so called legend of Rome's foundation, which critics tended to regard as a romance fabricated by Greek writers with the aid of folklore, showed itself trustworthy on two vital points: the initial date, which Fabius Pictor had not calculated at all badly, and the ethnic origin of the inhabitants, who were descended from the Latin people.

The most recent excavations have confirmed this rehabilitation of tradition. The last three kings of Rome were said to be foreigners from Etruria. Under their government the city had attained a great position, the whole of Latium was subject to her hegemony and the metropolis was covered with fine monuments. Researches in the Forum under the direction of a Danish scholar, E. Gjerstadt, have shown that from the sixth century this depression, drained by the construction of a sewer - the famous Cloaca Maxima - and paved with flagstones, had in fact become the centre of an important agglomeration. In the plain between the Palatine and the Tiber – the Forum Boarium, or cattle market – Italian archaeologists have discovered the bases of temples founded at this time; there is no longer any reason to doubt that they were built, as the ancients said, by the second monarch of the Tuscan dynasty, Servius Tullius, named Mastarna by his compatriots. Still following the same legend, this Servius Tullius had enclosed Rome with a wall, substantial ruins of which exist at several points; important fragments can be seen near the Termini station, for instance. This rampart was long believed to have been built after the Gallic invasion which ravaged the city at the beginning of the fourth century BC. A more careful study indicates that in reality it comprises two juxtaposed elements, one of which was indeed built in the fourth century, the other, however, dates from the sixth. Finally, we possess sufficient elements of the temple erected by the last of the Tarquins on the Capitol to enable architects to reconstruct on paper the design of this sanctuary, which must have remained to the end the religious symbol of Rome.

The last episode in this rediscovery of Rome's infancy occurred in the summer of 1964: an Italian team directed by the great Etruscan specialist, Professor M. Pallottino, was exploring the ruins of Pyrgi, an Etruscan port some sixty kilometres north east of Rome which served as an outlet for the important city of Caere; the bases of two closely neighbouring temples had been cleared, and in the narrow space which separated them a sacred deposit was found, containing decorative elements - terracotta statues especially - which could be stylistically dated to about 500 BC. While examining this hoard, the chief excavator was surprised to behold three crumpled leaves of gold. They were immediately transferred to Rome, where Pallottino noted that two of them carried an Etruscan inscription, the third a Phoenician text. Unhappily, since these records are rendered in imperfectly understood languages, they present great riddles to the linguists who are trying to decipher them. We can, however, establish that they were religious dedications consecrated by Thefarie Velianas in honour of the Phoenician goddess Astarte, assimilated with the Tuscan Uni, who was herself equivalent to Juno. This find proves that there was an important Punic colony in the port which played an essential part in the political life of the state of Caere, a neighbour of Rome. The Greek historian Polybius gives the year 509 as the date for the conclusion of a first treaty between Rome and Carthage which regulated

Most Roman art is derived from Greek originals, but one field in which their pre-eminence is unassailable is that of civil engineering. Massive walls, roads and aqueducts throughout Europe and Western Asia attest the skill of their designers. The Pons Aemilius, now known as the Ponte Rotto, was the first stone bridge over the Tiber. The piers of the bridge were begun in 179 BC but the connecting arches were not constructed until 142 BC. The bridge was in use throughout the Middle Ages but floods in 1575 and again in 1587 carried away the arches, which were never repaired.

The heroic mythology which grew up around the founding of their city was very important to the Romans. At the centre of their favourite legend were Romulus and Remus, children of Rhea Silva and Mars. Their usurper great-uncle, king of Alba Longa, ordered the babies to be thrown into the Tiber, but the flooding river carried them ashore near the Ficus Ruminalis, where they were adopted and suckled by a she-wolf. The figure of the she-wolf is from the fifth century BC and is Etruscan work: the twin babies are additions, made about AD 1500.

commercial exchanges between the two cities. We might hitherto have doubted whether the powerful African city needed to establish a proper legal agreement with a small peasant community lacking a coastline: we now glimpse the possibility of Caere's having served as an intermediary between its Punic ally and its Latin partner.

So, in two thirds of a century, archaeology has demonstrated its ability to guide history in solving particularly difficult problems essential to an understanding of the roots of our civilisation. In history, as elsewhere, the truth may very often look incredible, nor is criticism on purely formal lines competent to reveal mistakes. Archaeology, with its limited but scientific evidence, helps to avoid the shortsightedness of a hypercritical attitude, although we must avoid reading into archaeological evidence more than it can prove.

The temples of Largo Argentina

The most glorious era in Roman history began around the fourth century BC. The city on the Tiber, which had gone through a period of stagnation, if not of regression, after the expulsion of the kings, now committed herself to an expansion which was fast to become an explosion. She began by concluding a close alliance with the Campanian metropolis, the wealthier and more sophisticated but less warlike Capua, and subjugating her Latin sisters, who were outflanked by this pact. She then had to concentrate all her energies on defeating the mountain Samnites of the Apennines, who coveted fertile Campania for themselves. At the same stroke, the Etruscans in their turn had to accept her hegemony. But the magnitude of her successes attracted the attention of the great powers then dominating the Mediterranean; she had to withstand first the Greek king Pyrrhus, come to rescue his Italiot compatriots, then, soon after, the redoubtable western empire of the Phoenicians, whose capital was Carthage. Rome all but succumbed in this duel, but she was to emerge from it so strong that no other state could henceforward preserve its independence in the face of her imperialism.

This epic, which is unique in history, was worthily celebrated in literature, especially by Livy. Consequently, people anticipated the discovery of material evidence equal in magnificence to these splendid texts, but this expectation proved utterly delusive. Even in the heart of Rome, in the Forum, on the Capitol, practically no building survives from the fourth, third or second centuries BC; almost every

The civic heart of any Roman city was its market place or forum. Rome had several, the greatest of which was the *Forum Romanum* lying between the Palatine, the Capitol, the Quirinal and the Velei. The problems of studying this most important site have been infinitely complicated by its long history and many reconstructions.

At the south-east end of the Forum Romanum are the remains of the temple of the Dioscuri, Castor and Pollux. Very little survives except the sturdy grass-grown base and three fine Corinthian columns. Beyond the temple lies the arch of Septimius Severus, and the Victor Emmanuel monument on the left commemorates the rise of modern Italy.

thing has vanished beneath later structures without leaving so much as a trace of its foundations. Men once imagined that in the vast gallery of portraits amassed in museums it would be easy to find the great heroes of the conquest: Cincinnatus, Fabius, Scipio, Aemilius and the like. The development of iconography has corrected this misapprehension: the busts formerly attributed to these eminent men are in fact portraits of obscure persons of the imperial epoch. Even though archaeology can help to correct the errors of historians, it is incapable of restoring history. If we had only the results of excavations, nothing would lead us to suppose that Rome became the mistress of Italy in a century, and that in the next she subjected the entire Mediterranean coastline to her rule.

In this astonishing poverty we are glad that in the eighteenth century at least the hypogeum of the Scipios was found, with the sarcophagus and epitaph of the ancient Barbatus who conquered the city of Aleria, in Corsica, during the first Punic War. The brilliant excavations of M. Jehasse are currently in the process of restoring this city to us; however, imperial and early Christian structures predominate among the finds – very interesting ones, moreover – and the early centuries are principally evoked by an abundance of pottery.

It is precisely the study of small objects which will ultimately enable archaeology to put together a full and efficient history of the Roman conquest; this is thoroughly understood in the political and military sphere, its economic aspects remain much more obscure. The duel between Rome and Carthage was, for instance, accepted as the conflict between an agricultural and military state, on the whole very backward, and a maritime, industrial and commercial power of the first rank – a kind of battle between the elephant and the whale, as the rivalry between Russia and Britain was described not long ago. Today we can see that although Rome was indeed long content with the produce of her fields, augmented by plunder, her ally Capua had from the fourth century profited by the victory of the legions to disseminate everywhere the output of her workshops. Her triumph is measured in the spread of black Campanian ware, which is plentiful on all sites in the western Mediterranean, including the Carthaginian sphere of influence. Here we have a new element which explains many mysterious features of Hannibal's policy, for example, and notably the magnitude of the success which the winning over of Capua on the day after Cannae represented for the Carthaginian leader, since it would give his plans a far more positive purpose than a wish to sample the famous 'delights'.

While on the subject of Hannibal, it should be stated that it is only since 1956 that we have known what this great leader looked like. The credit for this discovery goes to E. G. S. Robinson, the British numismatist, who positively identified the coins struck by the Barcids in their Spanish territories, which they had virtually converted into an independent kingdom. Some of these pieces bear portraits of the three great Carthaginian generals Hamilcar, Hasdrubal and

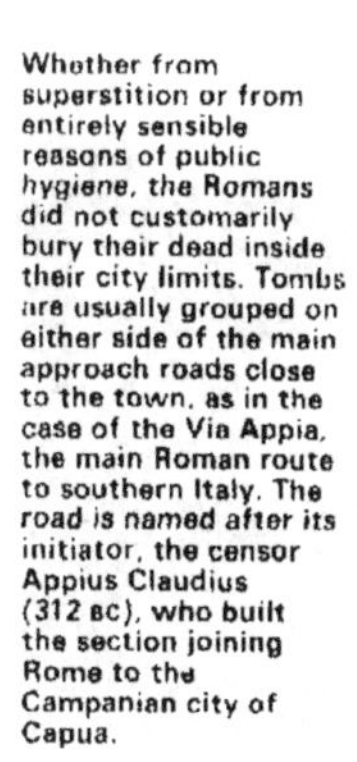

Whether from superstition or from entirely sensible reasons of public hygiene, the Romans did not customarily bury their dead inside their city limits. Tombs are usually grouped on either side of the main approach roads close to the town, as in the case of the Via Appia, the main Roman route to southern Italy. The road is named after its initiator, the censor Appius Claudius (312 BC), who built the section joining Rome to the Campanian city of Capua.

At the southern end of the Forum Romanum, in front of the Capitol, stands the triumphal arch of Septimius Severus, with its three arches and abundant reliefs of the emperor's victories in his campaigns against the Arabs and the Parthians. Inscriptions in Roman script, unrivalled for beauty and clarity, relate Severus' titles and triumphs, providing invaluable confirmation for historical records.

Hannibal. It has been recognised from a study of these coins that a bearded and helmeted bust in the Naples Museum, which (without any reliable foundation) was long believed to be a portrait of the victor of Cannae must be consigned to anonymity, while three other busts, one of which was found at Volubilis twenty years ago and the other two of which are now in the Ny-Carlsberg Glyptothek of Copenhagen and the Prado in Madrid, must represent Hannibal. They show the young general just after he was proclaimed leader by the army preparing for the great duel with Rome.

Rome and her monuments A district dating to the time of the Punic Wars still exists in modern Rome in the heart of the Papal city, two steps from the Piazza Venezia along the Corso Vittorio-Emmanuele which widens at this point to form the Largo Argentina. In ancient times this was part of the Campus Martius, a flat space outside the walls until about 300 BC, when overcrowding began to encroach upon it. During his censorship, Flaminius, a man who was to fall before Hannibal at Trasimene after a stormy political career, built a circus there rivalling the Circus Maximus, which had stood in the Murcian valley south of the Palatine ever since the monarchic period. These circuses were not just places of entertainment, but also centres of religious life, with temples clustered round them. Although nothing survives of the Flaminian circus, the Via delle Botteghe Oscure recalls the stalls set up in its ruins during the Middle Ages; several neighbouring shrines came to light in G. Marchetti's excavations around 1930. Four temples and the medieval buildings over them were recovered and restored. The nearest to the Corso was rectangular in shape and a large part of it

Unlike most of the people of the ancient world, the Romans had a thorough appreciation for solid bourgeois creature comforts. They realised that in so large and thickly populated a city as Rome efficient water supplies were essential and good drainage equally so. The result was the Cloaca Maxima, a huge system of sewers which drained the valleys between the Quirinal, the Esquiline and the Viminal. The original outlets survive and one can be seen in the lower left of the photograph.

Only the podium, a few columns and a fragment of wall survive of the elegant little circular temple believed to be that of Vesta, which lay on the Sacred Way through the Forum, opposite the Regia. The temple was dedicated very early in the history of the city (legend attributes it to Numa Pompilius, second king of Rome) but the podium is Augustan in date and most of the architectural fragments belong to restorations carried out by Julia Domna, wife of Septimius Severus.

colonnade is preserved around the whole cella above a podium three metres high. It was founded in the third century BC and took its final shape in the course of restorations carried out in the next century. It was repaired several times under the empire before becoming the Church of San Nicola dei Cesarine. Beside it is a round temple which suffered very similar vicissitudes at the same dates. Its southern neighbour, also rectangular in shape, is the oldest of all. It has been dated to the second half of the fourth century, at the height of the Samnite Wars. The colonnade on the facade and the long sides has unfortunately completely disappeared, and only the podium with its austere, strong lines preserves a record of the Roman architecture of this period which, according to the written evidence, was rich in an abundance of buildings. The altar which stood in the square in front of the podium survives intact, with its inscription of one Aulus Postumius, who restored it in the second century BC. But this all too-discreet magistrate never mentioned the name of the divinity who was to benefit from this dedication.

The attribution of the Largo Argentina temples is still a problem for the archaeologists, who have tried in vain to reconcile archaeological evidence with the written statements of the ancient authorities on the cults of this district. It is true that the remains of a colossal statue of a goddess were unearthed between the round temple and the one just described. Unfortunately the statue is a conventional type which could just as well represent Juno as Ceres or any number of personified abstract concepts. According to G. Lugli this image came from Temple D, the most southerly, and the largest of the four. This building, which was probably erected at the end of the second century BC, is the oldest example of a specifically Roman type of temple known as 'pseudo-peripteral' - a learned term meaning that the cella walls enclosed the columns of the porch, which are free-standing only in the rear part, or pronaos.

This sacred palace in the middle of the busy narrow streets of central Rome is pleasantly shaded by pine trees and is the only corner of the city where, without too great a strain on the imagination, we can reconstruct the familiar haunts of Cicero and even the great men of the generations before him whom he mentions in his orations. These ancient temples, however ruined, can easily be distinguished from the sumptuous buildings of the empire by their austere architecture, for which marble was not yet used. They clearly express the piety of these simple men who discovered the charms of the Hellenistic civilisations with a mixture of attraction and misgiving - charms which were hard to resist, but very apt to captivate or even to corrupt.

The town of Glanum

Southern France is scarcely less rich in Roman buildings than Italy itself. Everyone knows the Pont de Gard, the Maison Carrée at Nîmes, and the theatres and arenas of Nîmes, Arles and Orange, and the triumphal arches of Orange and Carpentras. The less celebrated temple of Vernègue and the portico of Rietz deserve to be better known. But what France lacks is a large ancient urban complex. Only two sites have been scientifically excavated. They are Vaison at the front of the Ventoux and Glanum near Saint-Rémy-de-Provence. Scarcely anything but a residential quarter and a theatre has been recovered at Vaison, and these have been too injudiciously restored. At Glanum, on the other hand, the very heart of the city has been revealed, thanks to the work of H. Rolland. The excavation has been conducted with exemplary care, with the result that it has been possible approximately to reconstruct the history of a Roman town which the ancient writers scarcely mention.

The site is very beautiful. The plain of Saint Rémy is closed on the south side by an offshoot of the Alpilles, which are cut by a narrow pass. A spring rises from the cliff on the west side of the defile, to which the native Ligurians, who fell under the political and cultural domination of the Celts in the fifth century BC, ascribed health giving and divine properties at an early date. The beneficent water was personified as a god named Glan, who was associated with the 'Mothers', ancestors of the fairies of our folktales. A barbarian sanctuary was dedicated to them, the remains of which have only recently been investigated. As in other high places in Provence, such as Roquepertuse and Entremont, the stone pillars of the building were carved with niches for the heads of decapitated enemies. But influence from Marseilles helped to civilise these uncouth people. In the second century Glanum became a communal republic where drachmae were struck imitating the Massaliot types. At the exit from the pass a council chamber, or *bouleuterion*, was constructed for the deliberations of the municipal assembly, or perhaps for the priests who administered the sanctuary. The region was conquered by the Romans in the last years of the second century.

The notables, their numbers probably increased by immigrants from Italy, built comfortable houses for themselves south of the pass; these were lighted in the Greek fashion from a colonnaded court in the middle and decorated with frescoes and mosaics which yield nothing to the finest homes of Rome and Pompeii. In 39 BC Agrippa, Octavian's closest colleague, passed through Glanum on his way to take possession of Gaul, which had just passed from Antony's control to that of his rival by the terms of the treaty of Brindisi. On this occasion a temple near the spring was dedicated to Valetudo, the Latin goddess of health. Apollo and Hercules had already come to join Glan and his Mothers.

In the subsequent years the citizens secured the services of a town planner of vision, who designed and carried out a total remodelling of the complex. The land to the south of the pass, where the new residential quarter was established, sloped sharply towards the north. The architect decided to level it with large quantities of fill on which two adjoining open squares were built. The one on the south side in line with the opening of the pass, was triangular in shape. It was flanked on the north west by a portico and adorned with a monumental *altar carved with captive barbarians* and trophies of arms in the middle. A little later it was finished by the addition of a sacred area *on the east containing two small temples*, probably dedicated to urban cults. The basic problem is the dating of the monuments; their identification comes next.

Because the Glanum excavations were carried out with meticulous care, H. Holland was able to make effective use of the essential stratigraphical information and paid particular attention to the large squares which the Roman *architects had built over the top of earlier* structures. Holland called Glanum III the city as it was before the construction of the squares, under it was Glanum II which consisted chiefly of houses buried in the fill which supported the squares. The first stage of the city before its Hellenisation was given the name of Glanum I.

To date Glanum II and Glanum III in principle, he had only to determine the age of the fill, which was clearly shown by the latest objects found in it. The most valuable 'fossils' from this point of view are potsherds, which are always very plentiful on ancient sites because almost everyday vessels and domestic utensils such as tableware, lamps, ornaments etc. were made of pottery. Around 30 BC the market was dominated by the Campanian workshops which distributed shiny black ware. From about 30 BC to AD 30 it was the Tuscans, especially the Arretine potters, who enjoyed what almost amounted to a monopoly. From the second third of the first century they were associated with and soon eliminated by the Gallic workshops which opened first in the Rouergue and then in Auvergne. These potters imitated Arretine ware, but were never able to equal the *high quality of the fabric nor the elegance of* the decoration. Most of the fill at Glanum contained no Arretine ware, while the latest Campanian types, which the experts call Campanian D, were found in abundance. The earliest Arretine wares were found only at points on the northern and eastern ends of the terrace, which were probably extensions of the squares. The terrace is rectangular in shape, surrounded by a portico and gives access on the north side to an adjoining basilica; an apsidal building beyond it was probably the curia, according to J. C. Balty's inspired conjecture, where the town council convened. As to the actual square, it is clearly the town forum. The Roman architects had not hesitated to bury beneath the fill several public buildings and private houses from the previous generation, in order to carry out this plan. North of the area, however, two roads flanked with houses dating to before 50 BC and a small bath house very like those at Pompeii were found intact.

Between 30 and 25 BC a monumental mausoleum on the north boundary of the city was built by a leading family who had served Julius Caesar and had been admitted by him to Roman citizenship, [illegible] thenceforward

The north entrance to the city of Glanum passed through a triumphal arch heavily and somewhat tactlessly decorated with reliefs of Roman victories over the Gauls. Just outside the gate stands a mausoleum which shows strong evidence of the town's *Hellenistic antecedents* in its Pergamene style. It is thought to commemorate the two grandsons of Augustus, Gaius and Lucius, who were heirs to the empire *until they both died in* their early manhood, but some evidence suggests an earlier date, about 48 BC.

Nowhere in the world is the Roman way of life so clearly illustrated as in the city of Pompeii on the bay of Naples. The first investigations in the eighteenth century were unfortunately little more than disorganised plundering expeditions, but subsequent excavations, conducted in a more responsible manner, have proved a mine of invaluable information.

entitled them to bear the dictator's *gens* name of Julius. It had a base decorated with large reliefs showing scenes of war and hunting. Nearby a triumphal arch, which served as a ceremonial gateway to the town, was dedicated a few years later. Although the rest of the town, which was destroyed in the middle of the third century AD by a Germanic raid, vanished beneath the assorted debris resulting from various floods, these two monuments were never lost, and the mausoleum at least survived comparatively undamaged. Known as the Saint-Rémy Antiques and admired by every traveller, they alone preserved the memory of Glanum's glory for seventeen centuries.

The dating of Glanum

In the interests of clarity we have described the history of the little city's monuments as if it had been all revealed at the same time without difficulties or problems, like the text of the fine inscriptions which epigraphists are sometimes fortunate enough to find. The controversial conclusions so far described about Glanum's monuments were the result of patient work and scholarship which combined the highly divergent facts furnished by nearly all the sciences at the archaeologists' disposal.

By considering other factors it is therefore possible to show that the basic transformation of the plan of the city took place about 30 BC. One of the houses buried by the terracing that was to support the basilica had been decorated with paintings and mosaics. Among the mosaics one was found which bore the name of the owner of the house - Co(rnelia) Sullae, (the house of) Cornelius Sulla. Although this cannot have been the famous dictator, it may have been a member of his family, perhaps a nephew, who had acquired property in Glanum, as many Roman nobles did in recently conquered provinces. The frescoes found in the debris were similar to those found in the Italian villas. A detailed study on the history of wall painting made it possible to date these to within a few years of 80 to 70 BC. Moreover, a fragment of painted plaster from Sulla's house bears a scratched graffito which, most fortunately, is exactly dated. 'Teucer was here', runs the text, 'in the consulship of Domitius and Sossius' (i.e. in 32 BC). Teucer's visit was therefore fifty years after the execution of the fresco and soon before the destruction of the house. In fact, as it is doubtful if Sulla allowed his guests to scratch their names on his walls and spoil his fine paintings, Teucer was probably one of the demolition workers who was trying – with considerable success – to inform posterity that he had taken part in the project.

Another indication of the exact date is furnished by the dedication of the temple of Valetudo, which shows the name of Agrippa associated with that of the goddess. Augustus' colleague made many visits to Gaul between 39 BC and his death in AD 12. With which of these trips should his stay at Glanum be associated? The answer is not supplied by the inscriptions, which do not include the date, but by the Corinthian elements in the architecture of the temple. The Corinthian style developed swiftly and its successive phases can be accurately dated to the second half of the first century BC. The least informed eye can discern at a glance some obvious difference in the style and proportions of the capitals of the Glanum sanctuary and those of the Maison Carrée at Nîmes, which Agrippa also built about 15 BC. Detailed analysis, based on comparisons with more precisely dated buildings, proves that the temple of Valetudo is clearly earlier than the building at Nîmes, so that its construction can be assigned to one of the illustrious Agrippa's first visits.

The date of the famous mausoleum of the Julii on the north side of Glanum can also be established by its architectural ornaments. Some of the capitals show features which are found only on the Arch of Augustus at Rimini and the temple of Apollo Palatinus, dedicated in 27 and 28 BC respectively. Moreover, the German scholar T. Kraus has shown that the carved scrolls on the frieze of the second tier of the tower are earlier than those of the Maison Carrée (about 15 BC) and the *Ara Pacis* at Rome, which was built and decorated between 13 and 9 BC. As the style of their ornaments is plainly more advanced than those of the temple of Valetudo, the chronological 'bracket' is narrowed to the years between 30 and 25 BC.

This discussion of the dating of Glanum shows how the archaeologist is able to calculate the age of a building by making use of all the factors at his disposal. In the case of Glanum, the excavators did not use any of the methods which physics has placed at the service of the modern explorer of the past. Indeed, these methods are of very little use to students of the 'classical' periods. They allow a margin of error which renders them virtually useless except for prehistoric periods, the early Middle Ages and cultures without any decipherable script.

Identifying Glanum's buildings

The identification of the buildings of Glanum has not raised any insuperable problems. The inscriptions from the sanctuaries of the past fortunately name the presiding deity. Some of the texts are written in Celtic and others in Greek, but the majority are in Latin. The case

One of the most meticulous and scholarly investigations of a Roman provincial town was the excavation of the city of Glanum, south of St-Rémy-de-Provence. Originally a Ligurian religious centre, a considerable Hellenistic settlement arose in the second century BC. It was badly damaged by the Romans during the troubled years around 49 BC, after which a massive rebuilding programme created the Roman city. The history of its many vicissitudes until its final destruction by the barbarians in AD 270 has been traced by the archaeologists who have been working on the site since 1921.

of the pairs of small temples to the east of the triangular square is not so simple because no inscriptions were found there. However, examination of a well sunk in the sacred area has yielded parts of statues which Rolland recognised as being representations of members of Augustus' family. Two of the heads, for instance, portray Livia, the emperor's wife, and his unhappy and notorious daughter Julia. These identifications are extremely difficult to make and are frequently disputed. Although hundreds of early imperial portraits are to be found in our museums, the greatest experts are often unable to identify them. The artists of this period idealised their subjects so much that the characteristic features of each face have vanished. Many of Rolland's identifications have recently been questioned in Germany, although they are supported by no less an authority than J. Charbonneaux. By no means everyone still accepts Balty's identification of the forum, the basilica and the curia, but the objections to the Belgian scholar's conclusions do not seem convincing. Glanum was too small a town and its flowering was too short (there is nothing worth mentioning later than the reign of Augustus) for it to have had two public squares like those found in the large cities, particularly in Africa.

The liveliest controversies have centred around the mausoleum of the Julii. Although it bears a perfectly legible epitaph some scholars have suggested that it is a cenotaph to Caius and Lucius Caesar, the grandsons of Augustus who died about the beginning of the Christian era. This theory is contradicted by indisputable epigraphic and archaeological evidence but it does show how many aspects of the celebrated monument are still obscure. It has also been very difficult to discover the precise meaning of the great reliefs on the plinth. These were once interpreted as describing episodes in Caesar's conquest of Gaul. But by 1840 Prosper Mérimée had realised that the general disposition of three of the scenes was borrowed from traditional Greek motifs—the battle for the body of Patroclus, the death of Troilus and the Calydonian boar-hunt. The fourth panel, which shows a cavalry battle, was inspired by a painting of an episode from the Punic Wars. The fact that his models were Greek had not deterred the artist from clothing and arming some of his characters in the style of his own day. We may conclude from this that he wished to refer to real episodes in the life of the dead man. The sculptor seems to have studied in the studios of northern Etruria and Italy. Like the decorators of the Tyrrhenian funerary urns but with a more advanced technique, he tried to give his reliefs a feeling of spatial depth, while the Greek sculptors placed all their figures in the same plane. By and large the decoration of the Saint Rémy monuments is closely related to those of north Italy. Artists from this district taught the Provencal Gauls, thus starting an original school which flourished for a hundred years. The arch at Orange, built in AD 26 in the reign of Tiberius, is the last noteworthy product of this school, the baroque pathos of which differs markedly from the classical taste which prevailed in Rome at that time.

During the Republican period, Rome's greatest enemy and the one which came nearest to destroying it was the north African city of Carthage. Founded by *Phoenician colonists*, the city preserved many oriental traditions and rites, some of which were extremely barbarous. In times of emergency sacrifices of children were made to the god Moloch, while dancers, wearing wrinkled grimacing demon masks, performed in honour of the god and the great goddess Tanit.

A street in Pompeii

The first real archaeological investigation in history started in 1719 when the Austrian Prince d'Elboeuf began to excavate the site of Herculaneum. Although his first attempt completely wrecked the theatre of the Campanian city, from the nineteenth century onwards archaeologists began to understand that it was possible to recover valuable pieces without totally destroying the buildings containing them. The houses and public buildings were gradually cleared while at the same time they were repeopled, more or less romantically, with the life that formerly animated them. For a long time, however, no one believed that anything but bare walls could be preserved *in situ*. All the portable objects found their way into museums or private collections along with the finest paintings, while the rest of the decoration was abandoned to certain destruction. It was not until 1861 that

The latest excavations at Pompeii have concentrated on a street which has been named the Via dell'Abbondanza. It is not the site of any outstanding public *buildings or fine* architecture, but it provides an immensely useful insight into Roman domestic life. New techniques are being used to replace *destroyed woodwork* as the walls are disengaged, so that the fronts of shops and houses, complete with their upper storeys, are preserved exactly as they were at the time of the eruption.

Titus, son and heir of the emperor Vespasian, destroyed Jerusalem and brought the sacred treasures of the Temple back to Rome for his triumph, which is commemorated by the arch standing between the Via Sacra and the Palatine.

excavations in horizontal strata instead of trenches were started under the direction of Superintendent of Antiquities C. Fiorelli, with all the finds preserved *in situ* as far as possible. Fiorelli thought also of injecting plaster into the hollows in the solidified volcanic ash left by organic matter, thereby making casts of corpses. An improved variant of this method made it possible to reconstruct parts of the excavated buildings and furniture. Fiorelli's methods have been continually improved by his successors, particularly V. Spinazzola, who was in charge of the work at Pompeii from 1910 to 1926. The cities obliterated in the eruption of AD 79 have therefore become not only the world's best known sites (the investigation of Herculaneum was abandoned in the eighteenth century and reopened in 1927) but veritable laboratories where the latest techniques are put into practice.

This is obvious to anyone who walks through the most recently excavated sector, the Via dell'Abbondanza, one of the main streets of Pompeii, which starts at a corner of the forum and runs in a north-easterly direction for roughly a kilometre to the Sarno Gate. About half of its length was excavated by Spinazzola. The façades of the houses along the street were carefully preserved, with the lively-coloured paintings decorating them, and on some of them electoral 'posters' were uncovered which referred to the town council which was about to be elected before the catastrophe occurred. Wooden door jambs have been restored and the bronze rings ornamenting them have been put back in their places. During the excavation, every part of the superstructure, or even its imprint in the ashes, which was found was preserved and put back into position. By means of this method the upper parts of the houses, which had been crushed by the weight of the volcanic debris without being totally destroyed, were restored to their original condition. Formerly the practice had been to finish an excavation before restoration, which meant that there was a large element of conjecture in the architect's solution to the problem of reconstruction. Spinazzola was able to prove that the Roman house was quite different from his predecessors' idea of it.

It used to be thought that the house had very few windows, and was lighted and ventilated almost exclusively from the inner courtyards. In fact, although there were no windows on the ground floor, the upper storeys were mostly open, and ventilated by balconies and loggias over the street. At the end of the day the inhabitants, especially the women, would take the air there, calling from one house to another and chatting with passers-by, as still happens in the crowded parts of Naples. This picturesque spectacle, which is depicted on a few frescoes, can easily be imagined by anyone who walks along the Via dell'Abbondanza. Spinazzola's excavations were basically concentrated on the street, which has always played such an important part in the life of the ordinary people of the Mediterranean. In this section the method of investigation initiated by Fiorelli, which consists of exploring each *insula* (a block of houses bounded by four streets) in turn, was also abandoned. The excavation was largely confined to clearing the section of each house along the road, and the only exceptions were a few particularly important complexes.

As we enter the section of the *scavi novi*, the first thing we meet is a fine house on the left-hand side of the street with an impressive doorway flanked, as usual, by two stalls, which were usually rented out by the owner. Popidius Montanus, who owned the house on the left, and had apparently been smitten by political fever. His electoral programme painted on the wall, states that he was the Games Club candidate! This kind of endorsement may scarcely seem serious, but it must have carried weight in a town where many leading citizens made their income by renting out property. His

A photograph taken before the First World War at the present Porta Maggiore. The walls of Rome were begun by the emperor Aurelian in AD 271 and completed in 275. They were continually improved and strengthened right up to the time of Belisarius in 536, and provide a permanent boundary mark of the imperial city.

neighbour, M. Vecilius Verecundus, was a busy manufacturer of cloth and sold his products direct to the public. A big fresco adorning his house showed on one side his labourers at work and on the other side the cloth being sold. These operations were placed under the protection of the popular deity Mercury, who is shown leaving his temple with his winged hat and caduceus, and Venus, who advances majestically in her chariot drawn by elephants. This Venus is thoroughly draped, a respectable Italian divinity, very different from the shameless Aphrodite whose amorous adventures often adorn the interior of houses. On the fresco there was enough space left above the temple of Mercury for another electoral manifesto.

On the other side of the street, Verecundus had a rival, a certain Stephanus, who devoted the whole ground floor of his house to his workshop and lived with his family in the upper

True Roman art, as opposed to work of Grecian derivation, tends to have a dual function. It is not only designed to adorn and magnify the city but to convince the citizens of the truth of the imperial 'party line' by its imagery and symbolism. A relief found under the Apostolic Chancellery in the Vatican in 1938 shows (below left) the arrival of Vespasian in Italy and (below) Domitian setting out on a campaign. Real personages are combined with allegorical figures such as the Genius of the Roman people and Victory, who are depicted giving their approval to the emperors' undertakings.

One of the finest surviving examples of Roman civil engineering is the enormous aqueduct, the Pont du Gard, which once brought a hundred gallons of water per day for every citizen of Nîmes from the spring at Eure, 49·75 kilometres away. The tiers of barrel-vaulted arches, built without cement from the local river-bed stone, rise 48·7 metres above the river. The aqueduct may have been part of Agrippa's public works programme when he was governor of Gallia Narbonensis in the reign of Augustus.

storey. Not far from the door the felt press can still be seen. The pool which usually occupies the middle of the atrium has been transformed into a rinsing vat. The chief workrooms, however, were at the end of the court. A little further along the street a man called Verus traded in ornamental and utilitarian bronzes. A surveyor's rod was found in his home. The whole of this district was given over to trade and industry after having been a middle class residential area.

Thus the House of Cryptoportico owes its name to an underground gallery beneath the garden which was planned as a refuge from the summer heat. Its last owner, however, was in the process of having it completely transformed into a storeroom, regardless of the magnificent painted stucco decoration executed a hundred years earlier. At the time of the catastrophe the unfortunate people who had tried to shelter in the cellar were slowly stifled by the ashes. The closely embraced corpses of a mother and her daughter were found there. Further on stood the mansion of Quintus Poppaeus, cousin of the notorious Poppaea, Nero's second wife. The walls of his house were covered with frescoes belonging to the last phase of Pompeian painting, Style IV, the baroque feeling of which reflects Nero's taste. Poppaeus' house has been named after Menander because of the portrait of the Athenian comic poet found there.

In the same block but served by another road was a small house furnished for a young couple. A verse scratched on one of the walls of the portico immortalises their too brief happiness: *Amantes ut apes vitam mellitam exigunt* (Lovers, like bees, live a honeyed life). One of the city's leading political figures, Paquius Proculus, lived next door. The front of his house was covered with electoral proclamations. The inside was distinguished by the exceptional quality of the mosaic floor. During this period mosaics were executed with a restraint which contrasted sharply with the exuberance of the wall paintings. However, the floor of Proculus' atrium depicts an aviary with many cages, each one containing the picture of a bird. At the entrance to the room was a picture of a tethered watch-dog, a very fashionable motif in the last days of Pompeii. A medallion set in the floor of another room was inspired by Nilotic themes, which had the same attraction for the Romans as the Far East did for the people of the eighteenth century.

We come next to the house of Amandus the priest, which has decorations from an earlier period. Over the entrance there is a drawing of a gladiatorial fight dating back to the time before Pompeii was Romanised, when the native people belonged to the Oscan nation. The Oscan settlement was one of the numerous branches of the great Italiot family which was closely related to the Latins. It was only in 80 BC, after the terrible civil war between Rome and her allies, that the dictator Sulla established a Roman colony in the Campanian city. The painting in the house of Amandus is a few years earlier than this event.

It is remarkable in that one of the combatants bears the actual name of Spartacus, which was that of the famous rebel of the years 73–71 BC. The rest of the painting belongs to Style III, which was at its height in the first half of the first century AD. Most of the subjects of the paintings derive from Greek mythology. While the great Hellenic painters and sculptors concentrated their skill on the modelling of a few beautiful figures, the artist selected by Amandus chose characters from fables, whom he painted in a romantic landscape enlivened by sharp contrasts of light and shade.

We next pass the front of a thermopolium, a wine shop where everything stands ready to serve the customers. Even the names of the barmaids—Aegle, Maria, Smyrna and Asellina—have been preserved. Their exotic names suggest that they were eastern slaves. The powerful 'psychological influence' of the imperial regime on the minds of the people can be assessed from some paintings on the opposite wall. One of them shows a sacrifice to the *genius* of the emperor by the *vicomagistri*, local leaders who belonged to the ordinary

The Roman bridge at the fortified town of Alcantara, in Caceres province not far from the Portuguese border in Spain, is unusually well preserved. In Roman times there was probably no settlement there, and the bridge, built in the name of Trajan in AD 105–6, was erected to provide a connection with Augusta Emerita. So impressive was the structure that the modern town was named after it (in Arabic al-kantara means 'the bridge').

people. Next to it is a picture of the twelve great Olympian deities. Elsewhere there is a painting of the great ancestors of the Roman people. Aeneas carrying his father Anchises on his shoulder and holding young Ascanius (the legendary forefather of Julius Caesar) by the hand, and Romulus bearing a victory trophy, stand out against an abstract background of lively red and yellow checks.

Naturally enough, patriotic propaganda also supplied the motifs for the decoration of a curious building interpreted by some scholars as a sort of club or military school for local youths. Two large trophies painted on each of the pillars framing a big door suggest the spoils of the nations conquered by Julius Caesar. Youth clubs at that time played a part which is not without parallels in modern totalitarian states. In addition, they supplied the only internal police force of the empire, where only the frontier provinces and the capital were permanently garrisoned by the army and the imperial guard.

The next houses belong to the middle class bourgeoisie. Pinarius Cerealis was a manufacturer of cameos and intaglio gemstones which were used as seals. The paintings in his house clearly demonstrate one of the features of Style IV, which was a passion for the theatre, an enthusiasm stimulated by Nero. The wall paintings show the fantastic and elaborate architecture of the stage where actors who seem to have emerged from behind the wall are performing *Iphigenia in Tauris*.

M. Epidius Hymenaeus, who probably began life as a slave, became an amiable if slightly sententious host in his old age. He could not resist having his dining room painted with judicious advice for his guests. 'There is a slave to wash and dry the guests' feet. There is a napkin to protect the cushions of the couch, please be careful not to dirty the cloth. Refrain from lecherous looks and do not make eyes at the women of the household. Converse on respectable subjects. Keep your temper and abstain from provoking others as much as possible, and if you can't, go home.'

The most attractive house in the Via dell' Abbondanza is probably that of Loreius Tiburtinus. Some think it belonged to Octavius Quartio, but this is unlikely. It is a fine home with an oaken garland above the door, which is framed by pictures of two bay trees painted on the pilasters. These same ornaments are found over the entrance to the house belonging to Augustus on the Palatine. At Pompeii the garland above the door indicates that the head of the household held the title of *sacerdos* of the first emperor, which was confined to the leading men. Inside the house we come to the atrium, the traditional heart of the Roman house. Usually there was an opening in the middle of the roof of the atrium for rainwater to fall through into the pool underneath. In Tiburtinus' house the *impluvium* has been transformed into a spouting fountain.

On the other side of the atrium there is no peristyle (an oblong court surrounded by a colonnade). Although there is a small court, it seems to be no more than a subsidiary of the long loggia dominating a delightful garden. Botanists were able to identify the trees and plants which once grew here, and have replanted the same species. This garden, like all Roman *horti*, was most artistically designed: a marble canal at right angles to the loggia divides it lengthwise into two. Cascades of water once flowed from a fountain on the loggia terrace; near by was a little dining room where one could dine in the cool air. This was decorated with frescoes showing the stories of Narcissus and Pyramus and Thisbe. The canal itself, which was named Euripos after the defile between Euboea and Boeotia, was flanked with statues and in the middle of it there was a small temple. Statues framed in the greenery were suitably placed throughout the garden. In the ancient world the private garden was virtually a sacred grave after it ceased to be merely a kitchen garden.

Two rooms which were cut off from the rest of the house opened on to the terrace. One was a small square room with a fine view from its windows looking towards the Lactarii mountains. The walls of this room are covered with frescoes. As everywhere else in Pompeii each wall is divided horizontally into three zones which in their turn are cut vertically into three parts by a sort of tall open pavilion made of graceful columns. Between these dream-like structures the central zone, the most important of all, has large white panels imitating tapestries. A single figure stands out in the middle of each panel. One of these is a priest of Isis with his shaven head, draped in a linen robe and holding a sistrum. This was a sort of rattle, the metallic clatter of which gave the rhythm for processions. A painted inscription, which is unfortunately almost illegible, once identified this character. However it still reveals that he was a native of Tibur, modern Tivoli. The likeliest explanation is that he was an ancestor of M. Loreius, who owned the house at the time of the catastrophe and whose surname, Tiburtinus, refers to the origins of the family. The devotion to the Egyptian goddess was hereditary in the family. The room with the portrait was a shrine with a tabernacle in its west wall which housed the sacred objects, forbidden to all but initiates. At the time of the eruption of Vesuvius Loreius snatched the sacred box, the palladium of his family, from the wall. Only the mark of it remains today, in the middle of the intact wall.

Even the garden was consecrated to the cult of Isis. No less than nine portraits of pharaohs

When a Roman legionary died at Chassenard in the ancient province of Gaul, he was accorded an elaborate cremation. A cast was taken from his face after death, as indicated by the half-closed eyes and slack mouth, and reproduced in iron. He was laid on the funeral pyre in full armour, the mask covering his face, and when the fire was lit the heat partly fused the metal. A fragment of his helmet can be seen above the forehead.

Leaders of the world in urban development, the Romans well understood the necessity for adequate water supplies for their chief cities. Rome itself received its first underground channel about 300 BC. In 144 BC the first overhead system, the Aqua Marcia, was installed and by the third century AD eleven major aqueducts were bringing in supplies from very considerable distances. The Romans believed in carrying the blessings of civilisation as they knew it to the distant parts of the empire. The remains of an aqueduct are still standing at Luynes in France.

In the Renaissance, when real interest in classical sculpture was first awakened, the Papacy, with its large land-holdings in central Italy, was in an excellent position to have first choice of the best Roman pieces, and even now that the temporal power of the popes extends only over the Vatican City, the papal collection of antiquities is one of the finest in the world. The stately galleries of the Vatican Museum are full of the Roman sculpture which also represented Greece to the men of the Renaissance, who knew nothing of original Hellenic work.

were found among the ornamental statues. An ingenious arrangement allowed the Euripos to overflow its banks, thus reproducing in miniature the fertilising floods of the Nile. A large marble figure holding a palm branch also personifies the sacred river. However, the Loreii's conversion to the Egyptian religion did not prevent them from remaining faithful to the cult of Hercules, patron of Tibur. A dining room adjoining the Isiac shrine was consecrated to him, and on its walls a large frieze showed one of Hercules' lesser known exploits, when he conquered the city of Troy to punish the treachery of Laomedon, and after killing the first king, put the infant Priam in his place.

This interpretation of the decorations of Loreius Tiburtinus' house which seems to have been established beyond a doubt by the work of M. Della Corte, A. Mauri and the Abbé Tran Tam Tinh, furnishes valuable confirmation for K. Schefold's theory. According to this scholar, who is Professor of Archaeology at the University of Basle, the motifs of the Pompeian paintings were not selected at random, nor even to please the aesthetic tastes of the owner, but to express religious and philosophical concepts. Although this theory has been hotly debated, some archaeologists believe it is the only possible explanation for the originality of Italiot painting in the late Republic and early Empire, in comparison with all the styles of wall decoration of every other time and place. The Roman fresco painter did not confine himself to decorating the wall. He tried to suggest an imaginary world beyond the wall, like the wonderland Alice found when she went through the rabbit's hole. This tendency appeared twice in the history of Roman painting, during Style II (60–15 BC) and Style IV which began about AD 60. There was a period of reaction known as Style III between the two. Schefold clearly demonstrates that this expressed the positive spirit of Augustus, that cold and Machiavellian ruler who established the reign of strictly rational order in all his domains. Style IV, on the other hand, reflects Nero's romantic enthusiasm which ended in madness.

The archaeologist's most difficult task is deducing and reconstructing from monuments the human psychology of bygone days. It is true that he must not let his imagination run away with him, since this can lead to some extremely peculiar aberrations. But an over-sceptical approach can be just as dangerous if it attempts to invest the ancients with a mentality very like our own. These two mistakes can only be avoided by the constant comparisons of Classical texts and monuments, which should never be studied separately.

This visit to the Via dell'Abbondanza will have indicated the wealth of information that a well-conducted archaeological investigation can

To the Romans of his own day, Trajan was *optimus princeps*, the best ruler. Clear-sighted, efficient, just and shrewd, he had, like most great administrators, an infallible sense of priorities and an unshakable coolness. His practical common sense and impatience with fuss are admirably conveyed by a portrait bust in the Capitoline Museum. During his reign, the idealism of the classic Augustan art was combined with a firm grasp of the vital need for political propaganda and the result is a representational style imbued with all the attributes of dignity and authority of the imperial power.

contribute to our knowledge of Roman civilisation. It has taken us into places almost completely overlooked in literature. The ancient historians in particular were interested exclusively in political events and mostly dominated by the moral code of a limited aristocracy and paid little attention to the daily life and problems of the middle and lower classes. A few rare storytellers like Petronius and a few *proselytes of Christianity are the only ones who* mention how ordinary people lived. But their evidence is not so explicit and complete as that of the streets and houses, the tools of daily life and the inscriptions of all kinds which have been dug up by the thousand. Intelligently conducted excavations reconstruct the setting of Roman life with indisputable accuracy and in the most minute detail.

A monument of imperial propaganda

It should not be assumed that archaeology is of no value in helping to compile the history of political power and the ruling classes. On the contrary, excavation often produces a great many objects connected with official propaganda, and this leads us to suggest comparisons with modern methods of 'psychological suggestion'. It is true that this comparison should not be carried too far, since the politics of ancient Rome were completely different from the modern system. For instance, too many historians are apt to forget that, though the empire was the first real state in history, the means at its disposal for controlling the individual were ludicrous in comparison with the most liberal modern democracy. The police force, for example, did not exist, nor did it begin to *appear in the most embryonic form* until the fourth century. There were few officials, but there were none at all in any other ancient community. However, the basic principle of the empire resembled modern totalitarian systems in that it was ruled by an all powerful leader – the word 'autocrat' is Greek for *imperator* – but he needed popular support. To obtain it he employed 'psychological suggestion', evidence of which has already been found in the streets of Pompeii.

Most of the official monuments of Rome were instruments of government propaganda. This is particularly true of great historical reliefs like those of the *Ara Pacis*, which has now been reconstructed on the banks of the Tiber near its original site, and the great triumphal columns of Trajan and Marcus Aurelius. Even the historians who are least disposed to admit the originality of Roman art are obliged to confess that these works have no Greek equivalent and are comparable with the monuments exalting the victories of the pharaohs or the kings of Assyria. The resemblance, however, is explained by the similarity of the aim in view, and not by any imitation by the Romans of eastern prototypes. In fact, the triumphal art of the empire originated in the electoral procedures of the Roman republic. Military commanders, who were also politicians, wished to keep their own achievements before the eyes of the electorate. From the third century BC onwards they had them immortalised in pictures, which were later replaced by carvings.

A large number of these reliefs have survived. Some of them still adorn the structures for which they were originally designed and others have been excavated. Two of the most recently dis-

The Julian founders of the Roman Empire were a popular subject amongst sculptors for many years to come, so that it can never be safely assumed that a portrait statue is contemporary with the man it depicts. There are, however, several other aids to dating. A figure of Julius Caesar in the Capitoline Museum is wearing an elaborate cuirass in the Hellenistic style, and detailed studies of armour can establish not only the date but the place of origin of such pieces with considerable accuracy.

covered examples have been selected to illustrate the problems this type of monument can raise for the archaeologist.

In 1938 the papal administration decided to carry out some work on the area beneath the palace of the Apostolic Chancellery, which stands in the northern part of the Campus Martius. To do this they had to dig eight metres below modern ground level, and it is seldom that anyone can dig as deep as this in Rome without finding ancient remains, but on this occasion the discoveries were outstandingly important. First of all they found the simple tomb of a well known historical character, Aulus Hirtius, one of the chief colleagues of Julius Caesar, who wrote the eighth book of the *Commentaries on the Gallic War* and who was killed in the civil war between Antony and

Very often the art of a great reign follows both the taste and the personality of the ruler, and that of Trajan's successor Hadrian is no exception. A more sensitive and sensual style with a more obvious relationship to Greek originals was developed. The emperor himself was an enigma even to his contemporaries, but something of the romantic spirit which influenced his patronage of the arts and fostered his extravagant devotion to the beautiful Antinous emerges in his portraits.

The father of classical archaeology was the great German scholar Johann Joseph Winckelmann (1717–68). He was one of the first to recognise the purity and idealism of classical Greek art and his passionate championship of the Hellenic spirit, expressed in his masterpiece *History of Ancient Art*, had a profound effect on the rising Neo-classical style. Curiously enough, he never even saw a genuine Greek piece of the best period, and all his knowledge was derived from Roman copies.

Octavian after being elected consul the year after the dictator's death. A deposit of marble had been placed near the tomb in the imperial period, containing fragments of monuments *which had been destroyed for one reason or* another. The pontifical workmen retrieved parts of two large relief decorations.

One of them, which apparently dated to the reign of Tiberius, showed one of the ceremonies of the imperial cult which we have already seen illustrated by the paintings in the Pompeian street. Two large panels came from a later monument. The characteristic features of Vespasian, who reigned from AD 70 to 79, were immediately recognised on one of them. It is an interesting detail that one of the pieces of these plaques was not found under the chancellery palace, which is papal territory by virtue of the Lateran Agreement, but under the Corso Vittorio-Emmanuele, which runs alongside the palace. It was claimed by the Com*mune of Rome and it was only later, after* diplomatic negotiations, that the work could be reconstructed in the Vatican Museum. It was then studied and published with admirable thoroughness by Signor Filippo Magi.

Each of these two friezes was carved on four adjoining marble slabs. In the frieze with Vespasian's figure, which has come to be called Frieze A, only the second slab from the right is intact. This is fortunately the most important. Vespasian is shown here in his toga with his hand on the shoulder of a young man who also wears a toga. F. Magi identifies him as Domitian, Vespasian's second son, who reigned from AD 80 to 96. Three other men are shown in this section of the relief. Two in the second row are symbolic personifications who can easily be *identified from coins, on which they are named.* The bearded old man in the toga represents the Senate and the half-naked youth with a cornucopia standing with one foot on an altar is the *genius* of the Roman people. The last figure is a lictor. Two other supernumeraries occupy the slab forming the right hand end of the relief, with a Victory coming to crown Vespasian. The left-hand half of the frieze shows the goddess Roma, helmeted and with one breast bare, en*throned amid the Vestals; they are turning to*wards the emperor to acclaim him.

The general meaning of this representation is clear. It refers to the historic events of the summer of AD 70. The empire had been ravaged by civil war for two years. Gaul and Spain first rose against Nero; Galba, his immediate successor, was overthrown by Otho, and he in his turn by Vitellius. In the end a coalition of eastern and Danubian legions established Vespasian. But this man, who was instrumental in suppressing the Jewish Revolt, did not personally take part in the struggle. Domitian, on the contrary, was in Rome and barely escaped with his life. It was he who came before his father when the new emperor landed in Italy at Beneventum and hailed him at the head of an official delegation. *Frieze A obviously shows this encounter in a* semi-symbolic form. It should be mentioned, however, that F. Magi's interpretation has been contested by the illustrious German scholar A. Rumpf, if only to show how minutely the smallest details must be examined in this kind of research. He bases his objection on a highly ingenious observation.

Roman senators used to wear a special shoe called a *mulleus*, which was characterised by a crescent-shaped ivory ornament. The so-called Domitian does not wear this shoe, but Vespasian does. Since Domitian was appointed praetor in AD 70 and therefore must have been a senator, A. Rumpf concludes that the relief does not represent him, but the *genius* of the equestrian order, the second rank of Roman *society, in association with the personifications* of the Senate and the people. This objection can nevertheless be rejected on the grounds that the *genius* of the Senate, too, is not wearing the *mulleus* which, for some unknown reason, has been reserved here for the ruling emperor.

One of the last great building programmes in Rome before the imperial crisis began to break up the solidity of the empire produced the baths of Caracalla, which must be one of the largest establishments of this type ever known. The baths are laid out symmetrically with the heating unit in the middle and a wing on either side, and are designed to supply 1,600 bathers at once. They include dressing-rooms, an exercise area, hot rooms, hot plunge baths, cooling-off rooms and a cold-water swimming pool, all of which were equipped with such an abundance of facilities they they must have placed a considerable strain on the city's resources.

The second frieze, Frieze B, shows a great procession. On the right is a group of soldiers in minutely detailed uniforms. These are the Praetorians, the emperor's personal guard. The *genii* of the people and the Senate go ahead of them, led by a young girl dressed as an Amazon in a short tunic leaving one breast bare and a helmet; she personifies either Rome or Virtus (courage). She is thrusting forward an emperor wearing campaign dress of a tunic and a purple *paludamentum*. This ruler has the features of Nerva, who succeeded Domitian, but F. Magi deduced that his face had been re-worked and must originally have represented another monarch. Minerva and Mars are conducting him towards a Victory, only the great outspread wing of which survives.

According to F. Magi these two reliefs were carved in Domitian's reign. This emperor attempted to impose authoritarian rule in spite of strong Senatorial opposition, and attached great importance to all kinds of propaganda. He used writers like Martial as well as artists, from whom he commissioned a large number of monuments such as statues, triumphal arches and trophies. However he was assassinated in AD 96. Suetonius describes the Senate's rejoicing at this news. Amid execrations of the dead man's memory ladders were fetched and his statues and the shields of honour on the walls of the Curia were thrown down. It was decreed that his name was to be obliterated from all the inscriptions, his portraits destroyed and anything which might perpetuate his memory suppressed. The law was implemented with the utmost severity. There are a large number of statues of Domitian on which the head was replaced by that of one of his successors and the inscription chiselled off. The Chancellery reliefs should therefore be 'corrected'; nothing was changed in the frieze where Vespasian, who was always regarded with veneration, holds the place of honour, but the condemned monarch was obliterated from the other and his place usurped by his successor, an opportunistic old senator who attempted to achieve some degree of national unity.

Magi's ingenious and convincing suggestion seems very likely to be right. If it has not been universally adopted by all scholars, it is because it upsets the accepted ideas about the development of Roman art in this period.

Until 1938 only one major example of Flavian political sculpture was known. This was the arch at the foot of the Palatine on the hill of the Velia which terminates the forum valley on the east and cuts it off from the Colosseum depression. This arch was dedicated to the memory of Vespasian's elder son Titus, who only reigned for two years, from AD 79 to 81. It has a unique bay in which two great relief panels celebrate Titus' triumph after his victory over the Jews. These reliefs are justly famous because they show the spoils from the Temple at Jerusalem, notably the seven-branched candlestick, being carried on barrows. Art historians are interested in it for other reasons. We have already seen in the case of the mausoleum of the Julii at Glanum that sculptors tried to obtain the same effects in stone as painters produced on canvas, i.e. to give the impression that their figures were moving freely in three dimensional space. This tendency is more marked in the Arch of Titus than anywhere else. For instance, the artist has shown the emperor's chariot obliquely in relation to the plane which forms the background of the relief, which is broken up by his movement. Unlike the Greek custom of placing all the heads on the same level, a large space has been reserved at the top of the field in which nothing is shown but the torso of the emperor and that of the accompanying Victory. The *fasces* of the lictors spread out behind the horses' heads, suggesting the presence of a vast crown in the background, like the spears in Velasquez' *Surrender of Breda*.

From the beginning of the twentieth century onwards this monument was regarded as the most perfect expression of the Latin artistic genius, the end product of a long evolution which included the Augustan and Claudian sculptors in its earlier stages. And now we have the Chancellery reliefs, which belong to exactly the same date but are composed according to different principles: the sculptor does not show the least regard for spatial depth, but basically conforms to the rules of the classical Greek style which Rome was supposed to have rejected! Rather than renounce or revise their theories, some scholars have attempted to dispute the date of these new plaques. A. Rumpf's argument, quoted above, aimed to date them to the reign of Hadrian in which a return to the classical style is generally believed to have occurred. This is surely unconvincing, and the theory of artistic unity under the Flavians must therefore be rejected and the coexistence of different styles admitted. These are probably not simply explained as the products of rival schools, as we see today, but as differences in kind, relating to the type of monument to be decorated and the spirit of the subject that was being celebrated.

The triumph shown on the arch was a completely historical occasion in spite of the presence of some allegorical figures. It was therefore suitably treated in a realistic manner. The Chancellery plaques, on the contrary, were dealing with an entirely imaginary and supernatural world, transgressing all rational laws of matter. A very interesting suggestion by a

The great engineers and master-builders of imperial Rome worked as much for posterity as for their own day. Their creations were meant to last. When Hadrian died in AD 138, a huge circular tomb was built for him at the end of a bridge over the Tiber. Both the bridge and the tomb are still there, and so sturdily did the tomb defy the passing centuries that it was adopted by the popes as a fortified citadel which they called the Castel Sant'Angelo.

When Diocletian finally resigned the imperial power, he had already prepared a retreat in which to spend his retirement. He was a native of Split (Spalato) in Yugoslavia, and there he built a palace which, in keeping with the warlike times in which he lived, was more than half fortress. A large rectangular space is enclosed by massive walls, with a walk for the sentries along the top and huge guard towers at the corners. The south wall gives directly on to the sea. The octagonal church almost in the middle of the enclosure is the emperor's tomb, which was converted in the early Middle Ages.

The original catacombs were those under San Sebastiano on the Via Appia. The underground burial passages have side niches for depositing corpses. The term ultimately came to be applied to all the burial complexes of this nature. In the early days of Christianity this vast labyrinth of tunnels supplied a refuge for the persecuted devotees. In due course of time, however, their numbers and influence had become so great that they were safe from official persecution and early in the fourth century the emperor himself adopted the faith, thus bringing an end to the classical Roman empire.

German scholar, Fraulein Erika Simon, proposes that plaque B is the apotheosis of Domitian, who is being led into the presence of Jupiter on Olympus by Minerva and Mars. The artist would have to resort to the exalted language of the epic to treat such a subject.

The historical value of archaeology

Is it possible to end by drawing up a chart of the contribution of archaeology to our knowledge of ancient Rome? We have seen that it has operated in a great many different fields, using a wide variety of methods ranging from stratigraphical excavations based on analysis of pottery styles to interpretations of great monuments, of architecture, sculpture and painting, and to the definition of the evolution of styles. The most tangible result of all this labour is that we have fundamentally changed our ideas on the evolutionary curve of Roman society. Historical texts would have us believe that the Republic reached its full maturity very early, perhaps in the fifth century BC or the fourth at the latest, and that moral decline set in during the second half of the second century with the great social wars. The imperial centuries were regarded as nothing but a prolonged decadence, the ancient virtues gradually smothered by despotism while Christianity was nourishing in its bosom a whole new system of totally opposed values.

The Tunisian city of Thugga (modern Dougga) was a hill town situated to the west of the road from Carthage to Theveste. Occupied from Neolithic times onward, it became a city of some distinction, strongly influenced by Carthaginian culture. In the time of Marius, Roman colonists settled there, and the native and immigrant communities lived side by side in their own ways till Septimius Severus united and Romanised the population. The town contains some of the finest Roman remains in north Africa, with a Capitoline temple, shrines to Caelestis and Saturn, and a theatre.

We now know that this doctrine is the result of the united labours of the reactionary historians, notably Livy and Tacitus, who supported the Senate and were regarded as champions of national tradition, and of the Christian writers who denounced the corruption of society under the empire. Philological criticism from the nineteenth century onwards has demonstrated (not without a few excesses) the inaccuracy of the traditional history of the origins of Rome. The primary function of archaeologists has been to illustrate the vitality of imperial society and to divert some degree of the attention hitherto focused almost exclusively on the capital to the provinces.

It is now apparent that Roman civilisation reached its full flowering only towards the end of the first century AD and in the course of the second, under the Flavians and Antonines, and that this period was one of the most productive eras the Mediterranean peoples ever knew. Furthermore, the researches we have discussed in the preceding pages show that although it was going through an extremely grave crisis, the empire in the third century AD still had plenty of reserves of strength, and that the fourth saw a positive renaissance, cut short in the west by barbarian invasions but continuing for hundreds of years in the east. In the overall development of mankind Rome therefore seems to be the true mother of the medieval civilisations, to whom she transmitted the heritage of Greece while transforming and adapting it to the spirit of each nation.

Wherever the Romans went they built roads linking their cities and towns and providing a swift means of moving their legions so that they could control their empire. In this picture is seen the road over Blackstone Edge in the Pennines - an indelible reminder of the precision of Roman engineering.

Europe in the Bronze and Iron Ages

A wooden cart complete with all its attachments was found among the wonderfully well preserved treasures of the burial hoard at Oseberg in Sweden, which dates to the ninth century AD. The solidity of the structure, with its thick-rimmed wheels, indicates that the local roads were not of the best. The most striking feature is the mass of carving which covers the body of the cart with mythological figures in flat relief.

The Celtic World

For a long time the earliest antiquarians knew the Gauls and Germans only through the writings of the ancient authors. They regarded the antiquities they collected as destined to throw light on 'the history of Rome in general rather than that of their native lands' (Peiresc, 1580–1637). The first sign of official interest seems to have been Elizabeth I of England's creation of the post of Antiquary Royal at the end of the sixteenth century. In the next century Gustavus Adolphus of Sweden (1594–1632) also appointed an Antiquary Royal, the distant ancestor of the present post of Director General of Antiquities. However, it was not until the eighteenth century that any institution made room for 'national antiquities'; indeed, the very phrase was not used in France until 1790. The creation of the Society of Antiquaries in London in 1718 was followed in 1780 by the Scottish Society of Antiquaries. Tsar Peter the Great published an edict requiring the preservation of all unusual objects found 'in the earth or under the earth or under the water', including 'old inscriptions on stone, iron or copper, arms, vessels and every other very ancient curiosity' (1718). In Sweden and later in Denmark the royal claim to buried objects of unknown ownership was made law, with compensation for the 'finders'. It was also in Denmark that the first 'official' excavations were conducted in 1776 at Jaegerspris by Prince Frederick.

Under the influence of the Romantic movement collectors became interested in the medieval and prehistoric periods under the all-embracing name of 'Celtic'. On both sides of the Channel the Celts excited much enthusiasm, and this enthusiasm gave rise to some rather extravagant publications (Cambry, *Monuments celtiques*). This 'Celtomania', however, was the beginning of several branches of western archaeology. For instance, one of the chief founders of French archaeology was the Société des Antiquaires de France, which was established in 1804 under the name of the Académie celtique.

In the eighteen thirties and forties European archaeology embarked on a positive phase. In Denmark, where a royal commission for the preservation of antiquities had existed since 1807, Christian J. Thomsen (1780–1865) founded in 1816 the Old Norse Museum, the original name of what is now the National Museum of Copenhagen. Thomsen adopted a method which had been in the air for some time of arranging and displaying the collections in his museum, classifying the objects by three technical stages: stone, bronze and iron. In his *Guide to Scandinavian Antiquities* published in 1836 he presented the general framework of the system, then his immediate successor J. J. A. Worsaae (1821–85) clarified the 'three ages' system and divided the ages into successive periods, two for the Stone Age, two for the Bronze Age and three for the Iron Age. He demonstrated the accuracy of this comparative chronological system by stratigraphical observations on the Danish peat bogs. In 1846 the excavation of the necropolis of Hallstatt in Austria was begun by Colonel Schwab. In 1856 Varga started to excavate the site of La Tène on the shores of Lake Neuchatel in Switzerland. These supplied the basis for reclassifying the Iron Age material, and in 1847 the Swede Hildebrand was able to suggest two divisions for the Iron Age, the first known as Hallstatt and the second as La Tène.

The enthusiastic interest of Napoleon III in the history of Caesar inspired a whole new series of investigations. There were excavations at Bulliot on Mont Beuvray, excavations or rather, raids, on the cemeteries in the Marne district, investigations of the tumuli on the Alaise (Doubs) highland, and excavations at Alesia and Mont Rhéa (1861–65), which resulted in the recovery of traces of the last battle of the Gallic War and brought to light weapons identical to those from La Tène. The eighteen sixties were particularly important. In France, where the *Revue archéologique* had been appearing since 1844, these years saw the creation of the Commission de topographie des Gaules, the first stages of the *Dictionnaire archéologique de la Gaule* and the departmental archaeological records, and finally and most important, the opening of the Musée des Antiquités Nationales in the chateau of Saint-Germain-en-Laye (1867). The names of Alexandre Bertrand (1820–1902) and Gabriel de Mortillet (1821–98) are associated with this enterprise. The Romano-German museum at Mainz, the German Archaeological Institute in Berlin and the National Museum of Zurich in Switzerland had already been in existence for several years. In England, where 'field archaeology was a national pastime in early Victorian times' (Glyn Daniel), the British Museum opened the new Department of British and

Opposite page: the village at Jarlshof in the Shetlands consisted of several contiguous stone huts, deeply countersunk in the sand near the shore. The islands attracted settlers by their good grazing and fishing, but the severe weather and shortage of timber called for particularly sturdy low-ceilinged houses. The people of Jarlshof not only fished but practised cattle and sheep farming, grew barley and worked bronze. The excavations have even yielded a few iron implements.

St Catherine's Hill near Winchester is one of the fortified mounds dating to the Iron Age which occur throughout Europe. The stronghold was once defended by a single oval rampart and ditch with an entrance on the north-east side. Sacked in the middle of the last century BC by the invading Belgae, the mound was later chosen as the site of a medieval chapel dedicated to St Catherine, the remains of which are hidden by the clump of trees on the summit.

Opposite page: the Temple of Poseidon at Paestum. The Greek city was founded by colonists in 600 BC and was soon one of the most flourishing centres of Magna Graecia. When all Italy came under Roman domination the city continued but the marshy site caused its gradual desertion.

Medieval Antiquities in 1866. Projects in Russia under the patronage of the Imperial Archaeological Commission assumed vast proportions often resembling organised looting, like the excavation of 7729 tumuli in the Vladimir region by Count Onvarov. From 1830 onwards they revealed the treasures of the Scythian tombs of south Russia and Kuban, and in 1865 the Scytho Sarmatian burials, which were wonderfully preserved in the ice, were found at Katanda in the Altai.

The tenuous relations between European archaeologists took firmer shape at the first anthropological and archaeological congress. The congress met in 1867 in Paris, where the Gallery of the History of Technology in the Great Exhibition showed French and European archaeology in two sections pre-metal Gaul and the Celtic, Gallic and Gallo-Roman periods.

From this time onwards the 'three ages' system was clearly inadequate to deal with 'the increasing complexity of archaeological achievement'. J. J. A. Worsaae himself recognised the regional limitations of this system, and in 1858 he suggested dividing Bronze Age Europe into geographical units. In 1873-74 Chantre established the first division of Europe into three regions: Uralian, Danubian and Mediterranean.

In view of the increasing number of systems, some kind of order became indispensable, and this was achieved by the Swede G. G. Montelius (1843-1921). The Montelius system, like the Linnaean, was based on an exact description of objects from closed deposits (tombs or hoards) of a known date. Comparison of these lists showed that there were always groups of identical objects characteristic of a period with which an isolated find could be associated. This system showed the technological advances of the objects, from which a typology specifying the evolution of the shape or decoration or geographical distribution of a given find could be drawn up. Montelius established the division of the Scandinavian Bronze Age into six periods on these principles and Otto Techsler suggested the division of the La Tène period into three periods: Early, Middle and Late. The extension of the absolute chronology of Egypt to Greece and Asia Minor by Flinders Petrie in 1889–91 enabled Montelius to suggest dates for the European metal ages, and he was approximately followed by the German Reinecke (1902) and the Frenchman Déchelette (1908).

Following Schumacher of Mainz and Kossinna of Berlin, German scholars evolved 'ethnic archaeology' (*Siedlungsarchaologie*) based on topographical studies which, with the help of finds, 'defines the areas of occupation, era by era, and notes variations in them, i.e. population movements' (H. Hubert). In England General Pitt Rivers and Flinders Petrie were 'the leaders of the revolution in archaeology which led it away from the contemplation of *art* objects to the contemplation of *all* objects' (Glyn Daniel). In 1880 Petrie applied the principle of total excavation at Cranborne Chase, devoting as much care to the study of living quarters as to tombs. He stressed the importance of everyday objects and their stratigraphical position, drawing detailed plans and sections. Similar methods had been pioneered by Worsaae's successor, Sophus Müller, when the National Museum of Copenhagen extended its excavation programme in 1897.

By the start of the twentieth century the stage was set for a new departure in European archaeology, which was now based on sound scholarship, the volumes of *Manuel d'archeologie prehistorique, celtique et gallo-romaine* by Joseph Déchelette (1862-1914) eloquently pointed the way. The new states like Poland, Hungary

A carved stone from the graveyard of Sanda parish in Gotland shows the runic script which ended the prehistoric period in Scandinavia. The runes appear to have developed in Gotland.

and Czechoslovakia, which were created after the First World War, added their many excellent findings to the archaeology of the Scandinavian and Western countries. After 1934, and even more after 1945, Soviet archaeologists achieved some remarkable successes which are insufficiently known in the West. In most countries the governments took over the supervision and financing of investigations so as to reserve and protect for the nation the archaeological wealth which forms part of their communal heritage. The publication of works on excavation methods and the exemplary quality of the results achieved by some scholars, for instance G. Bersu, Sir Mortimer Wheeler and L. Leroi-Gourhan, underlined the necessity for a systematic study of ways and means. The training of professional archaeologists first took shape at the Archaeological Institute in London. Archaeologists themselves encouraged public interest in archaeology by visits to excavation sites, television appearances and popular books and reviews.

During the last forty years interest in sociology and ethnography has led to exhaustive study of habitation sites, concentrating on traces of everyday life and economic activity, with excavations of high settlements like hill forts and *oppida* in countries from Great Britain to the Soviet Union, and bog villages in northern Europe. The concept of the era replaced that of the culture. Chronological groups no longer stood for successive periods but for cultural phases which were sometimes contemporary. Backward forms of a culture might persist in a region where a new civilisation had already begun to develop at a date ascertained from parallels with Western Asia and the Mediterranean. It became possible to recognise and date the movements of the peoples who spread them by studying the diffusion of cultures. The study of cultural interactions is closely connected with this branch of research. Outstanding discoveries in western Europe alone, such as the Vix tomb, the Celtic hill forts of Heuneburg and the warrior statue from Hirschland, have spotlighted the connections between the classical and the barbarian worlds. It is significant that there was a congress of classical and barbarian archaeology (Paris, 1963) on the theme of the influence of Greek and Roman civilisation on the peripheral cultures and its unifying effect on usually divergent branches of archaeology. Thus the aims of a new branch of research designed to contribute a series of increasingly accurate and better co-ordinated pictures of European history were established, in which man will take his place alongside his tools and his abandoned gods.

Dates, cultures and peoples

From the beginning of the twentieth century onwards the discoveries on which Montelius, Reinecke and Déchelette based their chronological classifications of the Bronze and Iron Ages have been augmented by new information. Many regional and local investigations have produced ten times as much material. Several scholars, including Schumacher, N. Aberg, Gordon Childe and C. Hawkes, have suggested new systems of classifying finds more accurately. Others, such as Holst, Vogt and Kimmig, have tried to define the limits of certain cultures like the 'Urnfield culture' which flourished in the Late Bronze and early Iron Ages. In more recent years J. Hatt has suggested the adoption in France of a more flexible system than the earlier

Runes undoubtedly had a completely magical function for the early Scandinavian people. The Old Norse word *rún* means 'mystery' or 'secret', and the characters are frequently accompanied by supernatural symbols. The hammer of Thor carved above the inscription on a stone from Laeborg (Denmark) invokes the god's protection against the ever-present powers of evil.

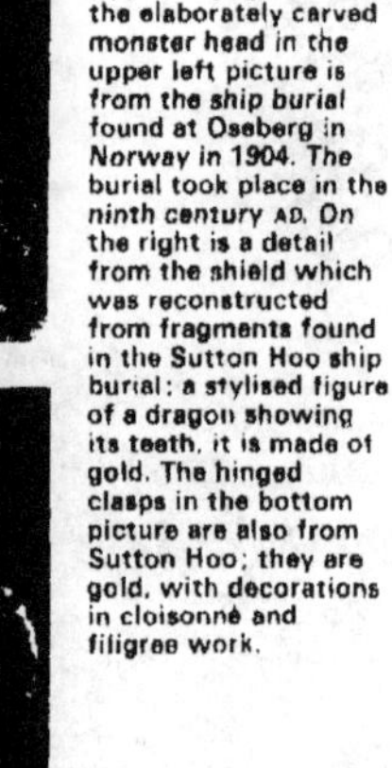

Opposite page: the elaborately carved monster head in the upper left picture is from the ship burial found at Oseberg in Norway in 1904. The burial took place in the ninth century AD. On the right is a detail from the shield which was reconstructed from fragments found in the Sutton Hoo ship burial; a stylised figure of a dragon showing its teeth, it is made of gold. The hinged clasps in the bottom picture are also from Sutton Hoo; they are gold, with decorations in cloisonné and filigree work.

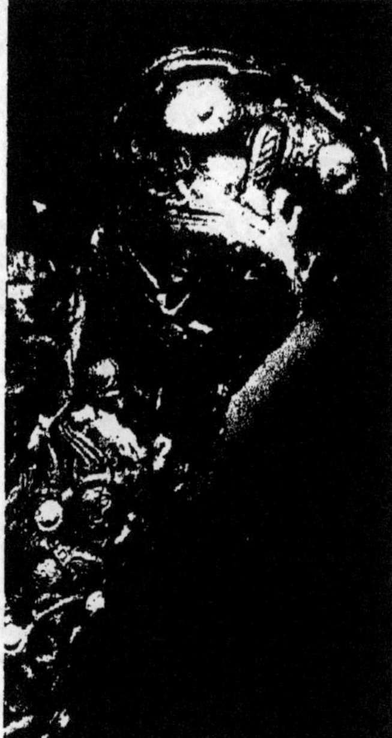

Viking art is savage and barbaric, but its effectiveness is undeniable. The cornel posts of the largest sled in the Oseberg burial are decorated with stylised animal heads staring outward at the observer with nightmarish ferocity. A writhing mass of humanoid and animal figures, interwoven with strapwork and embossed with metal adds to the overpowering impact of the work.

ones, as a 'temporary solution' to reconcile older data with the result of the latest excavations.

This system allows for the survival of belated forms of some civilisations, the dates of which overlapped from one period into another. There are many such survivals. For instance, after the appearance of the first typically Bronze Age products in Neolithic centres which A. Schaeffer believes came from the East (about 1800 BC to France), an increasing number of peoples, some of whom were already working copper in central and southern France, adopted the use of bronze, while brilliant homogeneous Neolithic cultures continued to survive in some districts (e.g. Morbihan and the Scandinavian countries). Stone tools were still being used as late as the Early Iron Age in poor communities such as those of Thiverny (Aisne) and Marmagne (Cher). Events followed similar courses in northern Europe at the end of the Bronze Age, which lasted until 500 BC there (hence Montelius' six divisions of the Nordic Bronze Age), while iron technology and its related cultures, the most widespread of which was the Hallstatt, were flourishing elsewhere. The Early Iron Age in these parts is contemporary with the Celtic La Tène civilisation. It persisted during the first millennium AD in a later phase sometimes known as Roman, which varied in length from place to place. The transition from the first to the second Iron Age in Celtic or Celticising regions is marked by the survival of belated or backward cultural forms such as the Hallstatt forms from Vix, Jogasses, Brittany and Aquitaine, which are broadly contemporary with La Tène.

Civilisations and cultures are generally classified by important sites and finds (La Tène), regions (Wessex), burial practices (tumuli or Urnfield) or characteristic products (food vessels in northern England). They are identified by typical objects, or preferably by groups of objects. Thus the spread of an important Early Bronze Age culture called Unetice can be recognised by the presence in tombs and hoards of bronze-handled daggers, torques with spiral terminals and pins with bulging heads or a ring on top. The first Urnfield phase is characterised by pottery shapes—large biconical urns with the high straight necks of burnished clay associated with small vases—and metal goods such as knives with the blade and handle cast in one piece, and poppy-head or collared pins etc., even more than by its burial customs (tombs with multiple incinerations). In

In the kitchen corner of the log cabins of Lake Constance, the walls and floor were lined with plaster to prevent fires. A shelf for cooking pots holds the rather crude household ware kept for everyday use at the end of the Neolithic period. An interesting span of equipment, ranging from the most primitive bone tools to imported metal objects, was found in the lake dwellings.

the same way, the three successive phases of the first Iron Age are marked by the great Hallstatt iron swords, horned daggers and swords with kidney-shaped or spherical pommels.

The relationship between the cultural groups and the distribution of their typical products illustrate migrations and spheres of influence. The invading waves of the Urnfield peoples, who might seem to have made 'unquestionably important contributions' to the basic elements of the Celtic civilisation, spread over Europe in the Late Bronze Age from a source which the latest evidence indicates to have been in the direction of Slovakia and northern Hungary. The distribution of their typical products now suggests two westerly movements: the first, which started between 1200 and 1000 BC, reached northern Lombardy, Champagne and Burgundy, while the second, which involved most of France, would have reached Languedoc and Spain in the eighth century. Another fundamental factor in the spread of the new metal technology was the arrival in the West during the eighth and seventh centuries BC of mounted warriors with an Eastern type of equipment, known as 'Thraco Cimmerians'. Their bronze and iron bits seem to appear in progressively later tombs from Hungary in the east to Languedoc (Cayla de Mailhac) in the west.

The combined evidence of archaeology, legend and history shows that Celtic expansion carried them far from their home in eastern France and southern Germany: Celt[illegible]lonisation at its maximum, [illegible] La Tène I and II, [illegible] Spain and northern Italy (settlements of the Insubres, Boii, Lingones and Senones in Lombardy), in La Tène II and III it extended to Poland, the Ukraine, Rumania and Bulgaria in

One of the most fortunate and illuminating finds of Scandinavian archaeology was the Oseberg grave, in which unusually favourable conditions had preserved tents, a wagon, four sledges, some beds, a quantity of domestic equipment and a beautifully carved ship, complete with all its gear, anchor, masts, gangway and oars.

One of the most striking landmarks of the Malvern hills is Herefordshire Beacon, on which stands the artificial mound of a medieval castle. Apart from its own earthworks, the stronghold was protected by a prehistoric wall which enclosed the entire summit of the hill and was adapted as a bailey in the Middle Ages.

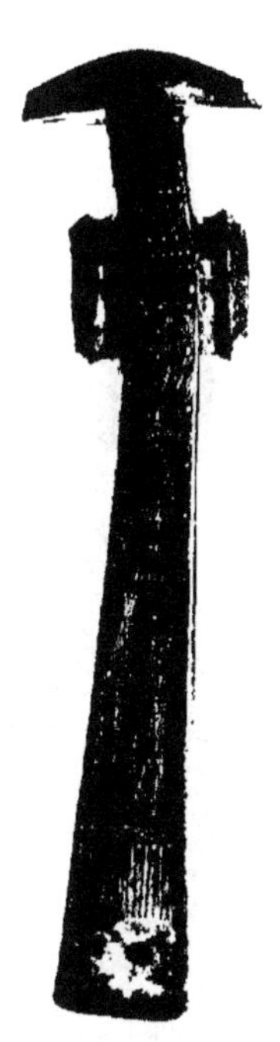

the east and Great Britain and Ireland in the west. In their easterly movement, which carried them as far as Greece – they attacked Delphi in 278 BC – the Celtic raiders clashed with the Scythians, who were migrating towards the west, and threw them back on their south Russian settlements. Some rich tombs in Hungary (Zedhalompuszta and Tapioszentmarton) and some isolated tombs in Prussia (Vottersfeld) and Schleswig (Plöhnmuhlen) mark the western limit of the Scythian advance. These contacts with the brilliant princely Scythian culture enriched Celtic art with elements from central Asia and the east (the so-called Basse-Yutz oenochoae). In the west several waves of Celts crossed from the continent to the British Isles. An original British Celtic art style was developed by the 'Marnians' who formed the second wave in the middle of the third century. The last arrivals after a major Belgic migration were refugees from north western Gaul, driven out by the Roman conquest. The peoples of northern Europe remained outside the Celtic world, but were strongly influenced by it. They traded with the Celts and bought products of Celtic craftsmanship, importing both goods – cauldrons and chariots, and technical methods – iron metallurgy, pottery and so on. The barbarian peoples of the Bronze and Iron Ages owed the homogeneous nature of their way of life from one period to another throughout the whole of Europe to these successive exchanges, as well as to the continuity of traditional techniques.

Habitation sites and dwellings

Iron Age settlements are not very different from those of the preceding period. Hills (Vix and Alesia) and spurs of land near the sea (Lostmarc'h-en Crozon, Finistère) were always preferred, or river banks (Heuneberg) and natural or artificial islands in lakes (Biskupin) or marshes (Glastonbury), but there were also many villages and isolated agricultural workings on sites of no particular character (Little Woodbury farm, the Germanic villages of Jutland and the Baltic islands). Many sites were occupied for a long time or several times in succession. On the fortified headland of Fort Harrouard (Eure) there are traces of occupation lasting from the Neolithic period to Iron Age II. The *terps* (mound) of Ezinge near Groningen in the Netherlands has superimposed remains of six rebuilding operations from the beginning of La Tène to as late as the thirteenth century AD. On the other hand, some sites were intermittently abandoned by their inhabitants for long periods. The Hallstatt fortress of Mont Lassois was only inhabited in the troubled times of La Tène III. Caves were often no more than a temporary refuge in time of unrest. The Mendip caverns in England, for instance, sheltered the Celtic population during the Belgic invasions of the late third century.

The wooden cities of Biskupin From 1934 onwards Polish archaeologists (such as J. Kostrezwski and Z. Rajewski) have been working on a complex measuring 2½ hectares on an island in the lake at Biskupin ninety kilometres north of Poznan. There were two cities in succession, the second of which was built on the same plan as the first after a destruction ascribed to Scythian raiders. The excavators dated this site to the beginning of the Polish Iron Age, or between 550 BC when it was first built and 400 BC when it was partially abandoned. In fact, occupation of this region dates back to at least the Neolithic period and Biskupin has also yielded finds from the Bronze Age and the Urnfield cultures. Ramparts, streets and houses were all built of wood. A palisade made of more than 35,000 piles driven into the water in rows from three to nine deep at an angle of forty five degrees protected the shores of the island. A rampart of square caissons of oak beams filled with earth and stones was erected on this. The only way into the enclosure was a gate six metres high in the south-west. This gate was a double door, three metres wide, surmounted by a watchtower and opening on to a bridge 120 metres long, which ended near a spring on the shore of the lake. The network of streets was paved with poles and planks laid on parallel stringers covered with sand and clay, and included a ring road following the rampart and eleven transverse streets flanked by rows of houses. A building with pillars, perhaps a sanctuary, stood in a small square between the seventh and ninth streets. The oblong houses, which usually measure eighty metres square, consist of an anteroom and living room, generally with a stone hearth, a bed (a board platform) and some shelves. The buildings were constructed by planting vertical posts with lateral grooves and filling in the spaces with overlapping boards with the ends bevelled to fit the grooves. The floor was boarded. The excavation yielded more than 5,000,000 different

When the inhabitants of the shores of Lake Chalain in the French Jura were driven from their settlements by floods, a branch of these people seems to have moved over the Alpine passes and settled, in the Early Bronze Age, among the north Swiss lakes. Forests of wooden piles were driven into the muddy shallows of Lake Constance as a base for their thatched log cabins, shown below in a reconstruction.

objects and fragments, many of which were made of wood, bone and horn.

The hill-fort of Vix The twin hills of Mont Lassois near Châtillon-en-Seine commanded the approach to the Seine at the point where it becomes navigable for small boats. Among other products, tin from Britain on its way to the Mediterranean passed through there. The settlement included a small city, perhaps inhabited by the workers, at the foot of the hill near the springs and a powerful fortress on the knoll which protected the wattle huts and the metal foundries on the slopes. The defensive system had three parts: first came a continuous ditch 5-7 metres deep and 19 metres wide, with only one zigzag way across, controlled by a guard house, then an earth rampart reinforcing the ditch on the inside and extending as far as the Seine on the east, and finally two enormous banks of earth which R. Joffroy believes must have been designed 'to ensure the protection of the springs and guarantee free access to water'. The rustic meanness of the buildings themselves contrasts with the high quality of the furnishings which included Attic black-figure vases and local painted pottery in abundance (1,500,000 sherds and 300 fibulae), and even more with the opulence of the other contemporary tombs in the neighbourhood, the tumuli of Vix and Saint Colombe. These bear witness to the extraordinary wealth of the 'rulers' of Vix between 550 BC and about 475 BC when the site was deserted. This picture is confirmed by several other sites of the period such as Camp du Château at Saline (Jura) which owed its prosperity to the neighbouring salt mines, Camp d'Affrique near Nancy and the nearby iron workings, and the great fortified castle of Heuneburg at Wurtemburg, which dominated the upper Danube valley not far from Sigmaringen.

Ensérune The *oppidum* of Ensérune (nine kilometres from Béziers in Hérault) was one of a double line of settlements from the Rhône to the eastern Pyrenees which commanded the route from the Cévennes highlands to the plain (Cayla de Mailhac) and the nearer passes to the Languedoc coast (Sainte-Anastasie, Mont Taures, Substantion and Ruscino). Three successive phases, dated by imported Greek and Italiot pottery, have been distinguished on this rocky spur, which rises to a height of 120 metres. At first the high ground was occupied by a modest village. There were no fortifications whatsoever to reinforce the natural defences. Many silos hollowed in the rock were used as storehouses to serve the mud huts scattered about the summit and the slopes. After about 425 BC buildings of crude masonry along streets intersecting at right angles replaced the primitive huts. A *dolium* (a large terracotta jar) buried under the house tended to supersede the silo. At the beginning of the fourth century a powerful cyclopean wall protected the inhabited area. About 230 BC a general reconstruction followed the destructions attributed to the last waves of Celtic invaders. Then the complex began to spread outside the enclosure and new houses, all one roomed, were built on the same chequer board plan. Monolithic columns, crudely imitating Greek models, replaced wooden posts. However, foreign influences from Greece and later from Rome were no more than a veneer. The way of life and the technical methods of the new Ensérune remained true to the local tradition. Unlike the coastal colonies, the hill-forts were never occupied by foreigners, who would have overthrown and probably obliterated the ancient indigenous culture. At the beginning of the first century AD the *Pax Romana* encouraged the inhabitants to desert gradually the uncomfortable high ground – Ensérune had only the water which accumulated in its cisterns – in favour of the lowland, which had been colonised for 150 years.

The Celtic hill-forts of La Tène III One of the outstanding characteristics of the late Celtic worlds is the close similarity or, to put it better in the words of W. Dehn, the 'astonishing likeness' between all the great fortified settlements of La Tène III, from Britain to Czechoslovakia. Although their sites – on a mountain or on high ground, in the confluence of two streams or the bend of a river – recall older sites, their fortifications show many new characteristics, most of which are common to them all.

The ramparts consist of a wooden frame filled with earth and stones. In the earliest, said to be Preist near Bitburg in the Rhineland, which was built at the end of the Hallstatt period, the woodwork consisted only of vertical stakes which could be seen above the stone facing from the outside, and some transverse beams dovetailed together or mortared. This arrangement has been recorded at Heuneburg and was common in Germany and most of the hill forts of Bohemia and Moravia. Another type, which Caesar mentions in his description of the siege of Avaricum (*Gallic War* VII, 23) under the name of *murus gallicus*, occurs in France, where there are twenty examples including Mont Beuvray, Alesia, Vertault, etc. In Germany there are three examples of this type; Belgium and Scotland each have one, and Switzerland has two. The framework was built in stages of transverse and longitudinal pieces of wood fastened at right angles by big iron pegs, the ends of the transverse pieces being visible on the outer face of the wall. The other characteristic features of these enclosures were walls and ditches running in long straight lines unrelated to the contours of the ground, giving a consequent loss of height, walls turning back towards the inside on each side of the entrance, to form recessed gates which were sometimes reinforced by posts (north-west entrance of Camp d'Artus in Finistère), and outworks protecting the gate by creating an obstacle (east entrance of Châtelier in Manche). Lines of multiple fortifications – two or three – enclose a number of hill-forts, as

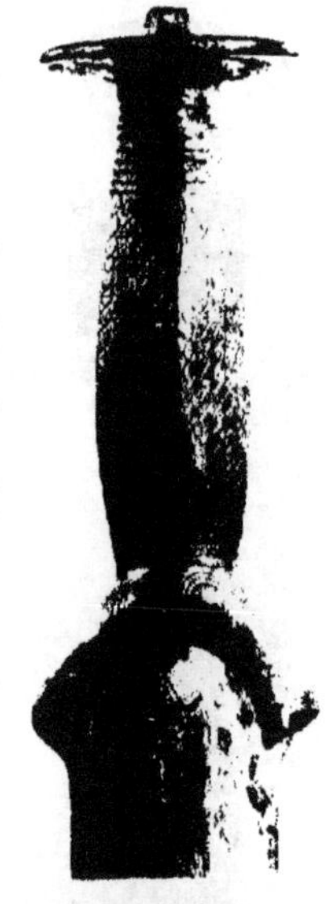

Bronze Age weapons from central Europe: above: decorated bronze sword hilt. Facing page: decorated bronze battle-axe.

The reconstructed interior of a pile dwelling from the shore of Lake Constance shows the sturdy box-shaped bed with its fur coverlet. A small axe hangs on one wall and on the other is the owner's bow and arrows. A long fish spear and a shelf with a few treasured metal vessels, just beginning to come into use, marks the transition from the Neolithic to the Bronze Age period.

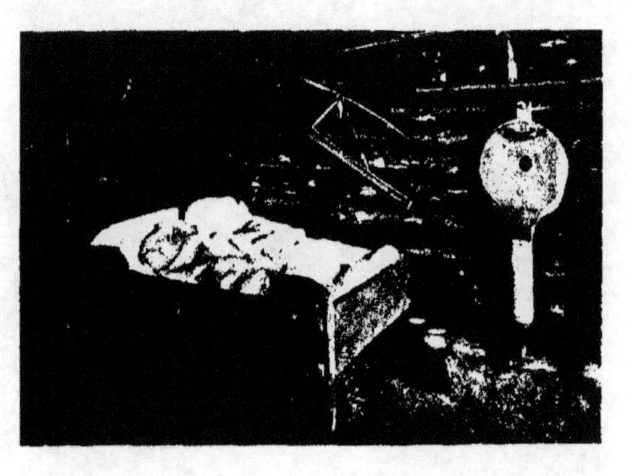

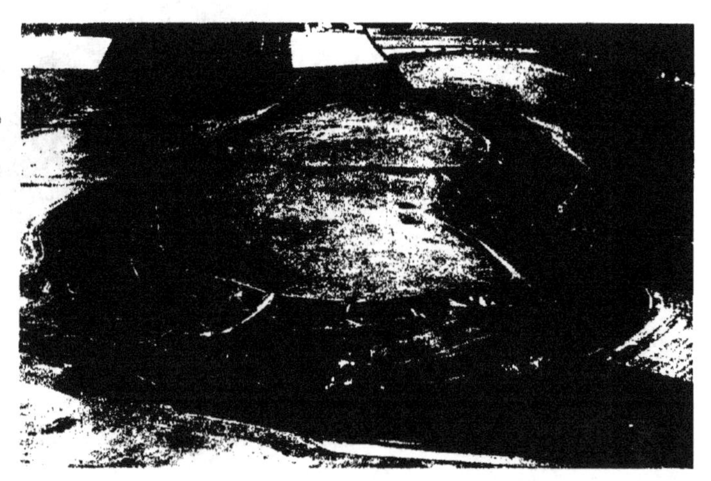

Maiden Castle is a colossal Iron Age earthwork on the road from Dorchester to Weymouth. The site was inhabited intermittently from the Neolithic period onwards. Its four ramparts and two entrances, on a scale hardly equalled in Celtic times, were raised as a defence against the Belgae in the last century BC. After the Belgae came the Romans. Vespasian's Second Legion sacked the fortification in AD 45 and moved on, leaving the local people to hasten their dead into crude graves near the eastern gateway, and from this time onwards Maiden Castle was never defended again.

south-west England such as Old Oswestry and Maiden Castle, and in Brittany a similar structure is seen at Kerkaradec-en-Penhars in Finistère, among others.

What was the function of these huge enclosures, which can measure up to several hundred hectares (Manching is 380 hectares)? Some of them, like Zarten near Fribourg-am-Brisgau, seem to have been used only as refuges, but in general, they were nearly always used for more than one purpose, being political (Bibracte, on Mont Beuvray, was the capital of the Edueni), strategic and economic centres at one and the same time. The huge expanses of open space, which can be seen behind the ramparts of Manching and Stare Hradisko, were useful both for defence and for the reception of refugees, and could also be used as pasture for the animals, or as large marketplaces. Forges and foundries and all kinds of specialised crafts enamel works at Mont Beuvray, glass bracelet makers at Manching, and official mints flourished in the industrial quarters.

Oppida played an important part in the defence of the Celtic world against the Roman invaders. In Gaul, Avaricum (Bourges) and Gergovia supported Vercingetorix in his defiance of the Roman legions. The decisive battle took place at Alesia, and traces of it have been found at Alise-Sainte-Reine. Maiden Castle in England was a fortified settlement with powerful defences reorganised about 50 BC to repel the invader. Near the remains of burnt huts, missiles from the Roman 'artillery' have been found, as well as the relics of the slaughtered inhabitants. The survivors seem to have returned to bury them when the danger was past.

Houses and villages The origins of the long house are very remote (there are Neolithic settlements in Germany, Poland and the Ukraine, and Bronze Age habitations at Wurtemburg), and the tradition continued through the Iron Age and far beyond, surviving until the present day throughout Europe. The round or oval hut, on the other hand, seems to have appeared later and only sporadically affected a smaller area. There are examples in western Germany and at the Scythian site of Nemirov in the Ukraine, at the La Tène site of Lochenstein in Swabia, in Alsace and especially in Great Britain. Another type of dwelling continued, until the Roman period, in Wales, Cornwall and northern England; this was a home and its subsidiary buildings inside a stone and earth bank enclosing an open court.

The main building material in most of these constructions was wood, whether they were made of posts and boards as at Biskupin, or whether they employed the more common method of posts and mud coated with clay. Even when the outer wall was made of stone and earth, the posts on the inside, set in parallel straight lines or circles, bore the weight of the thatched or tiled roof (the round house at Maiden Castle, the long houses of the Breme region in Jutland). The roof of the Little Woodbury farm, fifteen metres in diameter, had a square lantern to let the light in and the smoke out. In countries where wood was scarce, drystone houses were built, as in Ireland and the islands off Brittany, stone pillars were used instead of posts (the wheel houses of the Orkneys) and sometimes even megaliths were reused – the dolmens of the island of Geignog-

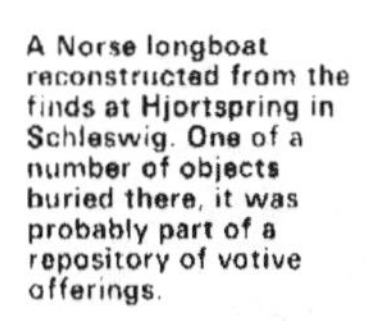

A Norse longboat reconstructed from the finds at Hjortspring in Schleswig. One of a number of objects buried there, it was probably part of a repository of votive offerings.

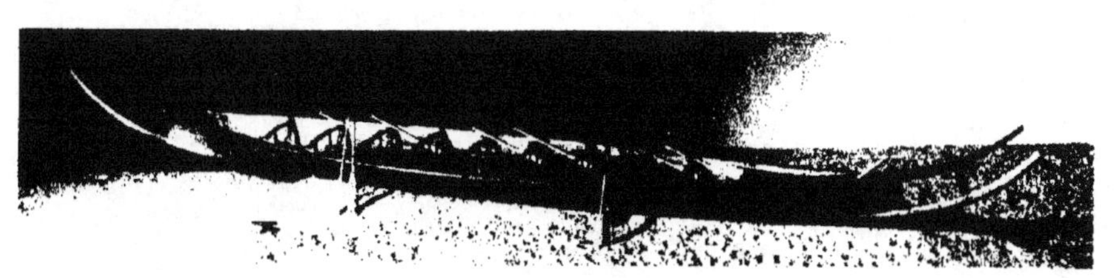

en Landeda in Finistère were converted into houses in La Tène II. Stone foundations and sub basements, either dry or cemented with clay, appeared at the end of the La Tène period (Mont Beuvray; Knelnice in southern Bohemia). The round houses of Glastonbury and Meare in the Somerset marshes stood on artificial platforms of beams, branches and stones laid on the marsh and retained by vertical stakes round the outside. Although the partial sinking of the construction in the ground, sometimes to a depth of half a metre or more, was not invariable it was common in central Europe (Czechoslovakia; Tuchlovice, Sobesuky and Hostosice) and eastern Europe (the Scythian hill fort of Kamenskoie on the Dnieper, the proto Slavic site of Novostroik 240 kilometres east of Kiev). Apart from fortified sites, there were also villages protected by their natural position, and even unprotected villages consisting of a few individual units up to several dozen dwellings; at Chrysanstén in Cornwall several courtyard houses are arranged more or less on each side of a road. The hamlet of Glastonbury, already mentioned, included about sixty cabins built without any apparent plan inside a palisade. In Russia the oblong huts of the proto-Slavic culture of Chernyakovo in the steppe country between the Dniester and Donets were arranged in severa[illegible] rows extending for a kilometre or more. Several isolated rural dwellings are also known. G. Bersu's excavations in Great Britain have produced information about the life of two rural settlements: Little Woodbury on Salisbury Plain (third to first centuries BC) and Ballacagen in the Isle of Man (first century AD).

Agriculture and Industry

Agricultural methods and products Although the Iron Age economy was primarily agricultural, archaeological evidence of agriculture is rare, and traces of this essential activity are hard to find. The peat bogs of northern Europe have yielded primitive wooden ploughs, some of which, from Jutland, have been dated to the pre Roman Iron Age by pollen analysis (ploughs from Hvorslev, Vebbestrup and Dostrup). A number of iron ploughshares have also been found. Rock engravings in Sweden show that these implements were drawn by oxen. Traces of farm workings have been preserved under ancient dunes at Saint Paul de Léon and Penmarch in Brittany, under the banks at the edges of old fields at Gurthian in Cornwall, and even under remains of buildings. There are traces of cross-ploughing under an road house at Alrum in Jutland. The formation of lynchets (small banks enclosing former lots of ground in southern England) has been attributed to these cultivation methods and to the practice of clearing the arable land of stones. The working of these square or rectangular lots, which vary in size from ten acres to half a hectare, can be approximately dated by sherds from the lynchets. At Fyfield near Marlborough these date to the beginning of La Tène.

The varieties of cereals grown can be discovered by studying the impression left by the grain on hand made pottery and by examining preserved calcined grain. For a long time there was only wheat and barley, then rye, which first appeared in Germany in the [illegible]lstatt period and did not become common [illegible] the rest of Europe until Roman times. La[illegible] sickles and scythes were clearly not general[illegible]sed until the end of La Tène; these implements have been found at Stradonice and the excavations of La Thielle in Switzerland. They are probably a little earlier than the reaper of the Treveri described by Pliny, which has been reconstructed from the reliefs of the late Roman period (Buzenol Arlon, Rheims). Post holes and pits in the enclosure of the Little Woodbury farm indicated buildings for parching the corn, a process necessitated by premature harvests caused by bad weather.

Storage pits for provisions and grain silos are common features of many sites, from Scythian forts such as Shirokaya Balka near Olbia, and Nemirov, to continental Celtic oppida and British Iron Age farms. Windmills with sails and turning wheels replaced hand operated mortars for grinding corn at the same time (mills from Hunsbury, Glastonbury and the Celles tumulus). Men and beasts usually lived together under one roof. The animals oxen, sheep, pigs and horses have been identified by the skeletons of sacrificed animals in sanctuaries and the kitchen debris and middens near the dwellings.

Carpenters and wheelwrights Wooden finds from a few exceptionally well preserved sites have provided an insight into the importance of this material in the making of [illegible] household utensils and tools. Br[illegible] has yielded parts of ploughs and [illegible] harrows, troughs and kegs, handles of tools and boats. Apart from various vessels and the remains of several chariots, the votive deposit of La Tène contained an ox yoke, a [illegible] complete with [illegible] shaft and a large [illegible] shield. Among the [illegible] from the villages of Glastonbury and Meare were pails, bowls and basins, a ladder, the supposed remains of a loom, a wheel with twelve spokes attached to a turned hub, and a boat made from a single piece of wood more than six metres long. Spades, forks and rakes were part of the equipment of the Germanic farms of Schleswig and Jutland.

Cooperage, which was regarded as a Gallic invention, its importance in Roman Gaul being attested by sculptures such as the reliefs of Langres and Cabrières d'Aigues, as well as by a few examples such as the barrel in the Mainz museum, originated in the containers made of staves attached to a countersunk base which existed as early as 1000 BC (the pail from

A reconstruction of the *Early Iron Age farm* at Little Woodbury shows the circular farmhouse inside which was a rectangle of posts, *perhaps surrounding* the hearth. The farm was equipped with threshing floors, granary platforms for seed and frames for parching corn, which was then stored in matting-lined pits. A fence with an outer ditch enclosed the farm. Among the animals kept there were cattle, a few pigs and sheep, indicating that the thick forests of earlier times were giving way to open grazing lands.

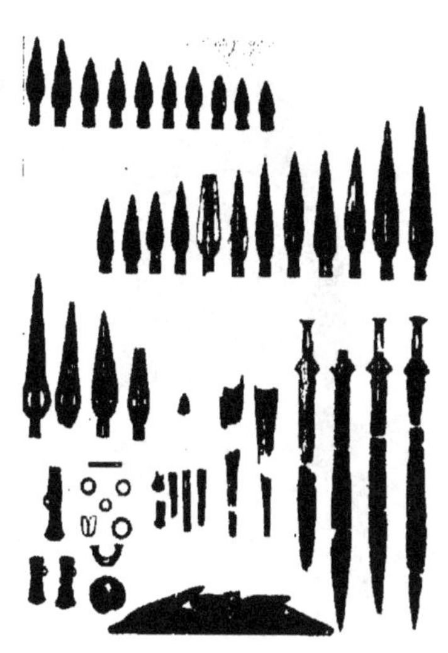

When archaeologists wish to establish a framework for dating a little-known period, the most valuable means at their disposal is a closed hoard in which all the objects were deposited at the same time and never subsequently disturbed, so that the issue is not confused by earlier intrusions or later additions. A hoard of this nature was the cache of Bronze Age weapons, some of which were worn or broken, but all roughly contemporary, found at Wilburton in Cambridgeshire.

Zurich Alpenquai in the National Museum of Switzerland). The use, in the second century BC, of metal hoops to hold the staves instead of wood or wicker hoops is established by the cylindrical buckets from Aylesford and Marlborough which are held by relief decorated bronze bands. Iron hoops were made in Roman Gaul (buckets from Alesia and Châteaumeillant).

Numerous finds confirm the Celtic wheelwright's reputation for skill. Several vehicles have been reconstructed from the metal ornaments from the fifth- and sixth-century chariot tombs; there are Hallstatt four-wheeled wagons from Vix, Ohnenheim and Bell im Hunsrück, and two-wheeled chariots from La Gorge Meillet and Somme-Bionne dating to La Tène I. The finest examples of these parade chariots of Celtic workmanship come from the votive deposit of Dejbjerg in Denmark. Although some of these vehicles must have been drawn by human beings, others were horse drawn, as shown by the decoration of the urns from Oldenburg in Hungary and Sublaines (Indre-et-Loire) and the presence of horses' bits in some of the tombs. Their excellent wheels consisted of a number of spokes—ten, twelve or fourteen—attached to turned hubs and iron-shod (or more rarely bronze-shod) rims. In the late periods (second and first centuries) spoked wheels were found alongside older types of solid wheels made in one or more pieces, which were probably kept for more utilitarian vehicles.

Celtic metal-working While the rare and very limited sources of raw materials (tin from Britain, copper from Ireland and central Europe) condemned many Bronze Age people to a costly import trade, the frequency of veins of iron ore near the surface favoured extensive development of iron working in the seventh century, and it was on this factor that Celtic power was based. Iron technology was introduced in Europe by the people of the Hallstatt culture. Their settlements often coincided with places where there was plenty of ore, and as a general rule the workshops and the mines were close together. The forges of All Canning Cross in England were supplied within a radius of eighteen kilometres. Burgundy, Franche Comté and Berry in France, all of which have abundant iron, have many cemeteries dating to the first Iron Age. The ore was reduced by adding charcoal in furnaces which were often partly sunk into the ground on hills or hillsides for good natural ventilation, sometimes assisted by the use of bellows. These installations were concentrated in the hill forts, or in specialised quarters. At Mont Beuvray in the Côme-Chaudron district the furnaces were grouped together in a single workshop—one building with sides ten metres long contained five of them. More than 1030 La Tène smelting furnaces have been counted at Zaglebia Staropolskia in southern Poland, with huge slag heaps nearby.

The earliest iron objects were modelled on bronze prototypes. The size and shape of the first iron sword were the same as those of the last bronze sword. The La Tène smiths added a mass of tools like those already known to the Hellenic civilisation to the repertoire of Hallstatt iron work, which was limited to a few ornaments such as torques, parts of chariots—hub-caps and wheel rims—and weapons—swords, daggers and spears. By the eve of the Roman conquest the shapes of these tools were perfectly adapted to their functions, and they remained unchanged for centuries to come. They included such agricultural implements as scythes, bill-hooks, hoes and ploughshares, woodworking tools like saws, rasps and files, and leather-working equipment—half-moon knives, paring knives, punches and borers—a fine set of which was found in the Celles (Cantal) tumulus. There were also metal workers' tools like those from Szalacska in Hungary, which included smith's tongs, anvils, hammers, hatchets, gouges, etc., and bundles of tiny goldsmiths' tools (La Tène).

Celtic goldsmiths were no less skilful. They worked in bronze, which was henceforth reserved for luxury goods and ornaments, as well as in precious metals—silver and gold. The smiths in workshops such as that of Mont Beuvray embellished their work with coral and later with coloured enamel, and some of them, for instance the Bituriges, had a high reputation for tinning and silvering. Metals were cast and chiselled or hammered and worked in repoussé. The most important products of Celtic art are goldsmiths' work, e.g. jewellery. There are bronze torques and bracelets from the Marne burials—Jogasse, Courtisols, Leury-le-Repos

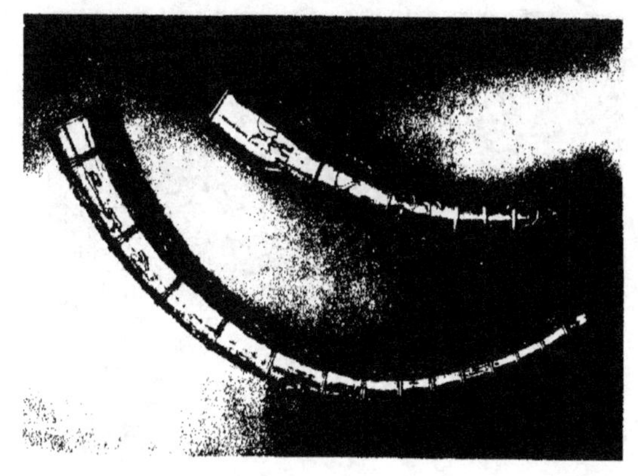

Among the largest and finest golden articles from Denmark are the two drinking horns from Gallehus in Schleswig, dating to the third and fourth centuries AD. Both horns are made with a smooth lining and a relief-worked outer casing decorated with bands of human figures, animals, stars, symbols and fantastic creatures. The runic inscription round the mouth of one of the horns gives the name of the goldsmith who made it.

In Bronze Age Scandinavia, heavy oak coffins were used to bury the dead, a custom which has proved highly advantageous to later archaeologists. The tannin in the wood often preserved the contents so perfectly that even the costumes have survived. Women wore a waist-length sleeved jacket and a short skirt consisting entirely of fringes with a narrow belt clasped by a large circular bronze disc. Men were more substantially clad in a knee-length kirtle strapped round the chest and shoulders with a longer cloak fastened at the neck by an ornamental pin.

etc., gold work from the royal tombs in Germany (Waldalgesheim, Reinheim, etc.), Belgium (Frasnes-les-Buissenal), eastern France (Apremont, Sainte-Colombe) and southern France (Las graisses, Fenouillet, Montans), as well as delicate, complicated fibulae (Vix) or highly elaborate examples (mask fibulae from the Rhineland and central Europe). There were also parts of harnesses—openwork phalerae from Somme-Bionne, enamelled bits and plaques from England and filigree ornamented horse yokes from Hradenin and Lovosice (Bohemia). There were ornaments, engraved helmets decorated with coral from the Marne tombs, a gold plated helmet from Amfréville, incised sword sheaths (England, Champagne, Switzerland and Bohemia) and shields from England (Whitham and Battersea).

The remarkable flowering of Celtic coinage, which began by imitating the Macedonian staters of Philip II and the coins of the Greek Mediterranean colonies (Thasos, Rhodes and Marseilles), was a result of the skill of these smiths, particularly those of Gaul.

The salt industry

The importance of salt in the ancient world is mentioned by a number of authors, beginning with Homer. Although a few of them (e.g. the older Pliny and Varro) supply details of the technical processes, it was left to archaeology to produce material evidence of the very early date of the 'industrial' working of this precious commodity.

The Hallstatt and Hallein salt mines When new salt pits were opened at Hallstatt in 1832 and 1836, shortly before the excavation of the famous necropolis, traces of very early workings were found. With the material discovered soon afterwards in the neighbouring Hallein (Dürrnberg) mines, they furnished a clear picture of extraction procedures.

Picks made of deer antlers, bronze (with flanges) and iron showed that mining had begun in the Neolithic period and continued throughout the metal ages. Pits were sunk into the veins of salt at an angle varying from twenty-five to sixty degrees; these could reach a length of up to 350 metres. Ascent and descent were effected by means of crude ladders, notched tree trunks or pieces of wood fitted across the tunnels. Narrow galleries followed the veins of salt, widening along with them to form veritable chambers. The salt, which naturally coagulated round the objects, preserved intact the pinewood torches attached to the walls, the stone and wooden wedges, the tool handles (picks and hammers) which were often broken in the process of extracting and shaping the blocks of salt, and the pails which were used to fill the goat-skin baskets strengthened with wooden laths employed for carrying the salt out of the mine. The remains of the woodwork which shored up the galleries has also been found, as well as the carpenters' hatchets and many fragments of leather, wood and linen cloth, and even parts of garments: leather caps, shoes and gloves. It is not known whether the salt was refined after mining, but it is certain that it was the object of important trading operations for the Celts from the Hallstatt period to the first

The earliest Iron Age European culture is known as the Hallstatt period, from the rich settlement found at *Hallstatt in Austria*. In addition to the iron artifacts, which marked the beginning of a new epoch in central Europe, some very fine bronzes were found in graves. A large bronze bowl with incised decoration has an unusual handle in the form of a cow, with a larger version of the same animal immediately above, its forelegs supported on a metal bar attached to the middle of the bowl

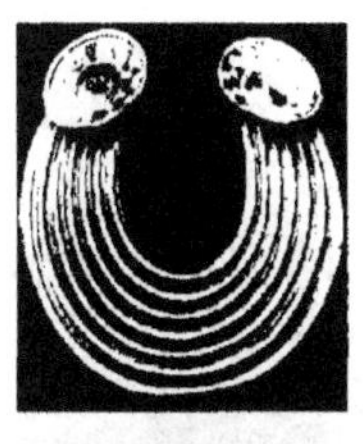

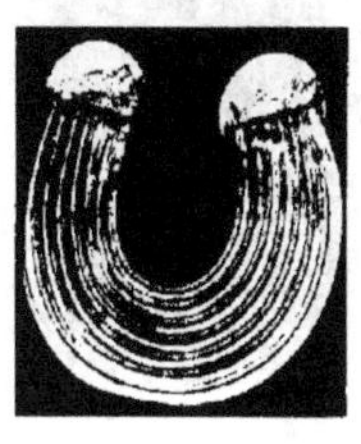

In the Early Bronze Age the main cultural movement was from Britain to Ireland, except for the craft of metal working, in which the Irish smiths had already established their superiority. The favourite ornament of this period, and certainly the most spectacular, was the flat, crescent-shaped collar of embossed sheet gold known as the lunula. A particularly fine example from Glenisheen, County Clare, is now to be seen in the British Museum.

Iron Age and the beginning of the second. This is proved by the richness of their tombs, with their imported Greek and Italian goods.

The processing of salt water People have long been intrigued by vast heaps of terracotta fragments and wood and charcoal ash near salt water springs or the sea shore. They were unable to suggest a satisfactory explanation. In 1740, for instance, Le Royer d'Artize de La Sauvagère believed that the 'brickwork' of Marsal in Lorraine was Roman road building. It was actually a dump of debris, estimated at nearly three million cubic metres, from the installations for processing the water from the Seille valley springs. Similar remains existed in Germany and Austria at Frankenhausen, Halle-Giebichenstein, Bad Nauheim, Schwäbisch Hall, Dammwiese near Hallstatt; in France at Vic, Saint Pierre sous Vézelay, and in the Cambridge fens in England. Others have been located on the coasts of the Atlantic, the English Channel and the North Sea. The best known are those of the eastern and southern coasts of England – the 'Red Hills' of Dorset, Essex and Lincolnshire – La Panne in Belgium and Nalliers (Vendée). Systematic exploration of the French coast has shown that there were many of these installations. Y. Coppens identified forty-two such sites in the Morbihan department alone. At the other end of Europe similar investigations are being conducted at Gruzii on the shores of the Black Sea.

The outlines of the process have been largely reconstructed from the finds, and seem to have been as follows: clay cylinders set up here and there on an area of bricks or beaten earth served as stands for shallow vessels of crudely fired pottery into which the salt water was poured. Fires were lit between the cylinders, causing water to evaporate, leaving a deposit of moist salt. This was then packed into containers of porous clay – goblets which were cup-shaped in the Bronze Age and truncated cone shaped in the Iron Age, or rectangular troughs sometimes divided into compartments – and finally dried near a hearth. The water from the salt springs was drawn from the wells through pipes of hollowed oak tree trunks (Saint-Pierre sous Vézelay) and conducted through open or closed wooden channels to wooden basins lined with clay, where it could be distilled. Salt was also gathered on the beach at low tide and dissolved in sea water which, after filtering and evaporation, also produced a comparatively pure product. The majority of these installations date to the Iron Age and the Roman period. Some of the German examples (Halle-Gebichenstein) go back to the Bronze Age.

Contacts and exchanges

The amber routes Amber from the North Sea (western Jutland) and Baltic coasts (Prussia) differs from southern amber by its high succinic acid content (six to eight per cent). Intermittently attested from the Paleolithic onwards in central Europe and from the Neolithic in France and Spain, it does not seem to have been the object of regular trading before the Bronze Age. It circulated in its crude form of drops and small pieces, which are found in groups; it was not worked until it arrived at its destination, when it was made into beads, pendants and various other ornaments.

The main trade routes crossed central Germany – the Weser valley, the Elbe valley and down the Moldau – Moravia and Austria, crossed the Alps by the Brenner Pass and met at the river Po and the north coast of the Adriatic before going on to the Peloponnese – Mycenae and Kakovatos – and Crete by sea. A secondary land route probably came down the Danube through the Balkans and another, starting in Germany (Saxe-Thuringia), turned towards England (Wessex) and Brittany (Morbihan). There are striking parallels between objects from England, Germany and Mycenae – amber plaques with converging perforations, amber pendants in the shape of miniature bronze halberds – which suggest that the Unetice metal workers were the central European middlemen in the metal-amber trade between the south, north and west in the second millennium.

Jutland amber was succeeded by the Baltic coast product in the Iron Age without any diminution in the trade, judging by the necklaces and beads from the Hallstatt cemetery and the great La Tène III deposits of crude amber found at Staré Hradisko in Moravia and a hoard weighing 2750 kilogrammes from Wroclaw-Parsenice in southern Poland.

In the Illyrian Hallstatt cemeteries beads of many different shapes were associated with discs incised with geometric patterns (Jezerine in Bosnia) or carved pendants (human faces from Kompolje in Croatia). In Celtic territory

One of the most fascinating and mysterious survivals of Celtic ritual art is the Gundestrup cauldron, a silver vessel found near Aalborg in Jutland in 1891. Gilded silver plaques covered with fantastic figures adorn both the inside and the outside of the piece, setting a problem of interpretation which has not yet been fully solved. One of the inside plaques shows a horned god who is holding in his right hand one of the torques which were so popular among the Celts.

amber beads, either singly or in necklaces, were found among glass beads, as in the necklaces from the Payre-Haute burial in Hautes-Alpes.

An eastern amber route between the Baltic and the Black Sea is attested by a few finds from the Vistula and Dniester valleys.

The Celts and wine The chief evidence of interchanges between the barbarian and civilised countries in the Iron Age is provided by the large numbers of products of Greek and Italian industry found in the territories occupied by the Celts. Exports of raw materials, particularly tin, from Britain (Cornwall) were exchanged for imports of southern wines, bronze and ceramic wine vessels, and brilliant and colourful products such as glass, coral and even ivory. The development of the Greek colonies, especially the Ionian settlements in the western Mediterranean, increased opportunities for contacts and trade. The intensity of trade and its chronological development can be estimated from the imported pottery in the local warehouses and neighbouring shops. F. Villard has shown that sherds of Attic vases at Marseilles touched a record figure at the end of the sixth century BC and diminished abruptly between 500 and 450 BC as if a period of economic depression had set in after a period of prosperity in the Phoenician colony. Thanks to recent German and French archaeological research the trade routes and their vicissitudes are beginning to be better understood.

From the seventh century onwards the Celts acquired Etruscan products, such as the corrugated cist from Magny-Lambert and the pyxis from Apponwihr near Colmar, and Greek goods, for example the Rhodian oenochoae from Vilsingen and Kappel am Rhein by the Alpine ways and perhaps by the old amber route. In the subsequent centuries these imports had to answer increased demands from Burgundy to Bavaria. The foothills of the Alps were always busy, as can be seen from the Illyro-Venetic finds from the Hallstatt necropolis and the Etruscan oenochoae from the Tessin cemetery, but another route had been found. After the founding of Marseilles about 600 BC, Massaliot traders, exploring the possibilities of the hinterland, set up relay stations and trading posts on the hills not far from the banks of the Rhône (Malpas, Le Pègue, Roquemaure) and extended their sphere of operations as far as Burgundy and Franche Comté. The early stages of the route are probably responsible for the finds of imported eastern Greek pottery – Ionian amphorae from Mercey – and Aeolian wares, which then gave rise to pottery made in Marseilles in imitation of Ionian vases of the second stage, e.g. plates with painted bands and amphorae (Mantoche, Mont Lassois). The great Celtic 'princely' tombs are contemporary with the peak period of Massaliot trade in the second half of the sixth century, but it is not necessarily responsible for all the Greek and Etruscan imports which enriched the graves of the Late Hallstatt period. Mediterranean influences were very strong north of the Alps at this time. The brick fortress of Heuneburg IV with its square bastions was perhaps modelled on Greek buildings, and the distribution of the finds indicates that trade between north Italy and Germany along the Alpine passes never slackened.

Massaliot trade received a setback after 500 BC from the Gallic invasions in the Rhône valley. Etruscan commerce, on the other hand, increased. Their outlets extended from Bohemia to Champagne with the spread of the Celtic La Tène culture, and reached Scandinavia and the British Isles in the north (imports of oenochoae, amphorae and other wine containers). Marseilles did not become active again until the end of the fourth century. The traditions recorded by the ancient authorities, Strabo and Diodorus Siculus, on the transporting of tin across Gaul to the Mediterranean, and the story of Pytheas' voyage of exploration along the Atlantic coast shortly before 300 BC belong to this period. The gold stater of Cyrene, struck between 322 and 313 BC, which was thrown up by the sea on the coast of Finistère at Lampaul Ploudal Mévrau, still 'caught in the grip of a deep water seaweed' (P. R. Giot), could be evidence of one of those Greek and Massaliot voyages.

Imports of Greek pottery in a few Languedoc sites in southern Gaul were still common in the late fifth century, for instance the Attic vases from the cemetery of Ensérune, but they were succeeded in the fourth century by Italian wares – Campanian and then Arretine pottery, which was also introduced into the interior by trading. In the strongly Gallic places (Le Beuvray, Essalois, Châteaumeillant, Basle) amphorae exactly like those from the wrecks on the Provençal coast show that Gaul bought lavishly of south Italian wines at the end of the independent period until the establishment of the Gallo-Roman vineyards, greatly to the profit

A magnificent Celtic shield, probably lost during a river crossing, was found in the river Thames near Battersea. It is an elongated oval in shape, with three conjoined circles of decoration carried out in relief work and highlighted with red enamel. The restraint, elegance and flowing line of the ornament is typical of Celtic workmanship at its best.

In the Late Bronze Age and Early Iron Age, sacrificial offerings of costly articles were made by dismantling the object and throwing it into a pool or bog. By this means a magnificent four-wheeled wagon was preserved at Dejbjerg in Jutland. The scattered pieces have all been reassembled, and it can be seen that the cart was richly decorated with bronze inlays in a local variation of the La Tène style. The rims of the wheels are made from a single piece of wood with an iron tyre for protection.

A strange bronze object from Balkaakra, near Ystand in Sweden, once believed to be an altar, is now thought to be a ceremonial drum. It is much earlier than the introduction of Christianity but apart from this its date is uncertain. The shape and the ten wheel-like supports suggest a form of sun-symbolism, but its precise use and significance are not known.

During the megalithic period of the Bronze Age, burials in the Hebrides were usually in the form of a passage and chamber covered by a cairn of boulders. Some of the outer stones of the cairn are occasionally taller than the cairn itself, and in the case of the tomb at Callanish on the island of Lewis, the uprights stand well away from the outer circumference of the mound, as if they had an independent function. A large avenue extends northwards from the circle and other alignments of stones radiate towards the other compass points.

of the Italian traders and also, no doubt, of their Massaliot agents.

Gods and the dead

Sanctuaries The evidence of the ancient authorities such as Lucan and Caesar leaves no doubt of the existence of *loci consecrati* (sacred places), woods and open-air enclosures where the Celts gathered to sacrifice to their gods, but their exact nature is not explained.

The location and arrangement of these favoured places was firmly dictated by a long tradition. It is accepted that the 'henge' type of monument, such as Stonehenge, was still a cult centre in the Iron Age. The great enclosure of Goloring between Coblenz and Mainz dated to the end of the Hallstatt period (fourth century BC) recalls these henges, with its circular bank measuring 200 metres in diameter reinforced on the inside by a wide ditch, and its central terrace.

A well preserved example of a Celtic sanctuary was discovered in 1959 during the excavations of the archaeological institute of the Czechoslovakia Academy of Sciences at Libenice near Kolin in Bohemia. A ditch reinforced on the outside by an earth bank bounded an oblong enclosure (seventy-four by twenty metres) with a female burial in the middle and a half buried place for sacrifices in the south-east. To begin with this quarter consisted of two unequal and overlapping ovals. A partially smoothed stela stood on a stone base in the smaller, as in an apse. At the point where the ovals intersected a post hole on a similar base marked the centre of a circular paved area, and near to this, on the north, two torques (rigid neckrings, symbols of the deity) of twisted bronze lay on the ground near the holes for two maple wood posts, to which they were once probably attached. Human remains (one adult and four children) and those of a dog and several bulls were buried in the earthen walls. Officiators and victims approached the sanctuary by a ramp between two stone walls. These installations date to the Late Hallstatt and Early La Tène periods. The tomb in the centre, dated to the third century BC by the fibulae pinning the dead woman's shroud, belongs to a second phase of occupation of the sacred enclosure. Another even larger complex is known in Bohemia at Msecke Zehrobice; this yielded the stone head of a god, or 'Celtic hero'.

Worship of the god and of the deified dead seem to have been associated with each other, or even intermingled in Gaul in the La Tène period. In the Marne district (Fère Champenois, Vert-La-Gravelle etc.) the tombs are grouped near or within the square enclosure bounded by ditches. In the middle of one of these enclosures at Ecury-le-Repos a shelter indicated by four post holes probably housed a stone or wooden image of the god, the position of which is marked by a fifth oval post hole. These enclosures were used right up to the Roman conquest.

The prototype of a sanctuary which became very widespread in Gallic territory appeared in the Roman period. It included a *cella* which was surrounded or simply preceded by a portico, and often set inside a vast square court flanked by a gallery. Remains of such wooden temples dating to the end of the La Tène period have been found under the foundations of the stone buildings of the same type which followed them, as at Saint Germain-le Rocheux (Côte d'Or), Mont Donon and the sacred city of Altbachthal at Trier, Germany.

Also Celtic but very different are the sanctuaries of the sites which mark out the Gaulish section of the ancient route from Italy to Spain between Aix-en-Provence and Montpellier. At a very early stage Hellenic influence profoundly affected this territory, which had been the lands of the Ligurii, who were conquered by the Celts in the fourth century BC. Mutilated and frequently reused fragments of buildings, destroyed in the Roman conquest between 150 and 120 BC evidence an interest in architecture which is unique in the Celtic world. Excavations at Gérin-Ricard have revealed the plan of a sanctuary established about 300 BC in a small rocky amphitheatre on a hillside at Roquepertuse (Velaux, Bouches-du-Rhône). The construction, which consisted of two successive terraces preceded by a roofed staircase with five steps, was entered from the front by a monumental portico. The square pillars of the *propylaea*, which were once painted with polychrome geometric and animal motifs, and later carved with sockets shaped like heads for displaying the skulls of conquered enemies, supported a lintel adorned with incised horses' heads and surmounted by a free-standing figure

of a bird on the point of flight. Inside were lines of human figures - perhaps gods, dead heroes or priests - double headed 'herms' and statues of cross-legged seated figures with their feet tucked under their thighs.

In the ancient Salyian capital of Entremont near Aix-en-Provence, a walk nearly 100 metres long, a true sacred way flanked by crouching figures led to the sanctuary, which housed many statues of half-naked warriors. The pillars of the portico were carved on three sides, with figures of armed horsemen and 'nude corpses in heroic poses' on the front face and piles of 'severed heads' on the sides. The existence of sanctuaries where the 'cults of the dead and the divinity were intermingled' in most of the oppida of the district is attested by numerous fragments of buildings and sculptures; there were friezes and stelae from Nouries, a lintel from Nages, statues of a lion from Les Baux, a man-eating monster from Noves and warriors from Sainte-Anastasie and Grézan, and pillars with niches and the 'crouching god' from Saint-Rémy-de-Provence.

The outer limits of Roman expansion are traced by the grave-goods buried with colonists. A Roman burial of Iron Age date at Bulbjerg in east Jutland contains clay vessels and a number of animal bones which suggest that a feast in honour of the dead man was held before the grave was filled in.

Representations of the gods

How did the people of barbarian Europe see their gods? Lucan mentions the 'sorry and inartistic' Celtic representations of gods: a Scandinavian saga refers to wooden figures of the Germanic god Freyr and his consort, the *Chronicle of Kiev* tells how Prince Vladimir had a wooden idol of the god Pe[illegible] with a silver head and a golden beard erected near his palace. Archaeological finds indicate that wooden posts and stones, wh[illegible] up in the sanctuaries as anicon[illegible]bols of the gods did not attain a human form until very late.

Some rare examples of wooden statues, the date and sacred nature of which is not always definitely established, have been preserved in exceptionally favourable places where conditions are humid. In the British Isles human figures from Dagenham in Essex, Shercock in Ireland and Ballaculish in Scotland, which vary in height from 45 centimetres to 1·5 metres, date at the earliest to the sixth century BC. They could be contemporary with the ithyphallic male figure found in the Brodbenbjerg bog near Viborg in Denmark and attributed to the Scandinavian Bronze Age. Just as primitive in technique, although they are much later in date (the end of the La Tène period) are the statue from Possendorf (Thuringia) and the couple from Braak (Holstein) near to which a heap of stones mixed with wood ash, cinders and sherds indicated rudimentary cult places. The large figure from Genfersee in Switzerland and the *ex votos* from Montbouy (Loiret) and the source of the Seine show that wooden religious carvings in the traditional rustic form but affected to varying degrees by classical influences persisted until the Roman period.

Celtic religious carvings in stone indicate by their proportions and their almost unworked surface that they belong to the same tradition as the 'menhirs'. They are either tentative attempts at the human form or else they are decorated in the La Tène style, with curves springing from one another and enclosing human masks and animal figures in their development. These two types of monument are evenly divided between south west Germany and Ireland. The presence of some of these stones in German burials indicates that they represent mythical ancestors or dead heroes. The fragment of a crudely roughed out figure from the tumulus at Stockach near Tubingen was associated with an Early Hallstatt cremation. A large free-standing carving of a helmeted,

An aerial view of the huge megalithic monument of Stonehenge on Salisbury Plain. The whole creation is not only an amazing feat of engineering; it indicates a surprising degree of political unity without which the work could never have been carried out.

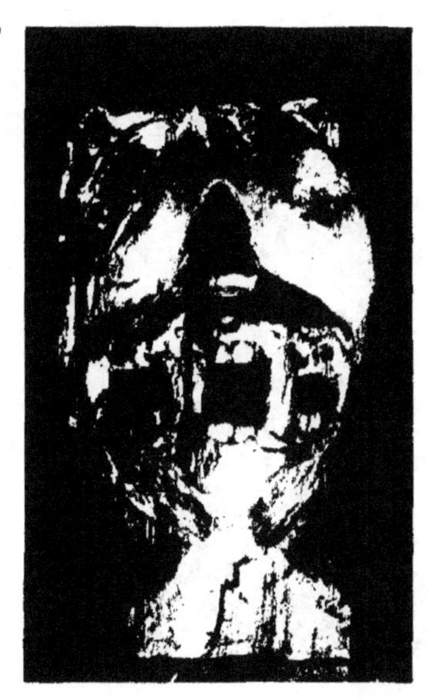

It is not very long since the British used to pray regularly for deliverance 'from the terror of the Norsemen', whose savage attacks spread a trail of death and destruction wherever they struck. The uncompromising ferocity of the Viking character is clearly expressed in their art, which invests even a domestic object like one of the Oseberg sledges with a demonic aspect by the addition of brutal carved heads on the posts.

armed and probably masked warrior surmounted in heroic nudity a burial mound in Hirschland, near Stuttgart. All the latest tombs here are contemporary with the beginning of La Tène. The janus-headed figure from Holzgerlingen, the Roquepertuse herms and the incomplete figure from Waldenbuch, whose ornamental scabbard relates him to the pyramidal pillar of Pfalzfeld, all belong to this same group. On each of the four faces of the last-named monument S-shaped plant motifs twine together, symmetrically framing a human face crowned with two large foils. The whole of the upper part is missing, but probably, like the Heidelberg head, topped by a head carved in the round showing the same stylisation as the masks.

In the Irish group there are also baetyls covered with curvilinear reliefs (the stones of Turoe and Killycluggin), and statues. These are in a strange crude style related by their shape (double-headed from Boa Island) and their pose (arms folded across the chest from Tandragee in Armagh) to the continental figures.

Apart from the Celtic-Ligurian temple carvings, a few late figures in Gaul dating from the end of the independent period and the early Roman empire should be mentioned. They include the 'gods with the torques' made of stone (Luffigneix, Rodez, Halle, Châteaumeillant) or repoussé bronze (Bouray), and the beaten or cast metal masks (Compiègne, Senlis, Notre Dame d'Alençon, Tarbes) attached to the wooden idols.

Period		Contemporary
BRONZE AGE	UNETICE, Adlerberg, Saxe Thuringia, Straubing	1750–1550 Cretan civilisation
Early (1800–1400)	Mainly flat tombs, a few barrows. Bronzes: flat axes, followed by flanged axes, flat triangular daggers, halberds, spiral ended torques, ingots in the form of curved bars. Pottery of the bell beaker type WESSEX. Individual burials under a barrow. Circular 'henge type' sanctuaries: Stonehenge, Avebury, Woodbury. Bronzes: tanged daggers, halberds. Imported Nordic amber (beads, miniature halberds, gold plated discs), Irish gold (lunulae), and Egyptian faience beads. ARMORICA. Barrows with burial chamber. Finely worked stone tools. Bronzes: swords and daggers, arrows, flat and flanged axes	
Middle (1500–1100)	NORDIC BRONZE AGE I. Flint imitations of bronze objects. Unetice-type bronzes. II. Many barrows with oak-wood coffins with well preserved organic material and corpses. Bronzes: daggers, straight-edged swords, axes with very narrow butts, belt discs with a point in the centre. Fibulae made in two pieces	1400–1200 Mycenaean civilisation
	Barrows, inhumations followed by incinerations. Bronzes: wide-shouldered swords followed by pointed swords, axes with flanges and butts, pins with bi-conical or trumpet shaped heads. Pottery with incised geometric patterns (Haguenau, Villhonneur)	1250 Dorian invasion in Greece
Late *(1200–700)* HA I–III	URNFIELD CULTURE *I. Inhumations and incinerations under a barrow.* Bronzes: straight sided swords (Rixheim type), knives cast in one piece with a ringed handle and hammered-back edges (Mels type). Collared poppyhead pins II and III. Incineration cemeteries (large bi-conical urns) with flat tombs. Burnished pottery decorated with rilling, *excision and applied ornaments. Bronzes:* swords with full grips (Moringen type), two-edged razors, knives with carved backs NORDIC BRONZE AGE. Incinerations. Many votive deposits. Bronzes: swords with massive decorated pommels; triangular razors, large serpentine trumpets (lurs)	1100 *Foundation of* Bologna. Villanovan civilisation

Sacrifices and offerings

Ancient authorities are unanimous in describing two religious practices common to most European peoples, but particularly the Celts and the Germans. These were bloodthirsty human and animal sacrifices and the depositing of offerings in sacred places, such as enclosures, ponds and streams. Although a similar interpretation is unavoidable in some cases (e.g. Libenice), it is sometimes difficult to tell the victim of a sacrifice from that of a murder or judicial execution, or a votive deposit from an itinerant craftsman's hoard.

Corpses preserved intact in the Danish and north German peat bogs raise a similar question. Tollund man still bears round his neck the leather thong with which he was strangled or hanged, and the man from the village of Grauballe had a deep wound in his throat. They had both been subjected to the penalty which Tacitus says was reserved for adulterers. But it is difficult to guess what could have been the crime of the partly scalped fourteen-year-old girl with bandaged eyes in the Schleswig Museum. Perhaps she was a sacrificial offering to some local god.

How are we to justify the votive character attributed to many of the deposits found in marshes and streams? The weapons and tools from La Tène and Port (Berne) were associated with human bones and the remains of a wooden platform from which they were thrown into the

Viking art, like nature, seems to abhor a vacuum. Every conceivable space is filled with busy, coiling garlands of interlaced carving, often carried out with masterly skill. The body of the Oseberg wagon rests on two curved beams which join it to the axles, and on the top of each beam is a fierce bearded head with the bared teeth and staring eyes typical of Norse art.

Period		Contemporary
	ATLANTIC BRONZE AGE. Incinerations. Many hoards of bronzes ([illegible] with pistillate and carp's tongue blades, axes with shafts and square sockets	753(?) Foundation of Rome
IRON AGE I		
Early Hallstatt U.F. IV (725–625)	Urnfield type incinerations and inhumations under barrows. Bronze swords, long iron swords. Imported Etruscan objects. 'Thraco-Cimmerian' equipment (bits)	c. 600 Foundation of Marseilles 688 Foundation of the Rhodian colony at Gela
Middle Hallstatt (625–540)	Inhumations under a barrow. Iron antennae swords, crossbow fibulae, boat fibulae and *sanguisuga* fibulae, metal situlae imitating pottery models, balloon-shaped pedestal vases. Imported Rhodian oenochoae and Italian bronzes	
Late Hallstatt (540–450)	Inhumations under barrows. Four-wheeled chariot tombs. Iron swords with kidney-shaped pommels, daggers with anchor-shaped chapes. Serpentine fibulae with disc catch plates, cup-shaped fibulae and fibulae with long springs and internal chords. Painted pottery (Vix) and vases with graphite decoration. Imported Greek bronzes and pottery, gold, amber and coral	530–500 Attic Black Figure vases 500 Attic Red Figure vases
IRON AGE II		
Early La Tène (500–250)	*Burial under barrows and in flat tombs.* Two-wheeled chariot tombs. Iron short sword. Fibulae with turned-up foot, short spring and external chord. Marne-type pottery with angular profiles. Gold and bronze jewellery. Imported Etruscan oenochoae with trefoil mouths. The 'severe' style of Celtic art (friezes of symmetrical motifs) followed by 'free' style (*curvilinear motifs*)	445–430 Building of the Parthenon 387 Rome captured by the Gauls 278 The Celts at Delphi
Middle La Tène (250–120)	Burials in flat tombs. Iron swords with bell-shaped ferrules. Fibulae with flattened bows and the foot attached to the bow. 'Plastic' style of Celtic art (reliefs and knobs forming human masks and animal shapes). *First Celtic coins and carvings in the round.* Imported Campanian ware	400–250 [illegible]
Late La Tène (120–30)	Incinerations. Fortifications of the *Murus Gallicus* type. Long iron swords with straight ferrules. Fibulae with strongly flattened bow and internal chords. Nauheim fibulae. Terracotta ox-head firedogs. Imported Italian wines and amphorae and Arretine pottery	118 Roman colony at Narbonne 118–113 Invasion of the Cimbri and Teutones 58–53 Romans in Gaul, Britain and Germany

Opposite page: Toltec figures at Tula. The Toltec culture dominated southern Mexico before the coming of the Aztecs, and its influence on the late Maya culture of Chichen Itza is clearly discernible. These gigantic figures may have supported the entrance to a temple of Quetzalcoatl. Eighth to twelfth centuries AD.

water. At Llyn Cerrig Bach in Wales animal skeletons, probably sacrifices, accompanied a large quantity of finds such as fragments of trumpets, cauldrons, chariot gear, tools and slaves' chains from the two centuries before the Roman conquest of Britain, as if the deposit had been made up of offerings repeated over a long period. The sacred nature of the Scottish deposit of Carlingwark and Blackburn Hill (first and second century AD) can only be confirmed by the fact that they had a cauldron for a container in view of the leading part played by this vessel in the mythology of Celtic Britain. Cauldrons also appear among the innumerable objects deposited in the Danish peat-bogs during the Iron Age. These include the Skallerup cauldron, which was mounted on wheels, the Braa cauldron, which had an iron rim and a decoration of bronze ox and owl heads, and the silver Gundestrup cauldron, the dating and ornamentation of which have roused so much controversy, although its religious function has never been doubted.

Like the Gundestrup cauldron, the little Trundholm chariot mounted on six wheels with a bronze horse drawing a solar disc, and the Dejbjerg chariots must have been intended from the first for religious ceremonies. Scandinavian archaeologists assign another meaning to the deposit from Hjortspring on the island of Als in Schleswig (about 200 BC), in which coats of mail, wooden shields, spears, wooden vessels, the skeletons of a dog and a horse and bones of oxen, sheep and pigs were heaped round a large ship. These were thank offerings to a god for a military victory.

Cemeteries and tombs

Man's equivocal attitude towards his dead has always been dictated by two contradictory needs: that of ensuring the survival of the soul and that of eliminating any evil influences which these spirits might exert on the living. The former explains the precious vessels from the Scythian and Celtic 'royal' tombs as well as the tools buried with a Gallic master smith at Celles (Cantal). The latter explains why the dead were bound before burial and some warriors were armed with severely damaged swords and spears (Celtic Iron Age II tombs).

The early peoples of Europe buried or cremated their dead without a change in the ritual, unmistakably betraying the evolution of their religious beliefs. Some of them practised only inhumation: Early Bronze Age cemeteries (Wessex, Alderberg, Straubing), Nordic Bronze Age cemeteries (the Jutland tumulus of the second phase), and La Tène I and II (Marne and Bohemia). Others practised cremation (Urnfield cultures, Gallic tombs of La Tène III, German and Slavic Iron Age burials). These two burial practices were sometimes found together, for instance in the necropolis of Polepy near Kolin in Bohemia, which belongs to the Unetice culture of the early Bronze Age, female burials sometimes contain the ashes of infants in urns. Numerous Hallstatt barrows in Gaul contained both the remains of cremations and skeletons (Burgundy and Berry).

The beliefs which led these people to bury the remains of their dead in pits in the ground (flat tombs) or to cover them with a mound of earth or stones (tumuli or barrows) are no better understood. The use of the barrow became the distinctive feature of central European Middle Bronze Age civilisation (Hügelgräber Kultur). It spread to all the peoples of Europe (Celts, German and Scythians), surviving until various dates in the Iron Age and some places, even to the Roman period (Trevires, Lemovices, Belgium). Flat tombs, which belong to a few cultures of the Early (Unetice) and Late (Urnfield) Bronze Age, superseded barrows among the Celtic people from La Tène III (dated about 300 BC).

The tombs were isolated from the world outside, sometimes by means of a wooden palisade, as in the bell-barrows in Holland and the Slavic barrows of the historic period at Borshevo near Voronezh, Soviet Union, or by ditches, as in the Early Bronze Age barrows in Wessex, the Celtic tombs in Champagne (Berry) and Slovakia at Trnovek and Holiare. This isolation caused the tomb to become a kind of sacred place, and this encouraged the grouping together of tombs: there are cemeteries with enclosures at Mulheim in the Rhineland, at Fcury le Repos in Champagne, and in Yorkshire, England.

The great Viking ship burials were furnished with every luxury the dead man could need, including a large quantity of very fine fabrics. Rich bedding, garments and wall tapestries were worked in gold thread, embroidery and appliqué designs in coloured silk. Pictorial scenes and abstract patterns varied the effect, which can be appreciated from a length of reconstructed tapestry in the University Historical Museum, Oslo.

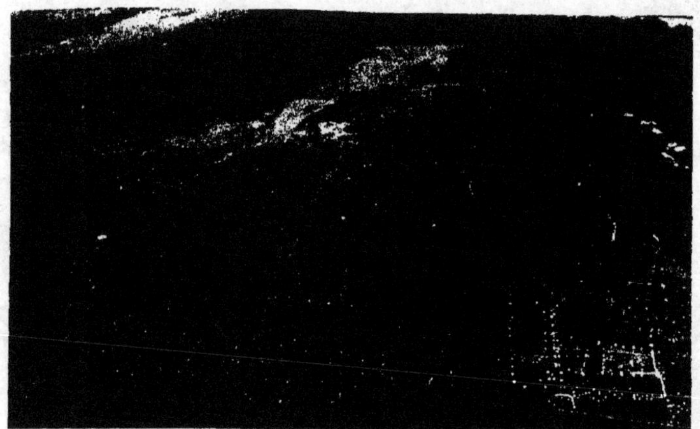

One of the most prevalent beliefs of the ancient world, occurring from Egypt to Scandinavia, is that the souls of the dead have to make a voyage across the sea. At Linholm Hills in Denmark, this belief was expressed by burying the dead in a grave monument consisting of a symbolic ship outlined in stones, a practice which was adopted in the Late Bronze Age and continued through the Iron Age almost until Christian times.

These cemeteries include dozens or even hundreds of tombs. Thousands of barrows are studded about the Russian plains. The development of the cemeteries has been traced by means of a chronological classification of the grave goods. Müller Karpe, for instance, has shown that the Kelheim urnfields spread from east to west between 1000 and 700 BC, and by means of a genuine 'horizontal stratigraphical investigation' in four stages. Hodson has demonstrated the continual shift of the Münsingen cemetery in Switzerland during La Tène I and II. Later burials in barrows were often added to the original tomb, and it is difficult to make out their sequence. The mound itself was sometimes enlarged for this purpose (Lantilly in Côte d'Or). Laid on the bare ground (Iron Age cremations deposits in Sweden and Denmark) or more commonly in a metal or pottery vessel, cremations were also grouped together in cemeteries. At Aylsford in Kent the pits which contained them were arranged in a circle (second century BC).

Some practices leave no doubt that the tomb was regarded as a dwelling for the dead. The Early Bronze Age barrows of Leubingen and Helmsdorf (*Thuringia*) concealed veritable houses of oak beams, with entrances and paved floors. The same concept dictated the construction of the wooden burial chambers of the Celtic rulers of the sixth and fifth centuries (Vix, Hochmichele) and the sumptuously painted and draped tombs of the Scythians and the related peoples (sixth to fourth centuries) of south Russia (Chertomlyk), Kuban (Kostromskaya) and as far as the Altai (Pazyryk). In the pre-Roman Iron Age in Denmark the excavations at Mandhoej in Bornholm show that a temporary conical hut above the incineration tombs was burnt down and the whole complex covered by a small mound. Many corpses were laid straight in the earth in the extended or contracted position, with no protection at all, or placed in crude boxes made of stone slabs. Pits were also lined with boards as at Libenice, or logs, as at Hirschland, Tomb 5. Bronze Age Danish barrows contained large coffins (external dimensions, 2.5 to 3.1 metres) carved from the trunk of an oak tree cut in half lengthwise. Some Scythian coffins were plated with valuable ornaments, such as the ivory plaques of Kul Oba in the Crimea.

The dead were generally *buried in their* clothes, but only the metal parts usually survive, such as fibulae, chains, belt buckles, jewellery and the warriors' weapons. Well preserved woollen garments, boxes and sheaths of wood have been found in Danish Bronze Age tombs. The garments destroyed by fire in incineration burials are evoked by fibulae or a few other accessories. Votive objects (late Bronze Age miniature swords in Denmark) might replace *weapons and jewels*.

The dead were provided with food – joints of meat, usually pork, boar or beef – and drink, with *metal and pottery vessels*. Among the Scythians, slaves and animals were sacrificed for the master's use. In the royal tomb of Chertomlyk five grooms were lying with their feet towards the chief burial, ready to use again. At Kuban twenty-two dead horses surrounded the burial chamber of Kostromskaya, and there were four hundred in a single tomb at Ulski. Among the Celts the custom of placing the bodies of important people such as chiefs (Vix, Vilsingen etc.) or priests (Ohnenheim) on a partially dis*mantled chariot, which also occurs among other* peoples, covered an enormous region stretching from Bohemia (Hradenin) to Gaul (Burgundy, Franche Comté and Poitou) at the end of the Hallstatt period (sixth century) and spread to Champagne (the regions of Epernay and Châlons-sur-Marne) and England (the Arras barrow in Yorkshire). Their wealth has caused these Celtic and Germanic tombs to be known as 'princely' or 'royal' because of the accumulations of jewels and golden ornaments found in them, along with imports from Greece and Italy. The tombs have yielded Greek painted pottery, Greek and Etruscan bronze vases (the Vix crater, the Grächwyl *hydria*, *oenochoae*, bowls, etc.) and Italiot vessels (the Aylesford oenochoae and paterae), while Roman silver goblets and glass cups were found at the Germanic incinerations in Denmark (Juellinge, Dollerup and Hoby). Valuable gifts were sometimes wrapped in cloth.

Under the influence of Christianity portable gifts gradually vanished from the tombs, but wherever paganism survived the dead were still equipped for life in the next world. The rich tombs of the Saxon kings of England (Sutton Hoo) and the Viking kings of Denmark and Norway (Oseborg) are brilliant examples of this *tradition*.

Opposite page: two great gods at Teotihuacan. The pyramid usually called after the Plumed Serpent (Quetzalcoatl) was in fact shared by the rain god Tlaloc, and symbolic representations of the two alternate on the *façade of the pyramid*. Originally the serpent masks were coloured, with eyes made of obsidian. The temple which surmounted the pyramid no longer exists. c. AD 300.

The Americas

The Haida people, who lived in the Queen Charlotte Islands, British Columbia, practised a religion based on a form of ancestor worship which required the performance of ceremonial mysteries and dances. For these rituals, the actors wore carved and painted wooden masks made with remarkable skill and representing nature spirits. The example shown here is a hybrid of a bear and a dragonfly, with a distinctly jovial expression

American and pre-Columbian archaeology

There is no branch of archaeology which occupies so equivocal a position as that of America. Until this century no one could establish any chronological divisions other than 'pre' and 'post' conquest because they believed that American history could not go back more than a few thousand years. For a long time everything earlier than Christopher Columbus' voyage of discovery was lumped together under the heading 'pre-Columbian' in teaching, research and museums, thus implying that later objects were quite different, and belonged to American colonial and post-colonial history. At the same time, because the notion of archaeology was allied to that of antiquity, remains dating to the sixteenth, seventeenth and eighteenth centuries were assigned to ethnological rather than archaeological collections.

These distinctions, established by force of circumstances rather than by preliminary thought, are both inaccurate and misleading, and before describing the amazing achievements of American archaeology, its limits and the broad outlines of its historical setting must be redefined.

The term 'pre-Columbian cultures' should be dropped, and indeed, it is scarcely ever used by scholars. In fact, although the heartlands of the great empires experienced a sharp break at the time of the conquest which justifies the idea of balancing everything that happened before the early sixteenth century against the subsequent periods, other parts of America were not reached by European influence until very much later. The word pre-Columbian, meaning the late fifteenth century and the early sixteenth, has an absolute chronological meaning and as a result, is only applied in the limited areas discovered or colonised by Columbus and his immediate successors.

It is obviously absurd to speak of pre-Columbian pottery of the eighteenth century, but nevertheless, in some parts of America pottery with no trace of European influence was still being made in the eighteenth century. Many more examples could be quoted. It is true that the term has been sanctioned by custom. It is used of the Incas of Peru and the Aztecs of Mexico, whose cruel overthrow did not occur until three decades after Columbus. It is no more appropriate when applied to the great cultures which were already declining before the conquest, than for the cultures which survived several decades or centuries after it. For the latter the term 'pre-European' can be used and it can be said, for example, that the Patagonian hunting cultures of the fifteenth and sixteenth centuries were pre-European in contrast with their later forms, in which the influence of the first Spanish settlements can be discerned.

The division of American cultures into two phases - before and after European influence - with their time lag of several centuries according to the locality, is of fundamental importance. It might seem to some people that everything 'before' comes into the province of archaeology and everything 'after' into that of ethnology or history. However, it would be completely wrong to assume that there is no historic archaeology.

Excavations were conducted in a former Jesuit settlement on the Parana where a Spanish community lived side by side with a group of American Indians for 200 years. Potsherds and the ruins of buildings were the main product of the excavation. There is no possible reason to regard such projects as other than archaeological although they deal with no very distant time. Techniques and customs in life and death, surviving practices, borrowed elements and cultural additions have been reconstructed by means of the excavation. This is certainly an instance of archaeology, although it is historic archaeology in the case of the Spaniards and protohistoric in that of the American Indians.

These terms are used here in their classic meanings, historic archaeology meaning that of a period in which writing was known to the group under discussion, protohistoric meaning that of communities who had no scripts but were in contact with literate peoples who left written evidence about them. The written evidence illuminates and completes the collected information on the culture in question, and this information, in its turn, makes for a better interpretation of the texts. The historic archaeology of the New World is not very spectacular compared with that of the Old. It covers a maximum of 400 years (from the sixteenth to the twentieth centuries) and only produces the sparse remains of the pioneers who made their way into these unknown regions and dared to establish contact with peoples whom they could not regard as other than 'barbarians'.

Opposite page: the strong influence of tradition on the central American cultures can sometimes raise considerable problems in dating. The Pyramid of the Magician at the Maya city of Uxmal, like many other structures of its period, contains a large number of elements salvaged from earlier buildings and reassembled in the obligatory form. These reconstructions can, at the closest, be dated to the seventh and ninth centuries.

One of the main centres of the Classic period was Teotihuacan, near Mexico City, which flourished from about 200 BC to AD 600. The city was founded long before the building of the temple complexes, which date to the first to third centuries AD. The religious centre, now partly reconstructed, includes three main temples: the Pyramid of the Moon to the north, the vast Sun Pyramid in the east, and the temple of Quetzalcoatl in the south, linked by a long road. Groups and lines of chambered buildings, formed of severe undecorated straight-sided steps, occupy the spaces between the three temples.

However, valuable evidence can be obtained from this branch of archaeology on the influence of Europe on American Indian culture, but it has not yet made enough progress to be discussed here.

American archaeology, then, can be divided into two highly uneven parts. The first part consists of prehistoric archaeology, which begins when America was first inhabited and ends between the sixteenth and twentieth centuries with the first texts and ethnological observations for each of the regions under discussion. The second part consists of roughly sketched protohistoric and historic archaeology, investigating the material remains from periods for which written evidence exists. Throughout America these documents are always written in Old World languages, since the rare texts in native American languages have never been satisfactorily deciphered.

The Incas knew how to perform complicated operations with the aid of an abacus and invented an elaborate mnemotechnic system for their calculations based on quipu knots and threads, but they had no writing. As early as 800 BC the Olmecs in Central America engraved signs on their monuments, but there are only a few of them and they are still undeciphered. Towards the fourth century AD Maya writing was well developed and used for books, or *codices*, and inscriptions on monuments, pottery and jewellery, but these have only been partially deciphered and most of them deal with astronomical observations. In the fourteenth and fifteenth centuries the Aztecs also had a script, but it was much more primitive than that of the Maya. And that is all. Although these groups can be said to have approached a literate stage in their cultural development, from the archaeological viewpoint these inscriptions and texts can only be treated in the same way as other material remains. Until new discoveries are made or the inscriptions deciphered we are still in the middle of prehistory when dealing with these cultures.

Finally, American prehistoric archaeology includes almost all the millennia between the present day and the time when man first set foot on the continent. It is a long story—thirty or forty thousand years, perhaps even more. It is divided into periods roughly parallel with those of prehistoric Europe, but it is advisable to adopt a different terminology for them in order to avoid arbitrary comparisons, either cultural or chronological.

There is nothing in America like the archanthropoids and paleanthropoids of the Old World, and it seems that the first settlers in America were *homo sapiens* with techniques parallel to those of the European Upper Paleolithic. For a long time this phase was known to scholars as 'paleo-Indian'. In recent years American scholars have substituted the term 'lithic', but this is not a fortunate choice. The lithic period starts with the origins. It includes all the primitive hunting and fishing cultures with their stone industries and ends with a post-glacial period which differs according to the region, during which the hunters met new climatic conditions by adapting their hunting methods to smaller game and giving greater importance to vegetable foods. These new hunting cultures are known as archaic. They are not technically very different from the preceding period and there is some argument as to whether or not they knew how to polish stone.

The archaic was succeeded by a fundamentally important period for the problem of placing the different American cultures at the time of the conquest. This is known as the formative period. At this time some American

The temple of the Warriors at Chichen Itza (seen here from the top of the Castillo) was probably built during the Toltec-Maya period, from the eleventh to the thirteenth century. The site is remarkable for the forest of columns which originally framed the temple on three sides.

The Zapotec culture evolved at Monte Alban in about the second century BC. and seems to have persisted with few checks until the Spanish conquest. The hilltop area of the city is the site of a magnificent temple complex of pyramids and sunken courts grouped round a central plaza. To the north of the plaza is an interesting mound which features the use of masonry columns grouped in fours between the oblong piers of a portico.

Indians discovered cultivation and settled in increasingly important villages. With squash, beans, maize, tomatoes, etc in the valleys, tubers in the tropical forests and potatoes on the high ground, the nomads settled down and the hunters became farmers. The first signs of this revolution are found in Mexico about the fifth millennium BC, but at the time of the conquest there were great areas of the Americas, the great plains of the North and the pampas of the South, for example, where it had not occurred. In the formative period during which pottery gradually spread through the most advanced centres, the first cities were established, and the first empires, with their remarkably highly developed techniques and crafts, followed. The American classification divides the history of the early stages into two parts: the classic period, which includes the period of development and peak, and the post-classic, which includes the period of decadence.

This classification has two drawbacks. Where terminology is concerned it is curious to classify groups of people by the chief raw material of their weapons and tools (lithic), and at the same time to introduce a notion of antiquity (archaic), and then the idea of cultural evolution (formative), while using terms borrowed from the Mediterranean civilisations (classic and post-classic). Perhaps the words are not particularly important so long as they define the concepts accurately and clearly, but the chief criticism of this system goes beyond the matter of names. It imposes chronological criteria on cultural criteria: 'the lithic peoples disappeared towards the beginning of the post-glacial period', but where should one locate the bison hunters of the great northern plains in the historic period and their immediate successors? 'The evolution of the archaic cultures is completely confined to the first millennia of the post-glacial period', but to whom should one assign the present-day food gatherers of the tropical forests, who are their direct descendants?

It is better to use a classification based on a single series of cultural features without any chronological implications which is broadly applicable everywhere. The first distinction will be between predators, who go out to look for their food, and producers, who produce it. The predators, hunters, fishers and food gatherers will be discussed according to their chief method of acquiring food, remembering that all these groups nearly always practised all three

The site of Copan (Honduras) belongs to the Classic period, when the art and culture of the Maya were at their height in the centuries between the first and the eleventh AD. The acropolis of the city is an extensive area, with several large concourses on different levels, the buildings of which mostly date to the sixth century. The main area, bounded on the south by stairs to the acropolis, is the setting for the ball court.

Religious ceremonial in central America included a ball game, the object of which was to knock a ball through a ring set in the wall of the court by using the elbow or hip. At first the game was played on fairly sporting terms, the winners being awarded the losers' clothes and jewellery as a prize, but by the time the ball court at Chichen Itza was built there were ritual sacrifices to the divinities who presided over the games.

methods, but in varying proportions. These predators can usually, but not always, be regarded as nomads. They may be sedentary (some groups of fishers), and they are often semi-nomadic (in some places food gatherers return at regular intervals to pick wild fruit etc.). These groups made up the whole population of North and South America from the origins to about the fifth millennium BC. After the fifth millennium they were gradually supplanted by groups who had learned to produce their subsistence by cultivating plants or breeding animals. The peoples nearest to the original scene of this discovery also adopted this new way of life and became producers in their turn. Those who lived farther away or in inaccessible regions were not affected until much later. The slow diffusion of agriculture was far from complete at the time of the conquest, for it had not then reached the far north (bounded approximately by the Great Lakes region) or the extreme south where the boundary lay on the north of the Argentine Pampas. Even today there are a few very rare groups surviving in the equatorial and tropical forests of Venezuela and Brazil who know nothing about agriculture.

The producers, especially the most primitive among them, generally practised a mixed economy in which hunting, fishing and food gathering still played an important part, along with the techniques of cultivation and stock-breeding. The discovery of cultivation in America is dated to about the fifth millennium BC, originating from a centre in Mexico; the cultivation of tubers probably originated independently in the forests of Venezuela. These discoveries and methods spread slowly like a patch of oil, leaving untouched areas which gradually shrank on the northern and southern fringes as well as within their boundaries. The movement was not complete in the sixteenth century; the Great Lakes Indians had just begun to cultivate beans when they made their first contact with Europeans. It is certain that this expansion would have continued as far as the limits imposed on primitive agriculture by climatic conditions.

While these customs were expanding far and wide and before the movement had lost its original momentum, something new was happening in the so-called central zone of the discovery. The most advanced groups entered a new phase of history characterised by urban concentrations, the formation of empires and the development of the arts. These were the high cultures which were destroyed root and branch by the Conquistadores. By following the chart in Plate I we can relate what is known of the evolution of these American prehistoric cultures from their origins onwards.

The problem of inhabitation

When it was first realised that America was not part of Asia and that the two were separated by the vast Pacific Ocean, the question immediately arose as to how the people who are still known as Indians reached the New World.

For the Christian world of the sixteenth, seventeenth and eighteenth centuries the problem naturally had to be solved in Christian terms. The Bible mentioned only one creation of mankind, so the Indians, like all other men, must be descended from the single basic human stock, Adam and Eve. This stock, however, was destroyed by the Deluge, and only Noah and his family escaped. The Indians must therefore be descended from the sons of Noah. Various theories were proposed in accordance with the historical view of the times. Some believed that one of the sons of Noah landed soon after the Deluge on the shores of America, his descendants founded a family and peopled the whole of both continents; some that a fleet of Trojans reached America after fleeing across the ocean after the sack of their city. Others believed that the remains of Alexander the Great's army, after many adventures, colonised the hitherto empty lands of America, and there were other hypotheses resting on equally tenuous foundations. No one attempted proof or offered discussion, and no one was deterred by the fundamental problem of the many variations in human types, languages, cultures and empires alleged to have developed among a few people from the Mediterranean in the space of two or three thousand years. Chronological perspectives were very short at this time, and the question of the time required for the development of the cultural variations of the American Indians was no more deeply felt than it was in the case of the peoples of Asia and Africa.

The problem was not investigated scientifically until the eighteenth century. No solution was found – this is still the case – but at least the problem could be assessed from the end of the nineteenth century onwards. The existence of fossil man, coeval with extinct fauna, the notion of prehistoric stone-working cultures and the establishment of a chronological framework for the evolution of the human race, were the achievements of the second half of the nineteenth century, without which it would be impossible to understand anything about the earliest history of mankind in America.

Apart from the fantasies of Ameghino, who believed that the origins of the whole human race were to be found in the tertiary and quaternary regions of the Argentine Pampas, two attitudes of mind divided the eighteenth- and nineteenth-century naturalists engaged in the study of man in America. Some were struck by the homogeneous nature of the communities with whom they came into contact, and by the predominance of Asiatic or Mongol features among them. The others were struck by the dissimilarities in physique, language, customs and techniques of the same communities. These two viewpoints were the basis of two systems of thought, or schools, with divergent opinions on American origins, which, even today, have not succeeded in completely reconciling their differences.

Those who emphasise the unity of the American Indian type, in which the Mongoloid features are very striking, believe that all American Indians sprang from Asiatic stock. They arrived via the Bering Straits, probably in successive waves, at a date which was once believed to be fairly recent but which archaeological discoveries have gradually pushed back. It was from this Asiatic stock, once thought to be Mongol but now known to have much earlier origins, that the various types and cultures of the American Indians developed on the spot by the action of interbreeding and widely different environments. The theory is now known as the American school. Its most inflexible form was expounded by an American anthropologist of Czech descent named Hrdlicka. In an adapted and [illegible] is generally accepted in the New World.

The French, Swedes, and Germans, on the other hand, were struck by the variations between the American Indians, and believed that these could not have arisen solely from environmental influences and the interbreeding of

The river valleys of the North American continent were the scene of the earliest agricultural settlements. While the people of the plains and the deserts were forced by hostile conditions to keep moving, the eastern Indians settled in villages and evolved fairly elaborate handicrafts and social structures. A bottle from Arkansas belonging to the Mississippi culture (about AD 1000) has incised decoration made after firing with a bone or stone point.

Traces of a human habitation site in north-west Mexico have yielded a radiocarbon date of 40,000 years ago, but the earliest known object from Mexico which can properly be described as a work of art only dates back 10,000 years. It is the head of a coyote carved from a spinal bone, and belongs to a Paleolithic phase which roughly parallels the cave art of western Europe of the same period.

About 12,000 years ago, elephants were being hunted in the swamps near Mexico City by the primitive people who inhabited the area. They knew nothing of farming or pastoral life, and little of building or the decorative arts, but the skull of one of the hunters, evidently killed in the struggle, shows that the physical characteristics of the original American people have changed very little.

Altar stones and stelae belonging to the Olmec culture of the La Venta area are carved in low relief with human figures. The legs are shown in profile and the torso in frontal view, which suggests a connection with the reliefs of Uaxactun. Most of the figures are seated, but the one above, on the so-called Alvarado Stele, is depicted in action. His elaborate headdress and jewels and the spherical objects on his shoulders indicate that he may be taking part in the ball game.

comparatively homogeneous emigrants from north-east Asia. Various explanations have been offered. The best-known in France is that of Paul Rivet, founder of the Musée de l'Homme in Paris. Rivet believed that America was inhabited in separate phases and waves over the course of ten millennia. According to his theory the first arrivals would have been Asiatic, coming through the Bering Straits and entering the empty wastes of America where they spread rapidly. A little later, perhaps about the anti-thermal period of the fifth and fourth millennia BC, proto Australians would have reached the southernmost tip of South America, probably following the chain of Antarctic islands. Later on Melanesians and then Polynesians from across the Pacific would have reached the South American coast, while in the first millennium AD North America was rediscovered once more by the Vikings. Other scholars have recently suggested that in the Magdalenian period men sailed to the coast of Canada, and some coins found in Venezuela indicate an extraordinary voyage across the Atlantic by Roman galleys.

In view of the numerous archaeological discoveries of the last forty years, it is now possible to produce a better evaluation of the arguments of the supporters of American Indian unity against the supporters of dissimilarity. Many details are still obscure, but the broad outlines of the inhabitation of America are now becoming clear.

The earliest migrations to America certainly took place through the Bering Straits. During the coldest part of the Wurm glaciation sea-level throughout the world was between 100 and 200 metres lower than it is today. Before the Bering Straits, which are nowhere more than 100 metres deep, emerged, Arctic Alaska was joined to eastern Siberia by a vast plain 1000 metres wide which could be called the Bering isthmus, between the Pacific and the Atlantic Oceans. The Diomede Islands were then a small mountain massif in the middle of this plain. The New and Old Worlds were united and divided again several times during the glaciations of the Quaternary period. Every time the straits emerged again more representatives of American fauna (e.g. the horse) crossed to Asia, and Eurasian animals (e.g. elephants and cervidae) crossed to America. During one of the last of these high dry periods of the Bering passage, when the human race had reached the cold regions of eastern Siberia, it was their turn to migrate to the New World, where no human foot had trod before.

When did this event take place? It is still not known. It was almost certainly more than 20,000 years ago, and perhaps less than 40,000. 35,000–40,000 seems to be a reasonable hypothesis. Why more than 20,000? Because some datings obtained in South America indicate that man had already reached the Straits of Magellan 10,000 years ago and was perhaps hunting in the forests and on the coast of Venezuela 15,000 years ago. Other datings have proved the existence of remains 19,000 years old in Nevada, 23,000 years old in California and even 37,000 years old in Texas. These figures are, admittedly, somewhat arguable, but even if some of them should be updated it still seems that 20,000 must be accepted as the minimum date for the arrival of the first men in America.

We regard 40,000 as the provisional maximum because so far no traces of any pre-sapient human being has been found in North or South America. There was nothing here like Neanderthal man, still less any archanthropoids like Pithecanthropus of south-east Asia. These beings, moreover, do not appear to have reached Siberia. Western Asian Neanderthal man never went any further north and east than Turkestan. South-east Asian Pithecanthropus stopped at Chou kou tien forty kilometres from Peking.

The city of Tula is the type-site for all the aristocratic warrior societies from AD 1000 to the conquest. Near the north face of the north pyramid in the ceremonial complex is a free-standing wall carved in relief on both faces showing stylised rattlesnakes swallowing human beings, a reference to the after-life of dead warriors.

Further north, the earliest inhabitants of the Siberian plains seem definitely to have been *homo sapiens*, whose powerful expansion across the world began about 50,000 years ago (Shanidar, Niah etc.) The most distant, inhospitable lands (eastern Siberia, the Bering isthmus, Alaska) could not have been inhabited before the centres of the expansion movement.

These first conquerors of the New World should be pictured as small groups of hunters venturing further and further eastwards in pursuit of their game, the rhinoceros the woolly mammoth, the reindeer and the horse. When they crossed the frozen steppe of the Bering isthmus they did not know that they were the first. When they settled in Alaska they were not aware of the bewildering discovery of the New World which opened before them. The first groups were followed by others. It seems that they all moved swiftly towards the east and then the south. Their speed should not be exaggerated, however. It took them 10,000 years to pass from the foot of the Rockies to Venezuela, and another 5,000 to reach the southernmost limit of South America. One and a half kilometres per year is the approximate speed of their diffusion southwards. In any case, it is certain that towards the beginning of the post glacial period (about 8000 BC) mankind had already traversed North America, taken the difficult route across the lower coasts and virgin forests of Central America, crossed the heights of Colombia and Venezuela, descended to the gigantic Amazon system, following or crossing its various branches, and finally reached the bare Pampas in the south.

Through forest, prairie and steppe they pressed on. In 8000 BC they camped on the shores of the Straits of Magellan which may still have been a lake at this time. The stages, routes, pauses and advances of this extraordinary epic are still unknown. There was probably no great wave of migrants, nor even several, but a succession of small groups, settling here, spreading their surplus population in ripples in all directions, always trying their luck farther east, farther south, but always farther on. The hunters and food gatherers pursued their way, more probably east than west of the Rockies and the Andes, more probably on the high ground than the unhealthy and dangerous lowlands. At the same time groups adapted to sea shore life must have followed and colonised the Pacific coasts and later, the Atlantic, but nothing is known of their earliest sites because they were destroyed when sea-level rose again after the glaciation.

Apart from the fact that they existed, we know virtually nothing about the first waves of settlers in America. Some rather crudely dressed stone tools are questionably attributed to them, and it has been suggested that they physically resembled Wadjak man, who had already spread where they were to follow, towards Australia. If Rivet supported the existence of prehistoric contacts between Australia and South America it was because he, like others, had recognised the occurrence of Australoid types among the earliest American Indian strata. It is now known that these American Indian 'Australoids' did not arrive by way of a highly unlikely voyage from Australia to America several millennia ago, but that they were descended from a very ancient Asiatic stock from which they broke away more than thirty or forty thousand years ago. Some of them were among the earliest migrants to America, others went in the opposite

The first Americans to enter the New World are now chiefly represented by the latest comers, the Eskimos. During the Ice Ages, the sea level dropped and the Bering Straits became a land passage to Asia, across which men and animals entered the north Americas, gradually pushing on south as others came behind. The Eskimos were the last to arrive, and have remained near their entry point. A caribou hunt engraved on a piece of bone (above) expresses their preoccupation with hunting.

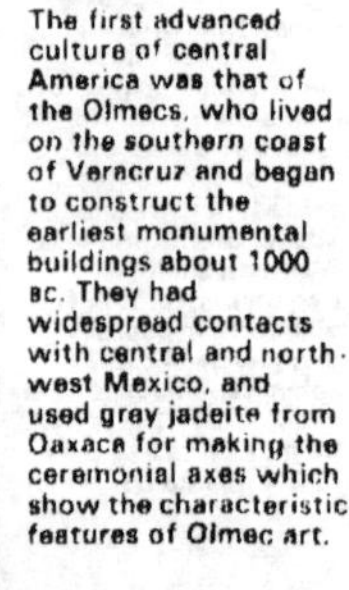

The first advanced culture of central America was that of the Olmecs, who lived on the southern coast of Veracruz and began to construct the earliest monumental buildings about 1000 BC. They had widespread contacts with central and north-west Mexico, and used grey jadeite from Oaxaca for making the ceremonial axes which show the characteristic features of Olmec art.

The fundamental rite which had the deepest meaning for the Aztecs was human sacrifice, and in order to obtain the numbers of victims needed a warrior caste devoted to the pursuit and capture of prisoners was necessary. The war god Huitzilopochtli, son of Coatlicue (see overleaf), presided over these expeditions and was specially favoured. His birth is shown on a relief panel of about 1507 from Tenochtitlán.

direction and were one day to populate the virgin lands of Australia. The American dolichocephalic types are not the great grandchildren of the Australians, but their first cousins.

Culturally and physically the Asiatic peoples slowly changed and slowly spread new, increasingly Mongol types and new cultural forms better adapted to the conditions of the Siberian steppe and the Bering isthmus in America. Perhaps ten or twelve thousand years ago the isthmus became a strait once more, but it seems that contact was not broken off because the human race knew how to navigate by that time, and canoe crossings succeeded the long trek by land. From this time forwards expansion towards the interior of America was more difficult. The way was no longer obstructed by glaciers, it is true, but the empty lands of early days had been superseded by country which had been continuously inhabited for several millennia and could not be crossed with impunity. The migrations of earlier times were gradually succeeded by slow diffusion movements. The last arrivals were the ancestors of the Eskimos, who never passed beyond the frozen north lands.

America was therefore populated almost exclusively by way of the Bering Straits from about 40,000 years ago until the arrival of the Europeans. This, however, does not exclude the possibility that other contacts took place across both the Atlantic and the Pacific. We shall describe these contacts chronologically as they occurred. A few years ago an American named Greenman tried to prove that there were Magdalenian settlements on the Atlantic coasts of Canada and America 15,000 years ago. His proof was chiefly based on the resemblance of the North American rock paintings and engravings to the French and Cantabrian examples.

An andesite statue of Coatlicue, the Mother Earth figure revered by the Aztecs, was found in the main plaza in Mexico City. The figure, far from maternal in appearance, has a double rattlesnake head and its necklace is formed from severed human hands and hearts with a skull as a central pendant. It once stood in the courtyard of the great temple at Tenochtitlán, where it served to remind worshippers that the maize crop would not be forthcoming without due meed of sacrifice.

The Aztecs believed that only through human sacrifices in large and increasing numbers could the continuation of the universe be assured. This belief perhaps explains why their art is so strongly permeated by a sense of imminent death, which their sculptors were able to convey in their powerful and expressive stone carving. A diorite piece, nearly four feet in height, probably from Tenochtitlán (far left), represents the head of Coyolxauhqui, sister of the war god. She is shown in death

The submissive fatalism of the Aztec people is shown by their great devotion to stoneworking, and the skill they achieved in pieces designed to last for eternity. They had no metal implements to shape hard stones, and a beautiful crystal skull, now in the British Museum, could have taken more than a lifetime to complete. Most of the working was done by patiently rubbing the surface away with strips of damp leather dipped in abrasive sand, a slow and intensely laborious process.

and on the identification of the Magdalenian 'ship' signs in an already complicated form of boat. It is not impossible that the coastal Magdalenians had boats like those of, for instance, the modern Eskimos, in view of their high technical standards, or that some of them may have reached the coast of America. But if there were contacts, they must have been sporadic since there are no cultural features common to both sides of the Atlantic in this period, and the resemblances mentioned above are also found in the rock art of other cultures, from which it could be concluded that all the cave art in the world is related.

The problem of Australians in America has already been touched on. It is probable that the physical, linguistic and cultural features common to Australia and South America, which suggest Australian migrations towards the west via the Antarctic islands, are due to a common origin in very ancient Asiatic stock. No one nowadays believes in migrants from Australia arriving in Tierra del Fuego via the Antarctic islands in the fifth and fourth millennia. Much migration could be conclusively proved by traces of human passage on the intervening islands, but no archaeological remains of any kind have ever been found on these islands.

The Melanesian elements, particularly those of a technical nature, recognised by ethnologists in America, have been interpreted by some scholars as evidence of Melanesian migrations across the Pacific and could be dated to the first or second millennium BC. There is no full study of the question, and in the present state of knowledge it is impossible to make any definite statement. A much more disputable point is the appearance about the same time of two important cultural features, pottery and cotton, on the north-west coast of South America. The earliest American pottery dates to the middle or end of the third millennium BC and appeared, without any formative stages, on the coasts of Ecuador and Colombia. It shows some similarities with Jomon ware, the earliest Japanese pottery. It can be deduced from this that American pottery was not an indigenous invention, but was introduced into Ecuador and Colombia by Japanese navigators, who must have arrived accidentally or, more probably, attempted to establish commercial links with distant peoples. This theory has an increasing number of supporters, but it has never been proved. The history of the introduction of cotton growing could be similar to that of pottery, but the evidence, especially on the botanical side, is still too scanty for definite conclusions.

The human race, therefore, evolved in America for forty, thirty or twenty thousand years independently from the rest of the world, with which their only contacts were through the inhospitable lands of Alaska. Some technical inventions and beliefs probably penetrated through this neck, but everything else was the result of internal development and contacts between the cultural zones established at comparatively early dates. Then, at a time when Siberian elements were slowly but steadily being transmitted, about 4000 years ago (this was when the ancestors of the Eskimos arrived), indications of contacts from across the ocean suddenly appeared. The Pacific itself was entering a new phase of its history. Its waters were scattered with canoes and junks. Chinese trade extended as far as the Sunda Isles and reached the first islands of Micronesia. (An early level in the Marianas, but not the earliest, gave a date of 1500 BC.) A little later farmers, who were also excellent seamen, began to settle in Polynesia. It would not be surprising if the impregnable barriers of the American coast were reached also when the [illegible] islands forming Polynesia were touched. It is not known if there were voyages in the other direction, but there are more and more arguments in favour of contacts with Japan and south east Asia and the Polynesian islands. Pottery and cotton may have come from Asia, but the ceremonial axe, or *patu patu*, would have been acquired from the Polynesians. On the other hand, the sweet potato, which occurs throughout Polynesia, is American in origin. All these questions have been asked, but never answered.

While mankind was thus discovering the virgin wastes of the Pacific and the peoples living near the shores were gradually adopting common cultural features, what was happening on the Atlantic coast? There were no islands to be colonised in the middle of the Atlantic, no relay stations between the coasts of Europe and Africa and those of America. The Atlantic was not sprinkled with all kinds of craft from the beginning of the Mediterranean civilisation. Setting aside the theory of Magdalenians in America—it could be set aside indefinitely—it was not until the beginning of the Christian era that the first signs of transatlantic influences appear. The Roman coins from recent archaeological excavations in Venezuela have been

The palace at Palenque chiefly belongs to the Middle Classic era, although some sections survived from the Early period and others were added in the Late. It consists of two courts bounded by parallel rows of chambers and is perhaps the earliest example of the closed-cornered quadrangle in Maya architecture. A square tower four storeys high has a *central stairway* connecting the corbel-vaulted levels. An inaccessible attic was placed between the storeys to preserve the outer proportions of bearing wall and vault zone.

One of the great achievements of the Mixtec people of southern Mexico was their production of the manuscripts known as codices. These are actually screens of deer hide folded in a series of pleats and painted with a continuous narrative told in pictures. The Zouche (Nuttal) Codex is, in effect, a family tree showing portraits of the various couples of ancestors and the exploits of a famous folk hero.

mentioned more than once. None of these discoveries has been published in enough detail to enable anyone to say whether they are real Roman coins, or only poor European pieces of the sixteenth or seventeenth centuries. The question, however, is open, and it is by no means unlikely that Roman galleys should have been driven to the coasts of America. In any case, these accidental contacts did not influence the American coastal cultures.

On the other hand, Viking settlements on the coasts of Labrador and Canada, sometimes extending as far as the latitude of New York state, are well attested. These Viking colonies seem to have been developed between the ninth and thirteenth centuries, but they vanished before they ever became really established, and their contacts with the northern Indians produced no technical or cultural innovations.

Three or four centuries later the massive wave of European immigrants was totally to transform the cultural and anthropological face of America. In brief, despite numerous obscurities, the history of the inhabitation of America can reasonably be summarised as follows.

1. Several waves of peoples of Asiatic stock, increasingly varied in culture and physique, penetrated through the north about 40,000 years ago. These waves spread through the interior of the two Americas, gradually slowing down and interbreeding while they adapted themselves to various extremely diverse conditions, evolving and making new discoveries.
2. From about the third millennium BC until historic times the inhabitants of the Pacific coast met with seamen from south east Asia, Melanesia and Polynesia, who introduced a few techniques and art forms among them, but made no profound changes in either the great cultural complexes or the anthropological stock.
3. From the sixteenth to the twentieth century the European conquest abruptly and completely changed the history of the people of America.

The principal stages

Predators A somewhat crude dressed stone culture consisting basically of worked cores with choppers, chopping tools, scrapers and bifaces (but not bifacial arrowheads) has been found in *apparently ancient* geological environments at various places in North and South America. Some experts have deduced from this that America had a lower Paleolithic period, the origins of which were connected with the Old World. This conclusion, founded on basically typological considerations, is highly suspect since the same cultures with choppers, chopping tools, scrapers and bifaces appear much later in a context firmly dated to an advanced stage of the post-glacial period. Nothing in the present state of knowledge either proves or even strongly suggests that the crossing to America was achieved before the second half of the Wurm glaciation by Old World archanthropoids equipped with typical lower Paleolithic tools.

A large group of buildings round a courtyard at Uxmal is known as the Nunnery from the cloister-like effect of its layout, but its true function is unknown. The corners of the courtyard are not correct right angles, which has sometimes been taken as a sign of architectural naiveté, but various refinements, such as the correction of the long horizontals of stair and architrave and the slight backward slope of the facades, suggest a high degree of sophistication.

On the other hand, it is quite possible that the first men to reach America some 40,000 years ago had an industry of tools made by striking a core, in which pebble tools perhaps played a comparatively unimportant part. They may not have known pressure retouching. To assign this group to the lower Palaeolithic would lead to inextricable confusion.

These men were the first 'paleo-Indians' of some authorities, or the early lithic peoples, as they have begun to be called in the last few years. In a climate several degrees colder than at present, dry in the high latitudes and very wet in the low latitudes, these basically hunting predators pursued prey consisting of fauna which is now partly extinct such as mastodons, mammoths, various animals of the camel species, bison, horses, cervidae of the Rockies in North America and the Andes in South America. Since the remains of their culture have been found in eroded open-air sites where nothing but stone has survived, nothing is known about them but this poorly documented industry. We do not know their physical type, their type of dwelling, their food or their burial customs. Neither do we know from which Asiatic groups they derived their technical traditions, nor, if these continued without a *break until the post-glacial period*, whether they were the ancestors of the core working craftsmen of the later era or whether the latter rediscovered types and techniques which had been abandoned in the meantime several thousand years later. Some groups must have settled along the Pacific coasts at the same time as these inland hunters, having followed the shores of the Bering Isthmus and gradually pushed on further south. However, because the scanty remains of their settlements are now submerged beneath many metres of water and

The Fejervary-Mayer Codex is a deer hide screen dating to the period before 1350. The strips of hide were fastened together with glue or stitching, prepared with a chalky varnish and painted with an astonishing clarity and precision of outline. The subjects are ritual, and some numerals are included. The pictures tell of the influence of the gods and the recurrent rhythm of their power over human life.

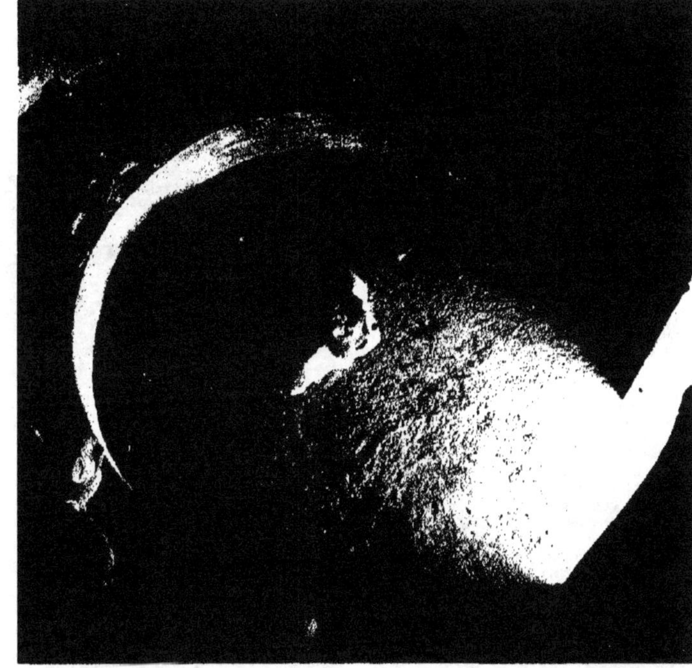

One of the latest Aztec religious structures, the rock-cut temple of Malinalco, was begun in 1476 and was still unfinished in 1520. On top of a platform approached by flights of steps is a circular shrine with an opening facing south. The sides of the chamber are lined with a horseshoe-shaped bench on which crouches the figure of a jaguar, with an eagle on either side. Another eagle, its wings half open, is set in the middle of the floor. The animals refer to the warrior orders of eagle and jaguar knights, to whose cult the temple was dedicated.

The so-called Pigeon Group at Palenque is less well preserved than some neighbouring buildings, from which some scholars deduce that it is earlier in date.

marine deposits, we can know nothing of them beyond the probable fact of their existence.

The evidence does not become clear-cut until the periods between 20,000 and 15,000 years ago for the hunters, and 7000 and 6000 for the coastal fishers. Towards the end of the Wisconsin period, which corresponds to the Würm glaciation in Europe, the glaciers retreated more swiftly towards the high land and upper latitudes. The Great Lakes region became accessible. In Patagonia and Tierra del Fuego in the distant south the tundra spread over the newly freed land. Plentiful game spread over the vast plains in the south (mylodons, horses, cameloids, cervidae) and north (mastodons, mammoths, cameloids, horses, bison and cervidae), hunted by groups of men whose numbers were still very limited. These new hunters were now armed with projectiles with stone heads with bifacial retouching carried out by the pressure method. These projectiles were cast either with a spear thrower or perhaps already fired from bows. It would be very interesting to know if these stone weapons and the means of propelling them were local inventions or imported from Asia.

However this may be, the new discovery developed, spread out and became the most typical product of the prehistoric New World hunters, and has made it possible to draw up a preliminary classification of them, the rest of their technical equipment not having been thoroughly investigated. The earliest known hunters made shouldered points slightly reminiscent of some Solutrean points, known as Sandia points from the cave of Sandia in New Mexico. They were still hunting archaic game

It seems that their numbers were still small, and that they gravitated only about the south western United States.

The shouldered Sandia points were succeeded by points with a central channel hollowed out of one or both faces by the removal of a long flake. These were, firstly Clovis points from the site of Clovis in New Mexico, then the smaller and finer Folsom points also from New Mexico. The appearance of these new forms was accompanied by an abrupt expansion of their distribution. The thickest centre is still the south-western United States, but the distribution of these shapes, which were probably the result of a new method of hammering, took place rapidly in all directions. Channelled points are found in Canada, in the United States, Mexico, Central America and various places in South America. They had already spread to the Magellan Straits 10,000 years ago. They have been found in Fell cave, one of the earliest known Patagonian sites, in association with knives and basalt scrapers and small hearths round which fossilised bones of horses, guanacos and perhaps mylodons were scattered. The physical type of these big game hunters is still unknown. Nothing whatsoever is known about their social organisation, their beliefs, their art forms and their dwellings, apart from the fact that they used caves as temporary shelters.

These channelled points were succeeded by points with a straight convex or concave base and triangular or foliate shap... n both North and South America. This was ... ther an immense wave of technical progress w... ich spread over both Americas or independent ... volution ending in the same results ... we cann... tell which. What is certain is that the scatter... hunters of the earliest times had become ... ore numerous. Some regions were more de... nsely populated, and in a more settled manner ... The diffusion of groups of peoples across ... he steppe and savanna, the high plateaux a... f the prairies did not meet only with natural obstacles as in former times, but with huma... opposition. The cultures began to be isolated ... rom each other.

In regions which were no longer amorphous man had to find ways of making better use of natural resources so as to get better returns. Game, moreover, was becoming scarce. Some of the large species, such as mastodons and mammoths were nearly extir... ...h horses, were on the way to extinction. It ... perhaps about this time that the appearanc... some groups characterised technically ... economically by more intensive use of ve... table food and the occurrence of grinding st... on some sites can be located. This is at the beginning of the post glacial period. Distinct contemporary cultures such as the Cochise culture in the western United States with its numerous grinding stones, the Plainview hunters in the south-west with their points, the men of the Tepexpan plains in Mexico, and the Ayampitin hunters in north-west Argentina with their points, can already be differentiated. We still have no remains of these hunters and food gatherers but a few weapons and tools, points, knives, scrapers, grinding stones and their favourite prey, which consisted chiefly of modern species. In Mexico, however, the last mammoths were still being killed and dismembered by the human race. It was also in Mexico that the skull of Tepexpan Man, which might be attributed to this period, was found.

The separate evolution and development of the groups of American predators continued

The Temple of the Warriors at Chichen Itza encloses a shrine of Chac, the Maya-Toltec rain-cloud spirit. His effigy reclines on a terrace. In the figure's lap is a shallow tray or platter, to catch rainfall.

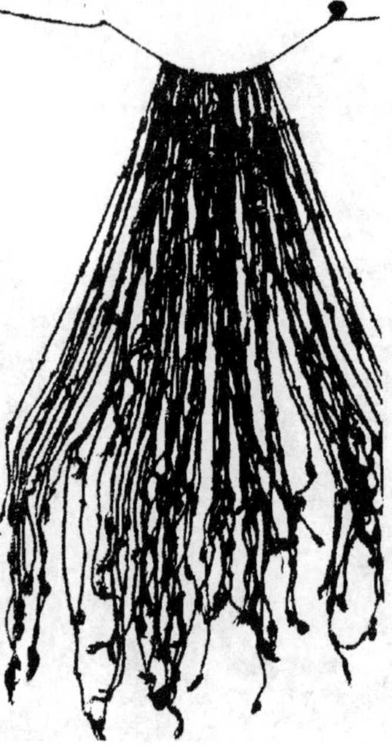

The Inca state was a complex organisation demanding a highly evolved civil service. In the absence of well-developed writing, this called for some form of simple but effective 'book-keeping' system for the regulation of accounts, and this was provided by the *quipu*. A length of string holds a fringe of knotted cords of different colours, the knots having a varying number of turns. Those nearest the string are units, next come tens, then hundreds and then thousands, while the colour indicated the subject.

The gods of central America were seldom realised in forms meant to attract by their charm and beauty. Religion ruled every aspect of daily life, and it was nearly always manifest in the form of suffering. Even the deities presiding over such pleasant functions as the growth of flowers or the provision of rain were usually hideous and frightening in appearance. A stone figure from Veracruz connected with the fertility of the maize crop is in the form of a sacrificial victim, his arms strained behind his back and his chest hacked open for the removal of the heart.

throughout the post-glacial period, with food gatherers, who are chiefly known by their grinding, beating and crushing tools, hunters, whose many types of missile points included bone, quill and wood but chiefly stemmed stone points, and fishers, whose sites are no longer submerged from this time onwards, sea level having reached its maximum about 5000 years ago. These groups now stayed within their traditional regions. Population movements became rarer and more difficult. Conversely, it is probable that contact between neighbouring groups, exchanges of goods and ideas, borrowings and trading became more frequent and continuous in the occupied regions.

A mosaic of peoples, still nomadic or semi-nomadic within the limits of their traditional areas and quite distinct from each other anthropologically, culturally, linguistically and technically was laid over the map of America. The still numerous descendants of the age-old traditions of hunting, fishing and food gathering still survived at the time of the conquest in huge tracts of America, e.g. the Eskimos in the far north, the Indian hunters of the great plains in the United States, the more primitive groups in the equatorial and tropical forests, the great Patagonians of the Pampas, the Fuegans of the Magellan archipelago and many others.

However, they were no more than survivals. A great evolution occurred seven or eight thousand years ago in Central America, and later spread like a patch of oil to North and South America. This took place several thousand years after the Old World, on the threshold of discovering agriculture and the Neolithic revolution, had undergone a profound change in its economy.

The discovery of agriculture Agriculture may have been discovered in America at several different centres. It is obvious that the tubers of the tropical forests such as manioc, the potatoes of the high Andean plateaux, the vegetables and the cereals of the great plains were not domesticated in the same regions. We still have very little information on the stages in the domestication of the various tuberous plants of the high plateaux and the hot forests. On the other hand, the history of man's control of the chief American vegetables and fruits about 7000 or 8000 years ago and their diffusion through North and South America is now beginning to be fairly well understood. The heart of this story lies in the Tehuacan valley in south-east Mexico and has been documented by Richard MacNeish.

From the beginning of the post-glacial period to about 6500 BC the inhabitants of the valley of Tehuacan consisted of small bands of nomads who lived by hunting and gathering wild plants, like their contemporaries in the rest of America. They hunted the horse and the antelope, but preferred small game - hares, rats, various small mammals, tortoises and birds. The climate was slightly colder and wetter than at present, and the plant growth formed an environment very different from the present arid condition. The industry consisted chiefly of variously worked flakes, bifacial foliate missile points. Polished stone was unknown. Vegetable food consisted entirely of wild species. This phase, known as the Ajuereado period, was succeeded without a sharp transition by the El Riego phase, which developed between 6500 and 4900 BC. The sites were small family encampments during the dry season and much larger encampments which must have been inhabited in spring and the wet season by several families.

In the daily menu cervidae replaced the horse and the antelope, rabbit the hare. Hunting methods do not seem to have changed, but

The largest pyramid at Chichen Itza is the temple of Quetzalcoatl-Kukulcan, known as the Castillo, which stands squarely in the middle of the site. The pyramid was rebuilt and enlarged on several occasions in conformity with the Maya belief in cyclical renewal. On top of the pyramid is a small twin-chambered temple with a wide architrave on which are reliefs of jaguars in procession.

vegetable food seems to have assumed much greater importance. They gathered many different plants. It is probable that a variety of squash (*cucurbita mixta*) and the avocado (*persea americana*) were already cultivated. Cotton, red pepper, amaranth and maize, all wild, were used. The Mexican Indians entered an intermediate phase between hunting and cultivation. Their equipment differed very little from that of the preceding period. Missile points were improved, knives and scrapers persisted and there were many large plano-convex graters as well as mortars, grinders, pounders and grinder handles, all made of pebbles, which were evidently used for crushing animal or vegetable foods. A few fragments of spiral basket work, threads, pieces of cloaks, wooden objects and bits of snares from this period have survived. Burial customs were complex: they included collective tombs, incineration tombs, and tombs in which the head was separated from the body.

The next phase, the Coxcatlan period, was more decidedly agricultural. Between 4900 and 3500 BC the inhabitants of Tehuacan were still nomads, but they gathered together in larger communities for longer periods, and the camps, one might almost say villages, became larger as the population grew.

Hunting and trapping were still practised, but gathering wild plants and cultivating domestic varieties became increasingly common. At the beginning of the period red pepper, which was cultivated from that time onwards, and maize were added to the squash vine and the avocado. Later on came amaranth, the bean, another kind of squash and the *sapodilla* (the sap of which now forms the basis for making chewing gum). There was no sharp change in the type of industry: missile points were finished with flanges, and the earlier simple mills became fully developed *metates*.

Like so many other peoples both before their time and after, the Maya were addicted to some curious notions of personal beauty. Small children had a ball of pitch fastened to their forelocks so that it hung between their eyes and thus encouraged a permanent squint; teeth were filed and the enamel inlaid with chips of jade or turquoise. One of the most persistent deformations shows clearly on a relief panel of a kneeling worshipper holding up a basket of offerings. His forehead has been flattened and reduced to a severe backward slope by pressing it in his childhood between two boards, one of which lay on his forehead and the other across the back of his skull.

The gr at ach ent of this period was the introduction of ma as a cultivated plant in the American Indi diet. Maize was probably the most importan ant in the history of prehistoric American agriculture. As the original wild species has never been identified, various theories have been put forward to explain its appearance in the ancient Mexican diet. The recent discovery in Mexico City of fossilized maize pollen seventy metres below ground level, dating t the last interglacial period,

The forms of early columns in Egypt and Western Asia nearly always take the shape of the natural elements, such as tree trunks or bundles of reeds, from which the stone prototypes were copied. This is not true of the architecture of Chichen Itza, where plant forms have no place in the decorative repertoire. The columns of the temple of the Warriors are carved to represent serpents, the head with its gaping mouth forming the base and the upraised body and tail being the shaft and capital.

proves that wild maize existed in the region hundreds of thousands of years ago and that it was neither Asiatic in origin nor the modified distantly related product of teosinte or *trip-sacum*.

The Neolithic movement gathered speed after the Coxcatlan period. From 3500 to 2300 BC we have the Abejas period. During the dry season hunting continued to be the main activity. During the wet season families gathered together in villages of from five to ten half-underground houses arranged along the *rio* terraces. These houses were not normally completely deserted during the hunting season; while the stronger and more skilful members of the tribe wandered about with their families, others stayed where they were. The first permanent villages were born. Cultivation developed. Several new species appeared in their gardens, such as the *canavalia*, a bean, several hybrid varieties of maize, and cotton. Dogs lived in the village. The old methods of stone working, basket making etc. persisted, along with some innovations like the making of polished stone vessels.

It was in the Purron period, between 2300 and 1500 BC, that the first fragments of pottery appeared in an environment differing very little from the preceding stage. In the Ajalpan period, between 1500 and 900 BC, subsistence was based almost entirely on agriculture, which included several more new species. Villages became more important, and housed between 100 and 300 inhabitants in mud dwellings. Pottery making was at its height, with human figurines and globular vases. This stage corresponds to what is known as the full Neolithic, with permanent well developed villages, cultivated plants forming the staple diet and widespread use of pottery.

The case of the Tehuacan valley has been discussed in detail here partly because this dry valley, where vegetable remains are remarkably well preserved, is fully documented, thanks to Richard MacNeish's systematic excavations, and partly because it is one of the earliest centres of the discovery of agriculture in the New World. However, it should not be assumed from this that all the inventions which appeared successively at Tehuacan were local. The use of beans and maize can be credited to Tehuacan, but squash appears earlier in the north-west of Mexico, the helianthus, a sunflower, originated in the south-western United States, a certain bean in Peru, the potato in the high ground of the Andes and the manioc root in the tropical forests. The discovery and diffusion of the cultivation of various plants in America should not be imagined as knowledge emanating from a pinpoint centre, but rather as the expansion and interaction of many discoveries from the sixth or fifth millennium BC onwards, and as a rather amorphous cultural movement extending perhaps from the south-western United States to Peru. The start perhaps occurred in Mexico, but after the beginning contacts, borrowings and interactions were such that it is impossible to disentangle their convolutions.

From this vast centre of discovery, which represented for the New World what Western Asia did for the Old, knowledge of agriculture was to spread in different directions. It reached the north-eastern United States in the ninth century – when communities belonging to the Hopewell tradition were drawing their staple diet from agriculture – and the south-western United States about AD 500. It spread towards the south more rapidly: in 2000 BC the inhabitants of Venezuela had a basically vegetable diet. The cultivation of manioc was definitely practised in 1200 BC, and that of maize in 1000. The first agricultural villages of the Amazon basin were established about the first millennium BC.

The Paraná–Paraguay basin was not affected until later, but agricultural techniques scarcely spread south of the River Plate at all, and the whole of the far South was still basing its economy on hunting, food gathering and fishing at the time of the Spanish conquest.

The great empires While these new ways of life were slowly spreading, new revolutions were beginning in the most dynamic centres: Peru, Central America and Mexico. Everything was then ready for the development of a new social and economic order, the most tangible signs of which, archaeologically speaking, are urban concentrations, the development and expansion of the crafts, the spread of art and architecture and the formation of empires with fluctuating frontiers and destinies. These transformations are similar to the slightly earlier course of Mediterranean history.

About 2000 BC most of the peoples from the south-western United States to southern Peru and Bolivia had reached a standard of living which enabled them to join in the great struggle for political and cultural supremacy. Three centres rapidly drew away from the others and made considerable advances, which were abruptly obliterated by the Spanish conquest. These were the Peru–Bolivia region, where the Inca empire originated, the central American region of Mexico, where the Aztecs rose to power and Central America, with the Maya territories. The last-named played a strikingly different part from the other two, since we only know the ceremonial centres, palaces, public buildings and works of art of the later Maya, but no community dwellings, workshop or city. This is a unique feature in the history of American civilisation – and perhaps in the history of the human race.

It would be impossible to review all the history suggested by the excavated remains of these mighty empires in a few lines, or even a few pages. But it is important to outline the essential facts.

The basic architectural element in Central America was a religious monument built on a platform. At first the platform was constructed of clay brought to the selected spot. Later on it was faced with slabs, then with stucco, and took the shape of a pyramid. The step pyramids of Central America are actually a series of truncated pyramids of decreasing size. On the top stood the temple, as it had stood on the original platform.

The first permanent villages of semi-underground houses in the Tehuacan valley date to 3500–2300 BC, in the Abejas period, and the first potsherds to 2300–1500 BC, in the Purron period, but it was not until the first millennium BC (the Santa Maria period, 900 BC–AD 200) that the first mud built villages with ceremonial centres occurred, and ceremonial centres were not built of stone with the classic group of pyramids, squares, ball court etc. until the Palo Blanco period (AD 200–700). The Tehuacan valley had already lost its head start. The oldest known pyramid is at La Venta in south-west Yucatan, and dates to the beginning of the first millennium BC. The custom of pyramid building spread rapidly and is found nearly everywhere in central America in the upper-pre classic period (the term is applied to the whole region, and covers the period between 600 BC and AD 300). The ceremonial centres, with their various buildings set round a central square, developed during the classic period, between the fourth and tenth centuries AD. In the post classic period from the tenth century to the conquest the religious buildings in the same ceremonial centres gradually lost their importance whilst buildings designed for the administration and the homes of the ruling classes developed.

Architecture, however, was not the most spectacular aspect of the movement, which had its roots in the settling of the early nomads. Trade, social organisation, religion, metal working, writing, mathematics and the calendar were other fascinating achievements of these cultures. Their history can be somewhat

Maya preoccupation with calendrical observances has helped with the dating problems which often beset the study of early cultures without a high degree of literacy. The Leyden Plate is a jadeite pendant worked on one side with a series of glyphs telling the date (AD 320, very early in the Classic Maya period) and on the other with a delicately incised drawing of a warrior chief trampling on a defeated enemy captive.

Facing page: despite the outstanding skill of the Classic Maya sculptors, they never seem to have attempted free-standing carvings. Their expertise was lavished on low relief work on altar blocks, drums and boulders, and especially on tall slabs of stone known as stelae. Some of the most beautiful stelae are found at Copan (Honduras) where Stela A, which dates to AD 731, seems to indicate an attempt to free the figure from the flat planes of the stone. The face is realistically modelled, but the treatment of the garments and headdress is purely ornamental.

The pottery from Chancay on the Pasamaya river shows a curious blend of independent tradition with provincial derivations. The workmanship is casual, and gives the impression (confirmed by the numbers of recognisable types) that a form of mass production for market-place sales was practised. A human figurine has impressed details and is painted in black on the white slip ground.

In 1952, when Dr Alberto Ruz was investigating the Temple of the Inscriptions at Palenque, he noticed holes in one of the floor slabs which led him to believe that it had once been removable. On closer examination it proved to cover the entrance of a stairway leading into the heart of the pyramid, at the base of which was a magnificent corbel-vaulted burial chamber. A huge carved slab covered the sarcophagus in which lay the remains of an unusually tall Maya nobleman of the late seventh or early eighth century, accompanied by a treasure of carved jade.

arbitrarily divided into four major periods: the archaic, the pre-classic, the classic and the post-classic. In central Mexico the archaic period extended from 1400 to 500 BC.

The later stages are studded with famous names: Teotihuacan, which lasted from just before the beginning of the Christian era until the ninth or tenth centuries, the Toltec period with its capital Tollan, from the tenth to the early thirteenth centuries, followed by the period of the Chichimec invasions and the rise of the Aztecs in the fourteenth century. Farther south the sequences are chiefly known from the excavations of Monte Albán, with its five periods extending from 650 BC when there was already a form of writing, as yet undeciphered, a calendar and a numerical system, until the Spanish conquest. In the Maya zone (Guatemala, Chiapas and Yucatan) the sequences were rather different. Pottery and agriculture were firmly established in the first millennium AD. This was also the period of the first architectural creations and of the Uaxactun pyramid. The early Maya empire developed from the fourth century AD onwards in Guatemala and the region round Lake Peten, with the sites of Tikal, Copan, and Palenque. In the tenth century all the early empire sites were abandoned. The latest known stele of this period is dated 909. These once-powerful groups migrated, for unknown reasons, to Yucatan. There they founded what is known as the New Empire (987–1697). Its history was punctuated by the rise of Chichen Itza, Mayapan and Uxmal, and by a period of decline beginning in 1461 – before the Spanish conquest – with the destruction of Mayapan.

Maya territory was not invaded by the Spanish until 1559 and the last Maya, who took refuge in the virgin forests, only surrendered to the Spaniards in 1697.

The first great urban civilisation of South America developed in the Andes valleys. The first architectural creations were built at Chavin Huantar at the beginning of the first millennium BC. They show obvious analogies with the stone buildings of Central America, with their superimposed square platforms with a small structure on top. Sculpture was already highly developed. It was dominated b. human and animal, mostly feline, heads. The same period in Chavin saw the development of luxury crafts with fabrics woven on looms, pottery and goldsmiths' work. The Chavin civilisation vanished rather abruptly and was followed by a period of cultural dissolution, with various small centres (Gallinazo, Paracas etc.) making rapid advances, which marked the beginning of the classic period.

The classic period in Peru lasted from the third century BC to the ninth century AD. Large numbers of local cultures (Mochica, Nazca, etc.) reached their apogee about the beginning

The deserts of the Peruvian coast do not look like an encouraging site for the beginning and development of a great culture, but the river valleys were remarkably fertile and encouraged the development of agriculture, and the sea supplied an unfailing wealth of food.

of the Christian era. The somewhat disrupted classic period was followed by the Tiahuanaco period, which like the Chavin period 1500 years earlier, influenced the whole of the central Andes and reached the small isolated centres. Tiahuanaco lies 4000 metres above sea level on the south shore of Lake Titicaca, 2000 kilometres south of Chavin. At present it is a huge complex of megalithic ruins of temples, monolithic gates, platforms and steps all built of immense dressed and polished blocks held together by tenons and mortices and decorated with low relief. Huge statues adorn the monuments. A remarkable system of water pipes and drainage has been found. The influence of the Tiahuanaco style (it was probably a religious centre rather than an imperial capital) extended a long way north and along the coast. The period of Tiahuanaco expansion was followed once again by a break up of this influence and the formation of small cities (Chancay, Chimu, etc.).

These small states, which had only just taken shape, then submitted to a new power, that of the Incas. This took place in the second half of the fifteenth century, and the Inca empire was one of the most ephemeral powers in the history of mankind, since it was hardly established and still at the height of its expansion when it was wiped out by the Spaniards in 1532. The best known Inca ruins are those of Cuzco and Machu Picchu, but many others could be named.

When the ancient civilisations of Europe embarked on the conquest of the New World in the sixteenth century they had to overcome a mosaic of groups of people who were still in the process of evolution. In the centre were the empires, Inca, Maya and Aztec, which *united the strongly hierarchical urban populations* with their specialised, highly developed craftsmanship, and the rural peoples who were almost all engaged in agriculture. On the perimeter of the empires a compact but scattered mass of peoples, anthropologically, linguistically and culturally very diverse, extended from the United States to the north-west of Argentina, practising a primitive form of agriculture. Finally, in the extreme north and south and in the most inaccessible depths of the forests, the high country and the savannahs, other sparser groups continued in a way of life with technical and economic foundations which had evolved very little in ten or twenty thousand years. In a few decades empr[illegible] and cities, rural peoples, hunters and fishers were all conquered, some massacred, some decimated by diseases against which they had no biological immu[illegible] disorganised and ready more or le[illegible] [illegible] and disastrous assimila[illegible]

What is left of this flowering of aspirati[illegible] and invention? There are many remains, both

The Temple of the Inscriptions at Palenque stands on the typical Maya pyramidal platform approached by a long flight of steps. It is now apparent that both the temple and the mound were contemporary with the princely burial they contain, and must be therefore thought of in the same light as the pyramids of Egypt, i.e. an unusually splendid funerary temple erected to honour the memory of a great ruler.

The Chimu people arrived on the coast of Peru from the north during the eleventh century. Their culture was more advanced than that of their Mochica predecessors, except in the field of pottery. A vase from the great frontier fortress of Paramonga, a few miles north of Lima, shows that the shapes had become formal and unimaginative, the colour reduced to an overall black, and the figures introduced in the surface modelling dull and stylised, though the dancing cat to the right has a certain charm (left). A spiky and elaborate double-spouted jar has a crenellated bridge surmounted by a crowned head, on either side of which are monkeys. On the upper part of the body, prostrate human figures are clumsily modelled in relief.

The unpropitious terrain of the Inca highland empire needed careful treatment in order to realise its full potential. On such precipitous sloping ground cultivation was difficult, and there was always a danger of the soil being swept away by erosion. In order to avoid this, dry stone terraces were built near all the main cities to conserve the land and increase the area available for agriculture.

humble and splendid, forming an important contribution to the history of human technical, social, political, religious and artistic experiment, which have been submitted only recently to scientific investigation. We also tend to forget that many 'discoveries' now in daily use throughout the world owe much to these civilisations. They include beans, tomatoes, potatoes, peanuts, avocados, maize, manioc, tobacco and many others — a whole gamut of tastes, scents or simply nutrition for which we are indebted to the distant prehistoric American Indians.

Special features

Prehistoric research has been fully developed for forty years in North America, and in South America has been developed since the Second World War. It has aroused interest outside the circle of scholars and students, and every year various works on prehistoric America in English, Spanish, Portuguese and other languages are published. The lure of the picturesque and exotic and the mystery of long-dead millennia partly account for this deserved popularity.

In fact, the fascination of prehistoric America goes beyond the mere reconstruction of the evolution of vanished peoples. It touches one of the basic problems of the study of humanity, which is the definition of the part played by determinism, as opposed to invention and originality, in human evolution, and finally, the identification of the laws of evolution, if there are any.

However, the study of humanity is fraught with difficulty. The physicist, the chemist and even the biologist to some degree use experiments as their basic research method, and this

Lake Titicaca, 3812 metres above sea level in the high Andean plateau, was inhabited at an early date by pastoral peoples who defied these inhospitable regions for the sake of the lake fish, the large mineral deposits of the surrounding mountains, and the pasture for their herds of llama, vicuna and alpaca. The lake also produced enormous beds of thick, strong reeds, which the people used to gather and tie in bundles for making the light boats which are still used today.

allows them to make comparisons based on an unlimited series of experiments. In the study of humanity experiment is impossible both for moral reasons (no community would wish to be the subject of an 'experiment' if it impelled them to some course they did not approve) and on practical grounds (an experiment would outlast the lifetime or lifetimes of the experimenters).

In the absence of experiment we would need to be able to follow the vicissitudes of several independent lines of human development. In the ancient world there were no 'independent' lines. The interplay of migrations, contacts and trade etc. involved every civilisation, and none of them, particularly not the most sophisticated, can be considered in isolation from the others. This continual intermingling dates back to very early times. We know, for example, that a homogeneous technical, religious and artistic culture extended from the Pyrenees to Lake Baikal 15,000 years ago. This culture, known as the Upper Paleolithic, illustrates the importance of human migrations and contacts, even among hunters who were mistakenly believed to live in small independent groups without any complex social structure.

Several millennia later mankind's most important discoveries took place in Western Asia. Man mastered the biological forces of nature by sowing seed and domesticating animals. Thenceforward he did not go out to look for his food any longer, but produced in his own fields the species of crop he had chosen as necessary for his subsistence. This process began nine or ten thousand years ago somewhere between the Mediterranean, the Red Sea and the Black Sea. But the question is, was it a unique discovery, the stroke of genius of a group of people, or a normal stage of human development? We may well ask this when we see the same story repeated two or three millennia later in China on the borders of Honan, and in Central America on the other side of the Pacific. The discovery of the cultivation of millet and rice in China was perhaps inspired from across the steppes and came through the medium of the caravans from Central Asia, from the cultivation of barley and wheat in Western Asia. But it is certain that the cultivation of the bean, squash and maize in America was discovered completely independently of the Old World.

The New World occupies a privileged position in the study of the development of human society. In the absence of different forms of human life such as inhabitants of other planets – and they would probably be too different to be valid for the study of humanity – our American Indians are the only means of comparison at our disposal. The Australians also evolved independently of the Old World, but they represent a very primitive technical and economic phase. America, on the other hand, went through stages comparable to those of the Old World, with her hunters and predators, her primitive agricultural communities and her cities and empires. What do these evolutionary similarities imply? Should we conclude from them that there are laws governing human evolution which allow very little latitude for originality? Or that the contacts we have tried to analyse were more important than might be apparent? Our knowledge is still very fragmentary, and the word 'law' is rather daring when applied to a series as incompletely understood and few in number as this. We can still do no more than point the way for investigation and thought.

The Inca were probably the finest stonemasons the world has ever seen. Their buildings were constructed of massive stone blocks with smooth bevelled faces, the contiguous edges ground away to a perfect fit, and even the many earthquakes of the region have totally failed to dislodge them. One of these indestructible Inca buildings is still in use in a street in Cuzco.

Chan Chan was the capital city of the Chimu people who preceded the Inca in Peru. The houses were all made of unfired mud-brick which was carved with fantastic animals and birds and painted in brilliant colours.

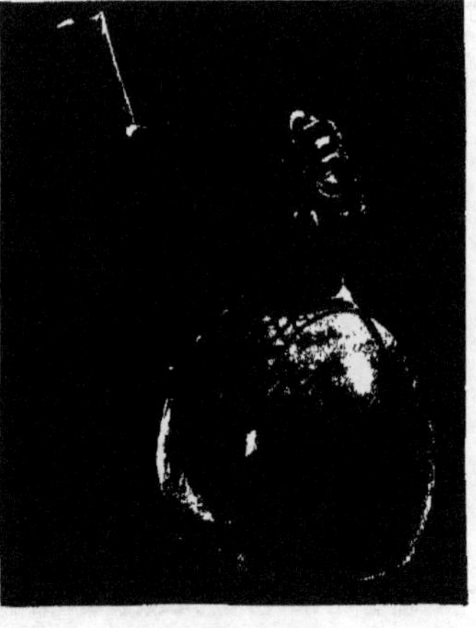

The most distinguished achievement of the Mochica people is undoubtedly their pottery. Zoomorphic or painted vessels reconstruct many details of Mochica life and appearances from the first to the late eighth century. A stirrup jar with the spout set in a large loop handle is crowned by a skull-like head, probably representing a beggar; the body of the pot is painted with a coca bag and a hand holding a pepper, carried out in reddish pigment on an attractive cream-coloured slip.

India, Pakistan and Afghanistan

A curious feature of the art of Mohenjo Daro is that the human figure is very rare, although the craftsmen of the period excelled in the modelling of animals. Bulls, dogs, sheep, elephant, rhinoceros, pigs, turtle, horses and monkeys were modelled. An example of a monkey found in a grave had an alignment of holes running through all four paws for the insertion of the stick up which the animal was shown climbing.

The Indian subcontinent in southern Asia, which includes the countries known today as Afghanistan, is the home of some extremely ancient civilisations. Although these great prehistoric cultures are fairly well documented today, it should nevertheless be remembered that this was not true at the end of the eighteenth century, when the spirit of enquiry was only beginning to turn towards new avenues. In fact, at this time the very history of India had been largely forgotten or radically distorted by legend.

Before embarking on the story of its rediscovery, which is in a large part due to archaeology, it would be helpful to mention the outlines of the geography, climate and history of this immense territory, whose frontiers have seen so many alterations from the earliest times up to the present day.

The geographical setting This Asiatic subcontinent, as big as Europe but smaller than the Soviet Union, can theoretically be divided into two halves by the Tropic of Cancer: the continental region lies north of the line, the peninsula to the south. The latter is triangular in shape, with its point jutting into the Arabian Sea, the Indian Ocean and the Bay of Bengal. The middle of the peninsula is dominated by the highland of the Deccan, furrowed by its numerous rivers and partially covered by vast forests. Its western side, dropping abruptly to the sea, has no space for coastal plains like those which form the eastern seaboard, with their alluvial deposits from the rivers flowing down from the plateau. The Vindhya mountains, which divide it from the Ganges plain, stand on its northern edge.

As to the continental region, it is bounded on the north by the high range of the Himalayas which extends as far as the Iranian highlands in the west, the Vindhya mountains in the south and the Bay of Bengal in the east. In this huge depression two rivers rising in the western Himalayas form two basins lying almost at right angles to each other. The Ganges waters the Indian plains from west to east as far as the lowlands of Bengal, where it joins the Brahmaputra in one of these basins; in the other the Indus flows from the north-east to the south-west, where its meandering course brings life to the sandy regions of southern Pakistan. These two valleys are separated by the Aravalli range and the dry regions of Rajasthan, the Punjab and the Thar deserts.

This land, with its immensely varied geography, has a climate which is equally excessive in its contrasts, ranging from extreme heat to the severest cold. The country is also subject to monsoons, to cold winter winds blowing from the north-east, and to wet south-east winds in summer. Rainfall, too, varies greatly according to the direction of the wind. Thus the mountains in the west (the western Ghats), like the southern foothills of the Himalayas which also meet the monsoon clouds, have very heavy rainfall, while the Deccan highlands remain extremely dry although they are fairly well irrigated by the rivers flowing across them. As to the Indus and Ganges plains with their continental climatic conditions, they suffer severe winters and torrid, dusty summers. They have little rainfall, but are well watered by the Ganges, the Indus and the Brahmaputra rivers and their tributaries.

The high plateaux of Afghanistan have an extremely harsh climate, with excessively hot summers and winters in which travel is prevented by snow and formidable winds.

Communications Although it may seem isolated, India, like Pakistan and Afghanistan, has never been completely cut off from other countries. There are only a few passes through the Himalayas to Tibet and China in the north, but the western and north-western approach routes have been used from a very early date in spite of their many hazards. These routes lie in the north and south: the former linking Iran with Kabul and the middle Indus basin, the latter joining southern Iran with northern Baluchistan and the lower Indus valley. It was by means of these roads that peoples from Western Asia and Europe came in, either for purposes of trade, or as invaders, so that the civilisations of Mesopotamia and Persia and those of Hellenistic Greece, Rome and later Arabia gradually penetrated the subcontinent.

The peoples of central Asia made use of the passes from Karakorum to northwest India.

The eastern coast of the Deccan, although somewhat inhospitable, was also accessible to navigation because of the monsoon winds which drove ships across the ocean. Roman navigators and merchants used this route and reached the coast of Coromandel.

Opposite page: one of the last great centres of late Indian Buddhism was the academic city of Nalanda in north-eastern India. Chinese travellers, who saw the town at the height of its glory in the seventeenth century, described it in words of almost ecstatic praise, but few of the surviving archaeological remains do anything to substantiate this eulogy. One of the exceptions was a stupa built on a podium, now almost submerged in the brickwork of a later building. The sculpture, in the style of early Mahayana Buddhism is similar but inferior to the Gupta work from Sarnath.

The Gandharan style of art flourished in north-western India from the first to the seventh century AD. Its founder was the great Kushan emperor Kanishka, who established his capital at Peshawar and pushed his victorious armies from Bengal to central Asia. He was a fervent devotee of the Buddhist religion, but although he was so important, very little used to be known of his reign. The writings of itinerant Chinese scholars, on which all the histories were once founded, are now supplemented by a few inscriptions, sometimes in corrupt Greek script.

Thus none of the contours, although they are so formidable, constitute an insuperable barrier to human movements in these regions and because of this movement of peoples, India, Pakistan and Afghanistan have achieved a considerable degree of cultural unity during several phases of their history. What is remarkable, however, is that India, entrenched behind her natural frontiers, never undertook military campaigns against her neighbouring countries. However it was by means of some of these same land routes that Buddhist monks from India travelled to these countries to preach the religion of Buddha, which by the fifth century AD had spread as far as China and Japan. In Ceylon and south east Asia Buddhism and the Hindu faith came via the sea route.

The historical setting The inhabitation of this huge region, which remained static for several hundred years, was the result of early ethnic migrations from the north west towards the Indus and Ganges valleys, in the direction of the Deccan and the distant southern regions. Little is known about these early inhabitants, and many questions still remain unanswered, including the inhabitation of the most scattered forest regions by tribal populations who are believed by some to have been driven out by more advanced newcomers. However, it is generally accepted that towards the middle of the second millennium BC the Dravidian peoples of the Indus valley were pushed into the east and south by Aryan invaders. Some evidence for this is found in the distribution of the languages, which belong to a Dravidian group in the south, while the languages which developed in the north are strongly marked by Indo-Aryan Sanskrit. In the heart of Baluchistan (west of the Indus) there are isolated peoples who still use a language belonging to the Dravidian group, which could be a survival from this ancient population movement.

A protohistoric civilisation, probably Dravidian by race, sedentary and city dwelling by custom, and familiar with the use of bronze and writing, then spread through the Indus valley, beginning in the middle of the third millennium and lasting less than 1000 years (we shall be discussing the history of its discovery later).

Aryan peoples, probably originating in Pamir or the Oxus region, then penetrated the Indus valley and India through the north-west frontier passes. It is not known whether they were responsible for the destruction of the culture they found in this region; in view of recent discoveries, archaeologists are inclined to think that this destruction may be attributable to natural disasters. In any case it is certain that the Aryans, who belonged to a linguistic stock with as many offshoots in the West as in the East, brought new beliefs and traditions with them. It was probably during the course of their slow conquest of Afghanistan, Pakistan and the borders of India that they evolved their sacred writings, the *Vedas*. These religious texts, which were only orally transmitted for a long time, tell of the pastoral origins of the invaders and the hardships they had to overcome in order to conquer the lands separating them from the Indus and Ganges valleys, where they succeeded, after many vicissitudes, in establishing themselves.

Other subsequent invaders of India altered and affected the country's artistic and cultural development, but never succeeded in destroying her national character. Thus Achaemenid influences, the result of the campaigns of the great kings Darius and Cyrus in the sixth and fifth centuries BC, can be clearly seen in the earliest products of Indian art. Later, in the fourth century BC, Alexander the Great also attempted a career of conquest which, although he never got further than the Punjab, had considerable influence on Indian art, especially on the schools of Buddhist art which developed in Gandhara (Pakistan) in the north west and Bactria (Afghanistan).

During the first centuries BC and AD Parthian Chaka and Kushan peoples invaded the borders of India from Kashmir to Bactria via the same routes and other passes further to the north before pushing on deeper into the Punjab, as far as the Ganges valley. The Kushans were the only ones who succeeded in founding a strong dynasty in India; this continued during the first and second centuries AD. From central Asia, too, came the hordes of White Huns who ravaged Bactria and the Punjab in the fifth century AD and were not halted until they had penetrated a considerable distance into Indian territory.

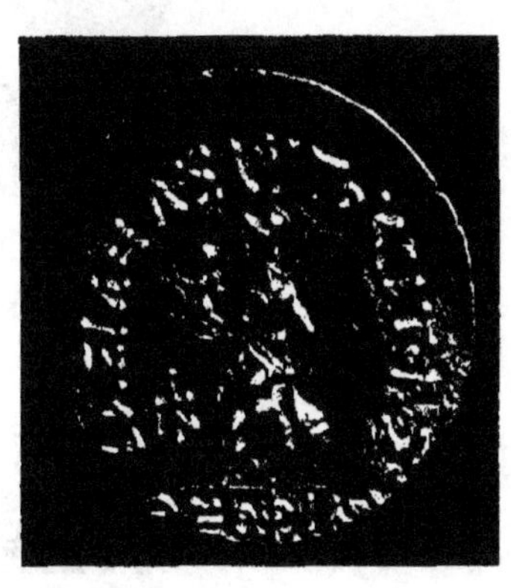

Alexander the Great's dream of extending his empire to the eastern shores of Ocean was never realised, but he was not forced back until he had reached central India, and his passage left an indelible mark on the arts and culture of central Asia. Nowhere was this more true than in Bactria, which was under undisputed Greek rule until well into the first century AD. A coin of Theophilus is entirely Greek in legend, type and workmanship.

These invaders, entering the Ganges lands at intervals of several hundred years, came into contact with the first independent kingdoms, some of which were already ruled by converted Buddhist monarchs. This period, starting about the sixth century BC, saw the development in India of the religious disciplines based on Brahminism, perhaps under the influence of new ideas from the West. Buddhism and Jainism evolved along the same lines, but the former, which was to vanish from India towards the eleventh century, had at the start a truly universal appeal. In the third century BC a Maurya king named Asoka, ruler of the kingdom of Magadha, and a convert to Buddhism, was instrumental in spreading the religion to central Asia and the Far East as well as to Ceylon and south-east Asia.

It was to this emperor, whose very name was to be forgotten until recently, that India also owed the first attempt to unite the land. The attempt did not succeed, but it was a profound aspiration justified by the common religious and moral codes as well as by the social and personal customs of the majority of the population, who were at that time only differentiated by their languages.

After the fall of the Maurya empire in 187 BC, India split up roughly into three regions: the south, the central Deccan and the north. Each in turn attempted to encroach on her neighbours to secure as large a realm as possible for herself. The whole of Indian history reflects this ceaseless competition, in which the Gupta Dynasty, King Harsha, the Pratihara Dynasty and the Chandella Dynasty, were successful in the north, the Satavahana, the Vakataka, the Chalukya and the Rashtrakuta dominated their neighbours for a time in the centre, while the Pallava, Chola and Pandya Dynasties in turn held power in the south.

During these years, however, Islam was slowly advancing, starting in the western provinces in the eighth century AD and eventually succeeding in overcoming the disunited people of India, to establish the Sultanate in Delhi in the twelfth century. The great Mogul empire of the sixteenth and seventeenth centuries made a further attempt to unite all the Indian kingdoms, including Pakistan and Afghanistan under one rule, but the project was never fully completed.

Finally, when in the nineteenth century Britain outstripped her continental rivals in the conquest of the Indian states, she too tried to unify the country by imposing a common economic, social and political system, but the differences of opinion between the two major religious groups, Hindus and Muslims, continued to grow. In 1947 the Indian empire achieved independence, and the country was divided into two states India and Pakistan, the Hindu population settling in India and the Muslims in Pakistan. Afghanistan has always preserved her autonomy and independence.

It should be clearly understood, therefore, that the phrase 'Indian civilisation' is used here to encompass all the elements which make up the common cultural stock of these three countries, at least up to 1947.

India discovers her history

Sources Although Indian culture is *rather* less ancient than those of Egypt and Mesopotamia, it has the advantage from the historians' point of view of surviving in a more or less continuous tradition into the present day. From north to south and from east to west, the orthodox Hindu still recites daily the Vedic hymns composed more than three thousand years ago. In the absence of political unity, which India has never completely achieved, it *was this literary, religious and social continuity* that ensured the moral and intellectual unity of the people.

This civilisation, founded on a much more extensive literature than the other ancient cultures, developed several religious and many philosophical systems over the course of the centuries. Her cultural and religious influence spread all over central and Western Asia and was *especially strong in south east Asia*.

Nevertheless, although it was piously preserved in oral tradition as well as written documents, the culture lost much of its historical context when the Europeans began to intervene in the eighteenth century. Story telling, so typical of the Indian mental process, transformed history into legends of an infinitely remote age, where the names of great men were only preserved in the Purāṇa writings and the sagas, where they were 'submerged in a mist of dreams or the contradictory fantasies of fiction', as Sylvain Levy puts it.

The emperor Asoka (about 268–233 BC) was the first ruler of any significance to make a determined effort to propagate Buddhism throughout his territories. Innumerable buildings and inscriptions attest the extent of his activities. A characteristic feature of his building programmes was the erection of columns always including the bell-shaped inverted lotus capital adopted from Achaemenid Persian architecture, a fine example of which was found at Sarnath near Benares.

A greyish limestone torso of a dancer comes from the second great Indus town, Harappa. The torsion of the figure and the telling simplicity of its surface treatment place it far in advance of the other stone statues of the Indus culture, and something in the movement of the body suggests the dancing figure of Siva from the later Chola period. Indeed, the figure has so little in common with most Harappan work that its authenticity has been disputed.

Inscriptions The history of early India began to emerge from these mists in 1784 when the Asiatic Society of Bengal was founded by Sir William Jones who, with some other scholars, devoted himself to the editing and translating of the Sanskrit texts. Then, in the early nineteenth century, people became interested in studying monuments and collecting inscriptions. The latter, incised on stone or metal, were carried out in Brahmi, Kharoshti, Devanagari and Persian characters. The earliest, written in Brahmi script, were indecipherable for a long time. It was not until 1837 that J. Prinsep succeeded in deciphering Asoka's edicts in this writing; the language was a dialect (Pakrit). A little later, the expansion of the power of Asoka, whose inscriptions were found on rocks and columns in all the countries over which he reigned, was traced by the deciphering of Kharoshti, a script of Aramaic origins, which was particularly common in the north-west of the peninsula. The chance discovery of a bilingual Greek and Aramaic inscription in the province of Kandahar in Afghanistan, and that of another version of one of Asoka's edicts shows that this monarch's empire extended far further than was once believed. Thus it is now evident that most of India was united under a single government in the third century BC.

Although Brahmi script remained in use in most of India until about the fifth or sixth centuries AD, Kharoshti script, the use of which was more restricted, disappeared at the beginning of the third century AD.

The inscriptions mainly consist of proclamations, panegyrics, foundation charters, dedications and pious sentiments. But the evidence on dating which accompanied them is rather difficult to interpret because, except in a few cases, they do not specifically mention the era to which they refer.

If the inscriptions make great contributions to the social and religious history of Indian territory, they are less valuable for art history, not only because their date is often uncertain but because they are rarely found *in situ*, so that no one knows which monuments they originally belonged to or the reasons for their dispersal.

Coins Whether punched, cast or struck, coins are a great help in illuminating the history of India in its obscure period. From the symbols and legends on them the realms of the various princes and the extent of their territorial conquests can be reconstructed. However, here too the lacunae are numerous, since coins are easily melted down for their metal content, and many have not survived.

Nevertheless, those struck by the Greek rulers of Bactria bear the royal portrait as well as the name, and this has made it possible to identify many of the princes reigning over these frontier districts whose names had not been preserved by tradition.

This is equally true of the Scythian and Parthian invaders who succeeded the Greeks in this region, and extended their power as far as Kathiawar. As to the Kushan kings, although they are well documented by excavations, their coins furnish interesting information on the precise extent of their territorial expansion and the length of their reigns.

There are also some rulers of Indian states in the first centuries AD who are only known by their coins. Finally, the splendour and riches of various other rulers, particularly those of the Gupta Dynasty, who made another attempt to unite the country in the fourth to sixth centuries AD, are attested by the remarkable beauty of their gold coins.

Greek and Chinese sources Historians and archaeologists sometimes call on Greek sources, the precisely dated information in which furnishes valuable pointers for dating the few *known facts of Indian history*. *Thus the date* of Asoka has been firmly established by means of references in one of his edicts to several contemporary Greek rulers of the second half of the third century BC.

These sources are also supplemented by Chinese records. These reveal the existence of memoirs written by the Buddhist pilgrims Fah Hien (fifth century) and Huan Tsang (seventh century), who journeyed from China to India to visit the Buddhist holy places. Huan Tsang gave such precise facts about the location of the holy places that archaeologists and geographers have been able, not only to undertake investigations based on these facts, but also frequently to rediscover totally forgotten sites which the two pilgrims do not name.

Above and left: the town of Mohenjo Daro consisted of two parts: the lower city to the east of the acropolis and the central citadel itself, with its massive artificial platform of mud and mud-brick on which stood all the main religious and ceremonial buildings. Long before the Buddhist stupa was erected at the highest point there was a granary, which suggests that the citadel, unlike the lower town, was defended in times of siege. A bath or tank of almost Roman dimensions may have provided ritual ablutions for the priests and people, and a living complex immediately to the north-east of the bath was perhaps a religious college.

Mohenjo Daro is believed to have flourished approximately from 2500 to 1500 BC. The site is subject almost annually to flooding from the Indus, but in spite of this obstacle, systematic investigations have yielded a wealth of information about the Indus cultures. The citadel was a large artificial mound with a number of mud-brick constructions. In the second century AD, a Buddhist monastery and stupa were built on the top.

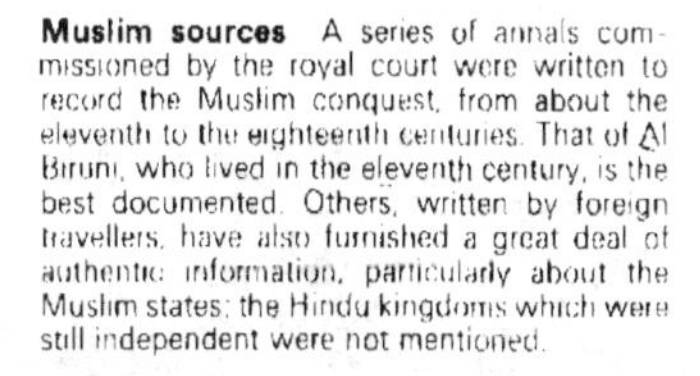

One of the most remarkable figurines belonging to the Indus civilisation is a small bronze dancer from Mohenjo Daro. The figure is nude apart from her necklace and the bracelets which cover one arm from wrist to shoulder. Her features are slightly 'aboriginal' in type, and the artist has skilfully conveyed the relaxed, provocative stance and the inviting sidelong glance of her heavy-lidded eyes.

Muslim sources A series of annals commissioned by the royal court were written to record the Muslim conquest, from about the eleventh to the eighteenth centuries. That of Al Biruni, who lived in the eleventh century, is the best documented. Others, written by foreign travellers, have also furnished a great deal of authentic information, particularly about the Muslim states; the Hindu kingdoms which were still independent were not mentioned.

Biographies Royal biographies are another source of information, but their historical value is severely limited by their panegyric character. However, that of King Harsha of Kanauj (seventh century AD) contributes some good evidence on the religion of the period.

Archaeology Although literary sources were the first to be studied, these lands owe the rediscovery of their remote past to the spade-work, intelligence and enthusiasm of the archaeologist.

Indo-Pakistani archaeology can be regarded as a model example of the part this discipline can play in the rediscovery of the early history of a civilisation. At first scholars concentrated on anything readily accessible, such as the monuments which were still visible and those which were comparable with Greek, Egyptian or Persian culture, and therefore gave points of reference for research.

As we have already seen, further impetus was supplied by the publication of the memoirs of the Chinese pilgrims whose descriptions of ancient cities and Buddhist holy places enabled early nineteenth century scholars to attempt a reconstruction of the early distribution of Buddhism.

This was a period of intensive study, exploration and cataloguing, soon to be followed by the clearing of a few monuments and by the excavations, which yielded quantities of finds. The excavators, however, were more interested in finding pieces which would make a good show in museums than in reconstructing the historical and cultural context which could bring these finds back to life. Serious excavations which accorded as much importance to historical evidence as to the works of art themselves began in the early twentieth century with the intensive researches of Sir John Marshall. It was at this time that the first protohistoric Bronze Age sites were brought to light. The science of archaeology made considerable progress after the Second World War under the leadership of Sir Mortimer Wheeler, who employed the stratigraphic methods developed in the great Greek and Western Asian excavations.

In recent years it has become possible to date the pre- and protohistoric civilisations accurately by means of carbon-14 tests on organic remains, but these tests are only useful for very early material, since there is always some degree of margin for error, which renders such results too inaccurate for the historic periods. In the Tata Institute of Fundamental Research India has a laboratory which specialises in these prolonged and meticulous studies.

Epigraphic evidence must be called on for the monuments of the historic period whenever inscriptions *in situ* are found; another alternative is the comparative method, but this can only furnish relative chronology in the absence of fixed historical dates.

Any research into the history of the art of these countries should therefore combine field investigations, which are indispensable for the assembling of new material, with academic study of the records.

Preservation The quantity and variety of monuments which have been catalogued and entrusted to the protection of the public authorities has called for the formation of a team of experts trained in modern methods of preservation and restoration. The latter process is being strictly confined to reassembling buildings, and is used with the utmost care and discrimination.

The preservation of mural paintings, which are particularly common in India, where they adorn temples (both rock-cut and built) and palaces, raised serious problems. The most fragile as well as the most precious, are those from Ajanta, in the caves dating from the first century BC to the seventh century AD In order to ensure their preservation, a particularly delicate task, the authorities in charge have called in the best international experts and studied the most appropriate procedures. The paintings were cleaned and treated in the nineteen-twenties, and have been reprocessed during the last twelve years.

A few paintings on palace walls which threatened to collapse have been removed and are now exhibited in museums. The best example of work of this nature was carried out in the palace of Rang Mahal (Chambal) by the National Museum of Delhi.

In addition, an ever increasing number of art and archaeology museums have been founded throughout the territory since the beginning of the century, especially since the achievement of independence.

However, the heavy responsibility devolving on these departments in the matter of protecting the buildings from theft and vandalism is by no means sufficiently alleviated by the laws regulating the export of works of art

Some contributions The contribution of archaeology in the Indo-Pakistani subcontinent is particularly spectacular in the realm of prehistory It has brought to light not only the chalcolithic Indus civilisation, but also the three previous Stone Ages (Paleolithic, Mesolithic and Neolithic), which affected the whole territory, in addition to the megalithic culture, later in date than the chalcolithic, which was particularly advanced in the southern part of the peninsula

Discoveries in the field of history are no less spectacular. Side by side with epigraphic and numismatic research and translation of texts, huge sections of history have gradually been rescued from oblivion The Maurya Dynasty of the Ganges valley, which attempted the unification of India under its most renowned monarch, Asoka has become an established historical fact once more. Furthermore, the sacred sites of Buddhism, which were completely lost for about 800 years, ever since the religion died out in the country where it had originated, have now been rediscovered, and have once again taken their places beside the legends of the Buddha and in the history of the

Mohenjo Daro yielded a considerable number of terracotta female figures which appear to have been connected with a fertility cult, although they lack the exaggerated emphasis on the genitalia which characterises Western Asian mother goddesses. Crudely made, with little or no pretence at artistry or detailed finish, they usually have a fantastic fan-shaped headdress with a curious cup fastened to each side with strips of clay. There are indications that these vessels were sometimes used for burning oil or incense.

The type-site of the Indus valley civilisation is Harappa in the Montgomery district of the Punjab. The ancient city covered an area not less than three miles in circumference, and many of its features have added valuable confirmation to the evidence obtained from Mohenjo Daro. Harappa too had a citadel mound, on which stood the main civic buildings and a lower town (Mound E, shown here) to the east. It has unfortunately, been much damaged by later inhabitants in search of bricks, and Mound E has been very little explored.

When Mohenjo Daro was finally destroyed by invaders about the year 1500 BC, they left very little of value in their thorough sacking of the city. A hoard which was buried in the lower city and overlooked by the plunderers shows that the Indus people, in addition to gold ornaments, wore lapis lazuli, turquoise, jade, carnelian, faience and steatite or beads made of ground steatite paste.

social and religious life of the first centuries AD These buried sites, among which are those of Nagarjunakonda, Ratnagiri, Sirpur, Devni Mori and Taxila, will be discussed later

The most striking stone sculpture from Mohenjo Daro is the head and torso of a bearded man wearing a cloak carved with relief trefoils originally filled with red paste. The fillet round the head and the presence of the trefoil, with its widespread symbolic significance, indicate that the figure may be a king or a priest. The hair and beard are rendered with exaggerated stylisation, but the face was given realism by the addition of shell inlays in the eyes.

The Indus or Harappa civilisation

When in 1920 N. R. Banerjee and D. R. Shani chanced to find some seals, then of unknown origin, on the sites of Mohenjo Daro and Harappa, which looked very similar to the examples from the same sites picked up fifty-two years earlier by Sir A. Cunningham, they had found nothing less than evidence of a new Bronze Age civilisation. This civilisation is doubtless not as rich as those of Egypt and Mesopotamia, but the sites where it developed are characterised by an extraordinary skill in town planning. It is now fifty years since researches first began, and the area affected by the culture is continually being found wider. Although its heart was in the Indus valley, evidence of it has also been found in Baluchistan, Afghanistan and India, from the Arabian Sea to the foothills of the Himalayas and the river Jumna in the east.

Sir John Marshall and the whole Archaeological Survey team began excavations in 1921 at both Harappa north of the river Ravi and at Mohenjo Daro on the Indus, about 650 kilometres to the south. This resulted in the re-appearance of two huge Bronze Age cities, where stone tools were used alongside bronze and copper, though no iron objects were found. Although no written history is known from this culture, it is designated protohistoric since it was already using a script, specimens of which, as yet undeciphered, appear on seals.

These great cities were built on strictly planned lines, with well designed houses along geometrically planned streets and avenues. The houses were often spacious and some were provided with bathrooms from which the water ran out to drains in the middle of the roads. These drains, like the houses, were remarkably well constructed of fired brick. Wells and latrines completed the sanitary arrangements. Sir John Marshall's excavations, reopened by Sir Mortimer Wheeler after 1944, revealed that outside the town there was a citadel enclosed within a defensive wall with bastions, and some inexplicable monuments set on high platforms, perhaps intended to protect them from floods. One of these buildings at Mohenjo Daro was excavated, and Sir Mortimer Wheeler realised that it was intended to store the city's grain supply. A similar building was also found at Harappa, but outside the citadel.

It seems that these people believed in an afterlife, as they generally buried their dead in the extended position, with the head pointing north, and surrounded them with their everyday utensils and with gifts.

Nevertheless, not a single building resembling a temple has been found, though at Mohenjo Daro a great bath surrounded by small rooms may have had a religious function. Only the numerous terracotta mother goddess figurines indicate that the inhabitants practised a fertility cult.

As for the arts, they are scarcely represented at all, although the steatite seals with animal motifs (mainly humped cattle, rhinoceros and crocodile) and the remarkable stone, bronze and even terracotta figures show great skill in execution, and point to the existence of artists able to capture the essence of a movement, the charm of a model or even a comical feature. Pottery is the chief of the minor arts. Shapes are varied, but it is usually of pinkish clay with a red slip, and black paint showing stylised animals, plant motifs (the most popular of which is the peepul tree) and geometric motifs of intersecting circles and scales formed the basic repertoire of the potters.

This pottery has provided valuable comparisons with Mesopotamian ware. The Indus populations evidently established trade contacts with Mesopotamia, proof of which is given by the discovery of seals and various objects of Harappan origin on the sites of Sumer and Akkad. The first dates suggested for the duration of the Indus civilisations, from 2500–1500 BC, were based on these finds. Carbon-14 tests conducted since 1949 on organic remains from recently excavated peripheral sites in India indicate a shorter period, approximately 2300–1850 BC. However, *Sir Mortimer Wheeler* believes that in the case of the great Indus cities, the occupation period was in fact much longer.

In spite of recent research, the origins of this culture are still obscure. Although the excavations of N. G. Majumdar and J. M. Casal at Amri, F. A. Khan at Kot Diji in Pakistan, and B. B. Lal and B. K. Thapar at Kalibangan have revealed the presence of cultures even earlier than that of Harappa in their lowest levels, it seems that they could not have been 'the direct ancestor of Harappa', as J. M. Casal wrote of Amri. What is more, the lowest flooded strata at Mohenjo Daro have never been excavated and only a few recent trials indicate that this ground contains evidence of civilisation.

As far as the expansion and evolution of this culture are concerned, a hundred sites, towns and villages which can be called Harappan have been recorded by means of surface finds and excavations in both India and Pakistan. Excavations at Rupar at the foot of the Himalayas and Alamgirpur north-east of Delhi have already *confirmed that it spread towards* the east. For the last fifteen years we have been confronted by veritable towns, with the coastal city of Lothal in Kathiawar excavated by B. B. Lal and B. K. Thapar. Although these towns are small, and have their own local features, they all share the basic characteristics of the Indus capitals, i.e. town planning, brick-built houses (the bricks were fired) with bathrooms, drains and wells of fired brick, an isolated citadel defended by massive walls with bastions, the use of seals with a script and animal motifs as well as pottery with the same characteristics as Indus ware. Burial methods, too, are similar. A few male and female figurines complete the scanty artistic output of these people. There is not a *single image which could have a religious* significance to lift them above the level of mundane life.

Lothal, which once lay close to the sea, is the only one of all the sites with any traces of port installations; here there is a large rectangular harbour with a sluice-gate across it.

The reasons for the disappearance of this culture from both Mohenjo Daro and Harappa as well as the minor towns of the Indus valley are still obscure. It was once attributed to the Aryan invasion because of the skeletons scattered about the ground at Mohenjo Daro, but nowadays excavators agree that if there was an invasion, it took place when decadence had already succeeded great prosperity, as Sir Mortimer Wheeler has clearly shown by his last excavations at Mohenjo Daro. Reasons for this decadence may be the successive floods of the unpredictable river, or the drying up of the farm lands caused by cutting down trees to stoke the brick kilns, and neglect of the irrigation system. The question is still disputed.

The decadence which is apparent in the upper strata of several Indus valley sites extends to the cultures of Jhuktar and Jhangar, two small centres to the north of Mohenjo Daro, but it is not certain that they were directly related to the Indus civilisation. These cultures are distinguished by the appearance of new pottery types and the abandonment of any attempt at town planning.

One of the most tantalising mysteries of the Indus cultures is the script, which has so far defied decipherment. It occurs on graffiti, tablets, pottery and seals, and changed very little throughout the history of the civilisation. The number of signs (nearly 400 have been recorded) suggests that it may be syllabic or even ideogrammatic, but it is unlikely to be alphabetic. The longest known inscription contains only seventeen signs and, unless a bilingual piece in a known language is found, there is little hope of a solution to the script. Some of the characters can be clearly seen over the figures of the bulls in the intaglio seals which were the outstanding achievement of the Indus valley artists. Apart from beautifully cut representation of animals, fabulous and real, which help to shed light on the fauna and climate of the period, a series of human figures, much less well executed, are also shown. The men of the seals are nearly always divine or ritually occupied. One of the gods was a strange figure associated with animals, perhaps a forerunner of Siva in his aspect as Lord of the Beasts.

The great painted caves at Ajanta contain some of the finest work of the Gupta and Early Chalukya periods. The walls were prepared by coating the rock with a blend of clay or cow-dung mixed with hair or chopped straw, and finishing the surface with gesso. The paintings themselves (the illustration shows the ceiling of Cave II) convey the profound charm of a period in which secular sensuality was closely interwoven with religious fervour.

Taxila

At the beginning of the twentieth century Gandhara (now in northern Pakistan), the meeting place of east and west, was known only from a few Greek and Indian texts, coins and inscriptions, including an edict of Asoka. However, Sir A. Cunningham's excavations and some chance finds had already indicated the highly distinctive nature of the culture which developed there.

Beginning in 1902 Sir John Marshall, newly appointed Director General of the Archaeological Survey, also undertook excavations at Charsada, the former capital of Gandhara, with the collaboration of P. Vogel.

His colleagues Spooner, Hargreaves and Aurel Stein, with the assistance of A. Foucher, a French archaeologist, set to work on neighbouring sites, and these excavations furnished the basic information on the Buddhist monasteries and the way in which they functioned.

In 1912 Sir John Marshall began to excavate Taxila (ancient Takshacila), which lies on the northern route between Afghanistan and India not far from the Indus. Alexander stopped there on his way to the East, and it is also where the Maurya domination of India started.

The findings of the excavations, recorded in the *Archaeological Survey of India* (the annual report of the Indian Archaeology Department), were considerable, and raised so many problems that present investigators are still divided over some basic points.

However this may be, Sir John Marshall and his assistants, who continued the work until 1931, revealed three successive cities on the site of Taxila. The earliest, Bhir Mound, which was still poorly laid out and built, belonged to the period of the Persian satrapies in the sixth century BC; it lasted until about the first century AD.

In the second century AD, however, a new town, Sirkap, had already been built on another bank of a small river. Sir John Marshall wrote that the excavations revealed five or six strata there covering a long period of occupation.

The lower strata attested occupation by Indo-Greeks, who were superseded by Scytho-Parthians in the first century BC and then by the first Kushans (earlier than King Kanishka, or about the second century AD).

The best preserved city belongs to the Parthian period. It is laid out in a chequer-board plan with a wide central road intersected by minor streets flanked by shops and religious buildings. A defensive wall of rough masonry with several bastions protected the town. This plan is somewhat reminiscent of contemporary Ganges cities, such as Kaucambi.

In the second century AD the inhabitants abandoned Sirkap and settled further north at Sirsukha, which was devastated by the White Huns in the fifth century AD, and has not yet been properly excavated.

In 1945 the Sirkap excavations were reopened by Sir Mortimer Wheeler, who made several trial trenches before attempting to establish the successive order of the strata and the cultures, which his predecessor's extensive excavations had done little to clarify. These trenches revealed that the first city near the river bank had been removed bodily to the south so as to include the hill of Hathial, where it was replanned and fortified.

There were many monuments, mostly relating to the Buddhist faith which, according to tradition, and confirmed by an inscription of Asoka, had been established in the country during the Maurya period. Some of these monuments were inside the city, but the majority, designed for congregations, were large structures outside the city dating to the Kushan period. The Dharmarajika, for instance, was a huge complex including a stupa, some small temples (often apsidal in shape) and some monasteries.

The wealth of decoration, both representational and architectural, which enlivens all these buildings, shows the extraordinary vitality of this new art style, which has been named Graeco-Buddhist. The style resulted from the fusion of Indian Buddhist art, which was still in a formative stage, with Hellenistic art, which had already been affected by Iranian influences. This basically religious art has various local characteristics which obviously arise from the use of two materials with profoundly different sculptural possibilities – schist and stucco. Other stylistic characteristics may be due in part to differences in weather conditions, but only the systematic excavation of a site similar to Taxila will be able to provide a thorough explanation.

The Graeco-Buddhist style was, as we shall see later, developed simultaneously in Afghanistan and the provinces of Bactria and Kapica. It then spread into central Asia, Kashmir and the northern Punjab. It even affected southern Pakistan, at Mirpur Khas in Sind, and reached Devni Mori in the province of Gujarat in India.

The whole of this complex style in which several civilisations meet is discussed in its main aspects in Alfred Foucher's *l'Art gréco-bouddhique du Gandhara* (1905–21).

Recent excavations on Indian Buddhist sites

The remains which are now to be discussed are either situated in inaccessible regions, or in areas threatened by dam-building operations.

Ratnagiri (Orissa) The Buddhist site of Ratnagiri near Bhuvaneśvar was recorded, but since it was cut off by irrigation channels, it was not easily accessible. The excavations of the Archaeological Survey of India, conducted by Mrs D. Mitra from 1957 to 1960, exposed a huge complex including a large star-shaped brick stupa surrounded by innumerable small votive stupas and accompanied by two monasteries and some small rectangular shrines a short distance away.

In addition to cells round a paved court, the two- or three-storeyed brick-built monasteries included a small shrine opening on to an end wall in line with a monumental gate. This gate, like that of the shrine, had stone jambs with very fine decoration, in which floral scrolls alternated with images of the gods of the Mahayana Buddhist Tantric (Vajrayana) pantheon.

None of the inscriptions from this site gives any indication of its date. On the basis of stylistic comparisons alone, Mrs D. Mitra has suggested the eighth century as the probable date of these buildings.

Sirpur The buildings on the ancient site of Sirpur, the former capital of a vanished empire, were of the same type as those at Ratnagiri. The site, which is buried in jungle east of Raipur, was already well known for its Hindu temple. However, the excavation of M. G. Diskshit's excavations, conducted for the University of Saugor in 1954, was to reveal the Buddhist remains. He found several Buddhist monasteries and temples but no stupas. The temples, like the Ratnagiri complex, followed a plan which seems to be related to the common northern Indian type of the sixth to eighth centuries, consisting of a porch and a hypostyle inner hall surrounded by cells, with a shrine-cell in the middle of the rear wall. Here, too, the bronze and stone carvings belonged to the Mahayana pantheon. Some of the bronzes show remarkable technical skill, and are related to a bronze-working tradition, examples of which have been found in several different provinces in India. In this case, Diskshit was able to date the buildings to the early eighth century by means of the inscriptions.

The terracotta animals of Mohenjo Daro are nearly all modelled by hand without the use of a mould, and some attain a degree of excellence which is seldom repeated in the other arts of this era. A figure of a bull, with its short legs, thick neck and heavy dewlaps, is admirably observed and the strong but simple treatment of detail shows a firm grasp of decorative principles.

Devni Mori The site of Devni Mori near the course of the river Sabarmati in northern Gujarat, which was due to be submerged by the waters of a new dam, was excavated from

At the right-hand side of the rear of Cave II at Ajanta are a pair of seated statues representing a ruler and his queen, surrounded by a wealth of smaller subsidiary figures and roofed in by a delicate floral ceiling. The enclosing niche gives a feeling of unity to the composition despite the multitude of characters, and the repose and dignity of the carving are typical of the classic Gupta style.

Among the largest and most magnificent of the early Buddhist cave sanctuaries, or chaitya-halls, is that of Karli near Bhaja, the interior of which is similar in scale to a Gothic church. The sixteen-sided columns rest on water-jars and are topped by groups of elephants with men and women riders, the elaborate carving of which forms a telling contrast with the severe treatment of the other architectural members. At the end of the hall is a model stone stupa with its umbrella-like canopy.

1959 to 1962 for the University of Baroda by the late R. Subbarao, N. H. Mehta and S. N. Chowdhary.

The monuments, a stupa and a monastery, belonged to a different style from those we have just discussed. They are the most distant indications of the spread of Graeco-Buddhist Gandharan influences in India, the best known example of which in this southern area is the stupa of Mirpur Khas in Sind in southern Pakistan.

The two square bases of the stupa were decorated with innumerable terracotta sculptures which Chowdhary compared to those of the last Gandharan period (third century AD). This hypothesis was confirmed by the discovery of a reliquary bearing a date corresponding to AD 205. If this was the foundation date, then the holy place must have been inhabited for several centuries. All the portable parts of these buildings have been housed in the museum of Baroda.

Nagarjunakonda Nagarjunakonda, one of the most typical southern Indian sites, was already known and excavated in 1927, but its position on the stretch of the river Krishna where a dam was to be built threatened it with total destruction. The Archaeological Survey of India undertook a salvage operation, like those on the Egyptian temples in Nubia, and since 1954 extensive stratigraphical excavations have been conducted there. These furnished information about the prehistoric and historic periods, and thus the history of a site which had an unusually prolonged occupation was reconstructed.

Men settled at Nagarjunakonda as early as the Neolithic period, and the megalithic culture subsequently produced several cist graves with stone enclosures as well as grave goods of red and black pottery and iron tools.

However, as far as one can tell from the brief reports published to date, the excavations succeeded in contributing supplementary information on the Ikshavaku dynasty which ruled this region during the third century AD. The names of several rulers belonging to this dynasty appear in some pious foundation inscriptions.

These rulers seem to have been ardent Hindus, as far as can be deduced from both the number of temples dedicated to the god Kartikaya (of the Shaiva pantheon) and from the installations destined, according to the inscriptions, for the great Vedic rites of royal consecration. However, the dynasties evidently also encouraged the establishment of Buddhist congregations, the vitality of which is attested by many foundations. The monasteries belonging to these foundations were radically different from those of the north. Two small apsidal shrines, one of which housed the cult image of

The most typical Indian monument is the stupa, or relic-mound, several of which occur at Sanchi. The proportions of these mounds were mathematically fixed, and every detail embodied a feature of cosmic symbolism. The dome itself represented the heavens enclosing the earth, the platform on the top was the abode of the gods, and the mast in the centre was the axis of earth and heaven, topped by the umbrella which symbolised the heaven of Brahma.

the Buddha and the other a miniature stupa (dagoba), were constructed inside the monastery. Stupas stood outside it. All these buildings were made of fired brick and covered with limestone reliefs with Buddhist motifs.

Other buildings, such as the stadium and the dock installations, probably had a more secular character. This stadium is the only known example in India; it consists of a central oblong surrounded by steps, all of which are faced with white slabs. The same care was lavished on the decoration of the ornamental stairs leading down to the river bank.

The findings of the fifteen years of work on this site have not yet been published, but the most outstanding buildings have already been dismantled and reconstructed by anastylosis on top of a nearby hill which is now a sort of island. They will gradually regain their original surroundings and appearance as the work nears its completion.

The story of the discoveries

Among the pioneers of Indian studies now to be discussed, are Sir William Jones and Charles Wilkins the former established the first fixed date of Indian history when he demonstrated that the Maurya king Chandragupta was contemporary with Alexander the Great, and the latter, who deciphered the Gupta script in the last quarter of the eighteenth century, can be regarded as the founder of this new discipline.

At the end of the eighteenth century and the beginning of the nineteenth scholars began to study the monuments in a more serious manner than previously. The discovery and the first accounts of a number of monuments which turned out to be landmarks in Indian art history date to this period. Sanchi, Amaravati and the caves of Ellora, Ajanta and Badami, as we have already mentioned, are the best known.

In 1837 J. Prinsep, who owned a collection of inscriptions in an unknown script including writings on stones, on columns and on the Sanchi reliefs, succeeded in deciphering this script, Brahmi, after long years of research. This discovery was an important one for ancient Indian studies, since it revealed the existence of the Maurya emperor Piyadasi. This Piyadasi, deciding to follow the Buddhist faith and urging his people to do likewise, was revealed as the man whom the Sinhalese annals called Asoka, who sent his daughter and son to Ceylon to preach Buddhism. We have also seen that the date of his reign was established by a reference in one of his inscriptions to contemporary Hellenistic rulers of the middle of the third century BC.

This discovery was confirmed by Masson's work on Indo-Greek coinage on which the names of Greek rulers appeared, and was shortly followed by the deciphering of the Kharoshti script, which occurs in north-west India and central Asia.

In the same period Prinsep organised the first excavations of a Buddhist site at Sarnath not far from Benares, which was to yield an enormous number of buildings and carvings dating from the first centuries AD to the early medieval period.

Prinsep died in 1840. His disciples continued his work both in India and in Europe, in the learned societies and institutes which were beginning to be established in the first half of the nineteenth century. In India itself Alexander Cunningham took up the torch, concentrating on the Indo Greek and Indo Scythian dynasties, while another group of investigators pursued their enquiries in the various fields of epigraphy, numismatics and the recording of the monuments, including the megalithic remains.

In 1844 the Indian government assumed the responsibility for setting up a council responsible for the preservation of the caves in Bombay province and for copying the paintings on the walls of the Ajanta caverns, but it was not until 1861 that the Viceroy of India, Lord Canning, ordered the foundation of an organisation for recording monuments and inscriptions, with Alexander Cunningham as inspector.

With somewhat slender means at his disposal Cunningham began the task of investigating the archaeological resources of the country. He was particularly interested in the problems

The view from Cave I at Ajanta shows the cliff face with the entrances to the carved and painted caves which are one of the most remarkable achievements of Indian art. The site was formerly a holy place associated with a Nagaraja who lived near the falls at the end of the valley. Buddhist monks began work on the cliffs as early as the second century BC, and by the sixth century AD they had hollowed out and decorated no less than twenty-six temple-caves and assembly halls

The vale of Kashmir, though closely connected with the empires of Asoka and Kanishka, always preserved a degree of individuality because of its geographical isolation high in the foothills of the Punjab Himalayas. In the fourth century AD and for some time afterwards, buildings at Harwan were decorated with terracotta plaques with impressed figures in oblong frames. They show a pleasing and highly distinctive blend of classical and Indian influences.

The village of Paya above Srinagar is distinguished for its elegant little Siva temple, which perhaps dates to the eleventh century. The trefoil enclosed in the pointed pediment holds a lively relief of a dancing Siva accompanied by a lute and a drum. The pilasters surrounding the temple have unusually fine floral capitals and the pyramidal roof is crowned by a fully open lotus flower.

of Buddhist topography raised by the publication of the journals of the Chinese pilgrims Fah Hien and Hiuan Tsang, who had travelled in India visiting legendary and traditional Buddhist sites in the fifth and seventh centuries AD. From 1861 to 1865 Cunningham scoured the whole of north and north-west India. His records appeared in the annual review which was founded shortly afterwards.

At the same time several geologists became interested in the prehistoric human remains from river beds in the Madras region, as well as in the megalithic tombs.

However, systematic records were impossible before the founding in 1870 of the Archaeological Survey of India. Alexander Cunningham was appointed Director General, and he and his assistants, J. D. Beglar and A. C. Carlleyle, tirelessly searched and studied all the provinces of northern and central India, where they located and described the chief sites of the ancient Indian civilisation. The most important sites were Taxila in the Punjab, Barhut in the central provinces, Khajuraho in the Bundelkhand and the Buddhist shrines of Malwa (Deogarh, Eran, Tigawa), Bihar and Uttar Pradesh (Boah-Gaya, Sravasti, Kaucambi, Ahichchhatra and many others).

All these discoveries were reported in an annual publication, the *Archaeological Survey of India Report*, which was published until 1885 when Cunningham gave up his post as director. The inventory compiled by Cunningham is still useful. His knowledge of epigraphy and history, allied to an extremely acute observation and profound intuition, enabled him to suggest stylistic relationships which are quite remarkable for his time, and are mostly acceptable to this day, even if a few of his conclusions have had to be revised. He also considerably helped epigraphic studies by publishing the *Corpus Inscriptionum Indicarum*, which led to the foundation of the Epigraphical Survey in 1883.

However, although the chief monuments were recorded and some of them were cleared, hardly anything was done about their actual preservation. This naturally had a disastrous effect on some of the sites, which were being destroyed by public works or systematically looted by collectors, antiquarians and even museum curators, who cared more about enriching their collections than about preserving the buildings, which were soon dismantled once they were stripped of their decorative carvings.

Cunningham's successor, James Burgess, was another of the great men of Indian archae-

Nagarjunakonda, one of the most important sites of southern India, has only been thoroughly investigated in recent years because its isolated position makes it extremely difficult of access. The third-century Ikshavaku rulers were devoted to the Hindu faith, yet they accepted a complex of ritual buildings, including two apsidal shrines, one for a statue of Buddha and the other for a dagoba (miniature stupa) inside the monastery. Outside stood another stupa with a curious star-shaped internal arrangement.

ology of the last quarter of the nineteenth century. He had already worked on the monumental inventory of southern India temples and was also interested in epigraphy. It was at his instigation that *Epigraphia indica* was founded. However, his chief interest was architecture, both Hindu and Muslim. Abandoning the annual publications inaugurated by Cunningham, Burgess adopted a system of monumental publications on specialised subjects. These gave place to a new publication, the *New Imperial Series*, the first seven volumes of which appeared under his direction.

During the next years the Indian government gave up archaeological research, and it was not until Lord Curzon, the new viceroy, expressed his indignation at the neglect of the monuments that the Archaeological Survey was resuscitated and took action once again. Sir John Marshall was appointed Director General in 1902 and reorganised this huge department with its varying functions, including both the administration and the practical maintenance and preservation of the monuments. In the latter field he laid down the rule that nothing was to be restored or reconstructed unless there was no other alternative, and that nothing was to be added to the decorations. Many monuments were thus saved from destruction, including the great Muslim complexes of Agra, Delhi, Ajmer and Bijapur, and the Hindu centres of Konarak, Chittor, Khajuraho and Bhuvaneshar, to mention only the most famous examples.

Marshall encouraged the independent states to organise their own archaeology departments, and museums, usually housing stone carvings, were established in several large towns as well as on the sites themselves.

The publications department was also reorganised. The *New Imperial Series* was continued in the earlier form of an annual report, and Marshall himself edited the magnificent series of *Annual Reports of the Archaeological Survey* from 1902 to 1937, covering all the department's activities. In 1919 he also inaugurated the new series of *Memoirs of the Archaeological Survey in India*. Epigraphy was not forgotten; he issued a new edition of the *Corpus Inscriptionum Indicarum*.

Sir John Marshall's diligence did not end there. He concentrated his own efforts and those of his colleagues on exploration and excavation, with the basic aim not of collecting works of art, but of discovering all the aspects of the ancient cultures of India – their towns, streets and furniture as well as their laws and customs.

The temple of Avantisvami-Vishnu at Avantipur near Srinagar is not large but it is well preserved and extremely ornate. It was built when King Avantivarman was a young man, shortly before he came to the throne in 855. Unfortunately, the defensive potential of the temple attracted military attention from an early date, and it suffered severe damage during wars and invasions.

Since 1954, the University of Saugor has been conducting the excavation of Sirpur, once the centre of a considerable empire which no longer exists. It was the site of a Buddhist community of monks, who built a monastery consisting of a court surrounded by cells. In the background of the photograph is the monastery shrine in the Mahayana Buddhist style, with a seated statue surrounded by heavily carved posts.

When the Greeks who were the descendants of Alexander the Great's generals established their rule in Bactria and Afghanistan, they brought their art and religion with them. These pieces form a curious contrast to the local style, as they are purely Graeco-Roman in inspiration and workmanship. A fine bronze figure dating to the first century AD represents a hybrid of the Greek hero Heracles with his club and the Egypto-Roman Serapis wearing the grain-measure symbol of fruitfulness on his head.

Opposite page: a view of the great central square of Monte Alban, looking to the east. One of the most spectacular of the Mexican sites, the city was the crown of the Zapotec culture.

The excavations at Begram in Afghanistan have yielded a wealth of varied material. In addition to Graeco-Roman finds, Chinese lacquer ware, bronzes and glass, were some fine incised ivory panels dating to the first century AD.

Overleaf, left: Ajanta, in Hyderabad, is the site of thirty caves hollowed out in the side of a ravine overlooking the Wagura river. They were discovered in 1817 and found to contain superb frescoes on Buddhist subjects dating from the sixth and seventh centuries AD. This painting is from Cave 1 and shows a Bodhisattva (one who will become a Buddha on earth). He is attired like a Gupta prince and holds a lotus in his right hand.
Right: a torso from Sanchi, of the post-Gupta style. This period in Indian art showed a mastery in the representation of the human form which was centuries ahead of the art of Europe. Ninth century AD, of red sandstone, the figure is now in the Victoria and Albert Museum, London.

Like Cunningham, Marshall was attracted by the culture of the Gandhara region where the heritage of Greece met that of India. He undertook the excavation of the site of Taxila, where he worked for twenty years. His excavations and discoveries, which are described more fully elsewhere, found their natural continuation in Afghanistan and central Asia.

Sir Aurel Stein's remarkable work in central Asia from 1901 to 1915 revealed the remains of a civilisation strongly marked by Graeco-Buddhist and Chinese influences dating from the first to the tenth centuries AD. He recorded texts in Kharoshti, Brahmi and Sanskrit, as well as paintings on silk, cotton and paper, and carvings in stucco, wood and stone.

Other investigators such as Paul Pelliot also gathered remarkable collections from oases like Tuen Huang, which now enrich European museums.

In India proper Sir John Marshall concentrated on the excavation of the Buddhist sites which he believed would yield a great deal of information and many finds. His hopes were not in vain; his findings from such sites as Sarnath, Kasia, Rajagriha, Basarh, Nalanda, Vaicali, Pataliputra, Bhita and Sanchi were remarkable, proving that there had been a veritable renaissance of Buddhist art and culture from roughly the third to the ninth centuries. The public marvelled at this hitherto unknown art, the aesthetic quality of which was comparable to the best of European art. Moreover, there was something in its range of expression, extending from barely modified Hellenistic to a profoundly original Indian style, which stimulated a great deal of research. To name only some of the earliest studies, there was Alfred Foucher's masterly book *l'Art gréco bouddhique du Gandhara* and Philippe Vogel's *l'Art de Mathura*, as well as Sir Aurel Stein's publications of his investigations. This period witnessed the reappearance of the early history of the entire Indus Ganges world.

In 1922, however, two chance finds attracted the attention of scholars to prehistoric India, which Sir John Marshall had been too busy to investigate. Excavations were immediately put in hand, and these resulted in the discovery of a protohistoric Bronze Age culture which evolved along the course of the Indus between 2500 and 1500 BC, the main features of which have already been described.

As a result of these discoveries investigations were divided into two streams: to find the origins of this culture, which was known as the Indus or Harappa culture, in the sites along the route towards Mesopotamia, and to attempt to define the area of the settlement and diffusion of this culture, the importance of which was instantly recognised by the whole academic world.

All these considerable labours and the publication of their findings occupied the end of Sir John Marshall's directorship. He retired in 1928, but continued to conduct the excavations of these sites until 1934. This resulted in the publication of *Mohenjo Daro and the Indus Civilisation* in 1931 and *The Monuments of Sanci* and *Taxila*, in collaboration with Foucher, in 1951.

Sir John Marshall's working hypotheses and methods of excavation have been criticised as being unscientific in comparison with subsequent research and the more scientific methods used later. There is some truth in this, but in his day methods based on the study of stratigraphy were not yet being regularly applied even in the Mediterranean excavations. The results he achieved can only be admired, and when in 1935 he left India, the country was considerably enriched by the knowledge of the past he had done so much to revive.

After this highly productive period the Archaeological Survey carried on similar investigations under the successive direction of Sir John Marshall's chief collaborators, but the world crisis of 1939 and the Second World War forced it to curtail some of its services, and its publications ceased.

During the next twenty years, however, archaeological policy for the north-west became more liberal, and foreign teams were authorised to work in these parts. This was the time of the Yale Cambridge expedition (1933), led by H. de Terra, in which Teilhard de Chardin took part. Tackling the problem of the Stone Age in the Punjab and Kashmir, he found an industry of 'pebble strikers' related to the Clactonian. It was named Sohanian.

In 1937 Indian universities and learned societies were also invited to take part in investigations and excavations. The results of this collaboration, which is still in progress, were excellent.

Despite the difficulties of this period work continued with the clearing of new sites such

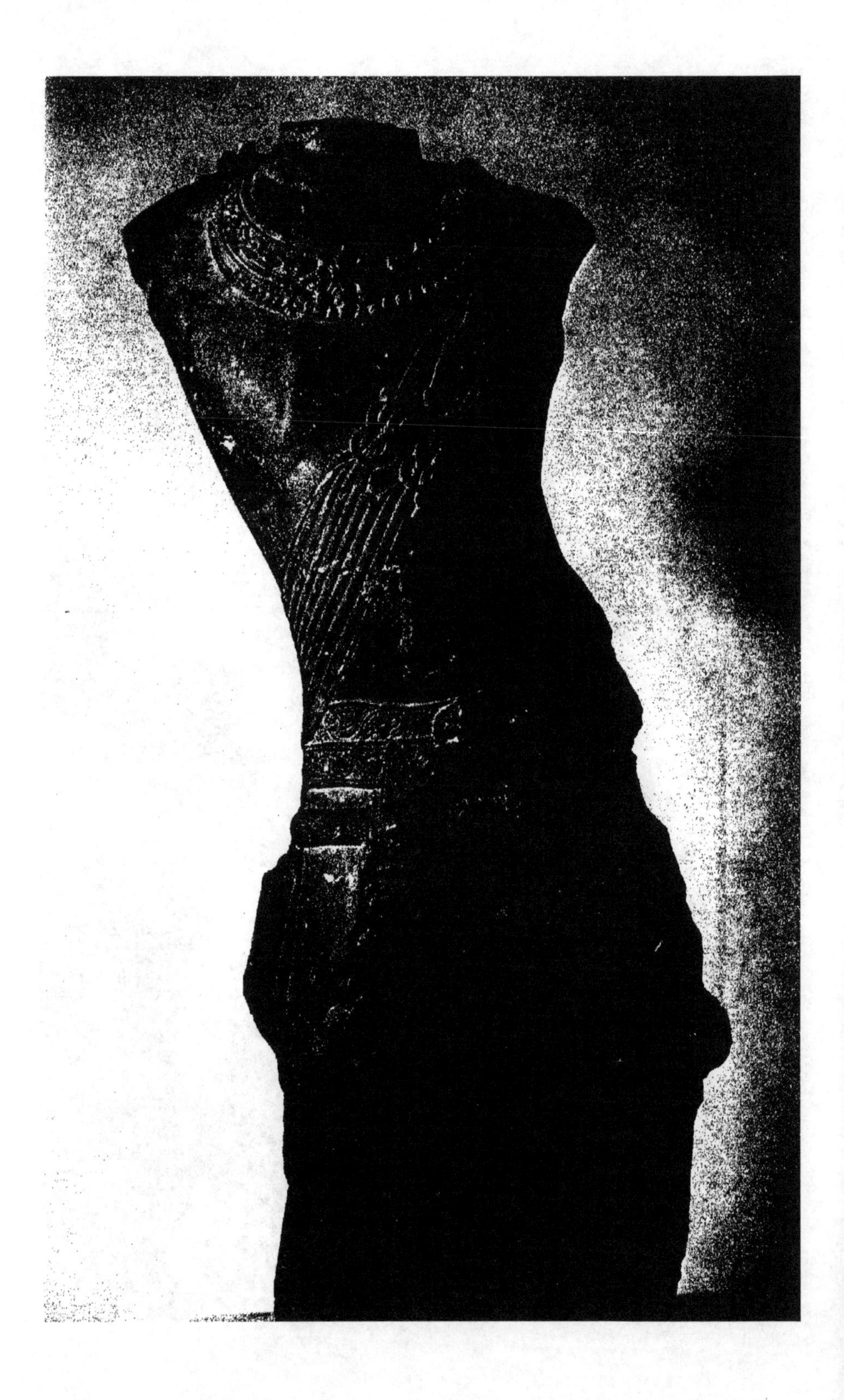

as the stepped temple of Paharpur in east Bengal. The discovery of an ancient Harappan city in the northern Punjab on the river Sutlej opened new avenues for later investigation by showing how far this culture extended from its centre.

A very large dig was conducted between 1940 and 1944 at Ahichchhatra on the upper Ganges in Uttar Pradesh. Besides an ancient Hindu temple, the lower strata of this site yielded grey painted pottery with black burnished ware above it. Both of these were later than the Harappan period, and subsequently served to define two cultures which spread throughout the Ganges valley during the first millennium AD.

After 1944, when Sir Mortimer Wheeler was appointed Director General of the Archaeological Survey, new methods of preservation, museum management and research were applied, particularly in the field of excavation, and new attention was given to the training of future archaeologists.

As to publications, the annual reports were not resumed as such, but were superseded by a new review, *Ancient India*, which was devoted to various branches of archaeology.

Excavations in the Indus valley, which had temporarily been abandoned, were reopened, and the stratigraphical method, adapted by Sir Mortimer Wheeler to Indian conditions and henceforth used for all the digs in this region, was employed.

The use of this method made it possible for the first time to establish continuity from the early strata to the full flowering of the Harappan culture and the late cemetery known as Cemetery H at Harappa itself. Sir Mortimer Wheeler was thus able to prove that the culture of these late strata was probably the result of infiltrations of new peoples whose origin could not be determined. Other excavations on this site and that of Mohenjo Daro showed that there had been a citadel next to the town. This citadel implied a martial aspect to these towns which had not previously been observed.

Sir Mortimer Wheeler also turned his attention to southern India, hoping that systematic excavations would prove able to establish the links between trade goods from the Roman empire and contemporary Indian local industries. Also in 1944, excavations were started in Pondicherry at Varampatnam (Arikamedu) where some Roman pottery had also been accidentally discovered in 1937 after some earlier finds. A French team led by J. M. Casal conducted three campaigns there, and Sir Mortimer Wheeler and the Archaeological Survey worked there for a season. These excavations produced Indian material associated with Arretine ware of the Augustan period (the end of the first century BC and the beginning of the next). These findings therefore confirmed Sir Mortimer's theory.

The information furnished by these discoveries encouraged him to return to work on the megalithic culture, which is widely represented in southern India. Excavations on the site of Brahmagiri in Mysore yielded important burials including porthole cists and red and black pottery, with iron tools, which he was able to date, by comparison with his earlier excavations, to the third century BC, before the beginning of Roman trade with India.

In 1947 archaeology took a new turn because of the separation of India from Pakistan. The immediate result of partition was that the Indus valley and Gandharan sites now belonged to Pakistan; consequently India was almost completely deprived of protohistoric and Graeco-Buddhist sites, while Pakistan had to organise her own archaeology department.

The valley of Kakrak is not far from Bamiyan in Afghanistan. The Buddhist sanctuary here consists of a number of rock-cut chapels dating to the fifth century AD, each of which has a cupola decorated with painted ornaments apparently deriving from earlier architectural forms of Bamiyan. In the central mandala is a painting of Buddha surrounded by a ring of contiguous circles, each of which in turn contains another picture of the same subject.

Indian archaeology today

In India the Archaeological Survey has more or less kept its former organisation and taken charge of the protection of ancient monuments, the formation and creation of archaeological museums, the publications service, and the new law forbidding the unauthorised export of 'any work more than a hundred years old'. However, the successive directors, N. P. Cakravarti, S. Vats and A. Ghosh, who were trained in Sir Mortimer Wheeler's school, have particularly concentrated on the following aims:

- Defining the easterly and southerly expansion of the Harappa culture.
- Investigating the 'Dark Age' between the end of the Harappan period and the first signs of the historical period as distinguished by grey pottery.
- Extending their researches in the field of Indian prehistory as a whole.

The expansion of the Harappa culture

Earlier explorations in the Punjab, Rajasthan and Kathiawar produced sites belonging to the Harappa culture. The investigations which naturally followed up these discoveries came to show that this culture spread over a considerable area. The site of Rupar which had already been dug in the nineteen-thirties and was now re-examined by the stratigraphic method, marked the southern limit of its expansion, while the eastern boundary could be found north-east of Delhi at Alamgirpur in the upper Jumna valley, perhaps reaching as far as Kaucambi further down the valley not far from Allahabad. Investigations in the south defined the spread of this culture along the coasts, across the Rann of Kutch and Kathiawar to the estuary of the Narbada.

S. R. Rao's excavations at Lothal in the Kathiawar peninsula revealed a Harappan city which was also equipped as a sea port. However, comparison and analysis of the pottery showed that this site was not occupied until late in the Harappan period and also that it was a local variant. Furthermore, this civilisation gradually declined and degenerated, as was

Opposite page: the causeway at Angkor Thom leading to the east gateway. The rearing head is of the serpent Vasuki, while the body is held by a long row of demons. The other side of the causeway shows a long line of gods, and the whole elaborate structure is an illustration of the Hindu myth of the churning of the soma from the sea of milk. The serpent was used to turn the churning stick, one end held by gods and the other by demons. Twelfth century AD.

The site of Bamiyan in Afghanistan demonstrates important links between the art of India and that of Iran and Central Asia. Here, too, are the first known colossal images of Buddha, the larger (shown here) standing no less than fifty-three metres high in its painted niche. An attempt has been made to reproduce the clinging folds of Gandharan drapery by attaching ropes to the stone core before coating the surface with clay.

clearly shown by the poorer quality of the seals from the later strata and the abandonment of any attempt at town planning. Also at this level, red and black pottery was found which archaeologists regard as the beginning of the red-and-black ware associated with the megaliths in the Indian peninsula.

Finally, the site of Kalibangan on the river Sarasvati in the half desert region of northern Rajasthan, now buried in sand, has been under excavation since 1961. B. B. Lal and B. K. Thapar, who directed the work, discovered the remains of a pre-Harappan occupation under the citadel, which they believe to be comparable with ancient Kot Diji (another Indus valley site), followed by a period of Harappa culture marked by careful town planning, with a separate citadel, as well as by pottery, script, seals and animal art. Carbon 14 tests, as we have seen, indicate an approximate date of 2300–1850 BC, after which the site was abandoned.

By revealing the importance of Harappan settlement in Indian territory, these discoveries increased the need for the investigation of sites where the last signs of Harappa culture might be followed by the first signs of the next culture.

The 'Dark Age' Evidence for this next period is scanty, and the archaeologist often has to use comparative methods to classify it. Methodical excavations conducted during the last fifteen years by B. B. Lal at Hastinapur and G. R. Sharma at Kaucambi, both ancient cities mentioned in the epic of *Mahabharata*, have supplied some information. Hastinapur, north of Delhi on the upper Ganges, was the site of an ancient occupation attested by poorly fired reddish ochre pottery dating to about 1500 BC, with which archaeologists are inclined to associate chance finds of groups of copper objects. After an interruption grey pottery painted with black motifs appeared, dating from 1000 to 700 BC, along with copper and even traces of iron. This ware had already been found in an early stratum at Ahichchhatra and an intermediate level at Rupar. The town was abandoned after a Ganges flood and the people who reoccupied it used the burnished black ware which is associated with wider use of iron, coinage and public works in the form of drainage pipes made of terracotta rings or jars set on top of each other. This burnished black ware is typical of the Ganges valley sites from 600 to 200 BC.

As to Kaucambi on the Jumna a short distance from Allahabad, the nature of the remains from the earliest strata is still uncertain, but they suggest traces of the Harappa culture. Then grey painted pottery appeared, to give way in later strata to black burnished ware, along with the use of iron and coinage. The city belonging to this level had a defensive system similar to those of other towns of the same period, but in this case it was unusually massive and well built, with bastions and gates. The dried brick of which the bulk of masonry is built was faced with fired bricks to strengthen it. Several monuments inside this enclosure have survived, including the Goshita-rama, a Buddhist monastery which, according to an inscription, was believed to have been visited by none other than the Buddha himself at the end of the fifth century BC.

Other towns belonging to this 'Dark Age' have also been excavated and found to have been occupied for about the same length of time. Most of these cities - Vaicali, Carvasti, Kumrahar and Rajfir, which have all yielded objects dating to the same two periods as the above were well known from the life of Buddha.

However, despite all these investigations, the nature of the culture which developed after the Harappan period is still very obscure. B. B. Lal and H. D. Sankalia, among others, accept the working hypothesis that grey painted pottery can be attributed to Aryan peoples from the west, and that it belongs to the period between the end of the Harappa culture and the historical period which is distinguished by black burnished ware and the use of iron.

A whole regiment of researchers from the central and provincial archaeological departments and the universities, all full of the same enthusiasm, are pursuing their enquiries along these lines, particularly in Rajasthan and the Deccan highlands. They have found numerous chalcolithic villages there, particularly along the river Narbada and in the Malwa plain. Their inhabitants used stone and copper tools and a fairly varied pottery, generally light red or black painted. Experts date this period to between 2000 and 600 BC. The site of Navdatoli on the Narbada, excavated by Professor Sankalia, is a good example.

This culture was swept away about 500 BC by the arrival of iron-using Gangetic peoples, who also brought black burnished ware and the use of coinage. They founded cities, the most important of which are Ujjain, the former capital of Malwa, and Mahecwar on the Narbada. This incursion of Gangetic peoples might correspond to the Maurya conquest.

Other explorations and excavations in the western parts of Bengal and Orissa have added some completely forgotten sites to the history of this period. One such is Tamluk in Bengal, and another is Cicupalgarh, which may have been the ancient capital of Kalinga (Orissa) dating to the time of Asoka, when he annexed this kingdom to his empire. The plan of this town is unusually regular and includes very carefully constructed enclosing walls with carefully spaced gates.

At Brahmagiri, further south in the state of Mysore, this was the period of the megalithic culture, which is characterised by cist graves and stone circles, but also by the use of iron and a red and black ware which is only found in these parts. The advance of the Maurya Dynasty is attested here, too, by an inscription of the emperor Asoka.

Prehistory The third object of archaeological research in the last fifteen years, that of further investigations into prehistoric sites, has also been extended to include the whole territory.

Scholarly investigations were focused on the sites in the peninsula and in Gujarat, some of which had been surveyed but not yet excavated, under the auspices both of the excavations branch of the Archaeological Survey and archaeologists on the staff of the major universities, particularly those of Poona, Baroda and Calcutta. The work was well planned and enthusiastically pursued, and the results were partially published in the annual reports of the archaeology department, such as *Archaeology in India* and *Ancient India*, and in reviews issued by learned institutions such as *Lalit Kala*, a new Indian review. Professor H. D. Sankalia has recently undertaken the task of collating all these findings in a fine analytical work, *Prehistory and Protohistory of India and Pakistan*. There are, as he says, probably still a great many points and cultural stages of ill defined origins or insufficient evidence, but in recent years the prehistory and protohistory of these regions have begun to emerge from the darkness in which they were shrouded.

The significant points can be summarised as follows. 'partition' deprived India of the Sohan sites in the northern Punjab where the Early Paleolithic pebble culture was found. Archaeologists then set to work to find other sites showing evidence of the same culture. The sub-Himalayan regions were surveyed with this aim in view, and soon yielded several similar sites. The investigation was then extended to central India, where no systematic research on the prehistoric periods had as yet been undertaken. Immediately after the Early Paleolithic, as distinguished by rather coarse bifaces, a Middle Paleolithic horizon was revealed, named Nevasa after a site near Ahmednagar in Maharashtra. Its diffusion in India was clearly defined, it was characterised by retouched flints, finer and smaller than those of the previous period including scrapers, borers and points.

The microlithic, or advanced mesolithic culture made such marked advances in India that for a long time it was believed to be much more recent, having survived in several backward regions until a later period. However, although its stratigraphy is not yet very firmly fixed this culture's intermediate position seems to be established in some cases and work is continuing on this assumption. The Neolithic culture seems to have varied in different parts of India, judging by the excavation results. It does not appear under the same conditions everywhere, sometimes already being associated with chalcolithic material. For the latter period we have already mentioned the site of Navdatoli, which yielded fine pottery.

In recent years archaeologists have also been clearing and organising sites belonging to a later but still unknown or poorly documented period. A brief review of the most important of these discoveries has already been given.

Since achieving independence India has also had to take the responsibility for a large number of monuments which were formerly under the control of independent kingdoms. In addition, the increase of tourists visiting all these regions has raised new preservation problems. Some of the most celebrated sanctuaries have had to be equipped for visitors without neglecting the matter of protection.

The mound of Mundigak near Kandahar has been under intensive investigation since 1951. It was found that about 2750 BC a dramatic change from village to urban life took place, typified by the building of an impressive palace. Despite severe erosion damage, a view of the west colonnade gives an idea of the appearance of this structure, with its row of massive engaged columns built, like the wall to which they are bonded, of baked brick.

Finally, all this huge quantity of evidence of vanished civilisations and all the masterpieces of art have had to be permanently housed, necessitating the creation of an ever increasing number of museums.

The Archaeological Survey of India is still responsible for the publication of *Ancient India*, which was started in 1946, and *Indian Archaeology*, which was founded by A. Gosh, the present Director General. This review is designed to give a brief account of each year's work in every branch of archaeology, including epigraphy and restoration.

No picture, however brief, would be complete without a mention of the importance of western research which, alongside the extensive achievements of the Indian experts, has contributed widely to the development and spread of knowledge of the various aspects of Indian civilisation. Among recent investigations in southern India the research and publications of the Institut Français d'Indologie at Pondicherry under its director, J. Filliozat, deserve special mention.

Pakistan

After separating from India in 19[illegible] Pakistan had to organise her own archaeology department. Sir Mortimer Wheeler was adviser for a time, after which A. Curiel became Director from 1954 to 1956, and was followed by Dr F. A. Khan. The Government, keenly aware of the amount of work which was needed, encouraged both Pakistani universities and foreign teams to help. The archaeology department was thus able to investigate the protohistoric period, Graeco-Buddhist art, Buddhist art and early Islamic art without neglecting the preservation of the monuments, all at the same time.

Since we have already discussed the discovery of the Indus civilisations we need only review the contributions made by recent excavations. In 1950 Sir Mortimer Wheeler reopened the excavation of Mohenjo Daro and found a huge granary full of corn inside the citadel, which has already been mentioned. Mohenjo Daro, which has certainly not yet yielded up all its secrets, was also investigated by two American teams at other times. One of these conducted a stratigraphic excavation in the heart of the citadel, and found the first traces of a calcined carved wooden door. The second attempted to penetrate to virgin soil by

The first city of Taxila lay in the area known as Bhir Mound. The soil has been badly eroded, but it has been possible to follow the layout of the city's plan and defences. First occupied in the sixth century BC or even earlier, the houses were built of sun-dried brick supplemented by timber, none of which has survived. The streets were narrow and winding and the houses themselves were very irregular, chiefly built on the plan of an open court with rooms haphazardly placed on one or more sides.

following the water table. This was discovered about ten metres below the present ground level, and the intermediate strata produced seals and other finds, emphasising the need for deep excavations.

Another American expedition led by W. A. Fairservis, looking for evidence of the origins of this culture, found rural centres in the Quetta district of Baluchistan, the pottery of which could have been the counterpart of Iranian ware. But they also found evidence of influences which could have come from the Indus at a date before the full flowering of this culture.

In 1959 J. M. Casal, at the head of a French expedition, continued the investigation started by Majumdar at Amri in Sind. The stratigraphic excavations Casal conducted there enabled him to distinguish the earliest levels, which were marked by a different type of pottery, certainly earlier than that of Harappa and not without similarities to the ware from the early strata of the nearby site of Kot Diji. At Amri this pottery was superseded by the typical Harappan red pottery with black painted patterns. However, in his monograph on Amri J. M. Casal has shown that the ancient culture of Amri could not be the ancestor of the Harappa civilisation. It was simply supplanted by a much more advanced culture, the early stages of which are still unknown.

As to Graeco-Buddhist studies, no new field investigations had been conducted since Sir John Marshall's last campaign at Taxila in Gandhara in 1938. In 1959 G. Tucci and an Italian team decided to reopen the excavation on the sites in Swat, formerly Uddiyana, on the borders of Afghanistan and Kashmir. This territory had been crossed by Alexander the Great, and the Kushan Dynasty later patronised Buddhist establishments there. Excavations on the site of Butkara shed light on the culture of the early strata, which were distinguished by primitive iron working. It is to be hoped that the publication of these finds, which included a monastic establishment heavily decorated with historical reliefs, will provide much-needed information on the true chronology of Gandharan sculpture.

G. Tucci also excavated other sites in this region, which disclosed rural communities still adhering to beliefs in the ancestral couple. G. Tucci believes that the Buddhists of Swat tried to assimilate local faiths by adopting divinities with demonic features.

Dr Khan's excavations since 1955 at Kot Diji south of Mohenjo Daro have revealed a fortified citadel and a lower town. As at Amri, Dr Khan found a new type of pottery in the lower strata, earlier than the Harappan ware which marked the next strata.

The archaeological department, which supervised all these researches, is also responsible for the protection of monuments and the organisation of museums, many of which have been created recently. Since 1964 it has published *Archaeology in Pakistan*, in which each year's work is reported.

Afghanistan

Afghanistan, which consists of high plateaux broken up by mountain ranges, the biggest of which is the Hindu Kush, lies on one of the ancient lines of communication, the civilising role of which has been discussed at the beginning of the chapter.

We need only recall the establishment of the Persian satrapies in the sixth century BC and the subsequent consolidation of the Greek dynasties during the third century BC, which were later overthrown by the coming of the Parthians. They were driven out in their turn in the first century AD by the Kushans from central Asia. These latter, who extended their sway throughout north-west India, adopted the religions of the country, especially Buddhism. About AD 480 the White Huns ravaged these kingdoms, but the Indo-Iranian culture nevertheless seems to have survived there until about the ninth century. Finally Islam overran the region in the eleventh century and used it as a base for the eventual conquest of India.

The importance of the inland routes had diminished with the development of maritime trade, and in 1840 Afghanistan cut off all contacts with the outside world so as to protect herself against her neighbours. From the archaeological viewpoint the country was very badly documented, because it had been impossible to conduct any large-scale excavations apart from a few rather hasty digs in the nineteenth century.

In 1922 an enlightened ruler, Aman Allah, decided to throw open the borders of his country and encourage archaeological research. He entrusted Alfred Foucher with the task of attempting to recover the national heritage, and with this aim in view the Délégation archéologique française en Afghanistan was created; it enjoyed special concessions for the next thirty years. A. Foucher, who had just published his masterly work *l'Art gréco-bouddhique du Gandhara*, set to work to retrace the footsteps of the Chinese pilgrim Hiuan Tsang, as Alexander Cunningham and Sir John Marshall had done in northern India. This was how he and his collaborators came to excavate the holy places of Buddhism which had already been located in the early nineteenth century: Bactria, Hadda, Bamiyan, Begram and Fondukistan, to name only the most important.

All these sites produced evidence of peoples of the first centuries AD who were strongly in-

fluenced by Greece and Iran, and had adopted the Buddhist faith under the auspices of their current masters, the Kushan kings. As at Gandhara, this influence could be seen in the arts of architecture and sculpture, in which western characteristics mingled with Indian features to produce a highly individual style. This is particularly clear at Hadda, which was inhabited from the third to the eighth centuries AD. The art of modelling was very commonly practised here. The stucco carvings, many of which adorn the monasteries and stupas, have been compared with work from Western Asia, *Central Asia and Kashmir, but they also seemed* to be related to contemporary Hindu and Buddhist monasteries in India, which also had stucco decoration, e.g. Nalanda, Sarnath, Devni Mori and many others.

The site of Bamiyan, north of Kabul, was investigated and written up by A. Godard and J. Hacklin. Tiny shrines and huge niches cut in the rock face holding gigantic figures of the Buddha were chiefly decorated with moulded, incised and painted stucco. These scholars have shown that here a very strong Sassanid *influence from Iran was united with elements* from India and the Hellenised west. The tradition of rock-cut temples, which also comes from Achaemenid Persia, had been adopted elsewhere in India (Lomas Tishi, Bhaja, Karli) during the last centuries of the first millennium.

In 1936 J. Hacklin, who worked first on these sites and then on that of Fondukistan, investigated Begram, (formerly Kapici) in northern Kabul. Early in the excavations he was fortunate *enough to find a treasure made up of Indian* ivory, Graeco-Roman glass, plaster and bronze, and Chinese lacquer which, chronologically speaking, indicated that the finds mainly dated to the first three centuries AD.

This precious collection, some of which now adorns the Kabul museum while some is in the Musée Guimet, was the subject of some brilliant studies both by J. Hacklin and his colleagues *and by a group of scholars including P.* Stern, J. Auboyer and O. Kerz, which were published in *Mémoires de la Délégation française en Afghanistan*. R. Ghirshman's stratigraphical excavations confirmed the date.

In addition to all these discoveries made just before the Second World War, we must mention the most important work carried out since that time. French teams directed by D. Schlumberger, who no longer enjoyed exclusive excavation rights, now conducted excavations on sites dating to different eras. In particular, the excavations of Surkh Khotal from 1953 to 1957 ad*vanced the sum of knowledge of the religious* foundations of the Kushan dynasty from the first to the fourth centuries AD.

D. Schlumberger excavated a huge building including a staircase and a succession of walks on which the rites of fire worship were possibly *conducted. Colossal images, probably repre*senting rulers of the Kushan dynasty, stood here, and Schlumberger believes that this may have indicated that there was an imperial and dynastic cult like that of the Devakula of Mathura. In addition, a plaque bearing an *Iranian inscription in Greek characters seems* to be a foundation plaque of King Kanishka himself. The inscription is fairly long, and defines the Takharian language spoken by these tribes, which was known only from coins.

We should recall also the recent chance discovery of a bilingual Aramaic Greek inscription setting out an edict of Emperor Asoka, the importance of which for knowledge of the expansion of Maurya power in Afghanistan has already been mentioned.

As the Afghan government encouraged sur*veys aimed at a better understanding of the* pre- and protohistoric periods in the country, J. M. Casal selected the site of Mundigak, *which offered possibilities of important informa*tion since it was located on a highway in southern Kabul linking Iran and the Indus. After several excavation campaigns from 1951 to 1958 he was able to establish a series of comparisons with the other sites recently excavated by Russian and American teams in Turkmenistan and Baluchistan, which blazed the trail from the Indus valley to the Mesopotamian territories.

Casal sees some similarities between the pottery from the earliest strata at Mundigak and that of Jemdet Nasr in Mesopotamia, Quetta in northern Baluchistan, and perhaps the early strata dating to before the Harappan culture at Amri and Kot Diji in the southern Indus valley. In the next period, known as 'period IV', a town surrounded by a buttressed wall, a palace and buildings probably intended for religious func*tions suggest a fairly prosperous urban people* who nevertheless knew nothing of writing.

Furthermore, this little city must have been a repository for exchanges between its eastern and western neighbours. Casal suggests that the vase shaped like a brandy balloon found there must have been inspired by the types used in north-eastern Iran, that the decoration of stylised goats and birds could have derived from the pottery of Susa II, and that the motif of groups of three peepul leaves could be evidence of contacts with Harappa pottery. Mundigak was abandoned about 2000 BC and there was no further activity there until about AD 800, when there was a brief settlement, *before it was finally deserted.*

In addition, Kabul museum, reorganised with the help of UNESCO, houses the collections which the archaeologists have rescued from *oblivion to shed light on the rediscovered his*tory of Afghanistan.

A monastic community in the Taxila district on a hilltop at Jaulian. The monastery consisted of two stupa courts with chapels and shrines for cult images, an ordination hall, a refectory and an open court surrounded by the monks' cells, shown in the *illustration*. There appears to have been accommodation for fifty-two monks in small cells, each with a wall niche, a window and a strikingly low doorway.

The Far East

The Chou Dynasty effectively came to an end with the period known as the Warring States, when China's vassals started the long struggle against the Chou's autocratic power. This bronze wine vessel (*hu*) is from Shensi and shows the influence of the Mongols who were a constant threat to the security of China. Fifth to second centuries BC.

South-east Asia

It would be impossible to give a full description of the history and geography of the countries to be discussed here, but it is nevertheless essential to give the broad outlines, because geographical considerations dictate the distribution and state of preservation of the remains, while the course of history influences their nature and variety. Despite its extreme diversity, which arises equally from its complex history and the varied geography, the archaeological material throughout south-east Asia raises the same problems of surveying, interpretation and conservation.

The geographical setting Geographically speaking, south-east Asia is a complicated area. It consists of a continental section, the Indo-China peninsula, and an island section, Indonesia, grouped round the China Sea. The former consists of alternate mountain ranges extending from the eastern end of the Himalayan massif and plains of various sizes connected by the great river basins. The latter is a succession of islands with high ground, usually volcanic, and a few short rivers. Although the peninsula, with its wholly tropical climate, seems to be unlike Indonesia with its equatorial climate, the monsoons give the area some semblance of unity with their seasonal changes of dominant winds. The archaeologist meets with the same problems everywhere, except for the Pagan region of Burma where the climate is comparatively dry. All the abandoned sites rapidly fall a prey to the exuberant vegetation, but this does not always play the destructive part generally assigned to it. All the perishable materials (wood, wall paintings, fabric, manuscripts, etc.) risk destruction by the combined action of the elements, the plants and animal life. Even stone does not escape unscathed, and recent studies have shown that some of the sandstone of Angkor suffered not only from the usual erosion factors, but from actual bacterial diseases.

The historical setting Several factors dominate the history and define the boundaries of the area known as south-east Asia. Outer India or Transgangetic India to the ancients, and Greater India to Indian scholars. The whole of this vast region, which nowadays consists of Burma, Thailand, Cambodia, Laos, most of Vietnam, Malaysia and Indonesia, apart from the Philippines, has various common features which, where the earliest centuries are concerned, can often be perceived only through the medium of archaeology. The political divisions of ancient times obviously have no relationship to the present boundaries. Not only have the frontiers, which were always fluid, varied considerably over the course of the centuries, but a number of the countries are comparatively recent creations, some, like Champa, have completely disappeared while others, like Cambodia, have survived despite centuries of unrest.

Quite apart from real prehistory, the study of which has generally scarcely passed the preliminary stage of exploration, it can be accepted that the Late Neolithic period almost everywhere lasted up to the beginning of the Christian era, and the Megalithic culture already belonged to the metal ages, in which bronze and iron technology seem to have arrived almost simultaneously. South-east Asia is a meeting point and a transit camp for the Mediterranean-like South China Sea, which lies like a wedge between India and China, and as early as this period it was subjected to the civilising domination of its neighbours.

Under the Han dynasty (212 BC to AD 220) China established commercial relations with the Indian Ocean (Arakan, Tenasserim) by land and sea, and pushing on to the Red River delta founded Nan-yue, the origin of the future Vietnam. Reliable historical sources narrate the process of the expansion of Chinese culture in the north-east of the peninsula, but the penetration of Indian civilisation can be studied only through the archaeological findings. Indian expansion, which took the form of a real colonisation, did not appear until the second century AD at the earliest. There were various reasons for this, including commercial and missionary activities and the emigration of cultured members of the community. The earliest contacts seem to have occurred at the same time as the fleeting extension of Roman trade attested by some scanty evidence from the lower Menam basin, Transbassac and perhaps as far afield as the neighbourhood of modern Hanoi.

Indian culture, incessantly renewed at its source, gradually spread throughout south-east Asia, producing kingdoms in the Indian style practising the Hindu and Buddhist faiths. It was not until the tenth century that it met its first setback, with the southerly progress of the

Opposite page: for 400 years, from the ninth century AD, the Khmer civilisation flourished in Cambodia. Immense temple and palace complexes were built in cleared areas – but after the decline of the Khmers the tropical forests began to close in, and the work of man to be submerged by nature.

future Vietnamese down the east coast of the peninsula. This was accompanied by the slow obliteration of the Indianised kingdom of Champa. Following the Mongol conquest, which had profound repercussions throughout south-east Asia, especially the central and western part of the peninsula, as early as the accession of Kublai Khan in 1260, the advance of Islam marked the halt and decline of Indian influence. Islam was established in Sumatra at the end of the thirteenth century and reached Java and Malaya by the early fifteenth. Apart from the island of Bali, where the ancient Hindu beliefs have persisted until the present day, only the lands which had accepted Sinhalese Buddhism (the kingdoms of Thailand, Burma and Cambodia) were to be able to resist the advance of Islam. With the capture of Malacca by the Portuguese in 1511 a colonial era which was only to end after the Second World War set in for the whole of south-east Asia except modern Thailand, and she was not entirely closed to Western influences.

It has not been possible to refer to the highly complex problem of ethnology in this brief survey of the historical evolution of south-east Asia. We should mention, however, that of all the migratory peoples of the historic period, only the Vietnamese took Chinese culture with them, the Thai and the Burmese having adopted the local Indian culture. It should also be noted that the advances of Islam were not accompanied by its frequently denounced iconoclasm, which was particularly prevalent in nearby India.

It is almost a commonplace to claim that our knowledge of the ancient history of Indian south-east Asia is, in a large measure, owing to archaeology, so obvious is it that nowhere else (except pre-Columbian America) has the progress of historical research been so closely bound up with that of archaeology. In spite of the disparity of the countries and the complexity of their history, the archaeologist finds a number of common features which arise from the fundamental unity of the cultures and the conditions in which the archaeological material survives.

A unique past The concept of the past assumes its widest meaning in south-east Asia, where the historic period embraces material which ranges from the prehistoric age (we have briefly mentioned that the last prehistoric strata are contemporary with the first evidence of Indian influence) to the present day. The necessity for considering ancient history as extending almost until the present arises from a combination of somewhat peculiar circumstances. On the one hand, the threat hanging over even the most recent objects from the climate and from the rapid Westernisation which is uprooting even the longest-established customs renders investigation, preservation and study urgently necessary. On the other hand, the recent 'prehistoric' period often includes some astonishing survivals, as we ourselves have recently seen, not only among a few hill peoples but even in highly civilised places like Thailand, where a few small groups have preserved traditional pottery techniques dating back to remotest antiquity.

In a Buddhist community the religious art copies images which enjoy particular veneration so frequently that it is sometimes difficult to distinguish between ancient works and 'copies' of every style and date. When we add that the manufacture — sometimes very skilful — and sale of forgeries is particularly flourishing, it will be clear that the archaeologist must always be on his guard against the many snares lying in wait for him, and also that archaeology cannot be as clearly defined here as it can in the West, since it frequently unites knowledge which should belong to the ethnographer or the prehistorian, if only for the requirements of interpretation.

Because the scarcity of dated monuments and the almost complete absence of works of reference are two of the basic features of Indian south-east Asia, the archaeologist has to resort to the comparative method to study his material. Comparison with firmly dated objects, the study of development and comparison with the arts of more or less neighbouring countries can reveal its affiliations, iconography and function. It follows that the south-east Asian archaeologist needs extensive knowledge, but this does not exclude a degree of specialisation, not only in matters native to the Indo-Chinese peninsula and Indonesia, but even in Indian and Chinese details. The day is past when the art historian could confine himself to the study of a single country and believe that every anomaly he met could be explained as a manifestation of 'local genius' or a resurgence of the ancient pre-Indian stock.

The methods Because of the almost universal absence of texts for reference, and the late and frequently suspect nature of the rare chronicles available in the best cases, everything remained to be discovered in south-east Asia. It is not surprising, therefore, that a preliminary phase, which was often very long, had to be devoted to investigating and cataloguing the above-ground monuments. The next step in the second phase was to take local possibilities into account and clear these monuments more or less completely, so as to study their general features and find any hidden statues and inscriptions.

Art historians and epigraphists took priority in making use of this information. The former, whose work is an end in itself, have furnished most of our information on the religion and daily life of each country. The epigraphers alone can gradually retrace and define the broad outlines of political and religious history and catch a glimpse of social organisation. This phase of the investigation meets with conservation problems, and these have always absorbed, and will continue to absorb, most of the attention of the scientific staff. The adopted measures, to which we shall revert later, all have to meet two requirements: to protect traces of the past, and to facilitate study of them.

The archaeologist cannot begin to investigate the less obvious remains revealed so abundantly by aerial survey or by trial trenches, or to begin real excavations until the end of this preliminary stage, when adequate preservation methods are readily available. Such excavations, which started only within the last twenty-five years, are far from being widely conducted, since so much attention is often claimed by the first phase.

It is always surprising to find how slowly methods used everywhere else, frequently with spectacular results, have been adopted in south-east Asia. Although it is true that some investigations have been conducted in a manner which failed to produce valuable information, it would be unjust to forget that conditions here are very unusual. Apart from the fact that there is no question of excavating without a detailed preliminary survey, there are so many standing monuments in need of study and conservation that the authorities naturally concentrate on the most pressing cases and those promising the least problematical results. Furthermore, it is well known that the application of the information furnished by the most recent research methods still includes a great many unknown quantities in south-east Asia, where none of the techniques of Western archaeology is immediately applicable.

As far as dating is concerned, nothing can be hoped for from the investigation of the residual magnetism of pottery until the decline of magnetism in the countries concerned has been tabulated. A few attempts at carbon 14 dating, which always has some margin of error for ancient material, have only produced deceptive results, but these may be explained by better understanding of local conditions. Smear tests, which have provided such remarkably precise information on the ancient cultivated and indigenous vegetation of India, can only be used when the native flora of each region has been defined. Even today the advance of research is subordinate to the progress of the preliminary work.

In every case academic research takes an important place in south-east Asian archaeology. Field work is obviously indispensable, but the essential object is the interpretation of the facts and finds. It is remarkable that not a single major theory has come to light in the field, and that they are often formulated a long way from the excavations. This arises solely from the fact emphasised above, that interpretation, being primarily based on the comparison of monuments with texts, cannot be done on the spot except in exceptional cases. It follows that centres of learning and field research are equally important and inseparable.

The contribution to archaeology Archaeology being the chief, and often the only, source of information on south-east Asia, the more abundant and varied the finds, the better chance we shall have of enlarging our knowledge. Apart from the concrete contribution it makes to the study of all south-east Asian art forms and their affiliations which were sometimes unknown before the start of archaeological research, and the historical information provided by the discovery of inscriptions, the great achievement of south east Asian archaeology is the revelation of unsuspected civilisations, for example the culture known as Dong Son, and of cultures hinted at by references in the ancient Chinese texts, like Fou nan (second to sixth centuries) and Dvaravati (towards the end of the sixth century and the beginning of the next).

The Dong Son civilisation Despite the situation of the earliest known sites (on the river Day in the province of Thanh Hoa) the Dong Son culture seems to be Indonesian in origin. Following the discovery on the site of Dong Son in 1924 of the abundant material (pottery, worked bone and many bronzes) in some very distinctive tombs (simple inhumation pits very different from Chinese burials) scholars succeeded in isolating the Dong Son culture. Its originality was most apparent in the bronzes, which included weapons for ritual and combat, jewellery, tools, vessels and cult objects, in which the bronze drums, usually most familiar from the late examples, play an important part. The style is marked by Chinese influences, but

Above and below: the great Buddhist monastery of Barabudur in Java. The building was begun in the ninth century AD, and consists of seven terraces. The walls are covered with fine reliefs depicting the life of the Buddha and provide an exercise in devotion for the faithful. There are seventy-two of the bell-shaped shrines called stupas (*far left, above*) and a gigantic one surmounts the whole. The restoration of Barabudur was the work of the Dutch architect Van Erp in 1907.

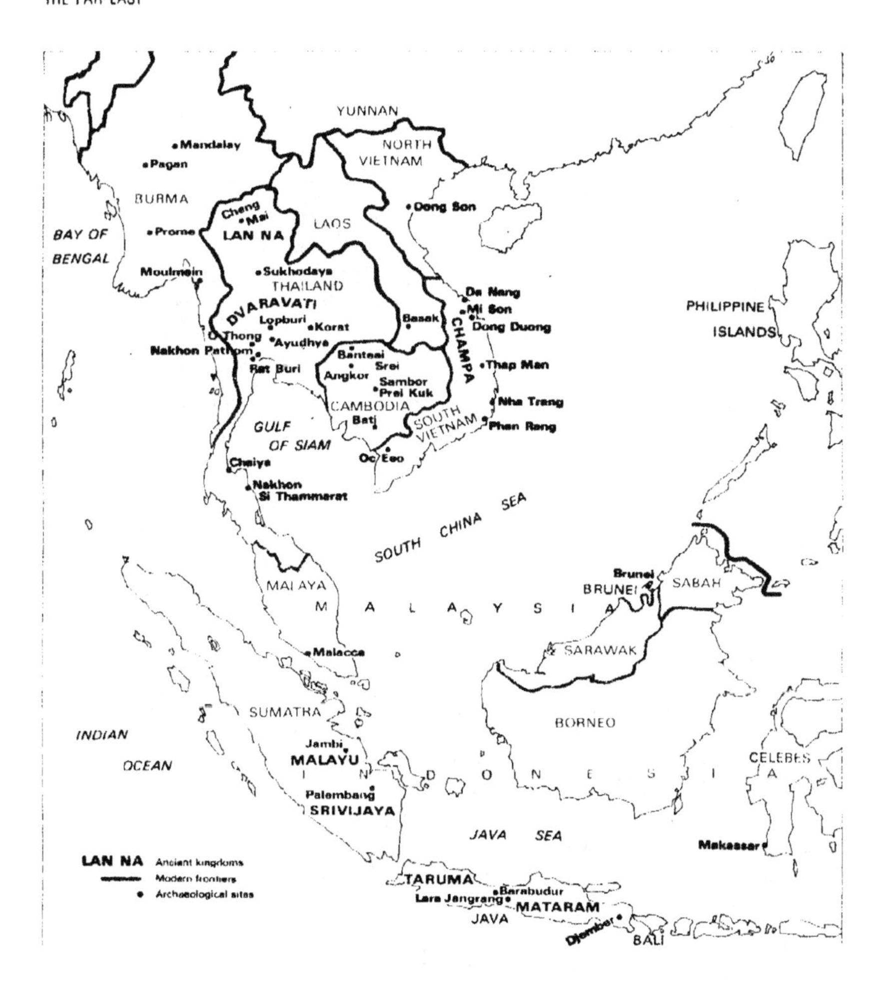

these by no means overshadow the native genius. It also shows Hellenic elements in its latest stages, and is characterised by geometric ornaments and systematically stylised figures.

This culture reveals complex beliefs somewhat reminiscent of certain shamanistic practices among agricultural and maritime peoples, and a fairly advanced military system. It flourished between the seventh century BC and the second century AD. It disappeared from Dong Son as a result of Chinese settlement in the Red River delta region during the Han dynasty. However, it achieved widespread expansion over an extensive area, starting from distinct centres. It is found in Malaya, Cambodia, Laos and Burma, is particularly rich in Indonesia (Sumatra, Java, Madwa, the Celebes) and penetrated as far as New Guinea. So far not a single text seems to be connected with it, but echoes of it seem to appear in a few scattered backward Indonesian communities of the Vietnamese highlands and the Indian archipelago.

Fou-nan and the Oc Eo culture For a long time Fou-nan, the earliest Indianised kingdom mentioned in the Chinese histories as existing between the second century AD and the end of the sixth, was only a ghost-realm. Textual criticism and the subsequent discovery of a few Sanskrit inscriptions gradually invested it with a degree of reality by detailing its position in the region extending roughly from Bassac to the Gulf of Siam. Various scholars tended to attribute to Fou-nan all the variants in Cambodian art which looked earlier, but nothing was actually known until the methodical investigation of Transbassac in 1938-45. This enabled L. Malleret, the honorary Director of the École Française d'Extrême-Orient, to identify several ancient sites in a region which had been dismissed, on unfounded geographical grounds, as recently formed and devoid of any trace of archaeological material.

The course of world events abruptly put an end to these excavations, which have never been re-opened, and thus it has only been possible to excavate part of the site of Oc Eo. However, the material from surface finds and trenches - dug with all the scientific precision that could be desired - illustrates the main features of a hitherto unknown civilisation. Air reconnaissance simultaneously showed a close, frequently connecting network of straight canals covering the whole area, probably to ensure drainage, controlled irrigation and communications between huge cities in geometrical enclosures. These installations in this naturally unhealthy and infertile low-lying country indicate the high level of the agricultural and maritime culture attained by Fou-nan, which evidently completely mastered the art of reclaiming land.

Archaeological finds, in their turn, show unsuspected art forms and commercial contacts. Some brick foundations and a few constructions related to the megalithic tradition, and rows of wooden piles supply what little information we have about their architecture. Except perhaps for a few terracotta reliefs, sculpture is practically unknown from the true Fou-nan period. On the other hand, the skill of their craftsmen is demonstrated by plenty of pottery with a variety of shapes and decorations, and tin and gold jewellery, sometimes very delicately made. Apart from these productions, which are usually based on imported Indian or Chinese models, epigraphic and symbolic amulets and seals establish the reality of the contacts mentioned in the Chinese sources between the third and sixth centuries and confirm the movement towards Indianisation. Some *intaglios* and medals of the Roman Antonine period and some *grylli* as well as a Sassanian *cabochon* attest the extent of their Western trade and echo Ptolemy's *Geography*. Traces of a similar culture were discovered in the U Thong and Dong Si Maha Pot regions of Thailand in 1964. These remains have not only made it possible to push back the boundaries of the historical period in the Menam basin, but have also confirmed the statements in the Chinese sources about the conquest of the neighbouring kingdoms by Fou-nan in the early third century.

The kingdom of Dvaravati Another striking example of the part played by archaeology in the study of early south-east Asia is the 'resurrection' of the kingdom of Dvaravati. This kingdom, according to the most recent research, seems to have arisen in the south of modern Thailand in the late sixth or early seventh centuries after the fall of Fou-nan, whence it derived a large part of its culture, and lasted until about the eleventh century. It is known exclusively through archaeological research.

In 1884 a British scholar set about re-translating some Chinese inscriptions into 'Dvaravati', and suggested that this was an ancient name for the lower Menam basin. As long as there was no field work to verify this theory it was widely accepted, and it was even suggested that the country was inhabited by the Mons, a people related by their language to the Khmers. In 1909, however, Commandant Lunet de Lajonquière, to whom the archaeology of the whole peninsula owes a debt for his ever active interest and for the compiling of a considerable catalogue, recognised that there was an art form in this region which was neither Cambodian nor Thai. At this time nobody thought of connecting the two.

It was not until ten years later that G. Coedès, a member of the Institute and honorary Director of the École Française d'Extrême-Orient, found the first evidence in the form of a pillar inscribed in the Mon language at Lopburi, from which, using stylistic comparisons, he was able gradually to define the original Dvaravati art which is chiefly represented by Buddhist-inspired sculpture. When research had progressed to the stage of reconstruction around this first nucleus of stucco and fragments of architectural decoration, two vast Buddhist sanctuaries of a hitherto unknown type at Nakhon Pathom were cleared by P. Dupont in 1939-40, and this enabled him to offer the first description of Dvaravati architecture and to publish a thesis entitled *l'Archéologie mone de Dvaravati (The Mon archaeology of Dvaravati)* in 1955, after the discovery of some fragments of inscriptions in the Mon language on the same site. This art style had now been definitely located and its essential features analysed, but its attribution was still partly hypothetical, since the actual existence of the kingdom of Dvaravati was not yet established.

The long awaited proof was not obtained until 1935 when study of two silver medals from Nakhon Pathom enabled G. Coedès to establish that the legend on them in seventh to eighth century characters mentioning a 'Lord of Dvaravati' finally confirmed this long-standing hypothesis. A systematic excavation campaign on the sites of the ancient cities by the archaeological branch of the Thai fine arts department simultaneously revealed the variety, richness and frequently classic beauty of the art of this kingdom. Its originality and cultural expansion are now established, but even the vague outlines of its history are still unknown.

Limitations Archaeology's part in south-east Asian studies is very important, *but it cannot* answer every question. Our examples have shown that archaeological research has sometimes revealed the existence of unsuspected civilisations, but without texts we are always reduced to conjecture as to the nature of the population of Dong Son, and although we are slightly better informed about Fou-nan, it is only because there are many references in the Chinese sources and a few surviving inscriptions.

The difficulties reappear in the case of Dvaravati, the only known historical information is occasional Chinese references to two embassies during the seventh century. This is nothing out of the ordinary. Archaeology does not usually furnish information on political and historical events. It rather plays the part of an auxiliary science to epigraphy. *No dates can be obtained* by studying the archaeological material in isolation. Despite the increasing possibilities of modern methods of analysis, a certain margin of error will always remain, growing larger where the historical outlines are more obscure. Except for a few purely material problems belonging more to sociology and ethnology than to archaeology, interpretation can be based only on a study of native or foreign texts, which furnish chronological references and explain facts.

We have stressed the fact that shortage of texts was one of the most striking characteristics of south-east Asia, and mentioned that we must look to the Indian sources for the possibility of new interpretations. This process has already borne fruit, and a number of architectural and *iconographic features which had been regarded* as original for several decades are now known to have been inspired by India. This does not in any way detract from the Khmer or Indonesian talent for adaptation, which sometimes amounted to genius. As a result of these advances in interpretation all the publications except those which are purely descriptive have quickly become obsolete. This process in its turn is accelerated by the fact that every new discovery in a region as little explored as south-east Asia may overturn the most apparently soundly based theories. Another consequence of the common lack of reference texts, or the all too-frequent ignorance of them, is the regrettable encouragement of a somewhat romantic interpretation which, disguised as a great theory or brilliant popular treatment, enjoys the perennial appeal of a historical or monumental mystery, but tends to offer the reader a variety of more or less attractive but usually imaginary speculations.

Offshoots The protection of archaeological material in its broadest sense also implies preservation of monuments and publication of findings. Remains should be preserved immediately after they are discovered. This is of the utmost importance in tropical countries. The preservation of objects from the excavations raises problems familiar to museum staffs all over the world, but the question of buildings is more pressing and calls for special measures. It is equally important to halt the process of their destruction and to maintain buildings in the most favourable conditions for study and, when appropriate, restoration. Contrary to popular belief, it is less dangerous for a building to remain buried or submerged in vegetation for a time than for it to be completely cleared but not given appropriate treatment for its condition. Far from breaking up the masonry, the network of the roots of big trees is less harmful than small growth, since it works its way into every crack and holds it together like a protective mesh. The shade of foliage, playing the part of a screen, a parasol, or an umbrella, saves the building from the most active agencies of erosion. After abrupt clearing, stone or brick which has been kept for a long time in a weakening state of humidity is suddenly subjected to the action of the sun, wind and rain, while stray fast growing vegetation will seize on the smallest crack, where its roots will dig in like a wedge. It is often difficult to arrest the speed of erosion, but the building itself can be saved by applying special processes in which archaeology and architecture are combined.

The temple of Siva Mahadeva at Lara Jangrang (Prambanan) in Java as it was in 1894.

The old system of temporary or permanent props, which were neither aesthetic nor effective, has been superseded by the anastylosis method, adapted from the procedures used on the Greek sites. Anastylosis consists of the restoration and repair of a building with its own fabric, using the original construction methods of the monument. It includes discreet and well-justified use of new materials to replace missing parts without which the ancient pieces cannot be put back into place. This method has produced spectacular results wherever it has been applied (Indonesia or Cambodia), but the quality of the reconstructions should not be allowed to obscure either their practical value or their scientific interest. Indeed, it is owing to these reconstructions that we can discover not only the original appearance of the rebuilt monuments, but the hitherto unsuspected existence of individual features such as the sacred deposits which conferred inviolability on them.

The prompt publication of all the findings from the excavations or investigations in the most widely available language is as urgent as the preservation of the monuments. We suffer nowadays from the brevity or absence of reports, and these are essential for interpreting some of the facts. The conditions under which a given discovery or excavation took place, if not published, will remain forever unknown, and it will be impossible to verify any hypothesis. This deficiency was forgivable in the early stages of research, but it continued until the Second World War. Its present widespread recurrence is the result of a variety of factors: for instance, the increase in the number of works published in their native languages by national archaeological services, which are seldom distributed abroad or are unavailable because of the large number of languages required by the south-east Asian specialist; the high, sometimes prohibitive price of extensively illustrated works, and the tendency of some scholars to keep their work nearly secret for as long as possible.

Historical survey

The whole field of archaeology in south-east Asia—except perhaps for that of Thailand—was originated by Western initiative. But it should be remembered that the first travellers, especially the missionaries, never paid much attention to ancient remains. Although some of them as early as the end of the sixteenth century left accounts which have contributed significantly to our understanding and interpretation of the facts, they were usually only interested in events taking place directly before their eyes. Indeed, it was not in the Indo-China peninsula but in Indonesia that the earliest interest in archaeology occurred. Since it is impossible to give a detailed description of the history and present state of research in each country, we must confine ourselves to a brief outline, reserving the longer explanations for a few typical examples.

Indonesia The oldest museum in south-east Asia, that of Djakarta (formerly Batavia), was founded as a result of the initiative of the Society of the Arts and Sciences of Batavia, formed in 1778. True interest in archaeology as such, however, began during the short British occupation of the archipelago under the leadership of the Lieutenant Governor Sir Stamford Raffles, and it was continued by the Dutch after their return to the islands. The foundation of the Leyden Museum in 1817 was the first attempt to organise a European research centre devoted to south-east Asian studies. In 1824 the first scholarly study of Javanese art appeared, and since then works of philology, iconography and art history have been published regularly. Perhaps because of the obvious connection between Indonesian and Indian art, which has resulted in the term 'Indo-Javanese' art, archaeologists here have come to extend their researches towards India proper in order to make swifter progress. In other countries, perhaps because of the originality of the material, the tendency is towards narrow specialisation, extending at the furthest to the whole of south-east Asia. Despite the substantial results obtained before the end of the nineteenth century, Indonesia had to wait until 1901 for the establishment of an official commission for archaeological research, which was transformed twelve years later into the Archaeological Service. This passed to national control when Indonesia achieved independence, and has been increasingly active ever since.

Indonesia can justly pride herself on being the birthplace of south-east Asian archaeology. She has also furnished an example of learned research for the whole peninsula, and has been the scene of the first use of anastylosis for the preservation and restoration of monuments. The first attempt, dating back to 1903, was concerned with the little Chandi Pawon. Although the results were and are questionable, the method was good and its rapid improvement allowed the repair of the admirable Barabudur (1907–11) and the majority of the sites. One of the most remarkable successes was undoubtedly the reconstruction of the great sanctuary of Shiva at Lara Jangrang, better known under the name of Prambanan, between 1937 and 1953. The painstaking methods employed here were also gradually to be adopted throughout south-east Asia.

There are a great many books on Indonesian archaeology embracing the whole archipelago, but they reserve a special place for Java, Sumatra and Bali, the most important islands from the historical viewpoint and, naturally, the richest in remains. They cover a vast period extending from the prehistoric era, which is represented by abundant material mainly from recent excavations, to about the sixteenth century. The most important period is the Indo-Javanese, or Indo-Indonesian, which is the most important from the historical viewpoint and the richest in monuments, including Mahayana and Tantric Buddhist features in its later phases and traces of Brahmanism: the Sumatran maritime empire of Srivijaya (second half of the seventh century to about the end of the thirteenth) which extended its domination from the archipelago to part of the peninsula, the central Java phase (eighth century to the first half of the tenth) with the Cailendra and Mataram dynasties; the eastern Javanese phase (second half of the tenth century to the middle of the fourteenth), with the Mataram, Kadire, Singhasari and Majapahit dynasties; and the kingdom of Bali, where the Buddhist rites have persisted from the eighth century until the present day. Most of the publications are unfortunately in Dutch, or more recently, in Indonesian, and are therefore out of the reach of most people.

Burma Apart from the remarkable publication on the principal monuments of Pagan by Sir Henry Yule in his book *Embassy to the Court of Ava* (1858), Burma was treated for a long time as a poor relation of Indian archaeology because of the country's completely arbitrary assimilation with the Indian empire (1886–89).

Nevertheless, the originality and diversity of Burmese art is undeniable, her wealth of monuments is considerable and the best chance of finding evidence of the earliest contacts with China and India lies in Burma because of the clues furnished by the texts. When we add this is one of those rare parts of south-east Asia where the climate has permitted wall paintings to survive, and that some of them, found in Buddhist monuments of a highly distinctive architectural style, go back to the beginning of the twelfth century, the importance of Burma's place in south-east Asian archaeology will be obvious. Unfortunately, there has been scarcely any research into epigraphy, although there is plenty of material, or into history which, unusually enough, is not directly dependent on undeveloped archaeology here.

Despite the profound researches of a few European and Burmese explorers and the foundation in 1910 of the Burmese Research Society, which has an immense research programme, there have been few archaeological studies and their subjects are limited. Archaeological research and the first preservation projects really started in recent years under the auspices of the Burma Historical Commission and the Archaeological Department. The destruction of most of the carved wooden buildings of Mandalay at the end of the Second World War has unhappily deprived us for ever of unique evidence on light construction methods, so important in south-east Asian architecture.

The main structure after restoration. The exacting task was begun in 1937, and resumed after the Second World War. It was not completed until 1953.

Malaya Encouraged by numerous local societies, particularly the very active Malayan branch of the Royal Asiatic Society, research started almost at the beginning of this century and has continually advanced, making a point of the production of useful publications. Although man-made monuments are rare, even if one

The sanctuary of the temple of Siva Mahadeva at Lara Jangrang.

ignores existing frontiers and regards the whole Malayan peninsula as a single unit, scholarly research is significant for two reasons: there is a possibility of illuminating numerous Chinese and Arabian ancient texts, mentioning the kingdoms and lines of communication in this country, which has always been a way station and a transit-point; and research can define the expansion and route of the influence of the Sumatran kingdom of Srivijaya on the mainland after the end of the seventh century. Research in northern Borneo has been expanding for some years now. Apart from some important prehistoric material (megalithic culture) this has yielded some evidence of Indian penetration into this large island from the fifth century onwards.

Champa Apart from two small ethnic groups driven back into the Phan Rang region and Cambodia, nothing of ancient Champa survived after the nineteenth century. Champa was an Indianised kingdom established at the end of the second century on the east coast of the Indo-Chinese peninsula, which was gradually absorbed by Vietnam in the course of her southerly advance from the end of the tenth century to the middle of the nineteenth. It could be said that knowledge of her history is entirely a reconstruction based on the comparison of foreign writings (Chinese, Vietnamese and European), epigraphy and archaeological material.

Late nineteenth century scholars produced some important philological and historical works, but discovered few monuments. Cham archaeology was born, like that of Cambodia and Laos, under the influence of the École Française d'Extrême Orient, founded in 1898. A summary catalogue of 229 groups of remains was begun in 1901, after which the first book, *La Religion des Chams* was published. From 1901 to 1915 Henri Parmentier, whose name is inseparable from the first stage of French archaeological research in south-east Asia, compiled an inventory of Cham monuments in Annam. This is still a basic work in spite of some ideas which are now obsolete.

The discovery of new archaeological sites by members of the school and its correspondents, and various excavation projects, considerably extended knowledge of Cham art and religion. Idols, foundation deposits and, at Thap Mam, an unsuspected style of sculpture, were found. Field work and even the most urgently necessary repairs to buildings were broken off by the events of 1945–46. Since these projects had never progressed beyond the stage of clearing the monuments, all the material capable of furnishing information on daily life is virtually absent, and in the present state of affairs it is doubtful if this serious hiatus will be filled in the immediate future. However, various interpretations have been offered on the basis of existing evidence, including a suggested history and chronology of the art style which extended from the sixth century almost up to the present day, iconography and the evolution of religion. These researches emphasise the importance of foreign influences (Cambodian, Chinese and Laotian) in the style, but it is for the most part profoundly original.

Laos The archaeology of Laos, the Thai kingdom founded in 1353, is highly complex because of the special geographical conditions and the varying influences to which the land has been subjected during its long history. It encompasses some very diverse elements – the strictly Laotian, the Khmer and the Burmese. Before 1900 all that was known of the art of Laos was contained in the book *Voyage d'exploration en Indo Chine* by Francis Garnier, published in 1873. Investigation of Laotian art – an original style characterised by a mixed type of predominantly wooden construction – was important and necessary, because this type of architecture is highly perishable. Apart from some limited work on prehistory (Tran Ninh, or the Plain of Jars), and studies of the Khmer monuments which naturally came into the orbit of Cambodian work, research has scarcely touched on anything but architecture and ritual objects. Without progressing much beyond the compiling of inventories and a few preservation projects (Vien Chan, That Luang, Wat Pra Keo, Wat Sisaket), Laotian studies have only resulted in the publication of works on the arts, the best of which is still Henri Parmentier's. Very little work has been done on Laotian archaeology and its subject seems very limited, but it would be unjust to minimise the work which has currently been accomplished. It has been very promptly published, has preserved the memory of many monuments which are now lost, and has awarded a frequently unappreciated art style its due distinction.

We have reserved the longest descriptions for Cambodia and Thailand because one clearly demonstrates how the advance of archaeology contributes to the knowledge of one of the most ancient kingdoms of the peninsula, while the other, because of its exceptional richness and variety, illustrates the complexity of south east Asian archaeological problems, and shows what can be hoped for from the progress of research.

Cambodia Work on Cambodia, or Khmer, has chiefly been carried out by the French. The account of the travels of the naturalist H. Mouhot, published in 1863 in *Le Tour du Monde*, attracted a great deal of public attention to the Cambodian ruins, which he called 'an architectural achievement which perhaps has not and never will have an equal in the world'.

Father Bouillevaux's *Voyage dans l'Indo Chine*, which appeared five years later, passed almost unnoticed. But it could be said that the first scholarly contacts with Cambodia did not take place until 1866–73, the first date being that of Doudart de Lagrée's expedition and the second that of the publication of Francis Garnier's *Voyage d'exploration en Indo-Chine*. Until then Western knowledge of the ancient Khmer civilisation was entirely derived from the first translation (1819) by Abel Remusat of Chou Ta Kuan's *Description of Cambodia*, a lively and highly informative picture of life in Cambodia in the last decades of the nineteenth century, and from the accounts of their various travels by British and German authors of the eighteen-sixties. The only local history was the royal chronicle, a late compilation in which everything that happened before 1350 was attributed to legendary times.

Research advanced side by side with exploration. Progress was closely allied to the foundation of the École française d'extrême-orient and the establishment in 1908 of the Conservation des monuments d'Angkor and the archaeological commission for Indo-China, which was dissolved in 1934.

The first phase of pioneers and initiators began with Doudart de Lagrée's expedition to explore the Mekong, the scholastic value of which arose from the great distinction of its members. This preliminary phase has never actually come to an end, since there is still plenty of unexplored territory in this area, which far exceeds that of the present kingdom. It can be regarded as continuing until the establishment of the first landmark, the creation of the Conservation d'Angkor, which was technically a turning point.

This was a period of investigations and enquiries which aimed to catalogue the monuments and collect inscriptions, and also succeeded in combing the whole country, often under appalling conditions and with highly precarious means of communication. There was no question of clearing the sites at this time, and even less of preservation. The first inventories were compiled, plans were drawn, often with surprising accuracy in view of working conditions, and restorations were attempted, generally less successfully. Casts were taken, and inscriptions reproduced. The photographs provide invaluable retrospective evidence, but unfortunately they were frequently distorted by draughtsmen who found it difficult to resist the temptation to heighten the romantic effect of the ruins in their illustrations for books.

The achievements of this period were considerable. Scholars not only assisted the remarkably rapid advance of Khmer studies but also provided much information which is still useful today. There is no space here to mention all the works of the pioneers of Cambodian archaeology, but we must quote a few unforgettable names. It is not generally known that in 1875 the Comte de Croizier drew up an early descriptive list of the monuments and in his book *l'Art Khmer* assessed the sculpture better than many modern authors. However, it was Louis Delaporte, a member of Doudart de Lagrée's expedition, who really introduced ancient Cambodian art to the West, especially to France. He was head of the 'Expedition for the investigation of the Khmer monuments' (1873, 1882–83), organiser of the Khmer museum from 1873–78 (which became the Museum of Indo-China in 1882) and of the former Trocadero Museum, which he directed until his death.

A guardian figure at the entrance to the eleventh-century AD temple of Ananda at Pagan in Burma. Since attaining her independence Burma has encouraged the investigation of her past—long neglected while she lay under the domination of a power that focused most of its attention on the antiquities of neighbouring India.

He was also the author of a remarkable book on Khmer architecture. The first generation of Cambodian archaeologists owed their training and often their vocation to him and to Etienne Aymonier.

Aymonier, who was also the founder of Cham studies, was directly or indirectly responsible for guiding research in the direction of true archaeology by insisting on the need for trained investigators and the use of material evidence. Aided by his thorough knowledge of the Cambodian language, of which he was the first professor, he laid the foundation of Khmer epigraphy with assistance from Bergaigne, the Indian scholar. His three-volume book *Cambodge*, which appeared between 1901 and 1904, sums up all the information available up to the beginning of this century. Despite some completely obsolete historical theories such as those on Funan or the date of Angkor Thom, it is still unquestionably valuable today, even if only for its full but summary list of monuments, which was not entirely superseded by Lunet de Lajonquière *l'Inventaire descriptif des Monuments du Cambodge* (1902, 1907, 1912).

This latter work, carried out by the École Française d'Extrême Orient, is still an indispensable work of reference fifty years after its publication. It can be criticised for its unscholarly approach, the inadequacy of its illustrations and the mediocrity of its plans, but no one can deny either the attraction of the method devised for describing the sites or the value of the evidence it contains. However outdated it may be, the book is still fundamental, and it needs annotation rather than revision. It will be obvious that the second stage of research could not have opened under more favourable auspices when we add that the same period saw the publication of L. Fournereau's great works on architecture and sculpture and the preparation of Dufour and Carpeaux's important monograph *Le Bayon d'Angkor Thom*, a complete review of the reliefs of the great temple.

The second phase is characterised by the application of documentary evidence and the clearing of the monuments accompanied by preliminary preservation measures. Even apart

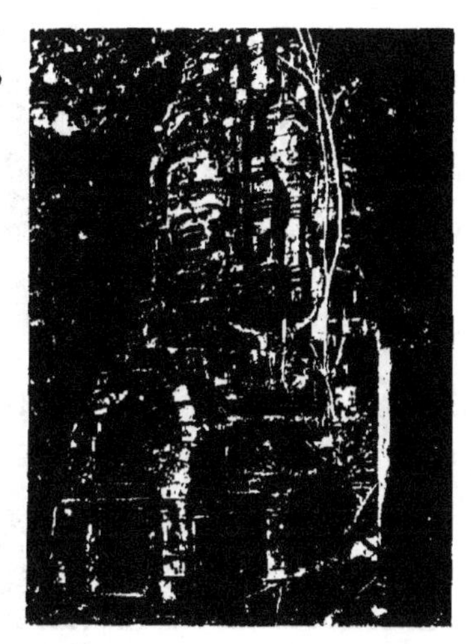

The temple of Prasat Lingpoun, which lies on the south-west side of Angkor Wat, has thirty-seven stone towers. Each one carries a huge stone face of the Buddha on all four sides. The encroaching forest threatens the survival of this great artistic treasure.

from epigraphic research bordering on archaeology which, thanks to the work of G. Coedès and L. Finot, developed remarkably swiftly and continuously, it is clear that the creation of the Conservation d'Angkor was a turning point.

The first explorers devised and shaped their own training, the new team, directly or indirectly attached to the École Française d'Extrême Orient, prepared for their task either by a university course or by field training, made use of research methods unknown to their forerunners, and could plan extensive programmes. L. Finot and G. Coedès' first attempts to interpret were devoted to iconography. Architects progressed beyond the stage of describing and set to work to study decoration and construction methods. H. Parmentier, head of the archaeological service of the École Française d'Extrême-Orient, finished Lajonquière's *Inventaire* and undertook the publication of precise and detailed monographs and overall studies, only part of which has, unfortunately, been published. J. Commaille, the first curator of Angkor, began to clear the ruins and published the first guide to the site of Angkor, which is still on sale to tourists. The work of clearing, which was bound to encourage people to visit and study the monuments, could only be accompanied by somewhat precarious preservation measures. Instructions were, however, to keep the ruins as they were rather than to make rash attempts at restoration, so as to avoid Viollet-le-Duc's type of fiasco. The first tentative reconstructions were only started in 1920, and included only the re-erection of the Giants flanking the approach road to the Victory Gate of Angkor Thom. As time goes by these scholars will perhaps be reproached for the limitation of their aims and methods, for concentrating on the park of Angkor at the expense of other sites, for paying too little attention to remains from a later date than the great Angkor period, and for ignoring everything but monumental art. It is true that there was much more clearing than excavating, and that no one in the archaeological service seems to have taken any interest in research on daily life.

But nowadays it is easier to criticise the work they did than to remember all the various difficulties they had to overcome, not the least being those emanating from the authorities—a Khmer museum had been requested and promised in 1908, but was not actually created until 1919. G. Groslier, a semi-independent scholar with a lively interest in Khmer life and history, was the first to assemble the records of Angkor civilisation. His book *Recherches sur les Cambodgiens* (1921) was unfortunately premature, but it could be regarded as the first real study of Khmer archaeology, using all the previously misunderstood evidence of the narrative reliefs, contemporary light architecture, pottery, ethnography, etc. Although so much attention was devoted to architecture and sculpture throughout this period, the chronology of Khmer art was regarded as a minor consideration.

Everyone accepted as an article of faith Aymonier's suggestion that Angkor Thom, the geometrical city enclosed in a wall twelve kilometres long with five monumental gates, centred on the Bayon, belonged to the town founded in the last years of the ninth century, as attested by an inscription. Even in 1924 when L. Finot showed that the carving of the Bayon and the gates was not of Shaiva inspiration, as the epigraphical evidence seemed to imply, but obviously Buddhist in character, no one was unduly disturbed. Except in the fairly rare case of a monument that could be dated by an inscription, everything reminiscent of the art of the Bayon was accepted as dating to the late ninth or early tenth centuries, and nobody worried about the apparently strange variations in the development of Khmer art.

The third phase, which was to raise Khmer archaeology to the front rank as a result of some spectacular discoveries, was marked by close collaboration between epigraphists, archaeologists and art historians, field workers and scholars in the research centres of Hanoi and Paris. It opened at the very time when art historians were becoming aware of the obvious discrepancies in this style, which could simultaneously produce monuments like the Bayon and others bearing dated inscriptions in an entirely different style. In the Indo Chinese Trocadero Museum, which met the demands of its founder L. Delaporte, a new theory which was destined to revolutionise Khmer studies was published by P. Stern. His book *Le Bayon d'Angkor et l'évolution de l'art Khmer*, which appeared in 1927, did not solve the problem of the date of Angkor Thom, but it laid the foundations of comparative study of the evolution of architecture and, by refuting Aymonier's theory, opened the way for a reconsideration of the whole question of chronology. G. Coedès' epigraphical studies enabled him to attribute Angkor Thom, the Bayon and all the monuments of the same style to their real creator, Jayarvarman VII—who was contemporary with Philip Augustus of France—a hitherto virtually unknown sovereign who was suddenly shown to be one of the greatest, if not the greatest, of early Cambodian history. Jayarvarman extended the country's borders and built an impressive number of monuments which are among the most astonishing creations of Khmer art—discounting the classic beauty of Angkor Wat and the royal temple of Suryavarman II, built fifty years earlier. However, this discovery raised another question: if Angkor Thom and the Bayon were three hundred years later, which was the first Angkor mentioned in the inscription consulted by Aymonier? This time the answer was supplied by archaeology. Topographical research and the excavations of

1931–34 enabled V. Goloubew to recognise the ditches, which had never been out of sight on the south and west sides, surrounding a city with sides four kilometres long, flanked on the east by a river in a diverted course and centred on the temple on top of the hill of Phnom Bakheng. Following up these investigations on the spot (air reconnaissance had already revealed a network of canals in the southern provinces in 1931), V. Goloubew was able to establish the importance of 'urban and agricultural hydraulics' in the life of Angkor, and simultaneously to extend the range of Khmer archaeology, which was still confined to monumental art.

Research on art advanced simultaneously. In 1927 Henri Parmentier's *L'art Khmer primitif* defined the profoundly original features, as well as the Indianising features, of pre-Angkor art, which flourished from the seventh century to the beginning of the ninth, and had tended to be eclipsed by the glories of Angkor. In 1931 P. Stern, working from a statue, some small columns and the lintel of a unique monument, suggested the definition of the transitional style marking the change from pre-Angkor to Angkor art. Five years later the discovery of twenty new archaeological sites in the same complex during a campaign lasting several months confirmed the original theory and enabled this scholar to expound the 'Kulen' style (from the name of the plateau thirty kilometres north of Angkor where these buildings stood), and to date it to the first half of the ninth century, despite the absence of epigraphic evidence.

The analytic method which had just been so triumphantly vindicated 'is based on a principle which seems extremely simple. When the decoration of one or more monuments shows similar features to that of a dated building, it is reasonable to assume that the monuments in question are roughly contemporary with this building; they must obviously be earlier if the decoration is less advanced and later if it is more highly developed.'

By a strict application of this principle Gilberte de Coral Remusat was able to publish *L'Art Khmer. Les grandes étapes de son évolution* in 1940, which twenty-five years later is still fundamental for the study of the history of Khmer art from the seventh century to the beginning of the thirteenth. Its only drawback is that it is a little too academic, being based more on the study of records than of the monuments themselves, and that it has somewhat neglected epigraphic evidence.

This period was rich in research, and the work of preservation advanced equally rapidly. The Conservation d'Angkor was still mainly responsible for clearing and organising the sites, but surveying, trial trenches and excavations added to the already extensive list new and sometimes very interesting finds, such as the pre-Angkor complex west of Angkor. Preservation methods were given a new direction by G. Coedès, head of the École Française d'Extrême-Orient from 1929 to 1946; reconstruction superseded reinforcement. Henri Marchal, who succeeded Commaille at the Conservation d'Angkor, was entrusted with the task of attempting the anastylosis of the Khmer monuments after a study and fact finding tour of Java.

Marchal's name is inseparable from that of Angkor, to which he devoted his discriminating care for nearly fifty years. His choice lighted on Banteay Srei, a small complex founded in AD 967 twenty kilometres north of Angkor. It had been decorated as a shrine, but was then almost completely in ruins. The operation lasted nearly five years (1931–36) and was completely successful – a veritable jewel rose from the ruins. All the previous restorations had only been able to give a very vague idea of its original appearance. The efficacy of the technique was proved. The safety of the ruins was now ensured and new possibilities opened up for research. Not only did the restored monument regain its original appearance, it revealed also unsuspected facts about construction methods, the use of sacred deposits etc., as well as yielding idols, inscriptions and cult objects during the course of the operation. Thanks to anastylosis, the results of which are perhaps too often assessed from the tourist's point of view, the actual history of the construction is subsequently known. After such an encouraging start this method was soon adopted by the Conservation d'Angkor, gradually improved and applied to increasingly ambitious programmes. Glaize, who has recently retired and who wrote an excellent guide to the Angkor monuments, was one of the most skilful successors of H. Marchal. We are indebted to him for some remarkable achievements such as the reconstruction of the central sanctuary of the Bakong temple pyramid, which literally rose again from a shapeless mound of earth.

The sanctuary of Phnom Bakheng as seen by Louis Delaporte in 1870. An illustration from Garnier's *Voyage d'exploration en Indo-Chine*.

While restoration techniques continued to advance and improve the fourth phase began. By now the archaeologists were not concerned simply with studying monumental art, but with looking for traces of former life by means of excavations, air surveys and analytical studies. L. Malleret pursued investigations in Transbassac which resulted in the discovery of the Oc Eo culture and the earliest evidence of life in Fou nan, published between 1938 and 1945. In 1944 Glaize began to excavate Angkor Thom and revealed part of the substructure of the ancient royal palace, which had only been investigated by a few hurried soundings until that time. Excavations were interrupted for ten years by international events, but they were subsequently re-opened in the Angkor region and extended to some prehistoric and protohistoric sites which L. Malleret had located by aerial survey. Unfortunately, none of the findings of these recent researches have been published. Work on interpretation is advancing side by side with field work, and new avenues are being opened up. These studies usually concentrate on working out the chronology of buildings and statues, and attach more importance to dated evidence, iconography and problems of the function and symbolism of the monuments. The result is a closer collaboration between archaeology and other sciences such as epigraphy and philology. Research is no longer limited to Khmer studies, but calls increasingly on comparison with foreign material and Indian parallels.

The progress and results of Khmer archaeology are spectacular, but this should not make us lose sight of the fact that these studies are still in their infancy, that there is still a great deal of work to be done, that a number of sites other than Angkor have been more or less neglected, and that many discoveries may always modify even the theories which seem nowadays to be the most soundly based.

Thailand Thailand, the only country in south east Asia to escape colonisation, remained both open to Western ideas and interested in her own past. She was indebted to her kings for the first steps in archaeology, which took place in the middle of the nineteenth century. In fact, King Mongkut (1851–68) first assembled specimens of archaeological and epigraphical material in Bangkok, brought them to the attention of Europeans and, after his accession, created a little museum in the style of the eighteenth-century amateur collection in his palace. This museum was gradually enlarged and was opened to the public by King Chulalongkorn in 1874. Other collections were subsequently created, the most important and scholarly being that of the Minister of the Interior, Prince Damrong, the true father of Thai (or as it used to be known, Siamese) archaeology. His example roused the provincial administration to emulation and led to the creation of archaeological storehouses, some of which had already assumed the proportions of museums by the early years of the twentieth century.

As Thailand's administration was very different from that of her colony or 'protectorate' neighbours, and archaeology there was based on the formation of museum collections, surveying and catalogues of monuments were neglected in comparison with the rest of south-east Asia. For a long time this side of the work was left to the discretion of European explorers, among whom some of the great pioneers of Khmer archaeology such as L. Fournereau, Lunet de Lajonquière etc. are to be found. The archaeological service is therefore fairly recent; it was only created, under the auspices of Prince Damrong and G. Coedès, in 1924. Indeed, apart from the foundation of the National Museum of Bangkok two years later, real archaeological research did not begin until the thirties, and only achieved its full growth about ten years later.

In these circumstances, the archaeological heritage of Thailand is so rich that the field research must be equally promising. In practice it is not concerned with the evolution of a single civilisation but, as G. Coedès puts it, 'with a species of stratification, the upper and more recent strata of which represent true Thai art from the eighteenth century onwards, while the lower and earlier strata consist of the archaeological traces of the peoples who occupied all or part of this region before the Thais established their rule here in the fourteenth century'. This scholar also stresses another peculiarity, this time of a geographical nature. 'Thailand is situated at a meeting point of various civilisations, and this explains the diversity of the archaeological remains and schools of art, which have nothing in common but the fact that they come within the influence of Indian art and are mainly Buddhist in inspiration.' The result of this combination of circumstances is that Thailand sums up within herself almost all the archaeology of south-east Asia.

Apart from her many rich prehistoric and protohistoric sites – those in the region of the Kve Noi (the famous 'River Kwai') have just yielded some interesting finds to a joint Thai-Danish expedition – we have an uninterrupted, sometimes contemporary, series of civilisations right up till the present day. Once the Fou-nan antecedents of the Dvaravati civilisation were recognised, as described above (this discovery was one of the most important contributions of Australo-Asian archaeology), a link was established between protohistory and the seventh century. By greatly simplifying the elements of this highly complex problem it could be said that Thai archaeology consists of the Fou-nan period (third to fourth century), Dvaravati (seventh to ninth centuries), Srivijaya (in the Indonesian tradition, especially in the peninsula region, eighth to thirteenth centuries), and Khmer (in the eastern provinces, from the ninth to the early thirteenth centuries, with its peak in the reign of Jayarvarman VII and continuations in the local school known as Lopburi).

Thai archaeology really includes the ancient kingdoms of Sukhodaya (thirteenth to fifteenth centuries), Ayudhya (fourteenth to eighteenth centuries) and Lan Na (or Cheng Mai, thirteenth to nineteenth centuries), which are distinguished by their individuality in using lessons from the borrowed elements from abroad, especially Ceylon and Burma in the Lan Na period. The most productive researches at present are studies of the civilisations of the first centuries, particularly the Dvaravati culture, but it is equally important not to forget the significant discoveries of the Thai period such as the treasures found at Ayudhya and Lan Na. These are of an admirable quality, and provide certain evidence of advanced cultures and techniques.

Some readers may be surprised by the importance of monumental archaeology and art history in south east Asian studies, but the reason for this is easy to see. We have seen that archaeology is a young science everywhere, even in Indonesia. It was only natural for scholars to concentrate at first on the most accessible remains and those which could provide the most concrete evidence, e.g. the historian's indispensable epigraphic texts and evidence of the artistic production which provides the most informative facts about any civilisation. As to art, research on dates was necessary because there were hardly any texts. This was the first step, without which research could have gone no further. It has taken more than fifty years to achieve it, but we should be surprised that such substantial results have been obtained so quickly in this unexplored terrain. There is obviously much to be done in every field, but only the uncharitable could blame the great pioneers who unassumingly cleared this unusually tangled terrain for the next generation of scholars.

China

Very little was known of prehistoric east Asia before the First World War, and serious archaeological research in China is fairly recent. The investigations of G. Andersson, the Swedish geologist employed by the Chinese government after 1920, uncovered the remains of a Neolithic culture in the village of Yang-shao (western Honan). In 1921 the discovery of the famous *sinanthropus* at Chou-kou tien south of Peking proved that there were hunters and food-gatherers in north eastern China 500,000 years ago. In 1928 scientific excavations were conducted in the An yang region on the site of the last capital of the Shang dynasty (1766?–1111? BC), and these recovered traces of the culture of the second royal dynasty, which ruled from the fourteenth to the eleventh centuries BC. In 1937 the war between China and Japan interrupted these investigations, but they were reopened by the government of the Peoples' Republic of China and extended to include an increasing number of new sites covering the prehistoric and archaic periods, the feudal age and the imperial era (221 BC 1911).

Even though this review is confined to the material and concrete evidence, it will necessarily cover a limited period, and will end with the foundation of the empire in 221 BC. At this date King Sheng of Chin annexed the kingdoms which were disputing for power, took the title of Huang-ti, 'August Monarch', and united all the lands of China for the first time under one ruler. The individual features of the various provinces faded away and a single culture spread throughout the empire.

The prehistoric period

The cultures of prehistoric China are among the most ancient in the world, and it is important for the student to realise that the human race itself evolved in China at a very early date. A discovery in 1960 indicated that *sinanthropus Pekinensis* may have had ancestors in northern China. The existence of an anthropoid from three to six times the size of historical man in southern China has been confirmed by a find at Kwang-si. The relationship between this *gigantopithecus* and Javanese *meganthropus* is interesting because it shows the great antiquity of the links between China and south east Asia. The continuity of human evolution in the middle and upper Paleolithic is now proved by the remains found in both the north (on the upper course of the Yellow River, in the Ordos and at Chou kou-tien) and in Szechwan and Kwang si. Contrary to recent theories, it is now known that the Mesolithic period was not a void, and that a microlithic industry appeared in China about 25,000 BC (excavations in Manchuria and Mongolia in the north, and in Szechwan, Kwang si and Kwang tung).

It seems that in this period the Mongolian steppe developed a different climate from that of the wooded valleys of north China and Manchuria. It also appears that a late Paleolithic way of life survived in southern China, where human remains show negroid characteristics, and that progress towards a less primitive culture was slower than in the north. The hunters and fishers of the north gradually learned to cultivate plants and to domesticate animals. The first use of polished stone and the first attempts at stock-breeding and cultivation can be dated to the fourth millennium. Village communities subsequently developed preserving a semi nomadic form of life in the early stages, and only achieving stability and permanence in the later periods. The culture named after the village of Yang-shao belongs to the first stage, the Lung shan culture, named after the site in Shantung where it was first found, belongs to the second. The earliest Yang shao site was found at Pao ki hien in the middle Wei valley in Shensi. This culture spread as far as the borders of modern Sinkiang in the west and Shantung in the east, but its centre was in eastern Shansi where the rivers Wei and Fen flow into the Yellow river. The influence of Western Asia does not seem to have been absent from the development of this culture. Indeed, it is characterised by pottery of fire-reddened fabric decorated with geometric patterns like those on the Neolithic pottery from Susa, Iran, the Ukraine and Rumania. This type of pottery, which seems to have persisted for a long time, included several different phases. The finest painted clay vases belong to the middle phase and come from the Pan shan necropolis on the western bank of the river T'ao-ho in Kan tzu. They are big jars with large bodies made by stacking rolls of clay flattened with a tampon. The decorative motifs (spirals, rhomboids, checks and lattices) were painted with red and black pigments before firing.

Pots from sites belonging to the Lung-shan culture are completely different. They are made of fine black clay fired at an appreciably higher temperature and thrown on a wheel. Unlike the red Yang shao pots, they have no decoration and are remarkable both for the striking effect of the burnished black surfaces and for the variety of their shapes. It should be mentioned that among the Lung-shan pottery are tripods with hollow legs (*li*) and solid legs (*ting*), cooking pots (*hsien*) and footed cups which probably served as models for the bronze-smiths of the Shang period. Although it sometimes occurs in association with the Yang shao culture, Lung shan seems to be later. The farming communities were already protected by earth walls, there is evidence of burial rites and some religious activities, and the practice of scapulomancy (divination by means of animals' shoulder blades exposed to fire) was extensive.

Where did this culture originate? Perhaps on the coast. It seems to have started from centres

A tree seems to be growing from the roof of the temple of Prah Khan. In fact the roots sink far below the very foundations, and many of the buildings are beyond salvation.

The great stone faces look down on the only nude statue known in the art of the Khmers. This figure, far right, from Angkor Thom wears the familiar enigmatic serenity, and identification has never been made. It could be a god, but the elaborate crown suggests an earthly monarch.

on the Shantung coast and northern and western Honan. It spread towards the lower Yangtze Kiang basin and the south-eastern seaboard. In Hsiao-t'un (An-yang region) it is closely associated with a transitional stage called the 'grey pottery culture', which immediately preceded the Bronze Age. The stratigraphic sequence throughout the excavated area at An-yang, on the site of Hou-kang and elsewhere, is as follows: red pottery is found in the lowest strata, black pottery in the intermediate levels, and grey pottery in the third layer. Vases made of a fine white pottery dated to the peak of the Bronze Age were found above the grey pottery and associated with it at Hsiao-t'un in particular on the site chosen as their capital by the Shang in the fourteenth century BC.

The cultures of the Neolithic period coexisted and followed each other without a break from the fourth millennium to about 1500 BC in China. At this date the Shang dynasty had ruled China for more than 200 years, and bronze working had reached its peak.

According to traditional history the Shang was preceded by the Hsia, the first royal dynasty. Still further back heroes and religious rulers taught the Chinese the major arts and inaugurated the rites upon which their civilisation was founded. Legend tells of the Three Monarchs (*San Huang*), then the Five Emperors, the first of whom, Shuen, had as his prime minister the Great Yu, founder of the future Hsia dynasty. So far prehistoric research has shed little light on the preliterate period. It is only with the advent of the Shang dynasty that it becomes possible to work out a chronology. We have been able to discover the names of the thirty Shang kings and give each of them a relative date from the divinatory inscriptions, which will be discussed later. Does the Hsia dynasty, which traditionally reigned from the end of the third millennium to the beginning of the second (2205–1767? BC), correspond to the Lung-shan period when the elements of political organisation appeared? According to legend the Bronze Age began with the Great Yu, and some versions indicate that the Iron Age could have begun between the ninth and eighth centuries under the Chou dynasty. The findings of current research are entirely provisional, but they indicate that Bronze Age technology began about 1700 BC, and the general use of iron about the year 600.

The archaic period

The Shang-Yin dynasty The accession of the Shang dynasty at the beginning of the eighteenth century BC therefore seems to mark the end of the Neolithic period in China, or at least in the central region (northern and western Honan, south-western Shansi and eastern Shansi). It is now known, however, that the transition from polished stone to metal-working was accomplished gradually and that the technique of bronze-working originated in China. It was known in south Russia and Mesopotamia in the second half of the third millennium. The knowledge of alloying metals could have crossed central Asia and the steppes to the Yellow River valley, but the theory that this technique was imported has now been abandoned. The appearance in the fourteenth century BC of magnificent bronzes, for which no antecedents can be found, could be explained by an influx of foreign peoples, perhaps invasion, which could have introduced the techniques and institutions which comprised the original Shang culture along with fully developed metallurgical techniques.

The horse drawn chariot, writing and the calendar appeared under the Shang Dynasty. Cities were established, offices became specialised, a hierarchy developed, and the differences between urban and village communities became more pronounced. Unlike the big rural family characterised by its unity, the aristocratic families assumed a patriarchal form with fairly complex organisation. The equipment of tombs, which will be discussed later, the abundance and splendour of the grave goods and the practice of human sacrifice all attest the importance of ancestor worship in the lives of the royalty and the nobility. The coexistence of two complementary structures–village and town–is perhaps the most remarkable feature of the earliest Chinese social order. However, it is not necessary to date the beginning of true Chinese civilisation back to the appearance of social segregation. The progress of the metallurgical arts certainly played a decisive part in this development.

Thanks to recent investigations, some previously unknown facts are now accepted. For instance it is known that the An-yang bronze workers had predecessors. Some Chinese archaeologists believe that the appearance of bronze could be slightly earlier than the accession of the Shang dynasty. The earliest foundries known to date, however, are those of Cheng-chou (northern Honan), where the Chinese opened an enormous excavation in 1950.

The excavations of Cheng-chou There are some Neolithic remains at Cheng-chou.

The excavations covered an extensive area, and the Shang sites, too, have a range of several periods; the earliest marks the transition to the Lung-shan Neolithic, and bronze does not appear until the third, which is known as Erh-li-kang. The foundations of a city surrounded by 1195 metres of walls date back to this period. This beaten earth enclosure seems to have protected the homes of the nobility, as well as the religious and civic buildings. Craftsmen and peasants worked outside the walls – an ivory and bone-working studio, a pottery factory and two bronze foundries were discovered outside the rampart. This city, which was occupied for a very long time, corresponds to the town of Hao, which was raised to the status of capital by the tenth Shang king. The population was very numerous, and must have overflowed out of the enclosing wall, which was built and then rebuilt in two different periods. House foundations, paved roads and tombs have been brought to light, with objects of stone, bone, ivory, jade and shell, enamelled pottery vases, bronze ritual vessels and three bone fragments with oracular inscriptions which could be contemporary with the inscribed bones from An-yang, the site of Hao only having been abandoned gradually.

It was, in fact, in a stratigraphical level corresponding to the latest phase of Cheng-chou in the early fourteenth century BC that writing appeared at An-yang. At this time (1384) King P'an-keng established his dynastic capital north-east of the present subprefecture of An-yang and changed the name of the dynasty, the Shang now became the Yin. Eleven kings followed him in this city until the fall of the Yin.

The architectural remains on the site of Hsiao-t'un where the administrative centre of the city was situated occupy the three strata above the Neolithic levels. The excavations so far have uncovered some underground dwellings in the form of pits at the deepest level, the stamped earth foundations (*hang t'u*) of some important buildings with drainage channels at the intermediate level, and similar foundations, but without the drainage system, at the highest level. The platform foundations of the intermediate level would be those built by P'an-keng when he established his capital at An-yang, and it was from the time that the city became the royal seat that the oracular bones, shells and tablets – in short, the whole body of inscribed objects, began to accumulate. It is important to note that *chia ku wen* (characters on bone and tortoise-shell) are too stylised to be symptomatic of the beginnings of Chinese writing. The excavations have not yet shed any light on the early stages of this script. However, epigraphists estimate that the pictographs engraved on ritual objects from An-yang could be evidence of an older script, and were saved from destruction because of their beauty. An epigraphist recognised the importance of the finds picked up by peasants in their fields by comparing inscriptions on bronze with the signs engraved on these bones. He realised that he had the oracle bones of the Shang Yin dynasty in his hands.

An-yang Despite the interesting investigations currently being conducted at Cheng-chou, An-yang is still the most remarkable excavation site of all. Here we shall mention only how the increasing scholarly interest in the deciphering of the inscriptions on bones determined the Academia Sinica to undertake systematic excavations at An-yang. From 1928 to 1937 fif-

teen excavation campaigns were conducted at Hsiao-t'un and in the surrounding regions. The investigations were interrupted by the war between China and Japan, but they were reopened in a limited form in 1950, when interest had shifted to Cheng-chou. The sites explored were on either side of the River Huan north-west of An-yang, covering an area fifteen kilometres square. Each of these sites formed a separate community, with its residential quarter, workshops and cemetery, but their evolution seems to have been dependent on the capital. At Hsiao-t'un, where the great Shang city was situated on the south bank of the river, archaeologists have been able to reconstruct the plan of the town.

The houses of the nobility probably stood in the north, the administrative and religious buildings (twenty-one platforms of beaten earth were aligned north/south in this sector) in the centre, and some ceremonial buildings and tombs in the south. The abundant human remains from tombs and foundations, from beneath gates and around buildings, are evidence of the practice of foundation sacrifices. Nothing is left of the actual buildings apart from the stone or bronze bases of the pillars which supported the roofs. They were huge constructions. The terraces measure an average of twenty to thirty metres long by ten metres or more wide (the biggest is eighty-five metres long by 14.5 metres wide). One of them still has two stairways in front and the bases of its thirty-one pillars. The gabled roof, supported by a wooden framework, was probably thatched. There were doors and windows through the beaten earth walls. Hsiao-t'un is not the only site where substructures have been found, but there are more than fifty of them in this centre of the Shang kingdom. Some underground dwellings near the great houses were probably the homes of the lower classes. They have been found further to the east, with bronze foundries, pottery kilns, and stone and bone workshops.

The palaces and temples of the capital symbolised the royal authority; death, after which the rulers were deified, only served to hallow their power. The royal cemetery has been discovered on the site of modern Hsi-pei-kang north of the river, and at Wu-kuan-ts'un next to Hsi-pei-kang. Eleven huge underground tombs have been investigated in this area. They all have a central chamber, usually facing north and approached at two of its corners (sometimes at the four cardinal points) by an access corridor. In the chamber and separated from the walls by a beaten earth partition stood the burial room. It was made of wooden beams and decorated with incised motifs and paintings which have left traces on the surrounding earth. The coffin rested in the middle of the room over a deep pit in which a man and sometimes a dog (the conductor of souls) was buried. With its two access corridors ending in a staircase to the north and south, the tomb of Wu-kuan-ts'un is forty-five metres long.

Human skeletons and grave goods were found in the chamber, on the ledges separating the burial room from the east and west walls. A male skeleton was extended face downwards in the pit under the coffin and kneeling men and dogs were buried at the foot of the access corridor. A smaller tomb at Hou-kang south of the river yielded a considerable number of burial gifts as well as human bones and twenty-eight male skulls. There were bronzes, jades and engraved stones, potsherds and gold leaf. Some smaller tombs on other sites have been examined, notably at Ta-ssu-k'ung-ts'un (north of the river opposite Hsia-t'un) and 22 out of the 166 tombs were found intact. These were more modest and had escaped the tomb robbers who, unhappily, did not spare the royal burials.

The finds will be described later. They included large bronze vases with strong decoration, weapons adorned with turquoise and jade, carved marble, white pottery and hard or enamelled pottery, and engraved or inscribed bone and ivory. Study of these treasures naturally pointed out the associated residential areas, and other archaic sites have been found in recent years.

The Western Chou The settlements belonging to the Shang kingdom were not ruled directly by the king, but by related lords or high officials appointed by him. *De facto* chieftains beyond the reach of the royal troops paid tribute to the central government which, in return, recognised their independent administration. Script and coinage established links between the capital and the minor centres. Intermarriages strengthened the ties between these groups of people who were united by their belief that they all belonged to a superior civilisation. Despite the conflicts which frequently divided them, these little states were invested with mutual solidarity by the hallowed sanction of family alliances.

The Shang civilisation, therefore, owed its originality to the aristocratic institutions of the walled cities. These towns were small. The wall of Hsiao-t'un has not yet been found, but it cannot have been more than 800 metres long. Despite their meagreness, these urban centres in the wild landscape exercised a certain fascination over the barbarians. This civilising virtue diminished with distance, but it extended as far as the modern subprefecture of Pao-chi in the middle Wei valley, and to the Chou, the power which was to overthrow the Shang-Yin dynasty. It was in the reign of the great-great-grandfather of King Wu who conquered the Shang, that the Chou adopted Yin civilisation. Their earliest settlement may have been in the lower Fen valley (in the south-west of modern Shensi). However this may have been, the excavations in south western Shansi and the middle Wei valley have produced the following sequence: Yang-shao, Lung-shan and Chou.

It is therefore probable that cultural exchanges cemented by marriage alliances gradually brought the Chou to absorb Shang customs and institutions. Their state was to become a nominal dependency of the neighbouring kingdom, but their power increased while that of the Yin decreased. This slow process of the integration of a semi-barbarous state into the Chinese civilisation would explain the difficulty archaeologists have in distinguishing Yin sites from early Chou. At a date assigned by Chinese scholars to 1122–1027 (nine different dates have been suggested between these two termini) the chief of the Chou, also known as 'Chief of the West', invaded Shang territory and penetrated into the heart of Honan. He crossed the river and destroyed the king's army. The body of the last Yin king was burnt, the conqueror assumed the royal title and the Chou dynasty was founded. The 'Chief of the West', now King Wu, then returned to his hereditary domain where he had just founded a capital, Hao (south west of modern Sian).

The stratigraphical sequence mentioned above was established on the site of K'o-hsing-chuang near Sinan. About nine places in the Sinan region have been associated with Chou stratigraphy. These sites are distinguished by deep cultural deposits. Rectangular pits and tombs attest the practice of human sacrifice and very dense occupation. Are they earlier than the conquest? Shang elements belonging to the Erh-li-kang phase have been found in tombs further east in the lower Wei basin, but Shang objects could have been brought to the west after the conquest. According to inscriptions on bronzes the four large tombs in the Wei valley belong to the end of the Western Chou. Whether or not the Chou possessed advanced organisation and techniques before the conquest, their civilisation after the conquest followed the route marked out by the Shang.

Like the Shang, the Chou built walled cities which protected the homes of the nobility. They included a central platform for the palace, the temple and its altars and the administrative offices. Also like their predecessors, they buried their dead in wooden coffins set in the bottom of a pit-grave over a small pit for the soul-conducting dog. They buried human victims, bronze weapons and vases, jade objects and sometimes horses and chariots with their dead. The use of tumuli does not seem to have been common before the eighth century.

The sites investigated in recent years are cemeteries. The most outstanding of them are situated in eastern and southern Shensi (Chang-chia-p'o and P'u-tu-ts'un near Sian; Li-ts'un near Mi-Hsien), in south Shansi (Tung-pao and Fang-tui-ts'un) and in north western Honan where the site of Hsin-Ts'un (subprefecture of Kiun) was discovered before the Second World War and a cemetery was recently found at Shang-ts'un-ling (Shen-hien). The tombs of Shang-ts'un-ling date to the end of western

Angkor Wat, the greatest work of architecture in Asia, built by King Suryavarman II in the twelfth century AD. The causeway is nearly a mile in length.

Chou and the beginning of the next era, those of P'u tu-ts'un, Li ts'un and Fang tui-ts'ui to the beginning of the dynasty. Some of the tombs at P'u tu ts'un are a little later, dating to the reign of King Mu (947 928 BC).

This period is not distinguished by a single innovation in either the shape or the decoration of the artifacts. The political centre, however, was moved further west and cultural influences seem to have spread further than under the Shang. Bronze vases dating to the western Chou have been found in the province of Jehol at Ling-yang in the north, in the lower Yangtze Kiang basin (Tan-t'u, Yi-cheng, Nanking) in the south, in northern and north-western Cho kiang and south Anhwei (T'un-hsi). However, the bronzes from these sites were associated with others which are not found any further north and seem to belong to a less advanced culture. *It is not impossible* that certain favoured centres were established in marginal areas.

The eastern Chou (Ch'un-Ch'iu period and the Warring States)

The expansion of the cultural influence of the Chou seems only to have reached its peak in the eighth century, after the transfer of the capital to Lo-yang in 770 BC. The feudal institutions which had first appeared under the Shang attained maturity in this period. Under the rule of the king, who derived his power from heaven, the lords *enjoyed far reaching privileges* on their estates. Some of them rebelled and invited the Western barbarians to attack the capital. Hao was sacked and the Chou fled to Honan under the protection of a vassal. Lo yang became *the seat of this diminished monarchy*.

From the eighth century to the middle of the fifth the feudal system was to disintegrate, great city states were established on the outskirts of the kingdom, and political power was no longer *vested in the king*, who arbitrated over disagreements but no longer wielded any force. The son of Heaven was obliged to give his support to the most powerful of the border chiefs in order to ensure domestic harmony and *to keep out the barbarians*. This was the period of the Warring States, dominated by the three states of Ch'i (Shantung), Ch'in (Shansi) and Ch'u (middle Yangtze basin), with two powerful states, Wu in the lower Yangtze basin and Yueh to the north of Ch'o kung, in the south.

The precarious barriers erected by the coalitions against the attacks of the great kingdoms finally collapsed in the fifth century. Uncontrolled anarchy reigned, and the rival principalities fought each other incessantly until the foundation of the empire in 221 BC. The period of Chinese history which saw the consolidation of the military and economic power of the great principalities is known as the Ch'un-ch'iu (Spring and Autumn) era, and the next period as the Chan-kuo (Warring States) period.

In these five hundred years of crisis Chinese society changed radically. After the sixth century fiscal and agrarian reforms were carried out and penal laws passed. In the fifth century centralisation became more pronounced and the feudal fiefs yielded to administrative controls. Battles ceased to be jousts for 'the defence of honour', and became expansionist wars in which different cultures clashed, east against west, north against south, Chinese against barbarian. As Granet writes, 'War became a necessity, a routine job'. Armies of mercenary troops were formed, with professional strategists and *bravos* instead of knights. They adopted equipment and arms from the nomads who opposed the Chinese social system in order to defeat them. The mounted archer replaced the nobleman in his war chariot.

Groups of barbarians in the north-west united to threaten the kingdom of Ch'in which achieved military supremacy in the fourth century by means of a series of severe reforms. The Yellow River culture encountered an indigenous culture on the Yangtze, which excavators have shown to be both original and sophisticated.

The development of written literature in the eighth century makes it easier to follow the course of this long evolution, but archaeological evidence is equally valuable. More than seventy-eight cities must have been built in the Ch'un-ch'iu period. Most of the towns identified so far seem to have been occupied for a very long time, until the end of the later Han (209–201 BC). The remains of these towns are also extremely varied, we have only space to mention the most important. At Lo-yang in Honan, the royal capital, more than 600 tombs were opened between 1950 and 1954, yielding more than ten thousand objects (the investigation of Wan-ch'eng is still in its early stages, but it is known that the beaten earth wall was nearly square and enclosed an area about eight kilometres square). The successive capitals of the Chin were Wo-kuo and Hsien-tien in south-western Shansi. Hsia-tu in northern Hopei was one of the capitals of the state of Yen, and Han-tau was the capital of the state of Chao in the south from 386 to 222 BC. Lintzu was the capital of Ch'i in Shantung.

Some tombs were dug up by farmers at Chin-ts'un, at the royal city of Lo yang. Since 1950 scientific excavations have been conducted in Hui Hsien at Liu-li-ko, Ku-wei-ts'un, Ch'ao-ko etc. A pit at Liu-li-ko yielded nineteen chariots; tombs at Ku-wei-ts'un produced bronzes inlaid with gold and silver. Some of the pieces from Hsin-cheng (Honan) could be earlier than the seventh century, but the bronzes from Chin-ts'un and Hui Hsien belong to the Warring States period.

The Eastern Chou towns are mostly square or rectangular, and their arrangement is reminiscent of those dating to the early years of the dynasty. An outer wall dating to the fifth century and enclosing industrial quarters, homes and commercial streets is found on some of the sites. At Lintzu this wall is four kilometres long from east to west and more than four kilometres from north to south, while the area inside the enclosure is no less than 1350 metres square. According to contemporary writings the inner wall delimited the citadel (*ch'eng*), the fortified area reserved for the ruler and his court. The outer wall protected crafts, industry, trade and everything that went to make up the wealth of the *kuo* (state), acquired by pillage and war. The farms on which the peasantry lived spread outside the walls. In the Warring States period, therefore, the cities housed a very large population enjoying a variety of skills and incessantly increasing in wealth. Commercial exchanges took place between the states, and a coin-using economy was established.

Agricultural methods kept pace with the development of the cities—great irrigation works and more advanced techniques made it possible to cultivate the marshy river valleys. The Chinese Iron Age began towards the end of the Ch'un-Ch'iu period, and the plough made its appearance. The earliest ploughs known to date come from Ku-wei-ts'un and belong to the Warring States period. Thanks to the spread of iron technology, agricultural methods improved and the yield of the farms increased rapidly.

Chou culture, strengthened by this progress, advanced rapidly in all directions. In Manchuria and on the borders of Mongolia in the north-east and north they clashed with nomadic warriors from whom they learned some practices. We know that barbarian strategy inspired far-reaching military reforms in the border states in the fifth century. The infantry became increasingly important and in the fourth century, when the threat of the Hsiung Nu from the steppes became apparent, the Chinese army enlisted mounted troops. New weapons, the crossbow and the catapult, were also adopted. The long tanged sword tended to replace the short Chinese sword. The shape and decoration of objects changed and a new style was gradually evolved from the wreckage of the Chinese kingdoms. Contacts between the nomads and settlers of the north and the shepherds, hunters and warriors became more frequent and exchanges more active, but tension also became greater and conflicts more widespread.

In the south, on the other hand, cultural expansion met with no major obstacles. Excavations in the middle and lower Yangtze valley, in the Red basin of Szechwan and on the Yunnan highlands prove that south China had attained a high level of civilisation in the fifth century BC. One of the most remarkable discoveries was a tomb found in 1955 at Shu Hsien (northern Anwei) on the site believed to be that of the city which was adopted as the capital of Ts'ai in 493 BC. With the lacquered wooden coffin which contained the remains of the Marquis of Ts'ai were rich burial gifts including pottery, lacquer, musical instruments and weapons. In 447 Ts'ai was destroyed by Ch'u, the most powerful of the southern states.

At the height of Ch'u power their territory extended to Shensi in the north-west, central Honan in the north, and Lake Tung-t'ing (southern Hopei) and northern Hunan in the south. This region was largely open to Chinese Yellow River influences, but it formed a distinct ethnic unit with an original culture. Its language was different from that of northern China and its script had individual features. The people were regarded by the northern Chinese as extremely superstitious, and they preserved myths and religious beliefs which the Confucian faith rejected. This land was classed as semi-barbarous in the central states, but it produced the most celebrated of the ancient Chinese poets Ch'ü Yüan (fourth century), whose mystical genius seems to have affinities with Chang-tzu.

According to the historical writings, the Ch'u administration and social organisations were very similar to those of the Chou. Their rice fields were well irrigated, and their agriculture was as prosperous as their industrial arts and crafts. None of their capitals have been found so far, but 1200 individual tombs have been explored in this region. The main excavation sites are situated in the Huai valley (Hsin-yang in Honan and Shou Hsien in An-yang) and in the Ch'ang-sha region of Hunan). They yielded rich and varied grave goods which show very advanced cultural development. There are iron tools and weapons, bronze ceremonial vessels, bells, weapons and mirrors, painted and enamelled pottery, freestanding wood carvings, bamboo mats, tooled leather, painted lacquer, silk and hempen cloth, slips of bamboo in bundles covered with written characters, and writing brushes in their containers. Among the most striking finds are the carved wooden tomb-guardian figures representing monstrous beings with long tongues and antler horns.

The Ch'u civilisation seems to have originated in the Huai valley during the Ch'un-ch'iu period and spread towards Hunan. It survived the overthrow of the state, which was destroyed in 223 by the founder of the Chinese empire. Its evolution can be followed up to and including the western Han period by means of epigraphy and study of the Ch'eng-tu region and the Chia-ling-chiang valley. It even reached Yunnan, where a distinctly important cemetery was investigated east of Lake Tien between 1955 and 1960. The excavation in Szechwan was important, but it cannot be discussed here; however we must not pass over the material found in 1955 at Shi c[illegible]han.

Th[illegible] investigated t[illegible]s belonged t[illegible] Dong Son culture th[illegible]ious site on the Gulf of Tonkin. Some similar[illegible] between Dong Son material (bronze drum[illegible]t shaped and egg shaped axes) and ob[illegible] [illegible] have led some prehistorian[illegible] to look for the origins of the Dong Son cu[illegible]ure in the west. Study of the funerary goods f[illegible]om Shi chi shan tends to prove that the [illegible] impu[illegible] originated with Ch'u and in western Szechwan. This does not exclude western influence, but it seems to have operated indirectly through the medium of Chinese culture. The earliest tombs at Shi chi-shan date to the end of the Warring States period and the latest to the beginning of the Christian era. Large numbers of Chinese coins and objects made in China during this period were found in the course of the excavation. By this time the empire had been established for more than two hundred years and the Han dynasty was ruling China.

The finds–stylistic evolution

The objects from the excavations cannot all be described here. The limited format of this chapter does not allow for detail, so we shall concentrate on the bronzes which are interesting from the point of view of both archaeology and art history. According to Tong Tso-ping's chronology, more than 1500 years lay between the Shang and the accession of the Han.

The Chinese archaeologist Kuo Mo-jo divides these fifteen hundred years into five periods:

1. stage of beginnings (late Hsia and early Shang).
2. stage of florescence (late Shang-Yin and early Chou).
3. stage of decadence (middle Chou to middle Ch'un Ch'iu, about 600 BC).
4. stage of renaissance (from the early seventh century to the end of the Warring States).
5. stage of decline.

Kuo Mo-jo assigns the vases to these five periods by their style, decoration and the inscriptions sometimes engraved on them. The vases played the part of *sacra* [illegible] and formed also useful reserves of metal for emergencies, but they were mainly used to hold the offerings presented to the ancestors at the great sacrifices. They were either cast in one piece in a mould or made in parts and fitted together, and their shapes vary according to their function. Archaeologists divide them into three classes.

I. Vessels for solid foods: tripods (*li, ting*), pots (*hsien*), bowls (*tou*, [illegible]), square boxes (*yi*), jars (*p'u*), round vessels ([illegible]).
II. Wine vessels: pots (*hu*), tripods (*chia, chueh*, [illegible]), pitchers ([illegible]), jars ([illegible]), cups (*tsun*, [illegible]), ewers (*kuang*), dishes ([illegible]).
III. Water vessels: bowls ([illegible]), basins (*p'an*).

These types of vases, especially the oldest (the *li* tripod in particular, which tended to disappear towards the end of the Ch'un ch'iu period) reproduce in bronze the shapes of the pottery and wooden vessels of the Late Neolithic.

Weapons (arrowheads, various axes, dagger axes, knives, plate armour and halberds), bells, tools, horse and chariot fittings, masks and mirrors, would all have to be described in order to give a true picture of bronze metallurgy. All these objects show a fairly homogeneous style

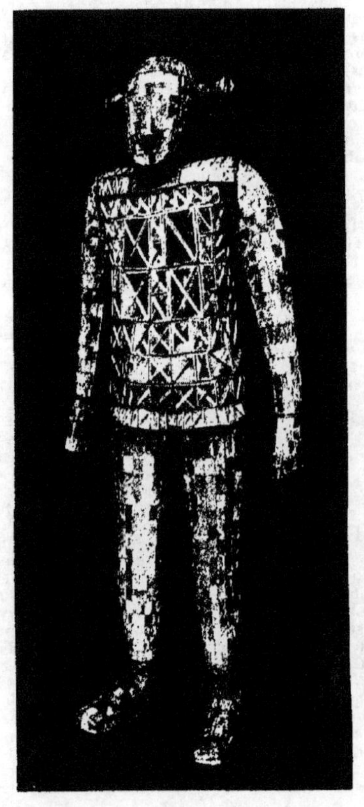

The funeral suit of Princess Tou Wan, consort of Prince Liu Sheng, shown in an upright position. It measures 172 cm (5 feet 7 inches) from the crown of the head to the soles of the feet. Alchemy was an active part of Taoist religion, which was quite different from Taoist philosophy, and there was a belief that jade would prevent a body from decaying; tablets of jade were used to seal the orifices of the body before burial. Prince Liu Sheng of Chung-shan ordered complete burial suits of jade for himself and his princess in the 2nd century BC and the one shown here was reassembled from 2,160 tablets of jade found in the Man-ch'eng tombs in the province of Hopei in 1968. Western Han dynasty (206 BC–AD 8).

A ritual vessel for holding wine, called a *hu*. A bronze of the middle Chou period, seventh century BC.

but the arrangement of the ornamental motifs varies. The decoration on vases from the stage of florescence is heraldic in character, placed in bands one above another and marked by strong toothed ridges. The metal is heavy, the appearance solid and sturdy. Fabulous animals (dragons, *kui*, and phoenix, *fong*), serpents and grasshoppers stand out in strong relief on a background of spirals (*lei wen*, the 'thunder motif'). A monstrous mask called the *t'ao-t'ieh* (wolverine) is one of the most striking motifs; it is often formed of two confronted dragons drawn in profile on either side of a central spine.

The human figure is not absent from this fantastic world, but animal themes are more common. Agriculture in the Shang period was not nearly as advanced as it was under the Chou, and hunting was still a basic necessity for survival. The aristocracy were warriors and hunters, every large estate included a hunting preserve, since the ancestors had to be provided with game and fish, meat was offered as a sacrifice, and the spoils of war were used to ornament artifacts. 'What is known as governing with the aid of symbols consists of borrowing the motifs for decorating the various insignia from hunting and fishing (*Tso Chuan*)'. These symbols and emblems appear on the dynastic talismans.

The bestiary of the vases is equally rich in a wide variety of species - rams, rhinoceros, elephants, goats, tigers, dogs, hares, fish, snakes, owls and various birds. Perhaps the owl, which can see at night, was to light the tomb with its gaze. The grasshopper was a symbol of immortality; the undershell of the tortoise was used for divination (it was put in the fire and the ancestors' response was interpreted from the pattern of the cracks). Sometimes carved wooden freestanding animal figures formed the handle of a vase or decorated its lid, and sometimes the animal formed the actual body of the vase.

The Shang also carved stone in the round. Buffaloes, toads, tigers, bears and even human statuettes were found at An-yang with marble sacrificial vessels. The oldest example of carved jade was found at Cheng-chou in the proto-Shang level, and the progress of the technique in the Shang period can be followed at An-yang and Hui Hsien. The Chinese have always spoken respectfully of 'true jade' (*chen yu*), the greenstone brought by caravan from Khotan and central Asia, but they also worked the jadeites. The Shang craftsman shaped his pieces with a saw (perhaps a circular saw), polished them with sand and water, produced contours by rubbing on a coarse-grained stone, and cut the decoration with a saw or a hard point. He also worked objects for ritual or ornamental use (rings, discs with a hole through the centre, sceptres and emblems of authority, ceremonial weapons, musical stones, appliqués and pendants). The art of jade working was not lost under the Chou, and in some tombs ritual jades

Under the Eastern Han dynasty (AD 24–220) China was prosperous and the landowners were the richest and most powerful class in the kingdom. The Wu-wei tomb in the province of Khansu dates from that period and excavations began there in 1969. A remarkable discovery was the group of bronze figurines, finely modelled and rarely exceeding 40 cm (16 inches) in height, of horses, mounted horsemen and dignitaries in carriages with a slave leading the horse. The number of carriages an official could command signified rank, and models of them were buried with him.

were even found inside the coffin: a disc (*pi*) representing the heavens rested on the back of the skeleton, which lay face downwards, a *t'sung* (a cylinder set in a square block symbolising the earth) had been placed under the abdomen, and to keep all evil influences within the corpse the 'seven openings' had been stopped up and the mouth sealed with a grasshopper, symbol of immortality. The decoration of the bone, ivory and jade objects was in the same style as that of the bronzes and the paintings, which had left their imprint on the walls of the burial pits.

Decadence and renaissance

The Shang style is remarkable for its coherence, strength and power to express fantasy. Its origins are problematical – there is nothing in China's Neolithic past to foreshadow the richness of these zoomorphic motifs. Scholars are fascinated by its seeming affinities with Indian, Alaskan and British Columbian art, but so far there has not been a single discovery which confirms their theories. It seems difficult to exclude the intervention of western and Scythian-Siberian influences in the phenomenon of stylistic mutation which took place in the middle of the Chou period. In this era ceremonial vessels lost the monumental character which they inherited from the Shang. The *ku* cup, the *chia* and *[illegible]* tripods and the square *yi* disappeared, the *ts'un* and the *yu* (wine pitcher) became very rare, the feet of the *[illegible]* tripod developed an inward curve, and new shapes appeared. The 'thunder motif' (*lei wen*) became rare and the *t'ao-t'ieh* was now [illegible] on the feet of tripods, purely decorative elements tended to replace zoomorphic figures, and the relief grew flatter. A transitional style developed and in the early Iron Age an entirely new style spread across the country. At Pu tu ts'un, for example, preliminary signs of this development appear, and can be followed until the seventh and fifth centuries by means of the bronzes from Hsing cheng and Chin-ts'un.

A pottery figurine of a horseman from the tomb of Princess Yung T'ai, excavated at Ch'ien-hsien in the province of Shensi in 1964. It measures 31.5 cm (just over 12 inches) in height and dates from AD 706 (T'ang dynasty). The rider is a huntsman and the animal clinging to the horse is believed to represent a cheetah. The horseman's dress and beard suggest that he came from Central Asia, the recruiting ground for many of those who served the rich families of the capital city, Ch'ang-an.

[illegible] find made between [illegible] th[illegible] [illegible] Li-yu, should be [illegible] period. The vases found [illegible] slide have oval or egg-shaped bodies with snaky interlaced dragons (*p'an lung*) in parallel friezes. The monotony of these twisted wreaths of zoomorphic motifs is relieved by light granulation, spirals, curlicues, commas and wings which gi[illegible] [illegible] vitality. A plait decorates the handle, base or neck. Animals in the round on the neck [illegible] treated with almost Western realism.

A considerable number of jades and small bronzes in the same style have been unearthed in the Huai-ho valley. Karlgren, the Swedish

One of the bronze horse-drawn carriages from the Wu-wei tomb at Khansu in north-west China. The driver sits behind a high dashboard and the horse is preceded by a slave; the umbrella or parasol is a sign of authority. (Another sign of authority, found in other carriages, is an axe with the blade turned outwards). The carriage is 43 [illegible] (17 inches) in height. Eastern Han dynasty (AD 24–220).

archaeologist, calls Kuo Mo-jo's stage of renaissance the Huai period. The typical feature of the material of this era is the fact that it includes elements of both the Chou culture and the Eurasian steppes. Among the small bronzes from the Warring States tombs are long swords decorated with geometric or zoomorphic motifs and inscriptions inlaid with gold and silver, and socketed *ko* halberds with terminal ornaments in the form of animal combats, keys, appliqués, buckles for belts or sword baldricks, disc-shaped mirrors decorated with dragons, stylised birds on a geometric background and T-shaped motifs on a background of hooks and spirals round a central handle. A glossy satiny green or greyish patina ('water patina') adds to the beauty of these objects, some of which are embellished with gold, silver, turquoise and malachite inlays. Men armed with spears, knives and bows battle with animals or great birds on the bodies of the bronze *hu* jars.

This mobile style which revives some elements of Shang decoration in a modified form (*t'ao-t'ieh*, grasshoppers and palmettes) seems also to be the result of a meeting of northern and southern influences. Its richness, daring, elegance and technical excellence are highly typical of this era, which was in love with luxury, power and novelty. It is still orderly, however, and even its baroque audacity is governed by a sense of balance.

From the fifth to the third centuries the Hui Hsien material grows poorer. Relief work tends to disappear from the vases from Chin-ts'un in the fourth and third centuries. Inlays of gold and silver form large scrolls, spirals, hooks and zoomorphic arabesques. The equipment in the Chin-ts'un tombs, which are probably royal burials, is splendid. The art of the kingdom of Ch'u alone deserves a long study. Influences from the north and perhaps from the south contributed to develop a brilliant, sophisticated and diverse civilisation in this powerful state towards the end of the ancient world. The most remarkable finds from the excavation are pots with a metallic coating, wooden carvings, fabrics and lacquers, with superb jades and beautiful bronzes.

These excavations proved that the Chinese were familiar with the properties of lacquer (a varnish obtained by refining the sap of the *Rhus vernicifera*) as early as the Shang period, at this time, however, it was only used for utilitarian objects. The craft of lacquer working and its use for works of art became widespread only under the Eastern Chou. Lacquers occur in Warring States tombs throughout the whole of China, from Kiangsu to Szechwan and from Jehol to Huna. The finest examples known to date, however, come from Ch'ang-cha in Hunan, where the humidity has helped to preserve them.

Lacquers were made in the imperial factory in Szechwan under the Han dynasty. A number of pieces from this factory have been unearthed at Lo-yang in Korea. Indeed, the Chinese were united at home, launched their armies against the Huns in Mongolia, and colonised north Korea under the Han Dynasty. They set up protectorates in the small central Asian kingdoms in the west and established links with Sogdiana and India. The Chinese had no direct contacts with the eastern provinces of the Roman empire at this time, but they would meet east Roman art imported by the Buddhist missionaries from Afghanistan in the oases of the silk route. We know that Chinese civilisation did not develop in isolation. China gave much, but she received much in return. From time immemorial ideas, techniques and shapes have circulated across the steppe and the northern taiga, between the Yellow River valley and eastern Eurasia.

Religious, intellectual and artistic influences from Western Asia renewed the springs of inspiration in China towards the beginning of the Christian era. During the troubled times which followed the fall of the Han dynasty contacts became more intermittent, but the routes to central Asia were never closed. When the empire was reunited under the Hsui (518-617) and the T'ang (618–907), Chinese expansion began once more. By the middle of the eighth century the Chinese were established in Pamir, whence they could dominate Tibet, Turkestan, Arabian Iran and India, and they seemed to be lords of all Asia. Commercial and cultural exchanges were more active and intense than ever before.

Excavations which are currently being conducted all over the country are providing material from all periods. Our knowledge of Chinese history, however, is still fundamentally based on the *Twenty-five Dynastic Histories*. Study of this vast corpus is unfortunately still only in its infancy, and excavations, which are increasing in number, help to throw light on it. The texts come to life through the finds. The soil of China conceals innumerable tombs and the treasures they contain are probably storing up some great surprises for tomorrow's archaeologists. Discoveries are taking place so quickly that archaeological research has considerable difficulty in keeping up with them. Several decades from now Chinese civilisation, which is so strongly original in spite of the contributions it has absorbed, will appear in a more vital light, comprehension of it will probably be transformed by this confrontation with the dead.

A bronze of the Chou Dynasty, c. tenth century BC. The decorative motifs are a development of the earlier Shang ideas but the Chou bronzes are generally more massive in effect. The tiger, one of a pair, carries a hollow chamber in the centre which may have held incense. The tiger was the creature of the west – the region of the Chou's origins.

Further Reading List

ARCHAEOLOGY AT WORK

Adams, Robert McC. *The Evolution of Urban Society: Early Mesopotamia and Prehispanic Mexico.* Aldine, Chicago, 1966

Bacon, Edward (ed.) *Vanished Civilizations.* Thames and Hudson, London; McGraw-Hill, New York, 1963

Bass, Georg F. *Archaeology Under Water.* Praeger, New York, 1966; Penguin, Harmondsworth, 1970

Butzer, Karl. *Environment and Archaeology.* Methuen, London; Aldine, Chicago, 1965

Celoria, Francis. *Archaeology.* Hamlyn, Feltham, 1970; Grosset and Dunlap, New York, 1971

Ceram, C. W. *Gods, Graves and Scholars.* Alfred Knopf, New York, 1951; Gollancz and Sidgwick and Jackson, London, 1952

Ceram, C. W. *A Picture History of Archaeology.* Thames and Hudson, London, 1958

Childe, V. Gordon. *Man Makes Himself.* Watts, London, 1936; New American Library, New York, 1962

Childe, V. Gordon. *What Happened in History.* Penguin, Harmondsworth and New York, 1942

Cookson, M. B. *Photography for Archaeologists.* Parrish, London, 1954

Cottrell, Leonard (ed.) *Concise Encyclopedia of Archaeology.* Hutchinson, London; Hawthorn Books, New York, 1960

Daniel, Glyn. *The First Civilizations.* Thomas Crowell, New York, 1968; Penguin, Harmondsworth, 1971

Daniel, Glyn. *A Hundred Years of Archaeology.* Duckworth, London; Macmillan, New York, 1950

Daniel, Glyn. *The Origins and Growth of Archaeology.* Penguin, Harmondsworth and New York, 1967

Edeux, Henri-Paul. *In Search of Lost Worlds.* Hamlyn, Feltham; World Pub. Co., New York, 1971

Clark, Grahame. *From Savagery to Civilisation.* Cobbett Press, London, 1946

Finegan, Jack. *Light from the Ancient Past.* Princeton University Press, New Jersey, [illegible]

Grant, Michael (ed.) *The Birth of Western Civilisation.* Thames and Hudson, London; McGraw-Hill, New York, 1964

Hawkes, Jacquetta and Woolley, Leonard. *Prehistory and the Beginnings of Civilisation.* Harper and Row, New York, 1962; Allen and Unwin, London, [illegible]

Kenyon, K. *Beginning in Archaeology.* Praeger, New York, 1953; J. M. Dent and Sons Ltd, London, 1964

Laet, Siegfried de. *Archaeology and its Problems.* Phoenix House, London, [illegible]

Linton, Ralph. *The Tree of Culture.* Alfred Knopf, New York, 1955

Oakley, K. P. *Framework for Dating Fossil Man.* Weidenfeld and Nicolson, London; Aldine, Chicago, [illegible]

Piggott, Stuart (ed.) *The Dawn of Civilisation.* Thames and Hudson, London; McGraw-Hill, New York, [illegible]

Plenderleith, H. J. *The Conservation of Antiquities and Works of Art.* Oxford University Press, London and New York, 1956

Wheeler, Mortimer. *Archaeology from the Earth.* Oxford University Press, London and New York, 1954

Wood, Eric S. *Collins Field Guide to Archaeology.* Collins, London, 1968

Woolley, Leonard. *Digging up the Past.* Thomas Crowell, New York, 1956; Ernest Benn, London, [illegible]

Woolley, Leonard. *Spadework.* Lutterworth Press, London, 1953

Zeuner, F. E. *Dating the Past.* Methuen, London, 1958

PREHISTORIC ARCHAEOLOGY

Bordes, F. *The Old Stone Age.* McGraw-Hill, New York; Weidenfeld and Nicolson, London, 1968

Breuil, H. *Four Hundred Centuries of Cave Art.* Hutchinson, London, 1952

Childe, V. Gordon. *The Prehistory of European Society.* Penguin, Harmondsworth and New York, 1958

Clarke Grahame *Prehistoric Societies* Hutchinson, London, 1965

Clark Grahame *World Prehistory* Cambridge University Press. Cambridge and New York, 1961

Clark, J. D. E *The Prehistory of Africa* Thames and Hudson, London, 1970

Clark J. G. D. *World Prehistory An Outline* Cambridge University Press. Cambridge and New York, 1969

Cole, J. M and Higgs, E. S. *The Archaeology of Early Man* Faber and Faber, London, 1969

Cole, Sonia *The Neolithic Revolution* British Museum (Natural History), London, 1960

Coon, Carleton S *The History of Man* Alfred Knopf, New York (rev. ed.), 1962. Penguin, Harmondsworth, 1967

Coulborn, Rushton *The Origin of Civilized Societies* Princeton University Press. 1959

Daniel Glyn *The Megalithic Builders of Western Europe*. Praeger. New York: Penguin, Harmondsworth, 1958

Day, Michael H *Fossil Man* Hamlyn, Feltham, 1969. Grosset and Dunlap, New York 1970

Oakley, K. P *Man the Tool Maker* University of Chicago Press. 1957. British Museum (Natural History), 1963

WESTERN ASIA BEFORE ALEXANDER

Albright, W. F *The Archaeology of Palestine* Penguin, Harmondsworth and New York, 1949

Braidwood, R. J *The Near East and the Foundations for Civilisation* Eugene, Oregon, 1952

Burrows, M *The Dead Sea Scrolls* Viking Press, New York, Secker and Warburg, London, 1955

Contenau, G *Everyday Life in Babylon and Assyria* E Arnold, London, St Martin's Press, New York, 1954

Frankfort H *The Birth of Civilisation in the Near East* Indiana University Press, Williams and Norgate, London, 1951

Gurney O. R *The Hittites* Penguin, Harmondsworth and New York, 1953

Pritchard James B (ed.) *The Ancient Near East*. Princeton University Press. 1958

Saggs, H. W. F *The Greatness that was Babylon* Hawthorn Books, New York, Sidgwick and Jackson, London, 1962

Woolley, Leonard *Ur of the Chaldees*, Ernest Benn, London, W. W. Norton New York, 1921. (New ed Thomas Crowell, New York, 1950)

THE NILE VALLEY

Aldred, Cyril *Ancient Peoples and Places The Egyptians* Thames and Hudson, London; Praeger New York, 1961

Cottrell, Leonard *The Lost Pharaohs* Evans Brothers, Ltd., London, 1950; Holt, Rinehart and Winston, New York 1961

Desroches Noblecourt, C *Tutankhamun* Connoisseur and Michael Joseph, London, New York Graphic 1963

Edwards, I. E. S *The Pyramids of Egypt* Penguin, Harmondsworth and New York, 1947

Gardiner, Alan *Egypt of the Pharaohs* Oxford University Press, London and New York, 1961

Shinnie, P. L. *Ancient People and Places Meroe* Praeger, New York, Thames and Hudson, London 1967

Ullendorff, E *The Ethiopians* Oxford University Press, London and New York, 1965

THE AEGEAN WORLD AND CLASSICAL GREECE

Blegen, Carl W *Ancient Peoples and Places Troy* Thames and Hudson, London, Cambridge University Press, New York 1963

Cook J. M *Ancient Peoples and Places The Greeks in Ionia and the East* Thames and Hudson, London, 1962, Praeger New York, 1963

Cook, R. M *Ancient Peoples and Places: The Greeks until Alexander* Thames and Hudson, London, 1961, Praeger, New York, 1962

Cottrell, Leonard *The Bull of Minos* Evans Brothers Ltd., London, Holt, Rinehart and Winston, New York, 1953

Cottrell Leonard *The Lion Gate* Evans Brothers Ltd., London, 1963

Graham James Walter *The Palaces of Crete* Princeton University Press, 1962

Hopper R. J *The Acropolis* Weidenfeld and Nicolson, London, 1971

Hutchinson, R. W. *Prehistoric Crete* Penguin, Harmondsworth, Peter Smith, New York, 1962

Philippaki Barbara *The Attic Stamnos* Oxford University Press, London and New York, 1966

Taylour Lord William *Ancient Peoples and Places: The Mycenaeans* Thames and Hudson, London, Praeger New York, 1964

Woodhead, A. G *Ancient Peoples and Places The Greeks in the West* Thames and Hudson, London, Praeger, New York, 1964

THE ETRUSCANS AND THE ROMANS

Bloch, Raymond *Ancient Peoples and Places The Etruscans* Thames and Hudson London Praeger, New York, 1958

Bloch, Raymond *Ancient Peoples and Places The Origins of Rome* Thames and Hudson, London; Praeger, New York, 1960

Charles-Picard, Gilbert. *Carthage*. Elek Books, London, 1964

Grant, Michael. *The Cities of Vesuvius*. Weidenfeld and Nicolson, London, 1971

Grant, Michael. *The Roman Forum*. Weidenfeld and Nicolson, London, 1971

Grant, Michael. *The Roman World*. Weidenfeld and Nicolson, London; as *The World of Rome*, World Pub. Co., New York, 1960

McDonald, A. H. *Ancient Peoples and Places: Republican Rome*. Praeger, New York; Thames and Hudson, London, 1966

Rostovtzev, M. *Rome*. Oxford University Press, London and New York, 1960

EUROPE IN THE BRONZE AND IRON AGES

Arbman, Holger. *Ancient Peoples and Places: The Vikings*. Praeger, New York; Thames and Hudson, London, 1961

Filip, Jan. *Celtic Civilization and its Heritage*. Publishing House of Czechoslovak Academy of Sciences and ARTIA, Prague, 1960

Gronbech, V. *The Culture of the Teutons*. Humphrey Milford, Oxford University Press, London and New York, 1926

Klindt-Jensen, O. *Ancient Peoples and Places: Denmark*. Praeger, New York; Thames and Hudson, London, 1957

Paor, M. and L. de. *Ancient Peoples and Places: Early Christian Ireland*. Praeger, New York; Thames and Hudson, London, 1950

Powell, T. G. E. *The Celts*. Praeger, New York; Thames and Hudson, London, 1958

Rees, Alwyn and Brinley. *Celtic Heritage*. Thames and Hudson, London, 1961

Shetelig, H. and Falk, H. *Scandinavian Archaeology*. Clarendon Press, Oxford; Oxford University Press, New York, 1937

THE AMERICAS

Burland, Cottie. *The People of the Ancient Americas*. Hamlyn, Feltham, 1970

Burland, Cottie. *Peru under the Incas*. Putnams, New York, 1967; Evans Bros., London, 1968

Bushnell, G. H. S. *Ancient Arts of the Americas*. Praeger, New York; Thames and Hudson, London, 1965

Caso, Alfonso. *The Aztecs: People of the Sun*. Oklahoma University Press, 1958

Hyams, E. and Ordish, G. *The Last of the Incas*. Longmans, London; Simon and Schuster, New York, 1963

Kubler, G. *Art and Architecture of Ancient America*. Penguin, Harmondsworth and New York, 1962

Lanning, E. P. *Peru before the Incas*. Prentice-Hall, Englewood Cliffs, New Jersey, 1967

Lothrop, S. K. *Treasures of Ancient America*. Skira, Geneva; World Pub. Co., New York, 1964

Mason, J. Alden. *The Ancient Civilisations of Peru*. Penguin, Harmondsworth and New York, 1957

Morley, Sylvanus G. *The Ancient Maya*. Stanford University Press, 1956

Thompson, J. E. S. *The Rise and Fall of Maya Civilisation*. Oklahoma University Press, 1954; Gollancz, London, 1956

Vaillant, G. C. *The Aztecs of Mexico*. Penguin, Harmondsworth; Doubleday, New York, 1944

INDIA, PAKISTAN AND AFGHANISTAN

Basham, A. L. *The Wonder that was India*. Sidgwick and Jackson, London, 1954; Macmillan, New York, 1955

Garratt, G. T. *The Legacy of India*. Clarendon Press, Oxford; Oxford University Press, New York, 1937

Marshall, John. *A Guide to Taxila*. Cambridge University Press, Cambridge and New York, 1960

Marshall, John. *Mohenjo-Daro and the Indus Civilisation*. Arthur Probsthain, Ltd., on behalf of the Archaeological Survey of India, London, 1931

Wheeler, Mortimer. *Ancient Peoples and Places: Early India and Pakistan*. Praeger, New York; Thames and Hudson, London, 1959

THE FAR EAST

Aston, W. G. (trans.) *Nihongi*. Allen and Unwin, London; Essential Books, New York, 1956

Chang, Kwang-chih. *The Archaeology of Ancient China*. Yale University Press, New Haven, 1963

Cheng Te-kun. *Archaeology in China: prehistoric China*. W. Heffer and Sons, Cambridge, 1959; Toronto University Press, 1970

Cheng Te-kun. *Archaeology in China: Shang China*. W. Heffer and Sons, Cambridge, 1961; Toronto University Press, 1970

Kidder, J. Edward. *Ancient Peoples and Places: Japan*. Praeger, New York; Thames and Hudson, London, 1959

Li Chi. *The Beginnings of Chinese Civilization*. University of Washington Press, 1957

Watson, William. *Ancient Peoples and Places: China*. Praeger, New York; Thames and Hudson, London, 1961

Watson, W. *Early Civilisation in China*. McGraw-Hill, New York; Thames and Hudson, London, 1966

Index

Figu

Acknowledgments

Colour illustrations

Front cover, Ann Hill. 17, PAF International. 18, ZEFA. 35, Hirmer Fotoarchiv. 36, Hamlyn Group Picture Library. 77 top, Camera Press. 77 bottom, Yigael Yadin/The Observer. 78–80, Camera Press. 105 top, Larousse. 105 bottom, A. F. Kersting. 106 top, Larousse. 106 bottom, Roger Wood. 123 top, W. F. Davidson. 123 bottom, A. F. Kersting. 124, ZEFA. 181, Barnaby's Picture Library. 182, Ashmolean Museum, Oxford. 183, Michael Holford Library. 184, Brian Brake/Magnum. 225 top, A. F. Kersting. 225 bottom, ZEFA. 226, Roger Wood. 243, ZEFA. 244, Fitzwilliam Museum, Cambridge. 285, Ann Hill. 286 top, W. F. Davidson. 286 bottom, PAF International. 287, ZEFA. 288, Ny Carlsberg Glyptothek. 329, ZEFA. 330 top left, University Museum, Oslo. 330 top right and bottom, British Museum. 347, Picturepoint. 348, Hamlyn Group Picture Library. 389, Barnaby's Picture Library. 390, Madanjeet Singh, Delhi. 391, Hamlyn Group Picture Library. 392, ZEFA. Back cover, Picturepoint.

Black and white illustrations

Front endpaper, Camera Press. 2, Hirmer Fotoarchiv. 7, Walter Drayer. 8, J. Bottin. 9, H. Roger Viollet. 10 top, Mansell Collection. 10 bottom, J. E. Bulloz. 11, J. Bottin. 12 top, Camera Press. 12 bottom/13, J. Allan Cash. 14 top, Hirmer Fotoarchiv. 14 bottom, J. Bottin. 15 top, Hirmer Fotoarchiv. 15 bottom, Popperfoto. 16 top, Giraudon. 16 bottom, Edwin Smith. 19 top, British Museum. 19 bottom, Hirmer Fotoarchiv. 20 top, Giraudon. 20 bottom, Paul Almasy. 21 top, Lucien Hervé. 21 bottom, Georges Viollon. 22, Giraudon. 23, Georges Viollon. 24, George Rodger/Magnum. 25 top, Popperfoto. 25 bottom, J. Bottin. 26, Mansell Collection. 27, J. Bottin. 28, H. Roger Viollet. 29, Giraudon. 30, Popperfoto. 31, Giraudon. 32, Metropolitan Museum of Modern Art, New York. 33, J. E. Bulloz. 34 top, Larousse/Musée du Louvre. 34 bottom, British Museum. 37, Giraudon. 38, Roger Wood. 39 top, Hirmer Fotoarchiv. 39 bottom, Popperfoto. 40, George Rodger/Magnum. 41 top, Popperfoto. 41 bottom, Georges Viollon. 42 top, Bibliothèque Nationale. 42 bottom, Popperfoto. 43, George Rodger/Magnum. 44, Giraudon. 45, George Rodger/Magnum. 46, Alison Frantz, Athens. 47, Hamlyn Group Picture Library. 48, Roger Wood. 49, H. Roger Viollet. 50 top, Hirmer Fotoarchiv. 50 bottom, Archives Photographiques. 51, Rapho. 52 top, Hirmer Fotoarchiv. 52 bottom, Larousse. 53, Roger Wood. 54 top, National Museum, Copenhagen. 54 bottom, Belzeaux-Zodiaque. 55, Universitetets Oldsaksamling, Oslo. 56/57, H. Roger Viollet. 58, Roger Clarke. 59, Aerofilms. 60/61, J. E. Bulloz. 62 top, J. Bottin. 62 bottom, Larousse/photo R. Goguey. 63, Larousse/photo B. Groslier. 64/65 top, Larousse/photo R. Goguey. 65 bottom/66 top, Larousse/photo R. Agache. 66 bottom/67 top, Larousse/photo I.G.N. 67 centre and bottom, Larousse/photo Bradford. 68 top, Larousse/photo Atkinson. 68 bottom, Larousse. 69, Larousse/photo Stierlin. 70 top, Larousse/Rapho Patellani. 70 bottom left, Larousse/Laoci. 70 bottom right, Larousse. 71, Larousse/Courbin. 72, Keystone. 73, The Art Institute of Chicago. 74, Hirmer Fotoarchiv. 75, Larousse/photo B. Groslier. 76 top, Larousse/Bibliothèque Nationale. 76 centre, Larousse/Weiss Rapho. 76 bottom, Larousse/Viollon. 81 top, Keystone. 81 bottom, H. Roger Viollet. 82, Keystone. 83, The Society of Antiquaries of London. 84, Larousse/photo Courbin. 85 top, Keystone. 85 bottom, Larousse/photo Courbin. 86 top and centre, Larousse/photo E. Pyddocke. 86 bottom, Larousse/photo Courbin. 87 top, The Society of Antiquaries of London. 87 bottom/88 top, Larousse. 88 bottom, Larousse/photo Courbin. 89, Larousse/photo J. Perrot. 90–92, Larousse/Courbin. 93, Camera Press. 94, Roger Wood. 95, J. E. Bulloz. 96 top, J. Bottin. 96 bottom, Giraudon. 96/97, Paul Almasy. 97 top, J. Allan Cash. 98 top, Giraudon. 98 bottom, J. E. Bulloz. 99 top, Giraudon. 99 bottom/100 top, J. E. Bulloz. 100 bottom, J. Bottin. 101, Alinari Giraudon. 102, Paul Almasy. 103, Research Laboratory, British Museum. 104, Popperfoto. 107, Hamlyn Group Picture Library. 108, J. Allan Cash. 109, National Museum, Athens. 110, Edwin Smith. 111, Popperfoto. 112/113, Hamlyn Group Picture Library. 114/115, National Museum, Naples. 116, Anderson-Giraudon. 117, J. Allan Cash. 118, Alinari Giraudon. 119, J. Bottin. 120 top, Hirmer Fotoarchiv. 120 bottom, Research Laboratory, British Museum. 121 top, Keystone. 121 bottom, Research Laboratory, British Museum. 122 top, National Tourist Organisation of Greece. 122 bottom, Rapho. 125 top, Paul Almasy. 125 centre, Camera Press. 125 bottom, Paul Almasy. 126 top, Central Office of Information, London. 126 centre, A.C.L. Brussels. 127 top, Keystone. 127 bottom, Paul Almasy. 128, J. E. Bulloz. 129 top, Benaki Museum, Athens. 129 bottom, Mansell Collection. 130, 131, Popperfoto. 132, Research Laboratory, British Museum. 133, J. Bottin. 134, Roger Wood. 135 top, H. Roger-Viollet. 135 bottom, J. Bottin. 136 top, H. Roger-Viollet. 136 bottom, Research Laboratory, British Museum. 137, Popperfoto. 138, J. Allan Cash. 139 top, Georges Viollon. 139 bottom, J. Bottin. 140 top, Hirmer Fotoarchiv. 140 bottom, Research Laboratory, British Museum. 141, Roger Wood. 142, H. Roger-Viollet. 143, Musée d'Aquitaine, Bordeaux. 144 top, Popperfoto. 144 bottom, H. Roger-Viollet. 145 left, Archives Photographiques. 145 right, H. Roger Viollet. 146 top, U.P.I. 146 bottom, H. Roger Viollet. 147 top, Giraudon. 147 bottom/148 top, Popperfoto. 148 bottom, H. Roger Viollet. 149, Popperfoto. 150 top and centre, Natural History Museum, Vienna. 150 bottom, H. Roger Viollet. 151 top, Werner Forman. 151 bottom, A. F. Kersting. 152, Universitetets Oldsaksamling, Oslo. 153, Jean Ribière. 154, Giraudon. 155, Foto Mas. 156/7, J. Bottin. 158, Dr. J. K. St. Joseph, Cambridge. 159, Central Office of Information, London. 160, Camera Press. 161, Hirmer Fotoarchiv. 162, Aerofilms. 163 top, John Donat. 163 bottom, British Museum. 164 top, Hamlyn Group Picture Library. 164 bottom, Rapho. 165/166 top, Giraudon. 166 bottom, Hirmer Fotoarchiv. 167 top, Giraudon. 167 bottom, The Oriental Institute, University of Chicago. 168 top. 170 top, Hirmer Fotoarchiv. 170 bottom, Foto Marburg. 171 top, Hirmer Fotoarchiv. 171 bottom, Foto Marburg. 172, Hirmer Fotoarchiv. 173, J. Bottin. 174–179 top, Hirmer Fotoarchiv. 179 bottom, J. Bottin. 180, J. Allan Cash. 185 top, J. Allan Cash. 185 bottom, Larousse. 186 top, Giraudon. 186 bottom, 188 bottom, Hirmer Fotoarchiv. 189, Mansell Collection. 190/191 top, Foto Marburg. 191 bottom, Hirmer Fotoarchiv. 192, J. E. Bulloz. 193 top, Radio Times Hulton Picture Library. 193 bottom–197 top, Hirmer Fotoarchiv. 197 bottom/198, Foto Marburg. 199 top, Camera Press. 199 bottom, Lucien Hervé. 200 top, Roger Wood. 200/201 bottom, Boudot-Lamotte. 201 top, Roger Wood. 202–205 bottom, Hirmer Fotoarchiv. 206, Roger Wood. 207, Editions Arthaud. 208, Roger Wood. 209, Hirmer Fotoarchiv. 210, Hamlyn Group Picture Library. 211 top, Roger Wood. 211 bottom, Michael Hetier. 212/213, J. Bottin. 214, Hirmer Fotoarchiv. 215, J. Bottin. 216/217, Hirmer Fotoarchiv. 218, Jean Bottin. 219 top, Hirmer Fotoarchiv. 219 bottom, J. Bottin. 220 top, Giraudon. 220 bottom, Foto Marburg. 221 top, Roger Wood. 221 bottom, Hirmer Fotoarchiv. 222 top, Michel Hetier. 222 bottom, Giraudon. 223 top, Roger Wood. 223 bottom, Popperfoto. 224/227, Hirmer Fotoarchiv. 228, Popperfoto. 229, Roger Wood. 230 top, H. Roger Viollet. 230 bottom, Giraudon. 231 top, Rapho. 231 bottom, Cairo Museum. 232 top, Paul Almasy. 232 bottom, Aerofilms. 233, Roger Wood. 234, J. Bottin. 234/235, Rapho. 236/238, J. Allan Cash. 239 top, Popperfoto. 239 bottom, Rapho. 240–241 top, J. Allan Cash. 241 bottom, Popperfoto. 242, Roger Wood. 245, Hirmer Fotoarchiv. 246 top and bottom left, Mansell Collection. 246 bottom right, Hirmer Fotoarchiv. 247, Josephine Powell. 248 top, Hamlyn Group Picture Library. 248 bottom, British Museum. 249 top, J. Bottin. 249 bottom, Hirmer Fotoarchiv. 250, H. Roger Viollet. 251 top, Radio Times Hulton Picture Library. 251 bottom/252 top, Hirmer Fotoarchiv. 252 bottom, Josephine Powell. 253 top–254 top, Hirmer Fotoarchiv. 254 bottom, Foto Marburg. 255 top, Hirmer Fotoarchiv. 255 bottom, Josephine Powell. 256, Hirmer Fotoarchiv. 257, Foto Marburg. 258 top–259 bottom, Hirmer Fotoarchiv. 260 top, H. Roger-Viollet. 260 bottom, Hirmer Fotoarchiv. 261 top, Popperfoto. 261 bottom, Josephine Powell. 262 top, Robert Descharnes. 262 bottom and 263 top, Hirmer Fotoarchiv. 263 bottom, H. Roger-Viollet. 264, Hirmer Fotoarchiv. 265, Foto Marburg. 266, J. Bottin. 267, Nick Stournaras, Athens. 268, Hirmer Fotoarchiv. 269 top, J. Bottin. 269 bottom, Hirmer Fotoarchiv. 270 top, J. Bottin. 270 bottom, Larousse. 271 top, Enrico Mariani. 271 bottom, Royal Institute of British Architects. 272, Roger Wood. 273, Picturepoint. 274, Hirmer Fotoarchiv. 275, Picturepoint. 276 top, Larousse. 276 bottom, National Tourist Organisation of Greece. 277 top, Hamlyn Group Picture Library. 277 bottom, National Tourist Organisation of Greece. 278 top, Mansell Collection. 278 bottom, A. F. Kersting. 279, Hirmer. 280, J. Allan Cash. 281 top, Stadtliche Antikensammlungen, Munich. 281 bottom, Mansell. 282 top, Alinari. 282 bottom, J. Allan Cash. 283/4, Hirmer Fotoarchiv. 289, Walter Drayer, Zurich. 290 top, Fototeca Unione. 290 bottom, Gerhard Klammet. 291 top, Mansell Collection. 291 bottom, Gerhard Klammet. 292 top, Hamlyn Group Picture Library. 292 bottom–294 bottom, Mansell Collection. 295, Alinari. 296 top, Mansell Collection. 296 bottom, Giraudon. 297 top–300, Mansell Collection. 301–305, Fototeca Unione. 306, J. Bottin. 307, Roger Wood. 308/309 top, Fototeca Unione. 309 bottom, Josephine Powell. 310 top, Fototeca Unione. 310 bottom, Mansell Collection. 311 top, J. E. Bulloz. 311 bottom/312, Mansell Collection. 313 top, Popperfoto. 313 bottom, Hamlyn Group Picture Library. 314, Rapho. 315 top, Roger Wood. 315 bottom, Fototeca Unione. 316 top, Hamlyn Group Picture Library. 316 bottom, Vatican Museum. 317 top, Mansell Collection. 317 bottom, Vatican Museum. 318 top, Jean Roubier. 318 bottom, Popperfoto. 319 top, Jean Roubier. 319 bottom, J. Bottin. 320 top, Mansell Collection. 320 bottom–321 bottom, Anderson Giraudon. 322, Mansell Collection. 323, Hamlyn Group Picture Library. 324 top, Yugoslav Tourist Office. 324 bottom, Vatican Museum. 325 top, Roger Wood. 325 bottom, James Mortimer. 326, Aerofilms. 327, Universitetets Oldsaksamling, Oslo. 328 top, J. Allan Cash. 328 bottom, Antikvarisk Topografiska Arkivet, Stockholm. 331 top, National Museum, Copenhagen. 331 bottom, Universitetets Oldsaksamling, Oslo. 332 top, Hamlyn Group Picture Library. 332 bottom, Mansell Collection. 333 top, Universitetets Oldsaksamling, Oslo. 333 bottom, Dr. J. K. St. Joseph, Cambridge. 334 top, Ashmolean Museum, Oxford. 334 bottom, Mansell Collection. 335 top, Ashmolean Museum, Oxford. 335 bottom, Mansell Collection. 336 top, Aerofilms. 336 bottom, Larousse. 337, Central Office of Information, London. 338 top, University Museum of Archaeology and Ethnology, Cambridge. 338 bottom/339 top, National Museum, Copenhagen. 339 bottom, Natural History Museum, Vienna. 340 top, National Museum of Ireland. 340 bottom, National Museum, Copenhagen. 341 top, British Museum. 341 bottom, Niels Elswing, Copenhagen. 342 top, A.T.A. Stockholm. 342 bottom, Aerofilms. 343 top, National Museum, Copenhagen. 343 bottom, Central Office of Information, London. 344/5, Universitetets Oldsaksamling, Oslo. 346, Universitetets Historiska Museum, Oslo. 349, National Museum, Copenhagen. 350, J. Bottin. 351, City Museum and Art Gallery, Birmingham. 352 top, J. Allan Cash. 352 bottom, H. Roger Viollet. 353 top, Rapho. 353 bottom, Mike Andrews. 354, J. Bottin. 355 top, Museum of Primitive Art, New York. 355 bottom left, Museo Nacional de Antropologia, Mexico City. 355 bottom right, Museo de Tepexpan, Mexico. 356 top, Fondo Editorial de la Plastica Mexicana. 356 bottom, J. Bottin. 357 top, James H. Hooper Collection. 357 centre, British Museum. 357 bottom, Eugen Kusch. 358, Hamlyn Group Picture Library. 359 left, J. Bottin. 359 right, British Museum. 360 top, Michel Hetier. 360 bottom, British Museum. 361 top, J. Bottin. 361 bottom, City of Liverpool Museums. 362 top, Henri Stierlin. 362 bottom, J. Bottin. 363 top, Michel Hetier. 363 bottom, British Museum. 364 top, Irmgard Groth Kimball, Mexico. 364 bottom, J. Allan Cash. 365 top, Irmgard Groth Kimball, Mexico. 365 bottom, J. Allan Cash. 366, Hamlyn Group Picture Library. 367 top, National Museum of Ethnology, Leiden. 367 bottom, Kemper Collection. 368 top, Henri Stierlin. 368 bottom, Mike Andrews. 369 top, J. Bottin. 369 bottom, Kemper Collection. 370, Mike Andrews. 371 top, Michel Hetier. 371 centre, Yvan Burlet. 371 bottom, Kemper Collection. 372, Conzett and Huber, Zurich. 373, Mansell Collection. 374, Josephine Powell. 375, Mansell Collection. 376 top, Hamlyn Group Picture Library. 376 bottom, Archaeological Survey of India, Government of India. 377 top–378 top, J. Allan Cash. 377 bottom, Paul Almasy. 378 bottom/379 top, Mansell Collection. 380 top, Mansell Collection. 380 bottom, 381 top and centre, Archaeological Survey of India, Government of India. 381 bottom, National Museum, New Delhi. 382, Eliot Elisofon. 383 top, Mansell Collection. 383 bottom, H. Roger-Viollet. 384 top, Conzett and Huber, Zurich. 384 bottom, Frederico Borromeo. 385–386 top, H. Roger-Viollet. 386 bottom, Larousse. 387 top, H. Roger Viollet. 387 bottom, Larousse. 388 and 393, Josephine Powell. 394, H. Roger-Viollet. 395, Larousse. 396/7, Paul Almasy. 398, Camera Press. 399, Private Collection. 400/1, Popperfoto. 402, Hamlyn Group Picture Library. 404, Larousse. 405/6, H. Roger Viollet. 407, J. Bottin. 408, Popperfoto. 409, Larousse. 411, Camera Press. 412, Picturepoint. 413/415, Paul Almasy. 417 top, Chen Collection, Hong Kong. 417 bottom, Freer Gallery of Art, Washington D.C. 418, Musée Guimet, Paris. 419, Freer Gallery of Art, Washington D.C. 420, Robert Descharnes. 421, Hamlyn Library. 422, Hirmer Fotoarchiv. 423, Alinari. 424, Musées Royaux d'Art et d'Histoire, Bruxelles. 432, Hirmer Fotoarchiv. Back endpaper, A. F. Kersting.

CPSIA information can be obtained
at www.ICGtesting.com
Printed in the USA
BVHW031601211022
649988BV00009B/587